Lecture Notes in Electrical Engineering

Volume 372

About this Series

"Lecture Notes in Electrical Engineering (LNEE)" is a book series which reports the latest research and developments in Electrical Engineering, namely:

- Communication, Networks, and Information Theory
- Computer Engineering
- Signal, Image, Speech and Information Processing
- Circuits and Systems
- Bioengineering

LNEE publishes authored monographs and contributed volumes which present cutting edge research information as well as new perspectives on classical fields, while maintaining Springer's high standards of academic excellence. Also considered for publication are lecture materials, proceedings, and other related materials of exceptionally high quality and interest. The subject matter should be original and timely, reporting the latest research and developments in all areas of electrical engineering.

The audience for the books in LNEE consists of advanced level students, researchers, and industry professionals working at the forefront of their fields. Much like Springer's other Lecture Notes series, LNEE will be distributed through Springer's print and electronic publishing channels.

More information about this series at http://www.springer.com/series/7818

Suresh Chandra Satapathy
N Bheema Rao · S Srinivas Kumar
C Dharma Raj · V Malleswara Rao
GVK Sarma
Editors

Microelectronics, Electromagnetics and Telecommunications

Proceedings of ICMEET 2015

 Springer

Editors
Suresh Chandra Satapathy
Deparment of CSE
Anil Neerukonda Institute of Technology
   and Sciences
Visakhapatnam
India

N Bheema Rao
Department of Electronics and
   Communication Engineering
National Institute of Technology
Warangal
India

S Srinivas Kumar
Electronics and Communication
Jawaharlal Nehru Technological University
Kakinada, Andhra Pradesh
India

C Dharma Raj
Department of Electronics and
   Communication Engineering
GITAM University
Visakhapatnam, Andhra Pradesh
India

V Malleswara Rao
Department of Electronics and
   Communication Engineering
GITAM University
Visakhapatnam, Andhra Pradesh
India

GVK Sarma
Department of Electronics and
   Communication Engineering
GITAM University
Visakhapatnam
India

ISSN 1876-1100   ISSN 1876-1119   (electronic)
Lecture Notes in Electrical Engineering
ISBN 978-81-322-2726-7   ISBN 978-81-322-2728-1   (eBook)
DOI 10.1007/978-81-322-2728-1

Library of Congress Control Number: 2015957125

Printed on acid-free paper

This Springer imprint is published by SpringerNature
The registered company is Springer (India) Pvt. Ltd.

Preface

This LNEE volume contains the papers presented at the ICMEET 2015: International Conference on Microelectronics, Electromagnetics and Telecommunications. The conference was held during 18–19 December, 2015 at Department of Electronics and Communication Engineering, GITAM Institute of Technology, GITAM University, Visakhapatnam, India. The objective of this international conference was to provide an opportunity for researchers, academicians, industry persons and students to interact and exchange ideas, experience and expertise in the current trend and strategies in the field of Microelectronics, Electromagnetics and Communication Technologies. Besides this, participants were also enlightened about vast avenues and current and emerging technological developments in the field of Antennas, Electromagnetics, Telecommunication Engineering, and Low Power VLSI Design. The conference attracted a large number of high-quality submissions and stimulated cutting-edge research discussions among many academic pioneering researchers, scientists, industrial engineers, and students from all around the world and provided a forum to researchers. Research submissions in various advanced technology areas were received and after a rigorous peer-review process with the help of programme committee members and external reviewers, 73 papers were accepted with an acceptance ratio of 0.33.

The conference featured distinguished personalities that include Dr. Lipo Wang (Nanyang Technological University, Singapore), Dr. V. Bhujanga Rao (Distinguished Scientist and Director General, Naval Systems and Materials), Prof. G.S.N. Raju (Vice Chancellor, Andhra University), Dr. Aditya K. Jagannadham (IIT Kanpur), Dr. Sanjay Malhotra (BARC, Trombay), Dr. Samir Iqbal (University of Texas, Arlington, USA), Dr. D. Sriram Kumar (NIT Trichy), Dr. P.V. Ananda Mohan (ECIL, Bangalore), Dr. G. Radha Krishna (IIT Madras), Mr. Rajan A. Beera (Global Director of Engineering, Cortland, NY), and Mr. G.S. Rao (Managing Director—Technology, Accenture). Separate invited talks were organized in industrial and academia tracks on both days. The conference also hosted a few tutorials and workshops for the benefit of participants. We are indebted to the management of Springer Publishers, GITAM University, for their immense support to make this conference possible at such a grand scale. A total of 13 sessions were organized as a part of ICMEET 2015 including ten technical, two plenary and one inaugural session. The Session Chairs for the technical sessions were Mrs. D.R. Rajeswari (Scientist-F, NSTL), Dr. V.S.S.N.S. Baba (Prutvi Electronics, Hyderabad), Dr. B. Prabhakar Rao (JNTU, Kakinada), Dr. S Srinivasa Kumar (JNTU, Kakinada), Dr. G. Sasibhushana Rao (Andhra University), Dr. P. Mallikarjuna Rao (Andhra University), Dr. P. Rajesh Kumar (Andhra University), Dr. P.V. Sridevi (Andhra University), Dr. A. Mallikarjuna Prasad (JNTU, Kakinada), Dr. N. Balaji (JNTU, Vizianagaram), Dr. Ch. Srinivasa Rao (JNTU, Vizianagaram), Dr. B. Tirumala Krishna (JNTU, Vizianagaram), Dr. Ibrahim Varghese (NSTL, Visakhapatnam), Dr. N. Bala Subrahmanyam (GVPCE, Visakhapatnam), Dr. M. Sai Ram (GVPCE, Visakhapatnam), and Dr. P. Ramana Reddy (MVGR College of Engineering, Vizianagaram).

We express our sincere thanks to members of the technical review committee and the faculty of Department of ECE, for their valuable support in doing critical reviews to enhance the quality of all accepted papers. Our heartfelt thanks are due to the National and International Advisory Committee for their support in making this a grand success. Our authors deserve a big thank you as it is due to them that the conference was such a huge success.

Our sincere thanks to all sponsors, press, print, and electronic media for their excellent coverage of this convention.

December 2015   Suresh Chandra Satapathy
                                                      N Bheema Rao
                                                      S Srinivas Kumar
                                                      C Dharma Raj
                                                      V Malleswara Rao
                                                      GVK Sarma

Advisory Committee

Dr. Tan Shing Chiang, Multimedia University, Malaysia
Dr. Matthew Teow, Yok Wooi, Head, KDU University, Malaysia
Dr. Lipo-Wang, NTU, Singapore
Prof. Fabrizio Bonani, Italy
Dr. A. Olatunbosun, University of Ibadan, Nigeria
Dr. Rajan A. Beera, Global Director of Engineering, Cortland, NY
Dr. K. Busawon, Professor, Northumbria University
Dr. V. Bhujanga Rao, DS & CC R&D (NS&IC)
Prof. R.V. Raja Kumar, IIT Kharagpur
Prof. G.S.N. Raju, Vice Chancellor, AU
Prof. B. Prabhakar Rao, Inc Vice Chancellor, JNTUK
Prof. S.K. Koul, Deputy Director (S&P), IIT Delhi
Prof. Shanthi Pavan, IIT Madras
Prof. Nandita Das Gupta, IIT Madras
Prof. Pallapa Venkata Ram, IISC Bangalore
Dr. Murthy Ramella Scientist-G, ISRO
Dr. P.V. Anada Mohan, ECIL, Hyderabad
Dr. C.D. Malleswar, Director, NSTL, Visakhapatnam
Dr. D. Sriram Kumar, NIT Trichy
Mr. Y.S. Mayya, Associate Director, BARC
Mr. V. Balaji, Head-Wireless Deployment, Reliance Corporate Park, Navi Mumbai
Prof. Ch. Ramakrishna, Director-UGC, GU

Principal Organizing Committee

Chief Patrons

Dr. M.V.V.S. Murthi, Hon'ble President, GITAM University
Prof. M. Gangadhara Rao, Ph.D., Vice-President, GITAM University

Patrons

Prof. K. Ramakrishna Rao, Hon'ble Chancellor, GITAM University
Prof. G. Subrahmanyam, Vice-Chancellor, GITAM University
Prof. D. Harinarayana, Pro Vice-Chancellor, GITAM University
Prof. M. Potharaju, Registrar, GITAM University

Organizing Committee

Chairman

Prof. K. Lakshmi Prasad, Principal, GIT

Co-chairman

Prof. C. Dharma Raj, Vice-Principal, GIT
Prof. V. Malleswara Rao, Head of the Department, Department of ECE, GIT
Prof. M.R.S. Satyanarayana, TEQIP Coordinator, GIT

Convener

Dr. P.V.Y. Jayasree, Assoc. Professor, Department of ECE

Co-convener

Dr. J. Beatrice Seventline, Assoc. Professor, Department of ECE

Proceedings Committee
Dr. GVK Sarma, Assoc. Professor, Department of ECE
Dr. K. Sri Devi, Assoc. Professor, Department of ECE
Dr. V.B.S. Srilatha Indira Dutt, Assoc. Professor, Department of ECE
Mr. B. Srinu, Asst. Professor, Department of ECE

Technical Programme Committee
Dr. T. Madhavi, Assoc. Professor, Department of ECE
Dr. S. Neeraja, Asst. Professor, Department of ECE
Dr. A. Satyanarayana Murthy, Asst. Professor, Department of ECE

Publicity Committee
Dr. G. Karunakar, Assoc. Professor, Department of ECE
Dr. I. Srinivasa Rao, Asst. Professor, Department of ECE
Mr. Ch. Raja Sekhar, Asst. Professor, Department of ECE

Finance Committee

Dr. Ch. Sumanth Kumar, Assoc. Professor, Department of ECE

Ms. K. Karuna Kumari, Assoc. Professor, Department of ECE

Mr. Ch.R. Phani Kumar, Asst. Professor, Department of ECE

Hospitality Committee

Dr. T.V. Ramana, Assoc. Professor, Department of ECE

Dr. A. Sreenivas, Assoc. Professor, Department of ECE

Ms. G. Radha Rani, Assoc. Professor, Department of ECE

Ms. D. Madhavi, Assoc. Professor, Department of ECE

Contents

About the Editors

Dr. Suresh Chandra Satapathy is currently working as Professor and Head, Department of CSE at Anil Neerukonda Institute of Technology and Sciences (ANITS), Andhra Pradesh, India. He obtained his Ph.D. in Computer Science and Engineering from JNTU Hyderabad and M.Tech. in CSE from NIT, Rourkela, Odisha, India. He has 26 years of teaching experience. His research interests include data mining, machine intelligence and swarm intelligence. He has acted as programme chair of many international conferences and edited six volumes of proceedings from Springer LNCS and AISC series. He is currently guiding eight scholars for Ph.Ds. Dr. Satapathy is also a senior member of IEEE.

Dr. N Bheema Rao is currently working as Associate Professor in the Department of Electronics and Communication Engineering, National Institute of Technology, Warangal, Andhra Pradesh, India. He has over 25 years of teaching experience. His areas of specialization include Design and Modeling of On-Chip Inductor for RF Applications and Device Modeling. He is an active researcher and has published more than 20 papers in various international and national journals. He has organized several MHRD sponsored short-term courses on "Analysis and Design of Electronic Circuits". He is a life member of INSEMIC. Under his esteemed guidance, two students were awarded with Ph.D. degree and three more are pursuing their Ph.D. He has also guided a number of postgraduate students.

Dr. S Srinivas Kumar is currently Professor in the ECE Department and Director, Research and Development, JNT University Kakinada, India. He received his M.Tech. from JNTU, Hyderabad. He received his Ph.D. from E&ECE Department, IIT Kharagpur. He has 28 years' experience of teaching undergraduate and postgraduate students and has guided a number of postgraduate and Ph.D. theses. He has published more than 50 research papers in national and international journals. His research interests are in the areas of digital image processing, computer vision, and application of artificial neural networks and fuzzy logic to engineering problems.

Dr. C Dharma Raj is currently working as Vice-Principal and Professor of Electronics and Communication Engineering, GITAM Institute of Technology, GITAM University, Visakhapatnam, India. He obtained his postgraduation degree from Osmania University and doctorate from GITAM University. He has over 25 years of teaching experience and 4 years of Industry experience (Electronics & Radar Development Establishment (LRDE), DRDO, Bangalore and MACE, Visakhapatnam). He held various administrative posts in the past, viz. Head, ECE Department; Director of Academic Affairs; Director of Student Affairs. His areas of specialization include Electromagnetic Interference and Compatibility. He is an active researcher and has published more than 60 papers in various international national conferences and journals. He has authored three text books with reputed International publishers (Electronic Devices & Circuits, Satellite Communications and Microwave Engineering). He has also guided a number of postgraduate students and is currently guiding a good number of Ph.D. students. He carried out a consultancy project with NSTL, Visakhapatnam, and was also co-investigator for a UGC Major Research Project.

Dr. V Malleswara Rao is currently working as Head of the Department and Professor of Electronics and Communication Engineering, GITAM Institute of Technology, GITAM University, Visakhapatnam, India. He is a postgraduate from Andhra University and received his doctorate degree from JNTU Kakinada. He has over 26 years of teaching experience. His areas of specialization include Signal Processing and Low Power VLSI Design. He is an active researcher and has published more than 40 papers in various international and national conferences and journals. He is currently guiding eight Ph.D. students and six other students were awarded the Ph.D. under his guidance. He has carried out a Major Research Project with UGC, India. He is currently working in Low Power VLSI Architectures for mobile applications.

Dr. GVK Sarma is currently working as Associate Professor in the Department of Electronics and Communication Engineering, GITAM Institute of Technology, GITAM University, Visakhapatnam, India. He has over 13 years of teaching experience and 2 years of Industry experience (Tejas Networks, Bangalore). His areas of specialization include MIMO Radar Signal Processing, VLSI Design and Underwater Communications. He has presented a number of papers at national and international conferences and has good publications in reputed journals. He has also guided a number of postgraduate students and is currently also guiding Ph.D. students. He delivers guest lectures on Radar and Sonar technologies at NSTL and at other academic organizations.

Subthreshold Operation of Energy Recovery Circuits

D. Jennifer Judy and V.S. Kanchana Bhaaskaran

Abstract This paper introduces a novel design methodology i.e. adiabatic subthreshold mode which inherits the features of both the subthreshold logic and the adiabatic (or the energy recovery) circuits. In this paper, analysis of essential digital gates is done for the three modes of operation namely, (1) subthreshold (2) adiabatic and (3) adiabatic subthreshold operation. The simulation results validate the benefits obtained in terms of the reduced energy consumption in the adiabatic subthreshold mode, making it suitable for ultra low power and medium throughput (10 kHz–1 MHz) applications. Specifically, it has been emphasized that the non-adiabatic dissipation that is prevalent in adiabatic circuits is almost discarded in the proposed mode. And, the challenges faced by the methodology are mitigated by the circuit level technique of upsizing the channel. Berkeley Predictive Technology Model (BPTM) 45 nm technology node has been used in the simulation studies on industry standard Spice tools.

Keywords Subthreshold · Device sizing · Adiabatic subthreshold · Nonadiabatic dissipation

1 Introduction

Generally, in a CMOS inverter, if $IN = 0$ and V is the supply voltage, only $\frac{1}{2}CV^2$ is delivered to the load capacitance for $OUT = 1$. The remaining $\frac{1}{2}CV^2$ is dissipated in the circuit transistors. Here, C represents the nodal capacitance. Then, when input signal IN becomes 1, the $\frac{1}{2}CV^2$ stored in the load capacitor is dissipated to the ground [1]. On the other hand, in the adiabatic logic, a time varying supply clock,

D. Jennifer Judy (✉) · V.S. Kanchana Bhaaskaran
ECE Department, VIT University, Chennai, India
e-mail: jennisteef@yahoo.co.in

V.S. Kanchana Bhaaskaran
e-mail: vskanchana@gmail.com

© Springer India 2016
S.C. Satapathy et al. (eds.), *Microelectronics, Electromagnetics and Telecommunications*, Lecture Notes in Electrical Engineering 372, DOI 10.1007/978-81-322-2728-1_1

called as *power-clock* is used, that reduces the voltage drop across the device at any time. The majority of the energy spent on charging the nodal capacitance is recycled back to the supply clock every time and is not dissipated to the ground as it occurs in CMOS. Hence, these circuits are called as *Adiabatic circuits*. This energy recovery property greatly reduces the switching power of the digital circuits. However, in the quasi-adiabatic circuits that consists of cross coupled PMOS transistors or cross coupled inverters at the outputs there is significant amount of non-adiabatic dissipation. This is due to the fact that the transistor helping in energy recovery remains *ON* only till the gate voltage is below its threshold voltage. When the output voltage reaches the threshold value of V_T, the recovery path transistor becomes *off* and no further energy recovery takes place. This contributes for the $\frac{1}{2}CV_{th}^2$ power dissipation, which happens every time when the input signal switches. This non adiabatic energy dissipation is a major setback in the Quasi adiabatic structures.

On the other hand, in the subthreshold logic, device sizing is done in such a way that they are operated with $V_{DD} < V_{th}$. In this weak inversion region, the leakage current or the subthreshold current is considered the computation current. It helps in surmounting the $\frac{1}{2}CV_{th}^2$ power dissipation drawback of the adiabatic circuits. Moreover, the exponential I-V relationship in the subthreshold region of operation increases the current gain. However, it is to be pointed out that, although the switching power is greatly reduced, subthreshold circuit operations are intended mainly for medium throughput applications. Additionally, due to the exponentially increasing current gain, the sensitivity of the circuit to the process and temperature variations increases. Nevertheless, the robustness of the circuit has been improved by different circuit level and device level techniques.

In this paper, it is proposed to utilize the advantages of both the subthreshold logic operation and the energy recovering adiabatic logic and launch a new mode of operation called *adiabatic subthreshold mode*. In adiabatic subthreshold mode of operation, the adiabatic circuit is device-sized to operate with the supply clock below the threshold value, i.e., in the weak inversion region. Here, the non-adiabatic dissipation discussed above is reduced to a bare minimum. Moreover, in this mode the process of near complete energy recovery property enhances the lower power operation capability even better than the conventional CMOS subthreshold operation. Further, the multi phase power-clocking in the adiabatic circuits assist in efficient pipelining due to their inherent nature of each clock lagging behind the other by 90°, with the successive power-clocks operating the cascade of circuits. The pipelining adopted in the adiabatic structures, may improve the throughput of the novel mode far better than the subthreshold circuits.

In this paper, the basic digital logic gates such as inverter, NAND, NOR, XOR, AND and OR shown in Fig. 1 are analysed for all the above three modes of operations to validate the claims. Each of the claim and the simulation outcomes are substantiated with necessary analytical support and conceptual clarification. Further, the proposed adiabatic subthreshold mode is illustrated with a 4 stage, 8 stage and 16 stage cascaded inverter chain to prove the operation of the adiabatic subthreshold operation with increased logical depth or higher latency.

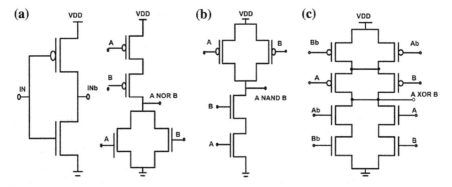

Fig. 1 Basic CMOS digital logic gates device sized for subthreshold operation **a** NOT, **b** NOR, **c** NAND

Section 2 discusses the device sizing and the impacts of subthreshold operation of the above mentioned basic digital logic gate circuits. Section 3 explains the circuit counterparts operated in adiabatic mode and detail out the features and benefits. Section 4 elaborates the adiabatic subthreshold mode of operation of the circuits listed in Sect. 3. Section 4 exemplifies a 4 stage, 8 stage and 16 stage cascaded inverter chain operated in all the three modes. Section 5 presents the simulation results, discussions and the analysis depicting the justification and validation. Section 6 concludes and the scope for the future work is also discussed.

2 Subthreshold Operation

2.1 Device Sizing

Scaling down of V_{dd} below the value of threshold voltage, for ultra low power designs require optimal sizing of transistors for accurate functionality, minimum delay, minimum power consumption and also symmetrical noise margin [2]. Thus, there is a need to obtain an optimal size ratio $\rho = (W/L)_P/(W/L)_N$. Changing ρ is done by changing the width of the transistors maintaining the length constant $(\rho = (W_p + \Delta W_p)/L_p/(W_n + \Delta W_n/L_n)$. PMOS transistor is made larger with respect to the NMOS device, maintaining the total size of the logic gate constant. Hence, in effect, C_L is maintained constant, to maintain the power dissipation of the gate also constant. Since the gate oxide capacitances contributing to the C_L is the same per unit area for both the PMOS and NMOS transistors, the energy does not vary so much. However ρ affects the delay, power and noise margin as follows.

- Increased width of the transistor will reduce the propagation delay due to the increased current. i.e., larger PMOS and smaller NMOS will give low t_{pLH}. However, it will increase t_{pHL}. Weaker PMOS and larger NMOS will result in reduced t_{pHL} and increased t_{pLH}.

- Changing ρ further changes the miller parasitic capacitance and hence the load capacitance C_L. This affects the total power consumption.
- Smaller PMOS and larger NMOS give reasonable noise margin low level *NML* and reduce t_{pHL} and the static power. However, larger PMOS and smaller NMOS improve *NMH* and reduce the t_{pLH}, however, at the cost of increased static power. Here, $NML = V_{IL} - V_{OL}$ and $NMH = V_{OH} - V_{IH}$.

Hence, in the presented work, an optimal size ratio ρ is designed for realizing minimum delay, minimum power consumption and reasonable noise margin characteristics. Thus, the device sizing is fixed as $\rho = (W_P = 135 \text{ nm}/L_P = 45 \text{ nm})/(W_N = 65 \text{ nm}/L_N = 45 \text{ nm})$ after analysing with various aspect ratios in the process corner models Fast PMOS/Slow NMOS and Slow NMOS/Fast PMOS for wide range of supply voltages below the threshold value $V_T = 0.61$ V.

2.2 Subthreshold Circuits

Subthreshold operation employs the leakage current as the operation current by maintaining the supply voltage $V_{dd} < V_{TH}$. The I-V characteristics of the devices in the weak inversion region are entirely different from the strong inversion region as shown in Fig. 2.

The weak inversion current has an exponential behaviour in contrast to the linear behaviour of the strong inversion region operation. The subthreshold current is given by the equation

$$I_{sub} = I_o \times e^{((VGS-VTH/nvt))} \times (1 - e^{(-VDS/vt)}) \times e^{\eta VDS/nvt} \tag{1}$$

$$I_o = \mu_o C_{ox} \times W/L(n-1)vt^2 \tag{2}$$

Here, V_{GS} is the transistor gate to source voltage, V_{DS} is the drain to source voltage, V_{TH} is the threshold voltage, v_T is the thermal voltage, η is the DIBL

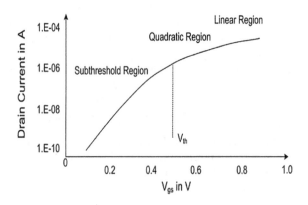

Fig. 2 Exponential behaviour of the drain current in the subthreshold region

coefficient, n is the subthreshold swing coefficient of the transistor, μ_o is the zero bias mobility, C_{ox} is the gate oxide capacitance, W and L are the width and length of the transistor respectively [3]. The equation signifies the exponential dependence of the subthreshold current on V_{DS}. This is the reason for the high transconductance gain of the circuit and the ideal VTC characteristics of the gate.

In the super-threshold region, a transistor enters the saturation region only when $V_{DS} > V_{GS} - V_T$ which gives a much narrower saturation region and thus an undesirable voltage transfer characteristic (VTC). However, the devices in the subthreshold region, the drain current saturates and becomes independent of V_{DS} for $V_{DS} > 3kT/q$ (~ 78 mV at 300 K). Hence, the devices are good current sources in the subthreshold region in the operating voltage range of *3kT/q* to V_{dd} [4, 5].

Furthermore, the exponential dependence of the drain current on the V_{GS} voltage increases the sensitivity of the circuit to the process and temperature variations and hence, the robustness of the device is declining which is one of the challenges faced in the subthreshold circuits. An added consequence of the subthreshold region is the decrease in the input gate capacitance, where the gate capacitance is given by

$$Ci = series(C_{ox}, C_d)||C_{if}||C_{of}||C_{do} \tag{3}$$

Here, C_{ox}, C_d are the oxide and depletion capacitances, C_{if} and C_{of} are the fringe capacitances and C_{do} is the overlap capacitance. The second order effects of the MOS devices are also less pronounced in the subthreshold region. The subthreshold currents are weaker and hence, the time taken for charging and discharging the nodal capacitance is longer, as given by

$$T_d = C_L V_{DD}/I_{on} \tag{4}$$

Thus, operating the devices in the weak inversion region exhibits several benefits such as 1. High current gain 2. Lower power dissipation 3. High noise margin 4. Low input gate capacitance 5. Reduced gate tunnelling current, Gate induced drain leakage and reverse biased diode leakage. However, the sensitivity of the sub-threshold circuits to the PVT variations and the low throughput of the devices operating in the subthreshold regime are major obstacles that need to be eradicated to acquire the benefits [6, 7]. The robustness of the circuits can be enhanced by various circuit level and device level techniques such as channel upsizing and body biasing [8].

Extending the subthreshold logic to adiabatic circuits can improve power efficiency of the energy recovery circuits for ultra low power operation besides deriving several benefits from each of the low power schemes. In order to illustrate the claims on the proposed adiabatic subthreshold mode, CMOS basic digital logic gates (AND, NAND, OR, NOR, NOT, XOR) are device sized to operate in subthreshold region. And also, the adiabatic designs of logic gates are evaluated in the strong inversion and weak inversion region of operation (adiabatic subthreshold mode) as discussed in the subsequent sections.

3 Adiabatic Circuits

Figure 3a–c show the (Efficient Charge recovery Logic) ECRL adiabatic structures of the digital logic gates. In these adiabatic circuits, it is noted that the constant power supply V_{DD} of the conventional CMOS circuits is replaced with a power-clock, which is trapezoidal in shape with four phases, as shown in Fig. 4. The cascaded successive stages of the ECRL adiabatic gates are driven by the four phases of the power clock PCLK1, PCLK2, PCLK3 and PCLK4. This pipelining feature of the ECRL adiabatic architecture lessens the energy dissipated per stage and enhances the throughput [9, 10].

The time varying power clock reduces the potential across the switching devices at any time, and this minimizes the dissipation in the pull-up and pull-down transistor networks. The energy delivered to the circuit nodes during *Evaluate* phase is recycled back to the supply during the *Recovery* phase. The output nodal value is cascaded to the subsequent stage in the adiabatic pipeline during the Hold phase [11]. This adiabatic dissipation of the power-clocked circuit during the evaluating/recovery phase is given by

$$E_{adia} = R_{on} C_L / TC_L V_{DD}^2 \tag{5}$$

$$R_{on} = [\mu_{effp} C_{ox} W_{effp} / L_{effp} (V_{DD} - |Vthp|)]^{-1}. \tag{6}$$

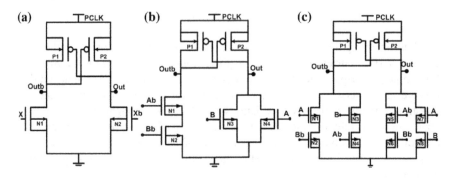

Fig. 3 ECRL **a** INV/BUF, **b** AND/NAND, **c** XOR/XNOR

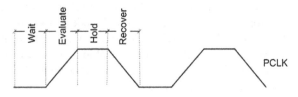

Fig. 4 Four phase power clock

where R_{on} is the ON resistance in the charging path, C_L is the load capacitance and T is the power clock period. However, there prevails a non-adiabatic component of power that is irrecoverable [12]. This is due to the fact that the PMOS transistors are maintained *on* only as long as its source voltage is more than the threshold voltage V_T [13]. The non adiabatic dissipation is the sum of energy components due to the V_{thp} and the leakage. i.e., $E_{nad} = E_{vthp} + E_{leakage}$.

It can be expressed by the equation as depicted below,

$$E_{nad} = 1/2 C_L V_{thp}^2 + V_{DD} I_{leak} kT \tag{7}$$

Then, the total energy loss E_{diss} per switching event of the adiabatic circuits is given by,

$$E_{diss} = 2\left(\frac{R_{on} C_L}{T}\right) C_L V_{DD}^2 + 1/2 C_L V_{thp}^2 + V_{DD} I_{leak} kT \tag{8}$$

Expanding,

$$E_{diss} = 2 C_L^2 V_{DD}^2 \left[T \mu_p C_{ox} \left(\frac{W}{L}\right)_p (V_{DD} |Vthp|) \right] + 1/2 C_L V_{thp}^2 + V_{DD} I_{leak} kT \tag{9}$$

where n, V_T, μ, is subthreshold slope factor, thermal voltage and mobility of the leaking transistors respectively [14]. It is inferred from the Eqs. (7) and (9), the adiabatic power dissipation component decreases with reducing frequency, while the nonadiabatic dissipation is independent on frequency. This detrimental feature of the adiabatic circuits is subsisting at all frequencies and hence is a design constraint and need to be sorted out. Thus, there come into view the novel approach i.e. adiabatic subthreshold mode that inherits the features of the subthreshold and the adiabatic styles of operation. Consequently, this proposed mode also exhibits most of the benefits of both the techniques as detailed out in the next section.

4 Adiabatic Subthreshold Logic

This section presents the adiabatic subthreshold mode of operation, i.e., the adiabatic structures are operated in the weak inversion region. The adiabatic structures are device sized for subthreshold operation and power clock *PCLK* voltage is held below the threshold value V_T for operation in the weak inversion region. This novel methodology acquires most of the merits of both the low power schemes which is listed as below.

- The dynamic power dissipation $P_d = \alpha C_L V_{dd}^2 f$ is minimized as the nodal capacitance C_L is reduced by device sizing due to the reduction in the gate capacitances, V_{DD} being less than the threshold voltage and the power-clock period T being longer. (meaning lower power-clock frequency)
- Besides, absolute energy recovery made possible through the use of time varying power clock improves the power efficiency to a greater extent.
- And, most crucially it is also justified that the non-adiabatic power dissipation is also brought down to an utmost low value in the proposed mode as given by the Eq. (10).

$$E_{nad} = \frac{1}{2} C_L V_{thp}^2 + V_{DD} \mu C_{ox} \left(\frac{W}{L}\right) V_T^2 \exp^{Vgs-Vth/\eta\, V_T} kT \qquad (10)$$

The non-adiabatic dissipation is dependent on 1. The power supply voltage PCLK that is now scaled down to a value below the threshold voltage 2. The nodal capacitance that is lessened by the reduction in the gate capacitance and 3. The value of threshold voltage that is come down to 3kT/q = 78 mV in the subthreshold region.

- Moreover, the gate tunneling current, reverse biased current DIBL, GIDL leakage currents are minimized due to their dependence on the supply voltage which is now below the threshold value.
- The weaker currents may increase the delay in the subthreshold mode and decrease the frequency of operation. However, the pipelining adopted in the adiabatic subthreshold circuits improves the throughput.
- Device optimization of the adiabatic circuits for subthreshold operation also improves the power delay product.
- Moreover, the current source property (drain current saturates at $V_{DS} > 3kT/q$) in the subthreshold region enables better pass gate logic or in other words reduces the voltage degradation in series connected devices. This is owing to the fact that the voltage drop across the devices is just the *3kT/q* drop ($\sim$78 mV) as against the full V_{th} value of around 700 mV.

The claims are exemplified with the basic digital gates and a 4 stage, 8 stage and 16 stage cascaded inverter chain in all the above three modes of operation analysed for power efficiency. A 4 stage inverter chain is depicted in Fig. 5 for better

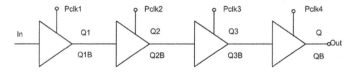

Fig. 5 Cascaded inverter chain

understanding in which the successive stages are driven by the four phases of the power clock. However, in the subthreshold mode, power clock will be replaced by a constant supply voltage V_{DD}.

5 Simulation Results and Analysis

The circuits are designed using the BPTM 45 nm process technology models and are simulated using TSPICE. The basic CMOS digital logic gate circuits shown in Fig. 2 are device sized (W_P = 135 nm; L_P = 45 nm) (W_N = 65 nm; L_N = 45 nm) to operate with a constant V_{DD} set at value 0.4 V for subthreshold operation. And, four-phase trapezoidal power-clocks with peak-to-peak voltage of 1.1 v (V_T = 0.61 v) are used to power up the ECRL circuits for the adiabatic mode of operation. Also, the ECRL circuits are device sized to operate with the time varying power clocks PCLK1, PCLK2, PCLK3 and PCLK4 fixed at peak to peak value of 0.4 V. Thus the power dissipation results so obtained in all the three modes of operation, i.e., subthreshold logic, adiabatic logic and adiabatic subthreshold logic for the basic logic gates have been depicted in Fig. 6.

The graph in Fig. 6 portrays the reduced dynamic power dissipation in the proposed mode due to smaller operating voltage, less nodal capacitance and reduced operating frequency.

Further, it has already been stated that the voltage degradation of series connected devices is reduced in subthreshold operation due to the current source property in the weak inversion region. Thus the logical depth that can be attained in the adiabatic subthreshold mode can defensed with simulation of the 4 stage, 8 stage and 16 stage inverter chains. The power results are plotted in Fig. 7. On comparison, the 16 stage inverter in the subthreshold operation consumes 570 nW, while the dynamic power dissipation incurred by the proposed subthreshold adiabatic counterpart decreases to 27 nW. Thus, a power efficiency of 95 % is realized in the novel mode of operation.

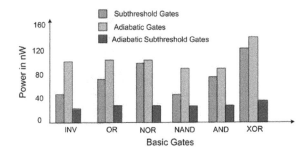

Fig. 6 Simulation results of the gates in the subthreshold, adiabatic, adiabatic subthreshold modes of operation

D. Jennifer Judy and V.S. Kanchana Bhaaskaran

Fig. 7 Simulation results of the power dissipation of the cascaded inverters

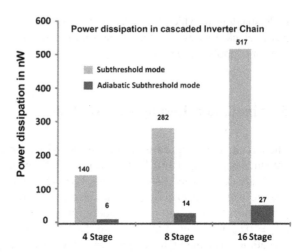

The frequency range of operation of adiabatic subthreshold circuits is held at 10 kHz–1 MHz and this improved throughput is mainly by the pipelining feature adopted in the ECRL architecture. This makes the subthreshold circuits suitable for medium throughput applications.

And also, it is clear that in the adiabatic mode when the power-clock reaches the threshold voltage during its negative ramping (or recovery phase); the recovery transistor is cut-off, leaving $1/2C_L V_{thp}^2$ of power dissipation unrecovered. This non adiabatic dissipation is narrowed to the least in the proposed mode. As inferred from Eq. (10), it is explicit that the scaled supply voltage, the threshold voltage which is now reduced to $3kT/q$ (=78 mV) instead of V_T (0.61 V) and the decreased leakage currents diminishes non-adiabatic component in the total power dissipation. The above analytical argument is proven with the simulation results of the 16 stage cascaded inverter chain shown in Fig. 8. It is validated that the inherent non adiabatic dissipation is 1600 nW in the adiabatic logic and reduced to 25 nW in the

Fig. 8 Non adiabatic power dissipation of a 16 stage inverter chain

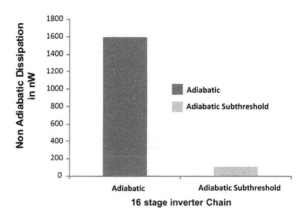

adiabatic subthreshold logic of a 16 stage inverter chain showing 98.4 % power efficiency.

However, the novel mode faces several challenges due to its sensitivity to the PVT. Due to these reasons, the noise margin is reduced and the robustness is declined. The above effects are addressed by upsizing the channel length to $L_P = L_N = 45$ n. The variability given by $\sigma V_T = \sqrt{WL_{eff}}$ is mitigated by increasing the channel length to several nanometers [15]. Due to the increased channel length, current gain is improved, reasonable noise margin is attained, and the robustness to variability is progressing. Moreover, the static power, which is larger in the subthreshold mode, is reduced by alleviating the leakage currents by channel length upsizing thus improving the energy gain.

Thus it is ascertained that the proposed adiabatic subthreshold mode reveals several advantages in terms of power efficiency discarding even the irrecoverable non adiabatic dissipation of the adiabatic circuits. Moreover, the throughput of the adiabatic subthreshold mode is escalated by opting the pipelining feature in the adiabatic circuits and thus fits well for medium throughput ultra low power applications.

6 Conclusion

This paper validates the benefits obtained in operating the adiabatic circuits in the subthreshold region. The main drawback imposed by the energy recovery design approach of the adiabatic circuits is the irrecoverable nonadiabatic dissipation that is surmounted by operating the same in the subthreshold region. The design methodology is proved using digital logic gate circuits and a larger circuit, namely, a 16-stage inverter chain with increased latency of 16 stages of logical depth to prove the capability of pipelining in the adiabatic circuits. It is also proven that the operating frequency of these circuits can range from hundreds of KHz to tens of MHz, with much lower power consumption compared to the traditional adiabatic and subthreshold counterparts. The future scope of the design may rely on incorporating the designed novel circuits to real time biomedical applications such as pacemakers, defibrillators and other applications such as RF IDs, sensor networks and battery operated portable devices that demands ultra low power and medium throughput.

References

1. J. Rabaey, *Digital Integrated Circuits* (Prentice Hall, 2003)
2. H. Soeleman, K. Roy, Ultra-low power digital subthreshold logic circuits. *Proceedings IEEE/ACM International Symposium Low Power Electron Devices*, pp. 94–96 (1999)
3. B.H. Calhoun, A. Wang, A.P. Chandrakasan, Modelling and sizing for minimum energy operation in subthreshold circuits. IEEE J. Solid-State Circuits **40**(9), 1778–1786 (2005)

4. A. Wang, B.H. Calhoun, A. Chandrakasan, *Sub-threshold Designs for Ultra Low-Power Systems* (Springer Publishers, 2005)
5. H. Soeleman, K. Roy, B.C. Paul, Robust subthreshold logic for ultra-low power operation. IEEE Trans. VLSI Syst. **9**, 90–99 (2001)
6. H. Soeleman, K. Roy, B.C. Paul, Robust ultra-low power sub-threshold DTMOS logic. *International Symposium Low Power Electron Design* (2000)
7. A.P. Chandrakasan, S. Sheng, R.W. Brodersen, Low- power CMOS digital design. IEEE J. Solid-State Circuits **27**, 473–484 (1992)
8. A.P. Chandrakasan, R.W. Brodersen, Minimizing power consumption in digital CMOS circuits. Proc. IEEE **83**, 498–523 (2005)
9. W.C. Athas, Low-power digital systems based on adiabatic switching principles. IEEE Trans. VLSI **2**(4), 398–406 (1994)
10. A.G. Dickinson, J.S. Denker, Adiabatic dynamic logic. IEEE J. Solid-State Circuits **30**(3), 311–315 (1995)
11. K.T. Lau, F. Liu, Improved adiabatic pseudo-domino logic family. Electron. Lett. **33**(25), 2113–2114 (1997)
12. Y. Moon, D.K. Jeong, An efficient charge recovery logic circuit. IEEE J. Solid-State Circuits **31**(4), 514–522 (1996)
13. V.S. Kanchana Bhaaskaran, J.P. Raina, Two phase sinusoidal power-clocked quasi-adiabatic logic circuits. J. Circuits, Syst. Comput. **19**(2), 335–347 (2010)
14. V.S. Kanchana Bhaaskaran, J.P. Raina, Differential cascode adiabatic logic structure for low power. J. Low Power Electron. **4**(2), 178–191 (2008)
15. B. Zhai, S. Hanson, D. Blaauw, D. Sylvester, Analysis and mitigation of variability in subthreshold design. *Proceedings International Symposium Low Power Electronics and Design*, pp. 20–25 (2005)

Design and Analysis of a Low Cost, Flexible Soft Wear Antenna for an ISM Band Working in Different Bending Environment

I. Rexiline Sheeba and T. Jayanthy

Abstract A low cost, flexible software antenna for ISM band is presented. The Novel antenna is proposed for ISM Band applications. Pure 100 % Cotton is used as dielectric substrate material with dielectric constant 1.6. This antenna is flexible and suitable for wearable applications. The designed antenna resonates at ISM (Industrial, Scientific, and medicine) band with a return loss of more than −25 dB. The simulated and measured results show the performance in terms of Return Loss, Radiation pattern which shows the efficiency of the proposed antenna and this flexible softwear antenna is measured in various bending environments are presented in this paper. Investigation focuses on an ordinary cotton cloth with 3 mm thickness, used as its substrate, and the patch and ground plane are made up of copper as conducting material together to form a flexible textile antenna. Proposed antenna is tested in various bending condition. Such Textile antenna designed for an ISM Band 2.45 GHz. Its radiation characteristics, return loss, gain, polarization have been examined which are the issues when it is used as a wearable antenna for medical purpose. Since it is a flexible textile antenna it bends for any condition. Observations were done for various diameter PVC pipes which is equivalent to the human body organs like arm, elbow, forearm, wrist or in the leg, ankle, knee, thigh and its resonant frequencies were noted. One of the advantages of these characteristics is once the antenna is flexible and bends in any condition then the specific absorption rate can be reduced, when this antenna is placed on the human body.

Keywords Softwear antenna · ISM band · Soft substrate · Bending environment · Flexible textile antenna · Human body physical structure

I. Rexiline Sheeba (✉)
Sathyabama University, Chennai 600119, India
e-mail: Sheebarexlin@gmail.com

T. Jayanthy
Panimalar Institute of Technology, Chennai 602103, India
e-mail: jayanthymd@rediffmail.com

© Springer India 2016
S.C. Satapathy et al. (eds.), *Microelectronics, Electromagnetics and Telecommunications*, Lecture Notes in Electrical Engineering 372, DOI 10.1007/978-81-322-2728-1_2

1 Introduction

Wireless communication devices and techniques are flourishing, convalescing and escalating nowadays. The improvement of such devices should assemble precise requirements miniature dimension, light weight, low cost with attractive appearance. To improve the characteristic of wearable microstrip antennas many techniques were developed. In recent years, wearable devices are getting popular and dominating in electronic industries. By using suitable materials such as textiles and foams, the electronic systems can be integrated into clothing. These smart textile systems can be deployed in different fields, and have been shown to function in garments. This low profile antenna is suitable for wearable applications and microstrip patch antenna topology is chosen. An Electronic Device Worn by a person said to be a Wearable contrivance. If it is a wearable one, then it should expect to be a light weight, low profile. One such device worn by a person for communication purposes such as navigation, monitoring health issues and is widely used in military and medical application said to be a wearable microstrip antenna. They enable the integration of flexible, robust conductive textiles to form the radiator and ground plane. Textile antennas already have been successfully implemented with satisfactory performance. Conductive textiles, metal foils can be used as the radiating element [1–3]. In 1993; FCC allocated 40 MHz of unlicensed band in the 1890–1930 MHz band. Several years later, the FCC also unlicensed the 5.15–5.35 GHz and 5.725–5.825 GHz frequencies considered as the existing 5 GHz ISM band.

The proposed Softwear antenna uses soft substrates in the microstrip patch antenna. Moreover this softwear microstrip patch antenna is used as a wearable one because of its compact size, light weight and ease of integration in clothes. So Textile substrates are used as soft substrates in this softwear antenna, wearable and textile antenna properties are in two dimensions. The Combination of textile antenna and wearable properties referred as softwear antenna in which it is 2-D flexible along two planes also it is optimized to perform proximity of the human body [1]. Researchers are focusing on such type of antenna because of its wearable system technology. User body and the characteristics of the antenna should be maximized for the coupling of antenna and human body interaction, treatment of malignant tumour can be found by using patch radiator and its operation is simple in microstrip patch antenna, strip line is separated with a separation which is flexible used to measure the human body temperature [2]. Several wearable antennas have been developed in the form of flexible metal patches on soft substrates which uses textile material. A new Hexagonal patch is proposed which is operating in industrial, scientific and medical frequency band at 2.45 GHz and was verified by the numerical techniques like Finite Element Method (FEM) and the method of moments (MOM) and the effect on the human body is known by its resonance frequency and gain. For Simulation human tissue is modelled as multi-layer's, for skin, fat, and muscle, various ε_r and σ value have been investigated to create a model of human tissue [3]. General scenarios of wireless body centric communication are namely, off-Body, in-body and on body. Also the Intra body and Interbody communications also explained in this review which deals Wireless Body Area Network (WBAN). IEEE 802.15.1 (Bluetooth) in (PAN) personal Area Networks (WPAN) is widely used, which extends the propagation range. Efficiency and gain of the antenna can be analysed in 3 parameters like Antenna-distance from the body, Location on the body and the type of the antenna, also the dispersive electrical properties of the human body is lossy at higher frequencies presence of human body changes the operating frequency of the antenna [4]. Normally human body is composed of water with dielectric constant and conductivity. When the metal based antenna placed on the skin, it reflects from the body. When EM waves coincide on the skin, a then there is a change in its resonant frequency. Electromagnetic Interference (EMI) between the human body and the antenna is calculated by Specific Absorption Rate, it is the rate of heat generated by the antenna and was sensitized as heat on the body surface. High dielectric constant increases surface wave losses and Bandwidth of the antenna decreases the impedance Bandwidth [5]. In medical application when a patient is to carry such radiator which constantly communicates the outside world can use this wearable technology working in various bending environments. In on body environment to keep the wearable antenna flat all the time is difficult when a patient worn on clothing. Due to the patient's body movement there is a possibility of bending the antenna in any condition. Also, this bending may modify the characteristics of the antenna like resonant frequency, Magnitude, return Loss etc. In general diameter of the human body organs are different; it also depends on the age factor. When this antenna is placed on the organs such as arm, elbow forearm, wrist or in leg, ankle, knee, thigh it is flexible and works in any pliable situation. This paper proposed some of the key features related to the wearable antenna design process include Textile material selection, material conductivity, antenna performance on various bending environment. Simulation and investigational interpretation were made on the performance characteristics of this Flexible softwear microstrip antenna are also explained rest of the section shows the bending performance of the antenna in various diameter using PVC pipes which represents the diameter of the human arm, forearm elbow, wrist, ankle foot, and also wherever bending is possible in human body according to the body movement.

1.1   Substrate Material

Pure 100 % cotton material is chosen with a firm and smooth surface and it is suitable for wearable applications. Thickness of this material is 0.3 mm. The electrical evaluations should be performed before the establishment of any kind of soft substrate material. It is important to know the dielectric permittivity of the chosen cotton material. If the dielectric constant is more, then gain, directivity, and efficiency increases. To Perform the effect of textile material Relative permittivity $\varepsilon_r = 1.6$ cotton material is selected. Comparing with other cotton materials like

Wash cotton, curtain cotton, poly cotton and Jean cotton the relative dielectric permittivity value is more [6]. If the dielectric permittivity value is more then performance of the antenna is more. Low dielectric constant in textile substrate reduces the surface loss and improves the impedance bandwidth of the antenna.

1.2 Conducting Material

To establish the communication system a conducting material with its electrical characteristics are required for the ground plane as well as the patch of the antenna. Material conduction should satisfy several requirements such as having a low and stable electrical resistance (1 Ω/square) in order to minimize losses [2, 7–9]. Variance of the resistance throughout the area should be small. Also conducting material should be flexible when it is worn, also when the antenna is deformed to any radius. The material used in such type of antenna should be in elastic because of bending, stretching and compression is possible when it is worn or integrated within the cloth [3, 4, 10–13]. Conducting properties of various materials plays major role in achieving the desired performance of antenna designs and also in fabrication. An impedance matching element controls the impedance bandwidth of the patch. In this flexible softwear antenna ground plane acts as an impedance matching element which create a capacitive load neutralizes the inductive nature of the patch to produce pure resistive input impedance. The proposed work focused on copper which is used as the conducting material because of its flexibility on the substrate. Copper has good flexibility and surface resistivity used in both patch and ground plane. One end of the patch is made up of copper and the other end conducting plane is also made up of copper. This is flexible for any bending radius which is possible when integrated in cloth or worn by the user.

1.3 Relationship Between Permittivity ε and Conductivity σ

Permittivity ε and conductivity σ are complex quantities expressed in real and imaginary parts as

$$\varepsilon = \varepsilon' - j\varepsilon'' \tag{1}$$

$$\sigma = \sigma' - j\sigma'' \tag{2}$$

Effective permittivity ε_e and the effective conductivity σ_e are defined as

$$\varepsilon_e = \varepsilon' - \sigma''/\omega \tag{3}$$

$$\sigma_e = \sigma' + \omega\varepsilon'' \tag{4}$$

Loss due to conductivity is expressed in dissipation factor or it is said to be tangent tan δ, which is defined as

$$\tan\delta = \frac{\mathrm{Im}[\varepsilon_e]}{\mathrm{Re}[\varepsilon_e]} = \frac{\sigma_e}{\omega\varepsilon_e} \tag{5}$$

Refraction Index of a substrate and patch includes both parameters

$$n = \sqrt{\varepsilon_r \mu_r} \tag{6}$$

ε_r Relative permittivity,
μ_r Relative Permeability

Ratio of space-wave radiation to surface wave radiation can be found for any small antenna mounted on the substrate and it can be applied to a patch. To achieve efficiency and high Gain dielectric constant should be decreased so that it increases the spatial waves which increase the bandwidth of the antenna. The Relative permittivity ε_r value changes as bandwidth changes. Dielectric constant thickness determines the bandwidth and efficiency performance of this planar textile softwear antenna. Low relative permittivity results in a wide patch and a thin substrate results in smaller patch.

2   Antenna Design Consideration

To have a low profile planar antenna this can be integrated into clothing. One of the familiar topology of microstrip antenna is preferred. This ensures radiation away from the body with sufficient bandwidth for a good coverage. Here the Use of 100 % cotton material is used as the dielectric substrate with a ε_r value of 1.6. Copper metal in the patch is acting as the radiating surface with a thickness of 0.1 mm. HFSS software is used for the design of the proposed antenna and the results were simulated for the antenna with the final dimensions in mm. L1 = 54.8 mm, L2 = 47.1 mm, h = 3 mm. Due to the larger physical area and higher bandwidth and ease of fabrication rectangular microstrip antenna is chosen. A 50 Ω microstrip feed line was provided for the antenna feed and SMA connector is the feeder for the microwave power.

$$w = \frac{c}{2f_r\sqrt{2/(\varepsilon_r + 1)}} \tag{7}$$

C   Velocity of Electromagnetic wave
ε_r Relative permittivity of the cotton Textile material
f_r Resonant frequency

Microstrip patch lies between dielectric material and air, thus electromagnetic waves related to effective permittivity (ε_{reff}) given by the expression introduced by Balanis [14], as shown in Eq. 8

$$\varepsilon_{reff} = \left[\frac{\varepsilon_r + 1}{2}\right] + \left[\frac{\varepsilon_r - 1}{2}\right][1 + 12h/w]^{-1/2} \tag{8}$$

where, h-Height of the substrate

Because of the narrow bandwidth of the patch, the resonant frequency depends on the length of the patch; design value of L is given by

$$L = \left[\frac{c}{2f_r\sqrt{\varepsilon_{reff}}}\right] - 2\Delta L \tag{9}$$

ε_{reff}-effective permittivity, additional line length ΔL and effect of fringing fields

$$\frac{\Delta L}{h} = 0.412\left[\frac{[\varepsilon_{reff} + 0.3]}{[\varepsilon_{reff} - 0.258]}\right]\left[\frac{w}{h}\right] + \frac{0.264}{\frac{w}{h}} + 0.8 \tag{10}$$

Effective patch length L_{eff} is given by

$$L_{eff} = L + 2\Delta L \tag{11}$$

Resonant Frequency of a planar rectangular patch antenna determines W and L, thickness t and permittivity ε_{reff} of the dielectric

$$f_{mn} = \frac{c}{2\sqrt{\varepsilon_{eff}}}[(m/L)^2 + (n/W)^2] \tag{12}$$

m, n-mode numbers when an antenna curved in an arc along its length

$$L = R\theta \tag{13}$$

R-Radius of curvature, θ-angle subtended by the patch length.

Equation (1) can be modified with the variation in the effective length by assuming the new length lies along an arc midway through the dielectric material, by ignoring the changes to the fringing field, new length can be given as

$$L_{eff} = \left(\frac{L}{\theta} - \frac{t}{2}\right)\theta = L - \frac{t\theta}{2} \tag{14}$$

Effective length L_{eff} becomes L in [14] and $\theta = 0$

A change in antenna performance reflects both increased and decreased frequencies depending on the bending environment. The curvature effect of the antenna on its resonant frequency was given by Krowne [15, 16] as

$$(f_r)_{mn} = \frac{1}{2\sqrt{\varepsilon\mu}} \sqrt{\left(\frac{m}{2\theta a}\right)^2 + \left(\frac{n}{2b}\right)^2} \tag{15}$$

2b-length of the patch antenna, a-radius of the cylinder, 2θ-angle bounded the patch width, ε, μ-permittivity, permeability.

2.1 Results and Discussion

Proposed softwear microstrip rectangular patch antenna has been modelled using HFSS (Fig. 1). HFSS is a commercial finite element method solver for EM structures. This software is provided with a linear circuit simulator with an integrated optimetrics for electrical network design. The geometrical construction and its material properties and also the desired output frequency should be specified. HFSS integrates an automated solution process HFSS automatically generates an appropriate and accurate mesh analysis for the given geometry [17].

Antenna geometry using HFSS have been shown in Fig. 1. Which shows the geometric construction of the flexible softwear antenna. Simulated $|S11|$ parameter in dB. (20Log10 $|S11|$) also said to be $|S11|$, the radiation pattern of the softwear antenna and the directivity are shown in Figs. 2, 3 and 4 respectively. Performance of the antenna depends on the return loss. As the return loss increases antenna performance also increases. More than −25 dB was achieved in the above observation.

Fig. 1 Simulated softwear antenna

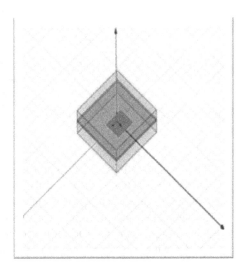

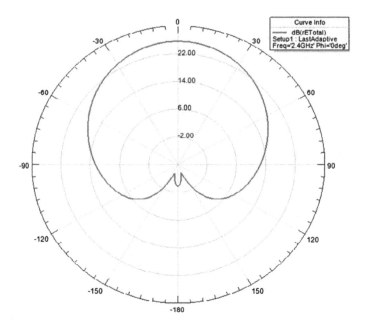

Fig. 2 Radiation pattern of the sofwear antenna

Fig. 3 Directivity

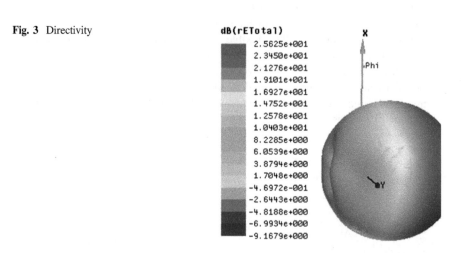

Above snapshot Figs. 5 and 6 shows the front and back side of the fabricated flexible softwear antenna. Measured S11 parameter using a Network Analyzer have been shown in Figs. 7 and 8.

The simulated and measured results shows the performance of the proposed flexible softwear antenna The proposed antenna is working in the frequency of 2.39 GHz, which is almost near to the simulated output of 2.4 GHz shown in

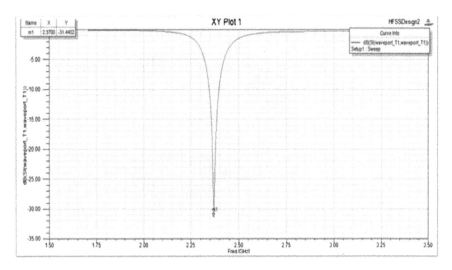

Fig. 4 Return loss of the proposed antenna

Fig. 5 Photographs of the
front and back side of the
flexible soft wear antenna

Fig. 6 Photographs of the
front and back side of the
flexible soft wear antenna

Fig. 7 Snapshot of the
measurement of S_{11} using
network analyzer

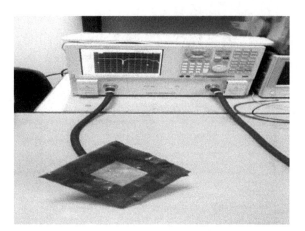

Fig. 8 Flexible softwear
antenna measured result

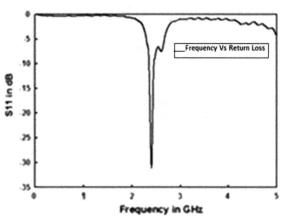

Table 1. When this flexible softwear is tested in various pliable condition different
frequencies were obtained which evidences, the proposed softwear antenna is
flexible and working in various meandering environment.

The measured return loss (s11) characteristics of the antenna under different
bending condition on the poly vinyl chloride (PVC) pipes of radii 11, 5, and 3 cm
respectively. Which is shown in above Figs. 9, 10, 11 and 12. When the flexible
softwear antenna is bent on the pipe of radius 11 cm, it resonates at 5.59 GHz with
the magnitude of (−12.414) dB as shown in Fig. 9. Similarly when the antenna bent
on the pipe radius of 5 cm the resonant frequency of the antenna shifted to

Table 1 Simulated and
measured output

Simulated output	Measured output
2.45 GHz	2.39 GHz

Fig. 9 Flexible softwear antenna placed on PVC pipe

Fig. 10 Antenna under bending on 11 cm dia PVC pipe

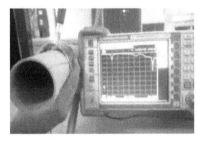

Fig. 11 Antenna under bending on 5 cm dia PVC pipe

(5.62 GHz) with the magnitude of (−36.9) dB and when the bending radius is about 3 cm then the resonant frequency is shifted to 5.7 GHz with the magnitude of (−29 dB). Experimental results shows in any bending condition resonant frequency of the proposed antenna increases as diameter decreases. It shows when more or less bending taking place while this flexible softwear antenna is placed on any of the above said organ of the human body then the resonant frequency oscillate in and around of the corresponding Resonant frequency. Also it was observed that when the pipe is kept in horizontal or vertical position same resonant frequencies and the corresponding magnitudes are obtained. The resonant frequency obtained from the sample diameters of the PVC pipes are coming under the ISM band between 5.5

Fig. 12 Antenna under
bending on 3 cm dia PVC
pipe

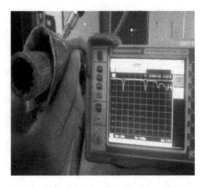

and 5.7 GHz. which shows the proposed flexible softwear antenna is working in ISM band.

Diameter of the circle is the longest distance in a circle, when the flexible softwear antenna is placed on PVC pipes with different diameter, shown Shift in its resonant frequency and its magnitude in dB shown in Table 2. The measured results show as the diameter increases resonant frequency decreases, and magnitude increases, it shows the resonant frequencies oscillates within the ISM band.

Figure 13 shows the geometry of flexible soft wear antenna on a PVC pipe. This shows the xz direction in which radiation is possible. By decreasing the radius of curvature, the strength of the Electric field increases [15], the magnitude value of the electric field depends on the effective dielectric constant and the same depends on the radius of curvature. Magnitude increases with decrease in curvature is shown in above Table 2.

Table 2 Measured result on placement of flexible softwear antenna on PVC pipes

Diameter of the PVC pipes (cm)	Resonant frequency (GHz)	Magnitude in dB
11	5.59	−12.414
5	5.62	−36.9
3	5.7	−29

Fig. 13 Geometry of flexible
softwear antenna on a PVC
pipe

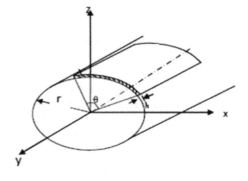

2.2 Experimental Observation and Concern on Bending Effects

Once the antenna is installed as an integrated part of the clothing on different parts of the human body like arm, elbow, forearm wrist, thigh, ankle and wherever human body movement is possible. This study examines the bending effect and the frequency resonance of proposed softwear antenna under various bending environment. In this experiment human body organs like thigh, knee, arm, forearm, wrist etc. are realised by the curved surfaces of 3 different diameter PVC pipes of its various internal radiuses. The diameter is almost similar to the above said humanbody organs. Hence different diameter shows different resonant frequency. As diameter increases frequency decreases which was tabulated above. Normally a cylindrical bend is more precise when compared to V-shape bends in clothing [12]. Proposed antenna is tested by properly bending it on the surface of PVC pipes.

While bending fringing fields should be taken into deliberation, due to this performance of the antenna has great consequence. At the Centre of the patch E-field is null. The fringing field between the margin of the patch and the ground plane guide the radiation. The thickness of the substrate plays major role for the amount of the fringing field. By using Eqs. (8–12) the resonant frequency of the antenna and all the parameters can be calculated.

3 Conclusion

The proposed Antenna performance depends on its structure. Using HFSS the proposed flexible softwear antenna was modelled with a soft substrate. The resonant frequency depends on the dimension of the patch shape, type of the substrate material and also the feed line technique. Variation of these parameters influences the resonant frequency. In this work when this novel antenna is placed on human organs with various diameter resonant frequency oscillates and shift to various frequency and this variation belongs to the ISM band was observed. This soft wear antenna is very cheap and is flexible for any bending Radiuses. By this observation the amount of heat generated on the human body can be reduced, while it was implanted on the human body organs with various diameters. Also the textile substrate used in this proposed antenna is 100 % cotton which absorbs water accordingly the performance also get changed and to be validated for future enhancement.

References

1. P.S. Hall, Y. Hao, *Antennas and Propagation for Body Centric Wireless Communication*, 2nd edn. (2012)
2. G.A. Conway, W.G. Scanlon, Low-profile microstrip patch antenna for over body surface communication at 2.45 GHz

3. A. Al-Shaheen, New patch antenna for ISM band at 2.45 GHz. ARPN J. Eng. Appl. Sci. **l.7**(1) (2012)
4. T. Yilmaz, R.F.Y. Hao, Detecting vital signs with wearable wireless Sensors. Sensors. ISSN:1424-8220 (2010)
5. R. Salvado et al., Textile materials for the design of wearable antennas: a survey. ISSN:1424-8220. Sensors (2012)
6. I. Rexiline Sheeba, Review on performance of soft wear Antenna. ISSN No. 0973-4562, vol. 10 issue 3, 6189–6196 (2015)
7. Y. Ouyang, W.J. Chappell, High frequency properties of electro-textiles for wearable antenna applications. IEEE Trans. Antennas Propag. **56**(2), 381–389 (2008)
8. M.A.R. Osman, M.K.A. Rahim, M.K. Elbasheer, M.F. Ali, N.A. Samsuri, Compact fully textile UWB antenna for monitoring applications. *2011 Asia-Pacific Microwave Conference (APMC)*, Melbourne, Australia, December 5–8 2011
9. M. Wnuk, M. Bugaj, Wearable antenna constructed microstrip Technology. *Progress in Electromagnetic Research Symposium Proceedings*, p. 67 (K.L. Malaysia March 27–30, 2012)
10. S. Sankaralingam, B. Gupta, Development of textile antennas for body wearable applications and investigations on their performance under bending conditions. Prog. Electromagn. Res. B. **22**, 53–71 (2010)
11. C.M. Krowne, Cylindrical rectangular microstrip antenna. IEEE Antennas Propag. **31**(1), 194–198 (1983)
12. A. Galehelar et al., Flexible light weight antenna at 2.45 GHz for athlete clothing. Centre for Wireless Monitoring & Applications Australia
13. A. Elrashidi et al., Resonance frequency, gain, efficiency and quality factor of a microstrip printed antenna as a function of curvature for TM01 mode using different substrates
14. C.A. Balanis, *Antenna Theory* (Wiley, New York, 2005)
15. Q. Wu, M. Liu, Z. Feng, A millimeter wave conformal phased microstrip antenna array on a cylindrical surface. *IEEE International Symposium on Antennas and Propagation Society*, pp. 1–4 (2008)
16. S.D. Murali1, B. Narada Maha Muni1, Y. Dilip Varma1, S.V.S.K. Chaitanya1, Development of wearable antennas with different cotton textiles DST-FIST sponsored. ECE Department, KL University, Vaddeswaram, Guntur DT, A.P., India
17. M.K. Elbasheed, M.A.R. Osman, A. Abuelnuor, M.K.A. Rahim, M.E. Ali, Conducting materials effect on UWB wearable textile antenna. *Proceedings of the WCE* (London, U.K. July 2–4 2014)

Bandwidth Enhanced Nearly Perfect Metamaterial Absorber for K-Band Applications

S. Ramya and I. Srinivasa Rao

Abstract In this paper, a nearly perfect metamaterial absorber is proposed with enhanced bandwidth and polarization insensitive. The proposed unit cell structure consists of outer split ring and inner asterisk shaped resonators printed on FR4 dielectric substrate. The designed metamaterial absorber gives broad bandwidth of 2.01 GHz and peak absorptivity of 99.98 and 99.94 % at 19.4 and 19.8 GHz, respectively. The simulation results prove the polarization insensitive behavior of the structure for oblique and normal incidence of the polarized waves at angles of 25° and 85°, respectively. The effective parameter is retrieved and the field distributions are studied. This metamaterial absorber is well suited for K-band applications like radar and satellite communications for uses in weather radar, imaging radar and air traffic control.

Keywords Metamaterial · Absorber · Polarization · K-band applications

1 Introduction

Due to unusual properties of the metamaterials like negative permittivity, negative permeability etc. along with its many advantages, extensive research was concentrated in various fields like antennas [1], absorbers [2] etc. The most promising application of the metamaterial is the microwave absorber. Compared to earlier days absorber, the metamaterial absorber due to its unusual properties, leads to nearly unity absorption with enhanced bandwidth, polarization insensitivity behaviour and compact size. These metamaterial microwave absorbers are widely used for electromagnetic wave absorption to reduce specific absorption rate in mobile

S. Ramya (✉)
Survey Research Center, VIT University, Vellore, Tamil Nadu, India
e-mail: ramya.sekar@vit.ac.in

I. Srinivasa Rao
School of Electronics Engineering, VIT University, Vellore, Tamil Nadu, India

© Springer India 2016
S.C. Satapathy et al. (eds.), *Microelectronics, Electromagnetics and Telecommunications*, Lecture Notes in Electrical Engineering 372,
DOI 10.1007/978-81-322-2728-1_3

phones [3, 4], Electromagnetic interference suppression [5, 6], Radar cross section reduction [7] and Electromagnetic compatibility. Various metamaterial absorbers were designed in microwave, terahertz and infra-red frequency ranges with dual band, triple band, multi-band absorption etc. Recent trend and the biggest challenge is the design of bandwidth enhanced metamaterial absorber to overcome the delimits of absorbers with narrow band absorption. The different forms of split ring based resonators which were used for ultra-wideband applications [8] are now replaced by electric field driven LC resonators. For achieving enhanced bandwidth for absorption, multi-layer structures were proposed but the structure was polarization dependant [9]. In terahertz region bandwidth enhanced structures has been proposed [10]. The dual-layer dual band absorber with enhanced bandwidth of 1.24 and 1.92 GHz was proposed in [11]. In [12] the single layer structure obtained the bandwidth enhancement of 940 MHz. The broadband absorbers in C and Ku bands were also proposed. This paper proposes a split ring with inner asterisk shaped structure for enhanced bandwidth and polarization insensitive metamaterial absorber. The structure gives wide bandwidth of 2.01 GHz with two absorption peaks of 99.98, 99.94 % at 19.4 and 19.8 GHz, respectively. Nearly perfect absorption is achieved by proper optimization of the structure. The surface current and electric field distributions are analyzed for good understanding of the physical mechanism of the proposed structure. The polarization insensitivity behavior of the structure for different normal and oblique angle of incidences are studied. The proposed metamaterial absorber has potential applications in K-band.

2   Proposed Design of Unit Cell Structure

The unit cell of the proposed structure is shown in Fig. 1. It consists of top outer split ring with inner asterisk shaped resonators and grounded bottom layer separated by FR4 dielectric substrate of height 0.635 mm with dielectric constant $\varepsilon_r = 4.4$ and dielectric loss tangent tan $\delta = 0.02$. The copper thickness of 0.035 mm is used for the bottom ground and top metallic patterns. The dimensions of the structure are a = 6 mm, b = 6 mm, c = 2.05 mm, d = 2.05 mm, e = 3.6 mm, f = 4.3 mm, $w_1 = 0.2$ mm and $w_2 = 0.2$ mm.

This structure resonates at 19.4 and 19.8 GHz and the absorptivity can be calculated using the reflection coefficient S_{11} and transmission coefficient S_{21} as,

$$\text{Absorptivity} = 1 - |S_{11}|^2 - |S_{21}|^2. \tag{1}$$

The S_{21} is zero due to copper ground which stops the wave transmission. The reflection of the waves should be minimized to make perfect metamaterial absorber which is achieved by impedance matching of the structure to the free space impedance. The input impedance is given as [12],

Fig. 1 Proposed unit cell structure

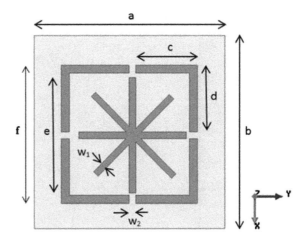

$$\text{Impedance} = \frac{1 + S_{11}}{1 - S_{11}} \tag{2}$$

A perfectly impedance matched structure should have real part of the input impedance as unity and imaginary part of the input impedance as zero [12].

3 Simulation Results

The proposed metamaterial absorber is simulated using Ansys HFSS with periodic boundary conditions. Figure 2 shows the reflection coefficient plot for the proposed structure with enhanced bandwidth of 2.01 GHz from 18.4 to 20.41 GHz. The enhanced bandwidth is due to fine tuning of the structure dimensions. The structure resonates at 19.4 and 19.8 GHz with peak absorptivity of 99.98 and 99.94 % as shown in Fig. 3, calculated using Eq. (1). The input impedance of the unit cell structure is calculated using Eq. (2) and is shown in Fig. 4.

The retrieved impedance parameter has almost unity real impedance and zero imaginary impedance at the resonating frequencies. Simulation results show that nearly perfect absorption with enhanced bandwidth is achieved and can be used for K-band applications. The electric field distributions and surface current distributions at 19.4 and 19.8 GHz are shown in Figs. 5, 6 and 7. In Fig. 5, the electric field distributions clearly show that the broad bandwidth absorption is achieved due to both split ring and asterisk shaped resonators. The Figs. 6 and 7 show the surface current distributions at 19.4 GHz and 19.8 GHz. The directions of the surface current on the bottom layer are in anti-parallel with the direction of the surface current on the top layer of the proposed structure and hence they constitute a circulating current loop.

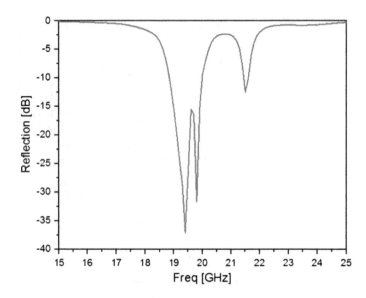

Fig. 2 Reflection coefficient of the proposed structure

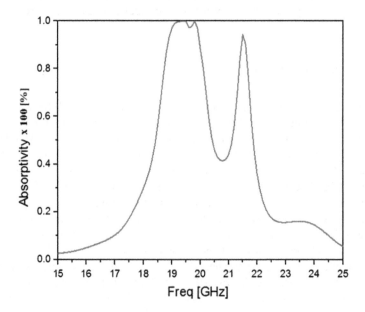

Fig. 3 Absorptivity of the proposed structure

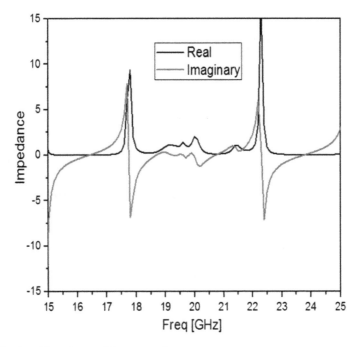

Fig. 4 Retrieved impedance of the proposed structure

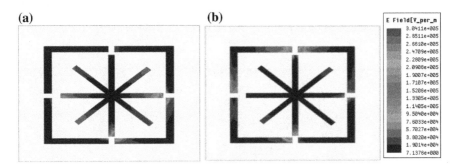

Fig. 5 Electric field distribution. **a** 19.4 GHz, **b** 19.8 GHz

4 Results for Oblique Incidence and Normal Incidence

The polarization insensitivity of the metamaterial absorber is verified by simulation under normal and oblique angles of incidence of the incoming polarized waves. Figure 8 shows the response of the structure under oblique incidence. The response of the structure under normal angle of incidence is shown in Fig. 9. The peak absorption is obtained only at zero oblique and normal angle of incidence. For normal and oblique incidence, at angles of 25° and 85° the structure is polarization

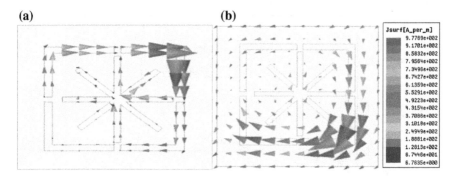

Fig. 6 Surface current distribution at 19.4 GHz. **a** Top layer. **b** Bottom layer

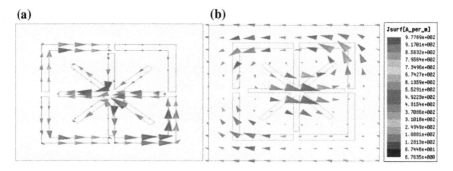

Fig. 7 Surface current distribution at 19.8 GHz. **a** Top layer. **b** Bottom layer

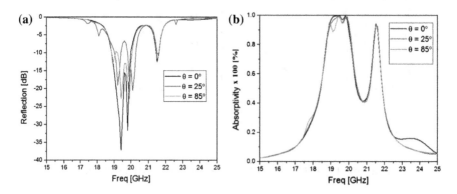

Fig. 8 **a** Reflection coefficient. **b** Absorptivity under oblique angle of incidence

insensitive. For other angles of incidence i.e. when the structure is rotated or when the position of the antenna is changed, the reflected power increases, absorptivity decreases and the structure becomes polarization sensitive.

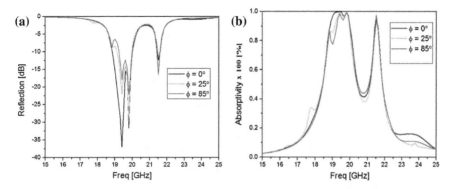

Fig. 9 **a** Reflection coefficient. **b** Absorptivity under normal angle of incidence

5 Conclusion

A bandwidth enhanced polarization insensitive nearly perfect metamaterial absorber is proposed. The split ring with asterisk shaped metamaterial absorber structure is optimized to give enhanced bandwidth and nearly perfect absorption at higher frequencies, making it as an promising absorber in K-band applications for satellite and radar applications for uses in weather radar, imaging radar and air traffic control. The simulation result shows the peak absorptivity of 99.98 %, 99.94 % at 19.4 and 19.8 GHz respectively. The bandwidth enhancement of 2.01 GHz is achieved between 18.4 and 20.41 GHz. The surface current and electric field distributions are plotted and analyzed. The proposed structure is polarization insensitive for both normal and oblique angle of incidence of the polarized waves at 25° and 85°.

References

1. S.D. Campbell, R.W. Ziolkowski, Lightweight flexible polarization-insensitive highly absorbing meta-films. IEEE Trans. Antennas Propag. **61**, 1191–1200 (2013)
2. N.I. Landy, S. Sajuyigbe, J.J. Mock, D.R. Smith, W.J. Padilla, Perfect metamaterial absorber. Phys. Rev. Lett. **100**, 207402 (2008)
3. J.N. Hwang, F.C. Chen, Reduction of the Peak SAR in the human head with metamaterials. IEEE Trans. Antennas Propag. **54**(12), 3763–3770 (2006)
4. M.R.I. Faruque, M.T. Islam, Novel triangular metamaterial design for electromagnetic absorption reduction in human head. Progr. Electromagn. Res. **141**, 463–478 (2013)
5. J. Sun, L. Liu, G. Dong, J. Zhou, An extremely broad band metamaterial absorber based on destructive interference. Optics Express **19**(22), 21155–21162 (2011)
6. S. Ramya, I. SrinivasaRao, Dual band microwave metamaterial absorber using loop resonator for electromagnetic interference suppression. Int. J. Appl. Eng. Res. **10**(30), 22712–22715 (2015)

7. T. Liu, X. Cao, J. Gao, Q. Zheng, W. Li, H. Yang, RCS reduction of waveguide slot antenna with metamaterial absorber. IEEE Trans. Antennas Propag. **61**(3), 1479–1484 (2013)
8. S. Ramya, I. Srinivasa Rao, Design of compact ultra-wideband microstrip bandstop filter using split ring resonator. *International Conference on Communication and Signal Processing*, pp. 102–105 (2015)
9. H. Xiong, J.S. Hong, C.M. Luo, L.L. Hong, An ultrathin and broadband metamaterial absorber using multi-layer structures. J. Appl. Phys. **114**(6), 064109 (2013)
10. V.T. Pham, J.W. Park, D.L. Vu, H.Y. Zheng, J.Y. Rhee, K.W. Kim, Y.P. Lee, THz-metamaterial absorbers. Adv. Nat. Sci. Nanosci. Nanotechnol. **4**(1), 015001 (2013)
11. S. Bhattacharyya, S. Ghosh, D. Chaurasiya, K.V. Srivastava, Bandwidth-enhanced dual-band dual-layer polarization-independent ultra-thin metamaterial absorber. Appl. Phys. A **118**, 207–215 (2015)
12. S. Bhattacharyya, S. Ghosh, K.V. Srivastava, Triple band polarization-independent metamaterial absorber with bandwidth enhancement at X-band. J. Appl. Phys. **114**, 094514 (2013)

An Approach to Improve the Performance of FIR Optical Delay Line Filters

P. Prakash and M. Ganesh Madhan

Abstract An approach to improve the filter characteristics of FIR optical delay line filter is presented. This scheme uses constrained least square algorithm to derive the optimum filter coefficients along with tuning of external phase shifter to reduce the overlap between adjacent pass bands. Based on this technique, a three port (1×3) lattice form FIR optical delay line filter is synthesized and compared with the existing results. The results of the filter designed is found to require only 18 stages, 50 dB stop band attenuation along with minimum attenuation of -24 dB in the overlap region. These values are significantly better compared to the literature reports available for similar filters.

Keywords Optical delay line filter · Constraint least square algorithm · Finite impulse response · Directional coupler · Phase shifter · Optical signal processing

1 Introduction

Optical delay line filters are widely used in the field of optical signal processing. They are implemented with phase shifters, directional couplers, optical delay lines and they are based on coherent superposition of incident fields [1]. The incoming signal, split into different signal paths are delayed with respect to each other, multiplied with appropriate weights and added together. Optical delay line filters are similar to conventional digital filters in electronics domain [2, 3]. FIR filters can be realized using fiber optic components as they posses periodic transfer function, which is used for filtering several adjacent channels simultaneously. Optical FIR circuits do not posses feedback paths and characterized by transfer functions with

P. Prakash (✉) · M. Ganesh Madhan
Department of Electronics Engineering, Anna University, MIT Campus, Chennai, India
e-mail: prakashp79@gmail.com

M. Ganesh Madhan
e-mail: mganesh@annauniv.edu

© Springer India 2016   35
S.C. Satapathy et al. (eds.), *Microelectronics, Electromagnetics and Telecommunications*, Lecture Notes in Electrical Engineering 372,
DOI 10.1007/978-81-322-2728-1_4

numerators alone [4–6]. Optical delay line filters can provide efficient RF filtering in optical domain itself, thereby avoiding optical to electrical conversion and processing in electronic domain. These filters find application in analog fiber optic communication applications.

The design of 1×3, 3×3 FIR filters have already been proposed in the literature [7, 8]. In the case of 1×3 filter, Sequential Quadratic Programming (SQP) optimization method has been used to obtain the approximate complex expansion coefficients. In their design example, the delay time was about 0.01 ns, which corresponds to a free spectral range (FSR) of 100 GHz. The number of stages used in the 1×3 filter design is 21 [7].

Kaname Jinguji and Takashi Yasui [9] in their work, have designed a $1 \times M$ ($M \geq 2$) optical lattice filter based on least square method with non linear constrains. In their approach, the number of stages was 39 and the stopband attenuation was 26 dB. Azam et al. [10] have proposed an optical delay line circuit which offers similar characteristics as $1 \times M$ FIR digital filter. In our earlier paper, we have designed a 1×5 optical band pass delay line filter based on the division of transfer matrix into canonical forms and the circuit parameters were obtained using Constrained Least Square (CLS) method [11]. This algorithm is found to increase the speed of execution and improves the numerical accuracy of the result [12]. In conventional delay line filter design reported in literature, the pass band overlap is significant and the same band of frequencies is passed through multiple ports, leading to cross talk. Filters with minimum overlapping of adjacent pass bands are always preferable.

In this paper, an approach to improve the performance of FIR delay line filter is reported. The improved characteristics of the filter results from the optimum filter coefficients developed using the CLS algorithm along with optimum external phase shifter value. A 1×3 lattice form FIR filter is developed based on the proposed scheme and found to result in reduced number of components (stages) and improved stop band attenuation. Further pass band separation is improved by tuning the external phase shifters to an optimum value.

2   Circuit Configuration and Synthesis

The circuit configuration for a three port (1×3) lattice form FIR optical delay line filter used in this work is shown in Fig. 1. The 1×3 structure has one input and three outputs along with of $2 \times (N + 1)$ directional couplers and $2 \times (N + 1)$ phase shifters and one external phase shifter. The delay difference in each path has a time delay $\Delta \tau$. The filter structure used in this work is based on the report of Azam et al. [7]. But they have used Remez algorithm, whereas, we deduce the filter coefficients based on constrained least square approach. The filter transfer function is the product of the transfer function matrix of directional coupler, waveguides and phase shifters.

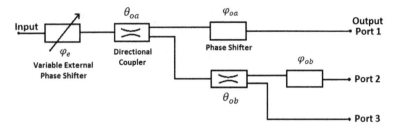

Fig. 1 Schematic of a generic, single stage 1×3 lattice for delay line filter

The transfer function of the first element in the structure shown in Fig. 1 is given as

$$S_D = \begin{pmatrix} e^{-jw\Delta\tau} & 0 & 0 \\ 0 & 1 & 0 \\ 0 & 0 & 1 \end{pmatrix} \tag{1}$$

The transfer functions of the directional coupler between the waveguides are given in the following expressions (Eqs. 2 and 3).

$$S_{cA} = \begin{pmatrix} \cos\theta_A & -j\sin\theta_A & 0 \\ -j\sin\theta_A & \cos\theta_A & 0 \\ 0 & 0 & 1 \end{pmatrix} \tag{2}$$

$$S_{cB} = \begin{pmatrix} 1 & 0 & 0 \\ 0 & \cos\theta_B & -j\sin\theta_B \\ 0 & -j\sin\theta_B & \cos\theta_B \end{pmatrix} \tag{3}$$

Similarly the phase shifters are characterized by their phase angles φ_A and φ_B, and their transfer function are given as

$$S_{pA} = \begin{pmatrix} e^{-j\varphi_A} & 0 & 0 \\ 0 & 1 & 0 \\ 0 & 0 & 1 \end{pmatrix} \tag{4}$$

$$S_{pB} = \begin{pmatrix} 1 & 0 & 0 \\ 0 & e^{-j\varphi_B} & 0 \\ 0 & 0 & 1 \end{pmatrix} \tag{5}$$

The total filter characteristic is expressed as the products of all these basic components. This approach is similar to that of Azam et al. [7]. The synthesis algorithm is used to determine the unknown circuit parameters such as optimum coefficients, phase angle of the phase shifters, the coupling angle of the couplers and one external phase shifter with phase value φ_e [7, 11]. The synthesis method has the following steps:

Step 1: Calculation of delay time difference $\Delta\tau$ from the desired frequency f_0

$$\Delta\tau = \frac{1}{f_0} \tag{6}$$

Step 2: Calculation of the optimum coefficients using constrained least square method [12].

Step 3: Obtaining the coupling coefficient angles of directional couplers and phase shift angles of phase shifters.

The recursion equations can be obtained by factorizing the total transfer matrix $S(z)$, which can be decomposed into the following form:

$$S(z) = S_N(z)S_{N-1}(z)\ldots S_2(z)S_1(z)S_0 = \prod_{l=N} S_l(z) \tag{7}$$

Each transfer function block has a delay time difference $\Delta\tau$, two directional couplers with coupling coefficients θ_{lA}, θ_{lB} and phase shifters with phase angles $\varphi_{lA}, \varphi_{lB}$. The transfer function of each block is given as,

$$S_l(z) = \begin{pmatrix} \cos\theta_{lA}e^{-j\varphi_{lA}}z^{-1} & -j\sin\theta_{lA}e^{-j\varphi_{lA}} & 0 \\ -j\sin\theta_{lA}\cos\theta_{lB}e^{-j\varphi_{lB}}z^{-1} & \cos\theta_{lA}\cos\theta_{lB}e^{-j\varphi_{lB}} & -j\sin\theta_{lA}\sin\theta_{lB}z^{-1} \\ -\sin\theta_{lA}\sin\theta_{lB}z^{-1} & -j\cos\theta_{lA}\sin\theta_{lB} & \cos\theta_{lB} \end{pmatrix} \tag{8}$$

The input–output relation of a three port optical delay line filter in response to an input (1 0 0) is given as

$$\begin{pmatrix} F(z) \\ G(z) \\ H(z) \end{pmatrix} = \begin{pmatrix} F^{[N]}(z) \\ G^{[N]}(z) \\ H^{[N]}(z) \end{pmatrix} = \prod_{l=N} S_l(z) \begin{pmatrix} 1 \\ 0 \\ 0 \end{pmatrix} \tag{9}$$

The key to solve the unknown filter parameters $\theta_{NA}, \theta_{NB}, \varphi_{NA}, \varphi_{NB}$ of the Nth block is to separate $S_N(z)$ from $S(z)$ is shown in Eqs. 10 and 11.

$$\varphi_{NB} = \arg\left(\frac{jC_N}{B_N}\right)$$

$$\theta_{NB} = \tan^{-1}\left(\frac{jC_N e^{-j\varphi_{NB}}}{C_N}\right) \tag{10}$$

For N = 0, the phase shift value φ_{NA} and the coupling coefficients θ_{NA} are given as,

$$
\begin{aligned}
\varphi_{NA} &= \arg\left\{ \frac{jB_N \cos\theta_{NB}e^{j\varphi_{NB}} - C_N \sin\theta_{NB}}{A_N} \right\} \\
\theta_{NB} &= \tan^{-1}\left\{ \frac{(jB_N \cos\theta_{NB}e^{j\varphi_{NB}} - C_N \sin\theta_{NB})e^{-j\varphi_{NA}}}{A_N} \right\}
\end{aligned} \tag{11}
$$

Here N is the number of stages and n represents the number of recursion of the corresponding equations. When nl = 0, additional equations can be derived, where l = 0, 1, 2... n − 1. $a_l^{[n]}$, $b_l^{[n]}$ and $c_l^{[n]}$ indicate the l_{th} expansion coefficients of $F^{[N]}(z)$, $G^{[N]}(z)$ and $H^{[N]}(z)$ respectively. At the first stage n = N, initial data for the recurrent equations are given as $a_l^{[N]} = al$, $b_l^{[N]} = bl$ and $c_l^{[N]} = cl(l = 0, 1, 2, \ldots, N)$.

$$
\begin{cases}
a_l^{[n-1]} = \left(a_{l+1}^{[n]} \cos\theta_{nA}e^{j\varphi_{nA}} + j b_{l+1}^{[n]} \sin\theta_{nA} \cos_{nB} e^{j\varphi_{nB}} - c_{l+1}^{[n]} \sin\theta_{nA} \sin\theta_{nB} \right) \\
b_l^{[n-1]} = \left(j a_l^{[n]} \sin\theta_{nA}e^{j\varphi_{nA}} + b_l^{[n]} \cos\theta_{nA} \cos_{nB} e^{j\varphi_{nB}} + j c_l^{[n]} \cos\theta_{nA} \sin\theta_{nB} \right) \\
c_l^{[n-1]} = \left(j b_l^{[n]} \sin\theta_{nB}e^{j\varphi_{nB}} + c_l^{[n]} \cos\theta_{nB} \right)
\end{cases} \tag{12}
$$

The external phase shifter value φ_e is calculated using the equation given below:

$$
\varphi_e = -\arg\left(a_0^{[0]}e^{j\varphi_{0A}} \cos\theta_{0A} + j b_0^{[0]}e^{j\varphi_{0B}} \cos\theta_{0B} \sin\theta_{0B} - c_0^{[0]} \sin\theta_{0A} \sin\theta_{0B} \right) \tag{13}
$$

Thus, all coupling coefficient angles $\theta_{nA}, \theta_{nB}(n = 0 \sim N)$ of $2 \times (N+1)$ directional couplers, phase shift values $\varphi_{nA}, \varphi_{nB}(n = 0 \sim N)$ of $2 \times (N+1)$ phase shifters and phase value, φ_e of an external phase shifter can be obtained from the above expressions.

3   Results and Discussion

This filter is aimed to filter out RF components, which are propagating in an optical carrier through fiber. The expansion coefficients, coupling coefficient, angles of directional couplers and the phase shift values of the phase shifters ($\theta_{nA}, \theta_{nB}, \varphi_{nA}$ and φ_{nB}) are determined using CLS method. The coefficients are shown in Table 1. The desired response is achieved with number of stages N = 18 and an external phase shifter value of $\varphi_e = -0.0110$. In order to evaluate the RF filtering function in the optical domain, the modulating RF signal of a 1550 nm

Table 1 Expansion coefficients and calculated circuit parameters for $\varphi_e = -0.0124$

Stage number	Expansion coefficients (a_k)	Expansion coefficients (b_k)	Expansion coefficients (c_k)	Coupling coefficient angle (θ_{NA})	Coupling coefficient angle (θ_{NB})	Phase shift value (φ_{NA})	Phase shift value (φ_{NB})
1	−0.0034	−0.0089	0.0096	−1.6980	0.9329	0.9809	0.3830
2	−0.0409	0.0334	0.0569	−1.3074	−1.9593	1.4889	0.9239
3	−0.0737	−0.0386	−0.0822	1.2252	−1.8130	1.5666	0.7375
4	−0.0955	−0.0632	0.0876	1.3656	0.3430	1.7211	1.1340
5	−0.0819	0.1029	−0.0594	1.6391	−0.9051	2.0834	2.7232
6	−0.0205	0.0683	−0.0033	−2.0604	0.4699	0.5186	32.3884
7	0.0736	−0.1752	0.0827	0.9788	−1.3811	1.2413	3.3292
8	0.1602	−0.0301	−0.1494	−1.1478	−1.4827	1.6812	0.3170
9	0.1953	0.2076	0.1753	−1.6227	2.2396	1.1644	1.8608
10	0.1602	-0.0301	−0.1494	−1.1478	−1.4827	1.6812	0.3170
11	0.0736	−0.1752	0.0827	0.9788	−1.3811	1.2413	3.3292
12	−0.0205	0.0683	−0.0033	−2.0604	0.4699	0.5186	32.3884
13	−0.0819	0.1029	−0.0594	1.6391	−0.9051	2.0834	2.7232
14	−0.0955	−0.0632	0.0876	1.3656	0.3430	1.7211	1.1340
15	−0.0737	−0.0386	−0.0822	1.2252	−1.8130	1.5666	0.7375
16	−0.0409	0.0334	0.0569	−1.3074	−1.9593	1.4889	0.9239
17	−0.0157	0.0071	−0.0290	−1.6980	0.9329	0.9809	0.3830
18	−0.0034	−0.0089	0.0096	0.4319	0.0889	0.5490	1.4576

laser diode is varied from 0 to 500 MHz. The response of the optical filter at different ports is shown in Fig. 2.

The frequency in the graph represents the modulating signal (RF) for the optical carrier. It is found that filtering action of RF components in optical carriers is efficiently carried out by the synthesized delay line filter.

The magnitude response shows almost 0 dB insertion loss at the center of each frequency band while the stop band attenuation is less than 50 dB. However the stop band attenuation varies for different ports. The results obtained shows that the optimum number of stages used to design a 1 × 3 filter is 18 (k = 18). This value is less than that reported by Azam et al. [7], where the number of stages used was 21, and the stop band attenuation was about 30 dB only.

The filter response is obtained for different value of φ_e and the attenuation value corresponding to the point of intersection of adjacent bands is determined. The point of intersection corresponds to the minimum attenuation in the overlap region between adjacent bands. Figure 3 shows the variation in minimum attenuation of overlap region for different external phase shift values. By tuning the external phase shifter, it is possible to achieve reduced bandwidth of overlap region, thereby reducing the cross talk between adjacent pass bands. It is observed that the optimal value of the external phase shifter is about −0.0124, which corresponds to an attenuation of 24 dB. Under optimum condition, it is found that the bandwidth of

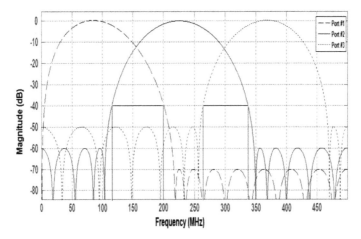

Fig. 2 Magnitude response of 1 × 3 FIR filter ($\varphi_e = -0.0110$)

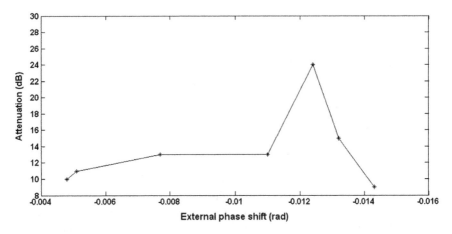

Fig. 3 Minimum attenuation in the region of overlap for different external phase shift values

the overlap region at 40 dB level for port 1 and 2 is 37 MHz. This is less compared to the unoptimized phase shifter value (φ_e) of −0.0110, which is about 85 MHz. Figure 4 shows the magnitude response of the filter with better separation in pass bands, which is obtained by tuning the external phase shifter value and corresponding circuit parameters. The bandwidth corresponding to overlap region of port 2 and 3 of the filter is around 60 MHz, which is less compared to the bandwidth reported in the previous case Fig. 2 (80 MHz). A minimum bandwidth and maximum attenuation correspond to region of overlap is determined for best filtering.

The variation of 3 dB bandwidth for different output ports from the filter shows almost constant bandwidth around 70 MHz. Table 2 shows the bandwidth of overlapping region at −40 dB level, for two external phase shifter values

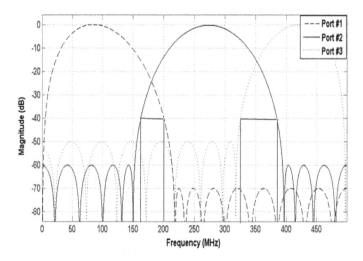

Fig. 4 Magnitude response of 1×3 FIR filter ($\varphi_e = -0.0124$)

Ports	Bandwidth of overlapping region at −40 dB level	
	$\varphi_e = -0.0110$	$\varphi_e = -0.0124$
1 and 2	85 MHz	37 MHz
2 and 3	73 MHz	59 MHz

Table 2 Bandwidth of overlapping region at −40 dB level for two different φ_e values

$\varphi_e = -0.0110$, and -0.0124, and it is found that for the optimum value of $\varphi_e = -0.0124$, the overlap region is less.

In the 1×3 report of Azam et al. [7], the stop band attenuation is around 30 dB; hence we compare the overlap bandwidth results at −30 dB levels. The overlap bandwidth is found to be 0.11 rad/sample, which is much higher than that achieved in this work (0.033 rad/sample). A 20 dB improvement is found in our case. As the overlap bandwidth is significantly less, it is expected to improve the crosstalk performance in the proposed filter.

The maximum insertion loss and minimum stop band attenuation in the proposed filter are found as 0.18 and 50 dB respectively. The stopband attenuation shows a periodic variation with the maximum value of 70 dB for the first and third output ports; however the attenuation is found to be 50 dB for the center band. Table 3 shows the comparison of Remez algorithm with CLS algorithm in terms of number of stages, stop band attenuation, overlap region and maximum attenuation in the overlap region. Since this paper focuses on synthesis of optical FIR filter with CLS algorithm, experimental results are not provided. However, practical implementation can be carried out, based on this approach.

Table 3 Comparison of CLS algorithm with different existing algorithms

Filter parameter	1 × 3 using REMEZ algorithm [7]	1 × 3 using CLS algorithm
Number of stages	21	18
Stop band attenuation (dB)	20	50
Bandwidth of overlap region at −30 dB level (rad/sample)	0.11	0.033
Attenuation level corresponding to intersection of two adjacent bands (dB)	3	24

4 Conclusion

An approach to improve the performance of FIR optical delay line filter is reported in this paper. This scheme uses constrained least square (CLS) algorithm to determine the filter coefficients effectively and also introduces an optimum external phase shifter value for minimum overlap between adjacent frequency bands. This concept is implemented in a (1 × 3) FIR optical delay line filter and found to provide significant improvement in terms of number of stages, stop band attenuation and reduced overlap between adjacent pass bands.

References

1. C.K. Madsen, Optical filter design and analysis, Chap. 3 (Wiley-Interscience Publication, John Wiley & Sons, Inc, 1999)
2. K. Jinguji, T. Yasui, Synthesis of one-input m-output optical FIR lattice circuits. J. Lightwave Technol. **26**, 853–866 (2008)
3. S. Azam, T. Yasui, K. Jinguji, Synthesis of 1-input 3-output optical delay-line circuit with IIR architectures. Recent Patents Electr. Eng. **1**, 214–224 (2008)
4. K. Jinguji, M. Kawachi, Synthesis of coherent two-port lattice-form optical delay-line circuit. J. Lightwave Technol. **13**, 73–82 (1995)
5. L. Benvenuti, L. Farina, The design of fiber optic filters. J. Lightwave Technol. **19**, 1366–1375 (2001)
6. G. Lenz, B.J. Eggleton, C.K. Madsen, R.E. Slusher, Optical delay lines based on optical filters. IEEE J. Quantum Elect. **37**, 525–532 (2001)
7. S. Azam, T. Yasui, K. Jinguji, Synthesis of 1-input 3-output lattice-form optical delay-line circuit. IEICE Trans. Elect. **90**, 149–155 (2007)
8. Q.J. Wang, Y. Zhang, Y.C. Soh, Flat-passband 3 × 3 interleaving filter designed with optical couplers in lattice structure. J. Lightwave Technol. **23**, 4349–4361 (2005)
9. K. Jinguji, T. Yasui, Design algorithm for multichannel interleave filters. J. Lightwave Technol. **25**, 2268–2276
10. S. Azam, T. Yasui, K. Jinguji, Synthesis of a multichannel lattice-form optical delay-line circuit. Optik. **121**, 1075–1083 (2007)
11. P. Prakash, M. GaneshMadhan, Design of lattice form optical delay line structure for microwave band pass filter applications. Prog. Electromagn. Res. C. **32**, 197–206 (2012)
12. I.W. Selesnick, M. Lag, C.S. Burns, Constrained least square design of FIR filters without specified transition bands. IEEE Trans. Signal Process. **44**, 1878–1892 (1996)

Development of a VLSI Based Real-Time System for Carcinoma Detection and Identification

N. Balaji, B. Nalini and G. Jyothi

Abstract Breast cancer is most common life threatening non skin malignant state in women. Accurate and early detection of breast cancer provides various chances for the survival of the diseased person. The proposed system also helps the clinical practitioner to diagnose the disease correctly. In this paper, an efficient VLSI architecture for carcinoma detection and identification is implemented. The implementation consists of three phases: pre-processing, feature extraction and disease detection and identification. Pre-processing removes the noise by arithmetic mean filters from the breast MRI image. Then the enhanced image is applied to an efficient dual scan parallel flipping architecture to extract the features using pipeline operation. The disease detection and identification can be performed better by using content aware classifier, which is based on Euclidean distance and positive estimation method. The proposed work aims at developing a VLSI system to diagnose on a particular breast cancer disease and is implemented on VERTEX-4 FPGA as it is a real time solution for disease detection and identification.

Keywords Breast cancer · MRI · Pre-processing · Feature extraction · Disease detection and identification

1 Introduction

Today, there is almost every area of technical Endeavour that is impacted in some way or the other by digital image processing. In the recent years, the most important diagnostic tool in medical applications is medical imaging, mainly MRI.

N. Balaji (✉) · B. Nalini · G. Jyothi
Department of ECE, JNTUK, University College of Engineering,
Vizianagaram 535003, Andhra Pradesh, India
e-mail: narayanamb@rediffmail.com

B. Nalini
e-mail: nalini.bodasingi@gmail.com

G. Jyothi
e-mail: jyothiece441@gmail.com

© Springer India 2016
S.C. Satapathy et al. (eds.), *Microelectronics, Electromagnetics and Telecommunications*, Lecture Notes in Electrical Engineering 372,
DOI 10.1007/978-81-322-2728-1_5

Mammography cannot determine whether an area is effected with cancer or not, but it can provide guidance for further screening or diagnostic tests [1]. One of the most powerful tool is breast MRI. The diseases that have been associated with structural changes in the breast are Ductal Carcinoma in Situ, Inflammatory Breast Cancer [2] Lobule Carcinoma and Metastatic Breast Cancer. The rate at which medical images are produced every day is increasing exponentially. Such images are the rich sources of information about shape, colour and texture, which can be exploited to improve the diagnosis and ultimately the treatment of complex disease. In this proposed work an attempt was made to improve the accuracy of disease detection and identification with less hardware utilization.

The following steps are involved in the disease diagnosis. The first step involves the reducing breast MRI image search space. Second step is the feature extraction using Dual Scan Parallel Flipping Architecture (DSPFA) and Memory Efficient Hardware Architecture on medical images. The third step is content aware classifier, which is based on Euclidean distance and positive estimation. The classifier parameters are trained for set of known breast MRI images. The parameters obtained during training phase and test phase from the breast images are used to identify the disease. The breast image data base is collected and an algorithm is implemented on Vertex-4 FPGA.

The rest of the sections structured as follows. Section 2 describes related works. Section 3 describes the VLSI architecture for disease detection and identification. Section 4 discusses technology and tools used for implementation. The results obtained are discussed in Sect. 5. Finally, Sect. 6 concludes the work.

2   Related Works

Mammography cannot determine whether an abnormal area is affected with cancer or not, but it can provide guidance for further screening or diagnostic tests. Magnetic Resonance Imaging (MRI) shows the most promise for improved breast cancer screening. Breast MRI is a developing technique for the evaluation of patients with primary breast carcinoma. MRI can be used to obtain three dimensional images of the inner parts of the human body, without using X-rays. MRI creates a detail picture of area inside the body without use of radiation and is a painless procedure. For the detection of primary invasive and non invasive breast carcinoma, breast MRI is very sensitive compared to mammography. MRI breast imaging is a supplementary tool, in addition to mammography, to help diagnose breast cancer.

Texture analysis places an important role for the characterization of biomedical images [3]. It can be classified as statistical, geometrical and signal processing types, out of which signal processing methods are used for texture filtering in the spatial or frequency domain to extract significant features [4]. There are many techniques available for feature extraction. In wavelet transforms, Discrete Wavelet Transform (DWT) is commonly used techniques for multiresolution analysis [5, 6].

For the characterization of biomedical images, multiresolution analysis is the most broadly used technique in signal processing.

The DWT takes an input image and decomposes into sub images that are used for characterization of horizontal and vertical frequencies. The DWT decomposition yields the approximation (LL), horizontal (LH), vertical (HL) and detail (HH) subbands. The process can be repeated for one or more levels if required. Various techniques to construct wavelet bases, or to factor existing wavelet filters into basic building blocks are known. One of these is lifting. The original motivation for developing lifting was to build second generation wavelets, i.e., wavelets adapted to situations that do not allow translation and dilation. Based on lifting scheme several novel architectures have been proposed. It has some additional advantages in comparison with classical wavelets like FT and STFT. It makes computational time optimal, sometimes increasing the speed of calculation by a factor 2. The 9/7-tap lifting DWT has reduced computational complexity proposed in [7], but it required delay of two multipliers and one adder. The extracted features like HL and LH are used identify the type of disease. Recently efficient architectures such as dual scan parallel flipping and Memory Efficient Architectures for a lifting-based 2-D DWT is proposed [8, 13]. Compared to memory efficient architecture, DSPFA has less critical path delay. In this paper, the Dual scan parallel flipping architecture is implemented for disease detection and identification and a real time system is developed with minimal critical path delay and 100 % hardware utilization efficiency.

2.1 Dual Scan Parallel Flipping Architecture

In recent years several VLSI architectures for lifting based DWT has been pro proposed [9–12]. These architectures intend either to improve the processing speed or reduce the hardware cost and memory efficient hardware. Lifting scheme entirely relies on spatial domain and it has many advantages as compared to convolution method such as fewer arithmetic operations, in-place implementation. The main drawback of the conventional lifting scheme for 9/7 tap filter is process the intermediate data in serial manner, thus results in longer critical path. In present algorithm lifting steps are realized in parallel and pipeline manner. The critical path delay is obtained to one multiplier delay of T_m by placing pipeline registers into the DWT unit that increase the control complexity [8]. The proposed dual-scan architecture only needs $N^2/2$ clocks to process an $N \times N$ image and only requires five registers for transposition. In the algorithm sequence x(n), with n = 0, 1 …, N − 1. The input sequence x(n) is divided into even and odd indexed sequences denoted by s_i^0, d_i^0. The steps involved in lifting algorithm are given by

$$s_i^0 = x_{2n} \tag{1}$$

$$d_i^0 = x_{2n+1} \tag{2}$$

$$\frac{d_i^1}{\alpha} = \frac{d_i^0}{\alpha} + \left(s_i^0 + s_{i+1}^0\right) \tag{3}$$

$$\frac{s_i^1}{\alpha\beta} = \frac{s_i^0}{\alpha\beta} + \left(\frac{d_{i-1}^0}{\alpha} + \frac{d_i^1}{\alpha}\right) \tag{4}$$

$$\frac{d_i^2}{\alpha\beta\gamma} = \frac{d_i^1}{\alpha\beta\gamma} + \left(\frac{s_i^1}{\alpha\beta} + \frac{s_{i+1}^1}{\alpha\beta}\right) \tag{5}$$

$$\frac{s_i^2}{\alpha\beta\gamma\delta} = \frac{s_i^1}{\alpha\beta\gamma\delta} + \left(\frac{d_{i-1}^2}{\alpha\beta\gamma} + \frac{d_i^2}{\alpha\beta\gamma}\right) \tag{6}$$

α, β, γ, δ are the 9/7 lifting filter coefficients. d_{i+1}^0, s_{i+1}^0, d_i^1, s_i^1 and d_{i-1}^2 are the intermediate data values obtained from internal memory [8]. These intermediate data are on different paths and this will be used for computing d_i^2, s_i^2, where d_i^2, s_i^2 are the outputs of current cycles. The intermediate data can be calculated in parallel with the current operation and then utilized in the subsequent operation. For example, in the operation in (3), during the computation cycle of $\frac{d_i^1}{\alpha}$, $\frac{d_i^0}{\alpha}$ is concurrently computed with the addition operation between s_i^0, s_{i+1}^0. In a similar manner Eq. (4) of first prediction stage is computed. During the second lifting step in (3) and (4), signals and are scaled by $\frac{1}{\beta\gamma}$ and $\frac{1}{\gamma\delta}$ respectively, instead of $\frac{1}{\alpha\beta\gamma}$ and $\frac{1}{\alpha\beta\gamma\delta}$ because outputs $\frac{d_i^1}{\alpha}$, $\frac{s_i^1}{\alpha\beta}$ produced from the first lifting operations are used in the first lifting step. The final outputs are scaled, as given by

$$d_i = \frac{d_i^2 \, \alpha\beta\gamma}{k} \tag{7}$$

$$s_i = s_i^2 * k\alpha\beta\gamma\delta \tag{8}$$

Thus, the proposed approach reduces the critical path delay to only one multiplier delay T_m from two multipliers delay of $2T_m$.

Block diagram of dual scan parallel flipping architecture as shown in Fig. 1. Mainly it consists of row processor, transposing unit and column processing. The odd and even samples from input data or signals are concatenated in Z-scan fashion. Z-scanning tolerates instantaneous row and column processing operation, ensuing

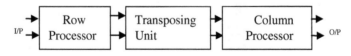

Fig. 1 Block diagram of dual scan parallel flipping architecture

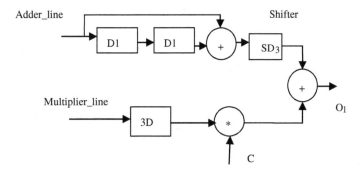

Fig. 2 Row processor element (RPE)

in a small fixed latency [13]. Thus, the transposing buffer size is independent of N, in which N indicates size of image and the corresponding pixels are read on the rising edge of the clock. The entire image segmented into an overlay of 2×2 pixels in order that column processing is able to start as shortly a 1-D DWT is produced by alternate row processing.

2.1.1 Row Processor

The row Processor Element as shown in Fig. 2. It has two inputs and one output. In addition to, the row PE consists of two adders, i.e., one multiplier and one hardware shifter.

In row PE operation the current and previous pixels are fed at the adder line are added, and the odd pixels fed at the multiplier line is multiplied by predetermined constant coefficients C, depending on the precise PE.

The above algorithm uses flipping. Due to this, the reciprocal values of the 9/7 tap filter coefficients are considered which are high enough to cause an overflow during multiplication operation. In case of two's complement arithmetic operations, the inverse coefficient values of 2^k is multiplied. With this critical path delay is reduced. Where k is an integer [14]. Therefore, (3), (4), and (5) are scaled by constants 16, 32, and 8, respectively. Finally, the output is recovered by multiplied (5) by 32 and (6) by 128 during scaling operation [8]. The second input coming from previous pipeline stage of PE1 delayed by five clock cycles through delay register 5D as shown in Fig. 3. Therefore, the critical path delay restricted to one multiplier delay only.

2.1.2 Transposing Unit

A transpose unit is essential among the 1-D row processor and the 1-D column processor to entire 2-D DWT operation. The TU has only five registers and one

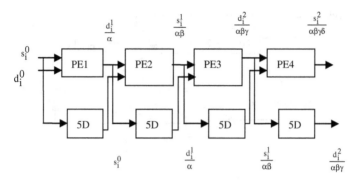

Fig. 3 Row processor

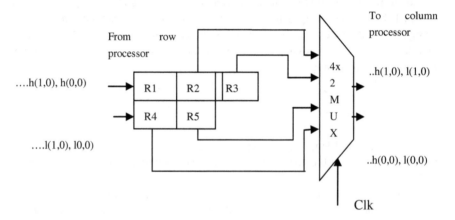

Fig. 4 Transposing unit

4 × 2 multiplexer, as shown in Fig. 4. The high pass and the low pass outputs from the 1-D row processor accumulated in transpose unit and concurrently feed to the 1-D column processor.

2.1.3 Column Processor

The aim of a column processor (PE) is same as the row processor (PE). Thus, the only difference is that the column processor acquires a line buffer and has two output lines, as described in Fig. 5. To produce output O'1 on output line O'2 is delayed by total three clocks to coordinate the two unit delays in two adders and one delay in the shifter. Subsequently, both the output lines are pipelined with the next PE. The entire column processor design is exposed in Fig. 6.

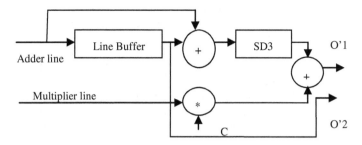

Fig. 5 Column PE

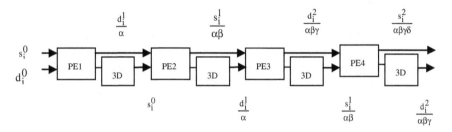

Fig. 6 Column processor

The DSPFA can also work out a multi level 2-D DWT by utilizing off-chip memory of size $N^2/4$ to accumulate the current level LL band coefficients for the next higher level DWT computations [8, 9]. The required clock cycle for j-level DWT is $\dfrac{N}{2^{j-1}\left(2+\frac{N}{2^j}\right)}$.

3 VLSI Architecture for Carcinoma Detection and Identification

The block diagram for carcinoma detection and identification is shown in Fig. 7. The implementation of carcinoma detection and identification has following three phases. Initially, image acquisition is done by giving a set of breast MRI images as input. The average (mean) filter smoothens image data, thus noise can be eliminated. For feature extraction, dual scan parallel flipping architecture is applied to obtain its high frequency image component as it often contains most of desired information about the biological tissue [3, 4]. Finally, Euclidean distance method is used to detect and identify the type of disease.

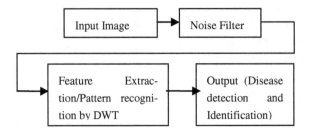

Fig. 7 Block diagram for carcinoma detection and identification

3.1 Noise Filter

Set of breast MRI images are given at the source input. The average or arithmetic mean filter is used to smoothen the data and eliminate noise present in the input image. The filter performs the spatial filtering on each gray level values in a square or rectangular window surrounding each pixel.

$$\hat{f}(x, y) = \frac{1}{mn} \sum_{(s,t) \in s_{xy}} g(s, t) \tag{9}$$

where m, n are the size of sub image centered at a point (x, y) and g(s, t) corrupted image, $\hat{f}(x, y)$ de-noised image.

3.2 Feature Extraction

Feature extraction is nothing but extract significant information from the input data in order to achieve required task based on this momentous data, instead of taking whole data for further processing. Also defined as transforming the input data into set of features is called feature extraction. In this, enhanced breast MRI image is applied to the presented dual scan parallel flipping architecture to extract relevant information.

3.3 Euclidean Distance Method

The Euclidean algorithm is a method for finding the greatest common divisor (GCD) of two integers. The divisor algorithm states that for any integers a and b, where abs (a) > abs(b) (abs is absolute value) and b ≠ 0, there exists a q1 and r1 such that,

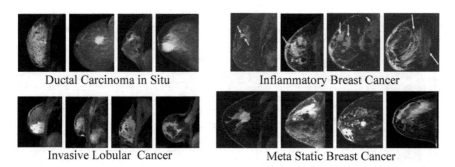

Ductal Carcinoma in Situ	Inflammatory Breast Cancer
Invasive Lobular Cancer	Meta Static Breast Cancer

Fig. 8 Four different diseases of breast MRI images

$$a = q_1 b + r_1 \quad \text{with } 0 \leq r_1 \leq b \tag{10}$$

The extracted features from dual scan parallel flipping architecture is applied to the Euclidean distance method by which it identifies type of disease it is. The projected approach is based on two stages, training and testing. In the training phase, the structure is trained with set of four known breast diseases associated to MRI image database. For each disease, set of four MRI images are taken as shown in Fig. 8. After feature recognition, all essential features are stored for additional processing. In the test stage, we will present one of the breast MRI images as test input to the dwt unit; subsequently the output is compared to the trained rest to identify the disease as shown in Fig. 10.

4   Technology and Tools

The architecture has been implemented in Verilog coding for feature extraction, feature recognition and disease identification and its synthesis were done with Xilinx synthesis tool. Xilinx ISE has been used for performing the mapping, placing, and routing. In behavioral simulation ISIM simulator has been used. The architecture is implemented on VERTEX-4 FPGA.

5   Design Results of Disease Detection and Identification

The behavioral simulation waveform for disease detection and identification is shown in Fig. 9. For different diseases, one MRI images for each of the diseases are taken in the test database. For all these MRI images, the real time system is

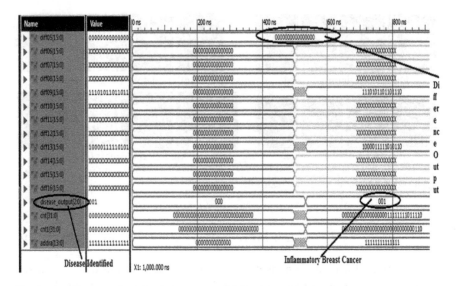

Fig. 9 Simulation result of the dual scan parallel flipping architecture output

implemented with minimal critical path delay and 100 % hardware utilization efficiency. For a particular input of breast MRI disease like ductal carcinoma in situ breast cancer, the corresponding output is shown in the simulation waveform. The entire system for breast disease detection and identification is implemented on Vertex-4 FPGA. The design implementation report of dual scan parallel flipping architecture is presented in Table 1. Also graphical user interface is created in the MATLAB environment to display the result in the image format as shown in Fig. 10. The implementation report shows that the proposed architecture has a critical path delay of 4.210 ns, and utilizes a total of 4498 four-input lookup tables for all the sixteen MRI images. From the device utilization summary as the architecture utilizes less hardware, the same FPGA is useful for detecting more number of diseases.

Table 1 Design implementation summary of dual scan parallel flipping algorithm

Logic utilization	Available	Used	Utilization
Number of slice Flip-Flops	12,288	3538	28 %
Number of 4-input LUTs	12,288	4498	36 %
Number of occupied slices	6,144	2645	43 %
Number of slices contained only related logic	2645	2645	100 %
Number of bounded IOB's	240	17	7 %
Minimum period		3.028 ns	
Maximum period		4.210 ns	

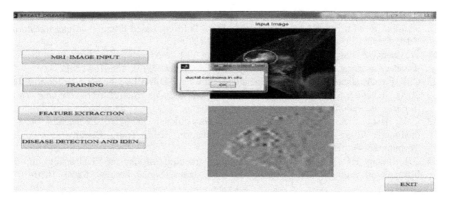

Fig. 10 Display of results using GUI

6 Conclusions

An efficient VLSI based dual scan parallel flipping architecture is implemented using both MATLAB and Verilog HDL program. The architecture is implemented on VERTEX-4 FPGA for breast cancer detection and identification and it is a real time application. The developed system is a portable device with advantages of high speed and low power consumption.

References

1. B. Verma, J. Zakos, A computer aided diagnosis system for digital mammograms based on fuzzy-neural and feature extraction techniques. IEEE Trans. Inf. Technol. BioMed. **5**(1), (2001)
2. American Cancer Society, *Breast Cancer Facts and Figures 2005–2006* (American Cancer Society Inc, Atlanta, 2005)
3. S. Lahmiri, M. Boukadoum. Hybrid discrete wavelet transform and gabor filter banks processing for features extraction from biomedical images. J. Med. Eng. (2013)
4. B. Li, M.Q.H. Meng, Texture analysis for ulcer detection in capsule endoscopy images. Image Vis. Comput. **27**(9), 1336–1342 (2009)
5. C.K. Chui, *An Introduction to Wavelets, Academic Press* (San Diego, Calif, 1992)
6. M. Vetterli, C. Herley, Wavelets and filter banks: theory and design. IEEE Trans. Signal Process. **40**(9), 2207–2232 (1992)
7. S.V. Silva, S. Bampi, Area and throughput trade-offs in the design of pipelined discrete wavelet transform architectures. in *IEEE Proceedings of the Design, Automation and Test in Europe Conference and Exhibition.* pp. 1530–1591 (2005)
8. A. Darji, Dual-scan parallel flipping architecture for a lifting-based 2-D DWT. IEEE Trans. Circ. Syst. **61**(6), (2014)
9. Wu, C. Lin, A high performance and memory efficient pipeline architecture for 5/3 and 9/7 discrete wavelet transform of JPEG2000 codec. IEEE Trans. Circ. Syst. Video Technol. **15** (12), 1615–1628, (2005)

10. Y.-K. Lai, L.-F. Chen, Y.-C. Shih, A high-performance and memory efficient VLSI architecture with parallel scanning method for 2-D lifting based discrete wavelet transform. IEEE Trans. Consum. Electron. **55**(2), 400–407 (2009)
11. W. Zhang, Z. Jiang, Z. Gao, Y. Liu, An efficient VLSI architecture for lifting-based discrete wavelet transform. IEEE Trans. Circ. Syst. II, Exp. Briefs **59**(3), 158–162 (2012)
12. C.-F. Lin, A memory-efficient pipeline architecture for 2-D DWT of the 9/7 filter for 30 JPEG 2000. in *2005 9th International Workshop on Cellular Neural Networks and Their Applications* (2005)
13. C.-H. Hsia, J.-S. Chiang, J.-M. Guo, Memory-efficient hardware architecture of 2-D dual-mode lifting based discrete wavelet transform. IEEE Trans. Circuits Syst. Video Technol. **25**(4), 671–683 (2013)
14. C.T. Haung, P.C. Tseng, L.G. Chen, Flipping structure: An efficient VLSI architecture for lifting-based discrete wavelet transform. IEEE Trans. Signal Process. **52**(4), 1080–1089 (2004)

A Novel Frequency-Time Based Approach for the Detection of Characteristic Waves in Electrocardiogram Signal

Kiran Kumar Patro and P. Rajesh Kumar

Abstract ECG is the electrical behavior of heart signal which is used to diagnose the irregularity of heart activity after visually inspecting the ECG signals but it is difficult to identify by physician's naked eye hence an effective computer based system is needed. One cardiac cycle of ECG signal consists of characteristic waves P-QRS-T. The amplitudes and intervals values of P-QRS-T segment determine the functioning of heart of every human. In this paper a novel methodology of Frequency-Time based approach is used to identify P-QRS-T waves. R-peak detection is the first step in characteristic waves detection, for identifying R-peak, wavelet transform (sym4) decomposition method (Frequency domain) is used. After R-peak detection other characteristic waves are detected by tracing to and fro from R-peak (Time domain). Standard ECG wave form is taken as base wave form and detects the waves in the estimated interval. MIT-BIH NSR database is taken and the methodology is implemented on MATLAB software.

Keywords Electrocardiogram (ECG) · Wavelet transform decomposition · Frequency domain · Time domain · MIT-BIH NSR · Matlab

1 Introduction

The Electrocardiogram signal is a recording of the heart's electrical activity and provides valuable clinical information about heart's performance. The electrical activity generated by Atria and ventricles by its depolarization and repolarization is depicted in both magnitude and direction in a graphical manner [1]. It provides

K.K. Patro (✉) · P. Rajesh Kumar
Department of Electronics and Communication Engineering, Andhra University
College of Engineering, Visakhapatnam, Andhra Pradesh, India
e-mail: kiranugcjrf@gmail.com

P. Rajesh Kumar
e-mail: rajeshauce@gmail.com

© Springer India 2016
S.C. Satapathy et al. (eds.), *Microelectronics, Electromagnetics and Telecommunications*, Lecture Notes in Electrical Engineering 372, DOI 10.1007/978-81-322-2728-1_6

Fig. 1 Characteristic waves
of ECG

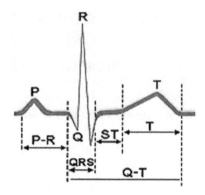

information about the morphology, rhythm, and. heart rate. ECG is not unique it
may vary from one person to other person as known fact that there is difference in
position, size, anatomy of the heart, age, relative body weight, chest configuration
and various other factors.

The electrical activity during the cardiac cycle is characterized by five separate
characteristic waves that are P, Q, R, S and T [1]. The amplitude of P wave is very
low (0.1–0.2 mV) and represent depolarization and contraction of the right and left
Atria. QRS complex having largest voltage deflection (1–1.2 mV) and represents
Depolarization of ventricles. T wave amplitude is also low (0.1–0.3 mV) and
represents ventricles Repolarization. The ECG signal is very sensitive in nature
therefore even the presence of small noise the various characteristics of the original
signal changes. Powerline Interference, Baseline wandering noise and EMG noises
corrupts the ECG during data acquisition, which makes detection of characteristic
waves difficult. So proper filtering is needed [2] for accurate detection.

Generally in ECG the major characteristic point is R-peak. Detection of R-peak
is a crucial task and it is the initial job in ECG signal analysis. After finding R-peak
location other components P, Q, S, T are detected by taking R-peak location as
reference and tracing to and fro from R-peak relative position. Figure 1 shows
normal ECG with all characteristic waves.

2 Frequency-Time Analysis for Characteristic Waves Detection

R-peak detection is the first step in ECG characteristic waves detection, by taking
R-peak as reference, the other characteristic waves was identified Pan Tompkins
algorithm [3] also used to detect ECG characteristic waves. In this frequency
domain analysis (wavelet transform) is used for detection of R-peak [4] and next
Time domain analysis is used for detection of other characteristic waves.

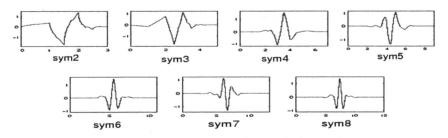

Fig. 2 Symlet wavelet transform

A. *Wavelet Transform Approach*

The Wavelet Transform is a time-scale analysis and used on wide range of applications, in particular signal compression. Now a day's wavelet transform is used to solve problems in Electro cardiology, including compression of data, ventricular late potentials analysis, and finally ECG characteristic waves detection. Wavelet transform decomposes the given signal into a number of levels related to signal frequency components and analyses each level with particular resolution [5].

In this the ECG signal decomposed into 4 levels using Symlet wavelet (sym4) transform for finding R-peak. In wavelet decomposition it down samples the original ECG signal, as a result the samples are reduced and QRS complex is retained. Symlet wavelet transform is the modified version of Daubechies wavelet with increased symmetry (Fig. 2).

Decomposed signals are noise free signals, and by making a threshold of 60 % of maximum value. The values which are above threshold [6] are invariably R-peaks. The decomposed signal can be reconstructed into actual signal by first multiplying the down sampled signal into 4, so that R-peaks are detected in actual signal.

B. *Time domain Analysis*

Frequency domain approach is used for the detection of R-Peak only and after time domain approach is used for other characteristic wave's detection.

R-R interval can be calculated by

$$T_{R-R}(n) = \frac{Rloc(n+1) - Rloc(n)}{f_s} (sec)$$

$$f_s = Sampling\,frequency$$

$$Rloc = location\,of\,R-peak$$

For identifying P-wave, a window in time domain is created with time gap limits from 65 % of R-R interval to 95 % R-R interval which is added to same R-peak location. In that window the maximum value will represents P-wave.

The Q-wave is identified by choosing minimum value in Time based window starting from 20 ms before corresponding R-peak. Similarly S-peak is detected by selecting least value in time based window after R-peak location.

For identifying T-wave, a window in time domain is created with time gap limits from 15 % of R-R interval to 55 % R-R interval which is added to same R-peak location. In that window the maximum value will represents T-wave.

The time domain windows are adaptive because they depend on R-R interval values.

3   Methodology of Frequency-Time Based Approach

The main aim of this methodology is for accurate detection of Characteristic waves P-QRS-T of Electrocardiogram. In this Frequency-Time domain approach is adopted [7]. Finding of R-peak is the initial step of identifying characteristic features of ECG which is done in frequency domain (wavelet decomposition) method [7, 8] and other features are detected by taking R-peak location as reference and thereby creating windows in time domain [9] (Fig. 3).

Steps wise methodology of proposed Frequency-Time based Approach:

Step 1: Load the ECG records from MIT-BIH NSR database [10, 11]

Step 2: Remove the different type of noises in ECG frequency range (Baseline drift noise, power line Interference, EMG noise) using cascaded based digital filters (Fig. 4).

Step 3: For finding R-peak decompose the signal using Wavelet (db4/sym4) at particular scale [7, 8].

Step 4: After decomposition identify R-peak in ECG signal keeping 60 % of the signal value as threshold.

Step 5: Reconstruction of ECG signal from decomposed signal find R-peak and R-location (Rloc)

Step 6: For finding Q-point by finding lowest value in the window range
$Rloc-X_1$ to $Rloc-Y_1$
Where $X_1 = 50 * t_s$ $Y_1 = 10 * t_s$ (t_s = sampling time)
$t_s = 1/f_s$ (f_s = sampling frequency)

Step 7: For identifying S-point create a window on the right side of R-peak in time domain in the range of
$Rloc + X_2$ to $Rloc + Y_2$
Where $X_1 = 5*t_s$ $Y_1 = 50*t_s$
$t_s = 1/f_s$ (f_s = sampling frequency)

Step 8: Detect T-point by finding highest value in the window range
$Rloc + X_3$ to $Rloc + Y_3$
Where $X_3 = 25*t_s$ $Y_3 = 100*t_s$
$t_s = 1/f_s$ (f_s = sampling frequency)

Fig. 3 Frequency-time based
methodology

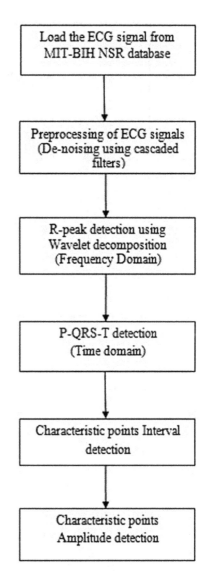

Step 9: Detect P-Point by finding maximum value in window range
Rloc − X_4 to Rloc − Y_4
Where $X_4 = 50 * t_s$ $Y_1 = 100 * t_s$
$t_s = 1/f_s$ (f_s = sampling frequency)

All the windows in time domain are created w.r.t Standard wave form of ECG shown in Fig. 5.

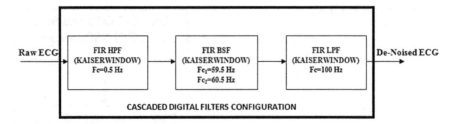

Fig. 4 Cascaded FIR filter configuration

Fig. 5 Standard ECG waveform

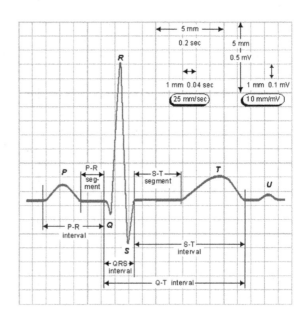

4 Results and Discussion

The analysis of Frequency-Time based approach was tested MIT-BIH NSR database [10]. The entire detection techniques developed in MATLAB software [12].

The above Fig. 6 represents original ECG signal taken from MIT-BIH NSR database.

The above Fig. 7 shows Symlet wavelet transform with 4 level decomposition (Frequency Domain). It is a down sampling process which reduces samples in each level) (Fig. 8).

The above Fig. 9 shows R-peak detected 4 level down sampled signal by taking this signal as reference R-peak is detected in actual signal by reconstructing down sampled signal (Fig. 10).

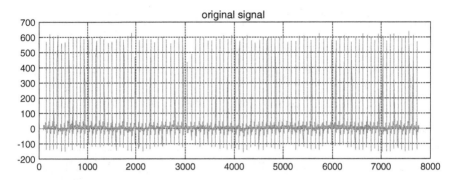

Fig. 6 Original ECG waveform

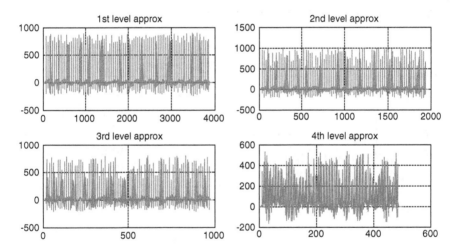

Fig. 7 Wavelet decomposition (sym4) of ECG

Finally the R-peaks are detected in actual signal using Wavelet decomposition and reconstruction methods.

The above Figs. 11 and 12 represents P-QRS-T detected actual ECG signal from MIT-BIH NSR database 16265.mat signal [10]. In that R-peaks represents by '*', P-peaks represents by 'O', Q and S peaks are represents by '+' and finally T-peak is represented by'Δ'.

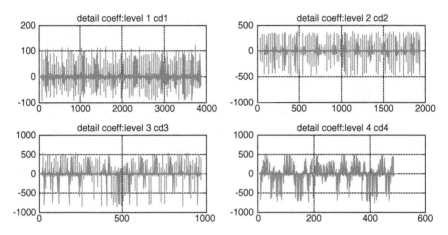

Fig. 8 Detailed wavelet coefficients (sym4) of ECG

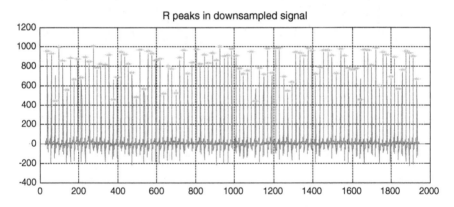

Fig. 9 R-peak detected down sampled ECG

The Characteristic waves detected ECG signal is taken from MIT-BIH Normal Sinus Rhythm database 16,265 mat signal (7500 samples with Sampling frequency $f_s = 128$ Hz).

The Tables 1 and 2 represents amplitudes and wave locations (samples) of actual ECG signal for first 10 samples only. The samples can be changed in time mode by multiplying the samples with 't_s' where $t_s = 1/f_s$ (f_s = sampling frequency).

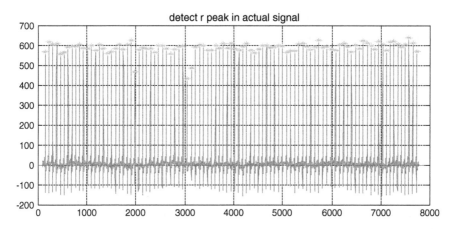

Fig. 10 R-peak detected in original ECG

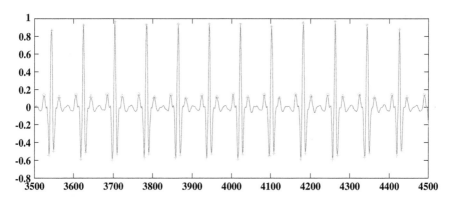

Fig. 11 P-QRS-T detected in original ECG

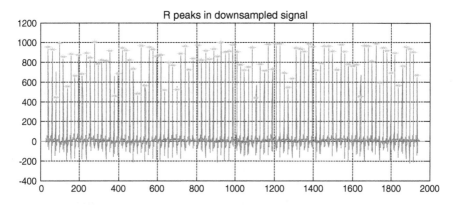

Fig. 12 Characteristic waves detected in original ECG signal

Table 1 Characteristic waves amplitude features in volts

Characteristic waves amplitude features				
P-wave amplitude (v)	Q-wave amplitude (v)	R-wave amplitude (v)	S-wave amplitude (v)	T-wave amplitude (v)
0.1483	−0.5898	0.9602	−0.5465	0.1123
0.1456	−0.5783	0.9522	−0.5412	0.1214
0.1412	−0.6117	0.9759	−0.5381	0.1222
0.1372	−0.5900	0.9304	−0.5275	0.1260
0.1405	−0.5705	0.9079	−0.5031	0.1159
0.1276	−0.5452	0.8924	−0.5043	0.1123
0.1412	−0.5548	0.9062	−0.5203	0.1135
0.1399	−0.5769	0.9486	−0.5420	0.1164
0.1304	−0.5763	0.9123	−0.5073	0.1272
0.1423	−0.5773	0.8789	−0.5334	0.1245

Table 2 Characteristic waves locations w.r.t. samples

Characteristic waves locations w.r.t samples				
P-wave locations	Q-wave locations	R-wave locations	S-wave locations	T-wave locations
136	149	155	161	174
214	227	232	238	252
291	304	310	316	329
369	382	388	393	407
448	461	467	473	487
527	540	545	551	564
606	619	624	630	644
682	695	701	707	720
760	773	779	784	798
838	851	857	863	877

5   Conclusion

In this paper a novel Time-Frequency based methodology has implemented, in that for R-peak detection Symlet4 wavelet transform decomposition technique is used and for other peaks detection Time based window technique is developed. The information about R-peak and QRS complex is very much useful for ECG analysis, diagnosis, classification, authentication and identification performance. The obtained amplitude features, interval features of characteristic waves of ECG are validated for MIT-BIH NSR database records. The results from MATLAB shows 100 % accurate detection rate for all characteristic waves (P-QRS-T) of ECG. Future scope includes calculation of all wave intervals, amplitude features for ECG applications like arrhythmia detection and Bio-metric tool.

References

1. R. Acharya, J.S. Suri, J.A.E. Spaan, S.M. Krishnan, *Advances in Cardiac Signal Processing*. (Springer, Berlin, 2007), ISBN: 3-540-36674-1
2. S. Banerjee, R. Gupta, M. Mitra, Delineation of ECG characteristic features using multi resolution wavelet analysis method. Meas. Elsevier **45**, 474–487 (2012)
3. J. Pan, W.J. Tompkins, A real time QRS detection algorithm. IEEE Trans. Biomed Engg. **32**, 230–236 (1985)
4. Lin, H.-Y., Liang, S.-Y.: Discrete-wavelet transform based noise reduction and R-wave detection for ECG signals. In: IEEE 15th International Conference on e-Health Networking, Applications and Services, 2013 (Healthcom 2013)
5. P.S. Addison, Wavelet transforms and the ECG: a re-view. Physiol. Meas. **26**(5), 155 (2005)
6. B.U. Kohler, C. Hennig, R. Orglmeister, The principles of software QRS detection. IEEE Eng Biol. Mag **21**, 42–57 (2002)
7. C. Li, C. Zheng, C. Tai: Detection of ECG characteristic waves using wavelet transforms. IEEE Trans. Biomed. Eng. **42**, 21–28 (1995)
8. J.S. Sahambi, S. Tandon, R.K.P. Bhatt, Using wavelet transform for ECG characterization. IEEE Trans Eng. Med. Biol. **16**(1), 77–83 (1997)
9. A.K. Manocha, M. Singh, Automatic delineation of ECG characteristic waves using window search and multi resolution wavelet transform approach. In: ACEEE International Conference of Emerging Trends in Engineering and Tech
10. http://www.physionet.org/cgi-bin/atm/ATM—MIT-BIH Normal Sinus RhythmDatabase/
11. http://en.wikipedia.org/wiki/File:SinusRhythmLabels.svg
12. Matlab help: MATLAB MATHWORKS

A Long Signal Integrator for Fusion Device Using Arm Controller

A. Nandhini and V. Kannan

Abstract Integrator is the basic circuit for measuring magnetic parameters like magnetic flux and magnetic field from the magnetic coil signal. In Nuclear fusion device the integrators are used to find the magnetic measurements which are significant for the long duration (signal > 1 s) device operation. The proposed integrator has been composed of input module, integrator module and processor module. It has been tested for long duration with different input signal. There is no conspicuous drift error has been noticed in the output.

Keywords Tokomak · Integrator · Fusion · ARM controller · Magnetic flux

1 Introduction

In modern nuclear fusion device like Tokamak, the fourth state of matter called plasma has been used for heating the radioactive elements. The temperature of this reaction is extremely high ($>10^6$ K). In order to tolerate the high temperature, the reaction takes place in the presence of doughnut shaped magnetic arrangement [1]. It is necessary to find the magnetic parameters like magnetic flux and magnetic field, in order to place the magnet and plasma in a safe distance. Hence, the integrator is mandatory to measure these magnetic parameters [2–4].

Electronic integrators have been considered as two types depend on the applied input and they are voltage integrators and current integrators. The time integration of voltage signal from the magnetic coil will give the total magnetic flux (ϕ) and the

A. Nandhini (✉)
Sathyabama University, Chennai 600119, India
e-mail: abnandhu@gmail.com

V. Kannan
Jeppiaar Institute of Technology, Kunnam, Kanchipuram 631604, Tamilnadu, India

© Springer India 2016
S.C. Satapathy et al. (eds.), *Microelectronics, Electromagnetics and Telecommunications*, Lecture Notes in Electrical Engineering 372,
DOI 10.1007/978-81-322-2728-1_7

69

magnetic field (E) [5, 6]. In similar manner the time integration of current signal from the magnetic coil will give the total magnetic charge (Q).

In order to accomplish the measurements during long pulses, precise systems and components level of stability of operation and performance. Measures such as active alignment systems, in situ calibration, special components (e.g. long pulse integrators) will be required [5].

In Tokamaks, the magnetic measurement system is based on passive coils used in association with analog integrators. However, analog integrators are subject to intrinsic problems. The two most important are firstly the integrator drift, due to the offsets of the operational amplifier which introduces an absolute error that increases with integration time, and secondly the saturation of the integrator in the case of a high flux variation such as a disruption [6, 7].

The main objectives of the proposed integrator system are as follows [8].

- Optimize existing integrators for fusion science and laboratory applications
- Configure integrators for long pulse applications
- Incorporate integrator system into widely used DAQ (Data Acquisition) modules for ease of use.

2 Functional Description

The integrator system is mainly composed of three modules namely input module, integrator module and processor module. Figure 1 shows the input module and integrator module and Fig. 2 shows the processor module. The input module has been composed of a buffer amplifier and a multiplexer. Similarly, the integrator module has been composed of two RC integrators and an instrumentation amplifier.

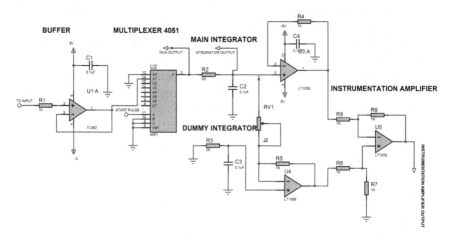

Fig. 1 Input and output integrator module

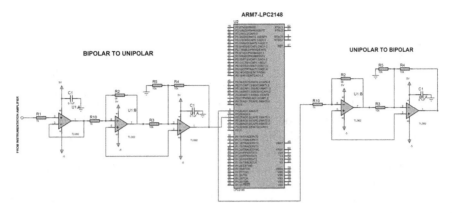

Fig. 2 Processor module

2.1 Input Module

The input signal (from the magnetic coil or function generator) has been applied to the buffer of the input module. TL062 is the buffer amplifier used in this method which is connected to give unity gain and also to avoid loading effect of input signal. The output from the buffer amplifier is connected to the multiplexer (Mux) which allows the signal up to 5 s. After that it switches to zero, so that the integration takes place only in that 5 s duration. 4051 Mux is used as a signal selector or used for switching the output from the buffer after 5 s.

2.2 Integrator Module

The integrator module has two RC integrators and an instrumentation amplifier. In which integrator1 acts as a original integrator and integrator2 acts as a dummy integrator. The input of integrator1 has been connected to the output of Mux and integrates the signal from the Mux. However the input of integrator2 is grounded and so it produces a reference signal for offset calculation. Both integrator output (integrator1&2) are connected to the two inputs of instrumentation amplifier which is actually used as a differential amplifier for eliminating offset in the integrated output.

2.3 Processor Module

The processor module has an ARM processor, UBC (Unipolar to Bipolar Converter) and BUC (Bipolar to Unipolar Converter). The integrated output from the amplifier is connected to BUC which converts the bipolar output signal into unipolar signal. Then the signal is given to the ARM processor for processing. The processor used in this method is LPC2148 which has in built ADC (Analog to Digital Converter) and DAC (Digital to analog Converter). As ADC accepts only unipolar signal, BUC has been used in this circuit. The integration formula has been dumped in the processor so that it eliminates the offset induced drift and saturation error. Then the digitally integrated signal has been given to the in-built DAC in order to get the analog output. Finally, the signal is send to the UBC in order to get the final bipolar output.

2.4 Operation

The buffer amplifier used for minimizing the loading effect at the input side. The integrator has been designed in such a way that the RC time constant is long compared to the ADC sampling interval Δt, but short compared to the pulse length (Fig. 3). The output voltage V_1 of the RC network is given by

$$RCV1 = (V_0 - V_1)\, dt \tag{1}$$

Fig. 3 A prototype integrator

Using samples V_i of the RC integrator's output voltage V_1, the real-time processor calculates the integral of the input voltage V_0, which equals the magnetic flux F through the pickup loop,

$$\Phi = \int V_0 dt = \int V_1 dt + RCV_1 \approx \sum (V_i \Delta_t) + RCV_i, \qquad (2)$$

where Δt is the sampling interval of the ADC. Here Vi is understood to represent the voltage at the amplifier input, so the amplifier gain G does not appear explicitly in the above Eq. (2). The passive RC integrator provides accurate integration of transients that are too rapid for the ADC sampling rate. For transients with a time scale shorter than the RC time, the RC integral initially contains most of the integrated value [9, 10]. The digital integral (first term on the right-hand side) can be thought of as a correction for the slow decay of the passive integrator. On the other hand, for slowly varying signals the digital integral dominates and the RC term becomes negligible.

Continuous measurement of the time-dependent baseline offset voltage is obtained by connecting the instrumentation amplifier input between the output of the RC integrator and a dummy integrator with resistance R. The instrumentation amplifier performing the role of differential amplifier in order to minimize the offset in the signal. The digitally integrated signal from the ARM processor tracks the offset change and used to eliminate that error.

3 Results and Discussion

The test signal from the signal generator has been given to the input module of the integrator. The buffer present in the input module has been sent the signal to the multiplexer unit without any loss in voltage. The multiplexer will allow the signal for 5 s since the reactor discharge happens for 5 s duration and after that it will retain in zero signal mode.

The output of the integrator system has been displayed in the Figs. 4, 5 and 6. It has been tested for triangular wave, square wave and pulse waveform. The main advantage of this system is to produce integration output at very low frequency.

The prototype integrator output for various signals has been shown below. The output response for triangular wave, square wave and pulse has been noted down. This output has been monitored up to 5 s duration using storage oscilloscope and there is no obvious offset induced drift present in the output.

Figure 6 shows the pulse signal which has been generated with the help of microcontroller followed by DAC. This is similar to the signal produced from the tokamak device [11, 12].

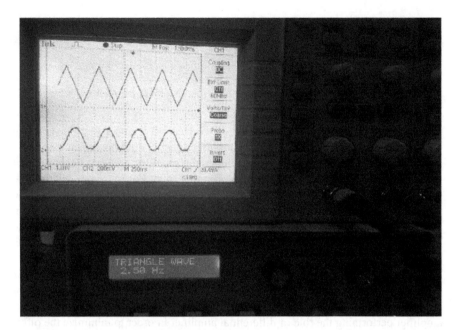

Fig. 4 Output response for triangular wave

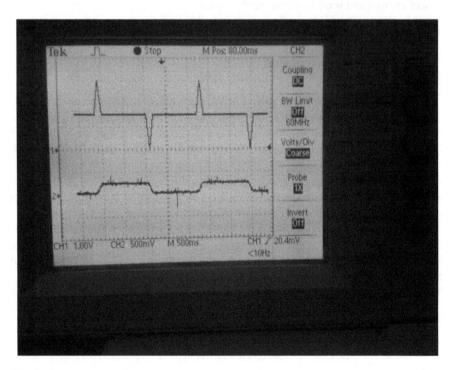

Fig. 5 Output response for square wave

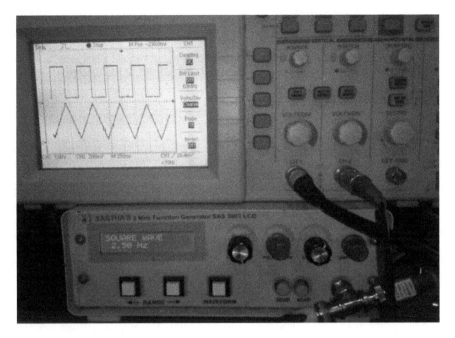

Fig. 6 Output response for pulse

References

1. S. Elieser, Y. Elieser, The fourth state of matter-an introduction to plasma science:, The Institute of Physics, London (2001)
2. E. Mazzucato, About the toroidal magnetic field of a tokamak burning plasma experiment with superconducting coils: Department of Energy, USA, (2002)
3. J.F. Schenck, Safety of Strong. Static Magn. Fields: J. Magn. Reson. Imaging **12**, 19 (2009)
4. Cordey, J.G. et al., in *Proceedings of the 28th EPS Conference on Controlled Fusion and Plasma Physics*, Madeira, Portugal, 2001
5. D.B. Jaffe, N.T. Carnevale, Passive normalization of synaptic integration influenced by dendritic architecture. J. Neurophysiol. **82**, 3268–3285 (1999)
6. T.H. Casselman, S.A. Hanka, Calculation of performance of magneto resistive permalloy magnetic field sensor. IEEE Trans. Magn. MAG **16**(2), 461–464 (1980)
7. A.E. Costley, et.al., Long pulse operation in iter: issues for diagnostics. in *30th EPS Conference on Contr. Fusion and Plasma Physics*, St Petersburg: vol. 27A, pp. O–4.1D. St. Petersburg (2003)
8. Y.K. Thong, M.A. Woolfson, J.A. Crowe, B.R. Hayes-Gill, R.E. Challis, Dependence of inertial measurements of distance on accelerometer noise. Meas. Sci. Technol. **13**, 1163–1172 (2002)
9. B.M. Liu et al., A new low drift integrator system for the experiment advanced superconductor Tokamak. Rev. Sci. Instrum. **80**, 053506 (2009)

10. K. Miller, et al., High gain and frequency ultra-stable integrators: Eagle harbor technologies
11. E.J. Strait, J.D. Broesh, R.T. Snider, M.L. Walker, A hybrid digital–analog long pulse integrator. Rev. Sci. Instrum. **68**, 381 (1997)
12. J.D. Broesch, E.J. Strait, R.T. Snider, M.L. Walker, A digital long pulse integrator. in *Fusion Engineering: 16th IEEE/NPSS Symposium*, vol. 1, pp. 365–368, IEEE, Champaign, 1995

Performance Analysis of Frequency Dependent and Frequency Independent Microstrip Patch Antennas

Naresh Kumar Darimireddy, R. Ramana Reddy
and A. Mallikarjuna Prasad

Abstract A miniature and low profile frequency dependent and frequency independent microstrip patch antennas are proposed in the paper. An idea of converting frequency dependent microstrip patch into frequency independent microstrip patch by altering the dimensions, relative spacing and with suitable arrangement of parasitic patches is presented and discussed. The performance analysis and design details of frequency dependent and frequency independent patch antennas are presented in the paper. To determine the mathematical calculations of critical microstrip patch dimensions and the corresponding optimized design of both antennas was carried out using Ansys HFSS solver. The parametric studies with respect to various substrate materials, substrate thickness and some design parameters of both the antennas are simulated and presented in the paper. The proposed antennas are fabricated and tested. The Simulation and Experimental results are compared. Based on these results, the proposed antennas operate in UWB (3.1–10.6 GHz) range of frequencies.

Keywords Frequency dependent · Frequency independent · Printed antenna · Parasitic patch · Microstrip patch

N.K. Darimireddy (✉)
JNTUK, Kakinada, A.P., India
e-mail: yojitnaresh@gmail.com

R.R. Reddy
Department of ECE, MVGR College of Engineering, Vizianagaram, A.P., India
e-mail: profrrreddy@yahoo.co.in

A.M. Prasad
Department of ECE, UCEK, JNTUK, Kakinada, A.P., India
e-mail: a_malli65@yahoo.com

© Springer India 2016
S.C. Satapathy et al. (eds.), *Microelectronics, Electromagnetics and Telecommunications*, Lecture Notes in Electrical Engineering 372, DOI 10.1007/978-81-322-2728-1_8

1 Introduction

In some wireless and broadband applications we need directional antennas such as Frequency Dependent and Frequency Independent antennas. To achieve the end-fire radiation, the parasitic elements in the direction of the beam are smaller in length than the feed element will be used. Typically the driven element is resonant with its length slightly less than $\lambda/2$ (usually 0.45–0.49λ) whereas the lengths of the directors should be about 0.4–0.45λ. However, the directors are not necessarily of the same length and/or diameter and the separation between the directors is typically 0.3–0.4λ [1, 2]. Mishra et al observed that, while the element widths are scaled-out log-periodically with scaling-in of operating frequency, the element lengths do not pursue this principle. However, almost all reported work [5] on Log-Periodic microstrip antennas use a log-periodic scaling for the parasitic element length.

In this paper, frequency dependent and independent microstrip patch antennas are considered. Basically frequency dependent antennas work at a fixed frequency of resonance and depending on that frequency the lengths of the directors, driven element (or) active element and the reflector are changed. These antennas cannot meet the desired features of broad band systems in a given band of frequency, so if the antenna need to operate at different resonant frequencies then separate antennas need to be designed for every operating frequency in case of frequency dependent antennas [3]. In view of the above disadvantage, and to overcome the shortcoming, frequency Independent antennas are considered with minor changes in the dimensions of parasitic elements, relative spacing and suitable arrangement of parasitic patches of frequency dependent antennas.

2 Antenna Design and Configuration

2.1 Frequency Dependent Antenna

Frequency dependent antenna, the word itself mentions that antenna depends on a particular band of frequency of the required system. One of the frequency dependent antennas is Yagi-Uda antenna and the design of six element frequency dependent antenna is proposed in the paper. The microstrip frequency dependent antenna consists of a driven patch along with several parasitic patch elements [6, 7] which are arranged on the same substrate surface in such a way that the overall antenna characteristics are enhanced.

2.1.1 Design Procedure

The design of frequency dependent antenna with six elements of desired gain at different frequencies is carried out using the following relations.

- Wavelength

$$\lambda = c/f \text{ in meters} \tag{1}$$

where
c velocity of the light
f resonant frequency

- Length of Directors

$$
\begin{aligned}
L_{d1} &= 0.44^* \lambda \\
L_{d2} &= 0.44^* \lambda \\
L_{d3} &= 0.43^* \lambda \\
L_{d4} &= 0.40^* \lambda
\end{aligned}
\tag{2}
$$

- Length of Reflector

$$L_r = 0.475^* \lambda \tag{3}$$

- Spacing between Reflector and Driven Element

$$S_L = 0.25^* \lambda \tag{4}$$

- Spacing between Director and Driven Element

$$S_d = 0.31^* \lambda \tag{5}$$

- Diameter of the elements

$$d = 0.01^* \lambda \tag{6}$$

- Length of the array

$$L = 1.5^* \lambda \tag{7}$$

The proposed dimensions and the design of frequency dependent antenna are shown in Fig. 1.

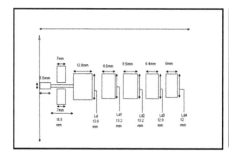 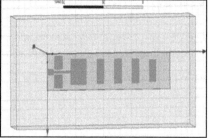

Fig. 1 Proposed frequency dependent antenna design with dimensions

2.1.2 Simulated Results and Parametric Study

The simulated design parameters of frequency dependent antenna of six elements at different operating fixed frequencies using MATLAB are recorded in Table 1.

The following results are recorded using HFSS. Figure 2 shows the return loss of the frequency dependent antenna and it resonates at the single band frequency of 9.5 GHz.

The Frequency dependent Antenna is designed using HFSS software whose gain, bandwidth and return losses are varied with the variation of ground length and substrate materials, substrate thickness and director width are recorded in Tables 2, 3 and 4 respectively.

2.1.3 Fabricated Antenna and Experimental Results

Figure 3 shows the fabricated structure and experimental setup of the antenna. Figure 4 shows the measured return loss with single band frequency of operation.

2.1.4 Comparison of Results

See Table 5.

2.2 Frequency Independent Antenna

The Frequency Independent Antenna consists of parallel linear dipole elements of different lengths and spacing. The lengths of the dipole elements, the spacing from the virtual apex to the dipole elements, the wire radius of the dipole elements, the spacing between the quarter wave-length dipoles are proportional with the geometric scale factor τ, which is always smaller than one. A wedge of enclosed angle

Table 1 Simulated design parameters of frequency dependent antenna using MATLAB

S. no	Frequency (MHz)	Length of directors (mm)				L_a	L_r	S_L	S_d	d	L
		L_{d1}	L_{d2}	L_{d3}	L_{d4}						
1	125	1.056	1.056	1.032	0.96	0.998	1.14	0.6	0.744	0.024	3.6
2	200	0.66	0.66	0.645	0.6	0.624	0.712	0.375	0.465	0.015	2.25
3	500	0.264	0.264	0.258	0.24	0.249	0.285	0.15	0.186	0.006	0.9
4	1000	13.2	13.2	12.9	12	13.8	14.25	7.5	9.3	0.3	110

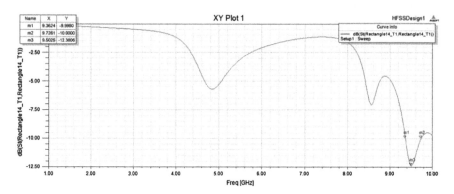

Fig. 2 Return loss of frequency dependent antenna

Table 2 Variation of gain and bandwidth with respect to various substrate materials

S. no.	Substrate	Full ground	Half ground	One-forth ground
1	RT/Duriod 5880	Gain = 5.8228 dB	Gain = 4.5071 dB	Gain = 4.498 dB
		No bands	No bands	No bands
2	RT/Duriod 6002	Gain = 6.0048 dB	Gain = 4.8578 dB	Gain = 4.2345 dB
		No bands	No bands	No bands
3	Glass	Gain = 6.8700 dB	Gain = 6.5108 dB	Gain = 4.9437 dB
		Single band at 7.5 GHz	Single band at 7.6 GHz	Four bands at 7.6 GHz, 7.8 GHz, 8.3 GHz, 9.2 GHz
		Bandwidth = 0.19 %	Bandwidth = 0.88 %	Bandwidth = 2.66 %, 1.14 %, 2.48 %, 6.47 %

Table 3 Variation of gain and bandwidth with respect to substrate thickness

S. no.	Thickness	Full ground	Half ground	One-forth ground
1	1.4 mm	Gain = 3.095 dB	Gain = 2.6596 dB	Gain = −6.4275 dB
		No bands	Single band at 9.6 GHz	Single band at 8.5 GHz
			Bandwidth = 2.98 %	Bandwidth = 4.69 %
2	2 mm	Gain = 3.7006 dB	Gain = 3.3456 dB	Gain = 1.0060 dB
		No bands	No bands	Single band at 8 GHz
				Bandwidth = 5.76 %

bounds the dipole lengths. The spacing factor, σ is defined as the distance between two dipole elements divided by the twice of the length of the larger dipole element. The relationship between the different parameters can be summarized as follows:

Table 4 Variation of gain and bandwidth with respect to director's width

S. no.	Directors width (mm)	Full ground	Half ground	One-forth ground
1	$W_{d1} = 4.5$	Gain = 3.3825 dB	Gain = 3.0518 dB	Gain = 8.5769 dB
	$W_{d2} = 4.5$	No bands	No bands	Single band at 8.3 GHz
	$W_{d3} = 4.4$			Bandwidth = 4.74 %
	$W_{d4} = 4$			
2	$W_{d1} = 6.5$	Gain = 3.3748 dB	Gain = 2.8366 dB	Gain = 8.5891 dB
	$W_{d2} = 6.5$	No bands	No bands	Single band at 8.3 GHz
	$W_{d3} = 6.4$			Bandwidth: 4.82 %
	$W_{d4} = 6$			
3	$W_{d1} = 8.5$	Gain = 3.3723 dB	Gain = 2.8334 dB	Gain = 1.1970 dB
	$W_{d2} = 8.5$	No bands	Single band at 9.5 GHz	Single band at 8.3 GHz
	$W_{d3} = 8.4$		Bandwidth = 4.68 %	Bandwidth = 4.37 %
	$W_{d4} = 8$			

Fig. 3 Fabricated structure and experimental setup of frequency dependent antenna

$$\tau = \frac{R_{n+1}}{R_n} = \frac{L_{n+1}}{L_n} = \frac{D_{n+1}}{D_n} \quad (8)$$

$$\sigma = \frac{1 - \tau}{4 \tan \alpha} = \frac{D_n}{2L_n} \quad (9)$$

Here,

τ Geometric ratio and $\tau < 1$,

L   Length of the dipole,

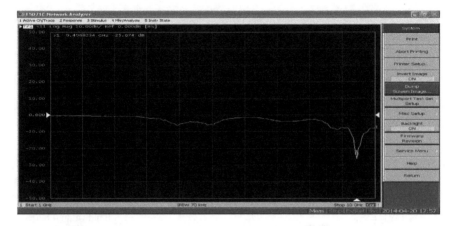

Fig. 4 Experimental return loss of fabricated frequency dependent antenna

Table 5 Comparison of simulated and experimental results

Parameters	HFSS	Network analyzer
Start and stop frequencies	1–10 GHz	1–10 GHz
No. of bands	Single band	Single band
Resonant frequency	9.5 GHz	9.4 GHz

R   Distance from apex to the dipole elements,
D   Spacing between the dipole elements.
σ   Spacing factor, which relates distance between two adjacent elements with the length of the larger element which is equal to half of the apex angle.

As the first step of the design procedure, fundamental design parameters τ and σ should be chosen for a given directivity. For a given directivity, corresponding σ and τ can be found. For a certain τ, if maximum directivity is desired, σ_{opt} should be chosen through the curves. Optimum σ can be formulated as

$$\sigma_{opt} = 0.258\tau - 0.066 \qquad (10)$$

$$\alpha = 2\tan^{-1}\left(\frac{1-\tau}{4\sigma}\right) \qquad (11)$$

After determining σ, τ and α, bandwidth of the system which determines the longest and the shortest dipole elements can be calculated. Active region bandwidth, B_{ar} can be related with the fundamental design parameters by the following equation

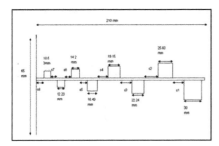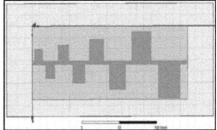

Fig. 5 Frequency independent antenna design with dimensions

$$B_{ar} = 1.1 + 7.7\left(1 - \frac{\tau}{2}\right)^2 \cot \alpha \tag{12}$$

In practice a slightly larger structure bandwidth, B_s is usually designed to reach the desired bandwidth, B. These bandwidths are related by

$$B_s = B^* B_{ar} = B^*\left(1.1 + 7.7\left(1 - \frac{\tau}{2}\right)^2 \cot \alpha\right) \tag{13}$$

Boom length of the structure is defined between the shortest dipole and longest dipole elements and is given by

$$L = \frac{\lambda_{max}}{4}\left(1 - 1/B_s\right) \cot \alpha \tag{14}$$

$$\lambda_{max} = 2L_{max} \tag{15}$$

The proposed dimensions and design of frequency independent antenna are shown in Fig. 5.

2.2.1 Simulated Results and Parametric Study

Based on the equations discussed in the above section, a frequency independent antenna is designed using HFSS software whose gain, bandwidth and return losses are varied with the variation of ground length, substrate, and its thickness and whose simulated parameters are also presented in Table 6 using MATLAB code.

The following results are recorded using HFSS. Figure 6 shows the return loss of the frequency independent antenna with two operating band of resonant frequencies. The analysis of gain and bandwidth of frequency independent antenna are recorded in Table 7 and Table 8 with respect to various substrate materials and substrate thickness.

Table 6 Design parameters of frequency independent antenna

S. no.	Desired directivity	Wedge angle (degrees)	Length of the structure (mm)	Relative mean spacing	Average characteristic impedance (in ohms)	Output impedance (in ohms)	Center to center spacing (in inches)
1	9 dB	24.2	110	0.001	328.1961	60	0.8457

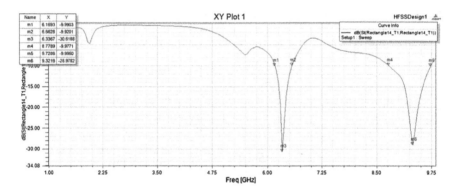

Fig. 6 Return loss of frequency independent antenna

2.2.2 Fabricated Frequency Independent Antenna and Experimental Results

Figure 7 shows the fabricated structure and experimental setup of the antenna. Figure 8 shows the measured return loss with dual resonant frequencies.

2.2.3 Comparison of Results

See Table 9.

3 Discussion of Results

The performances of miniaturized frequency dependent and independent patch antennas with respect to different parametric variations, where the gain, number of operating bands, bandwidth and desired resonant frequencies are observed and studied. With the substrate material FR-4 epoxy (εr = 4.4), the frequency dependent antenna operates at single band (9.4 GHz) and the frequency independent antenna operates at dual band (6.4 and 9.3 GHz) of frequencies. From the parametric study it

Table 7 Variation of gain and bandwidth with respect to various substrate materials

S. no	Substrate	Full ground	Half ground	One-forth ground
1	RT/Duriod 5880	Gain = 7.0422 dB	Gain = 7.0117 dB	Gain = 7.2209 dB
		Single band	Double bands	Single band
		at 8.6 GHz	at 1.7 GHz, 8.6 GHZ	at 8.6 GHz
		Bandwidth = 8.94 %	Bandwidth = 4.37 %, 9.47 %	Bandwidth = 8.94 %
2	RT/Duriod 6002	Gain = 8.9148 dB	Gain = 8.2243 dB	Gain = 8.9565 dB
		Single band	Double bands	Double bands
		at 7.7 GHz	at 1.5 GHz, 7.8 GHZ	at 1.9 GHz, 7.6 GHZ
		Bandwidth = 6.43 %	Bandwidth = 3.39 %, 8.12 %	Bandwidth = 2.2 %, 8.27 %
3	Glass	Gain = 7.9850 dB	Gain = 8.5121 dB	Gain = 7.6071 dB
		Nine bands	Seven bands	Six bands
		at 4.9 GHz, 6.7 GHz,	at 4.9 GHz, 5.9 GHz,	at 1.4 GHz, 4.9 GHz,
		6.8 GHz, 7 GHz,	7 GHz, 7.4 GHz,	7 GHz, 7.8 GHz, 8.3 GHz, 9.7 GHz
		8.3 GHz, 8.6 GHz, 7.4 GHz, 9.6 GHz, 9.9 GHz	7.7 GHz, 8.4 GHz, 9.9 GHz	
		Bandwidth = 1.48 %,	Bandwidth = 1.42 %,	Bandwidth = 5.98 %,
		0.34 %, 0.46 %,	0.62 %, 0.69 %,	1.61 %, 1.99 %,
		0.31 %, 0.83 %,	0.51 %, 0.71 %,	1.63 %, 4.74 %,
		0.74 %, 2.35 %,	0.26 %, 0.74 %	1.29 %
		0.62 %, 0.72 %		

is observed that using glass or RT-Duriod as substrate material multiband characteristics are obtained for both the antennas. The proposed antennas operate in UWB (3.1–10.6 GHz) range of frequencies with good radiation characteristics and peak gains of 4–9 dBi.

Table 8 Variation of gain and bandwidth with respect to substrate thickness

S. no.	Thickness	Full ground	Half ground	One-forth ground
1	1.4 mm	Gain = 4.2898 dB	Gain = 5.6992 dB	Gain = 7.0237 dB
		Double bands at 6.4 GHz, 9.4 GHZ	Triple bands at 1.4 GHz, 6.4 GHZ, 9.4 GHz	Four bands at 1.6 GHz, 6.4 GHZ, 7.9 GHz, 9.4 GHz
		Bandwidth = 6.22 %, 12.76 %	Bandwidth = 1.79 %, 6.31 %, 12.07 %	Bandwidth = 6.44 %, 5.23 %, 1.71 %, 12.55 %
2	2 mm	Gain = 5.7329 dB	Gain = 6.6087 dB	Gain = 6.7023 dB
		Double bands at 6.3 GHz, 9.3 GHZ	Double bands at 6.3 GHz, 9.3 GHZ	Triple bands at 1.6 GHz, 6.3 GHZ, 9.3 GHz
		Bandwidth = 7.19 %, 9.75 %	Bandwidth = 6.47 %, 9.8 %	Bandwidth = 7.09 %, 6.70 %, 10.67 %

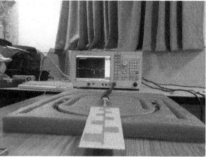

Fig. 7 Fabricated structure and experimental setup of frequency independent antenna

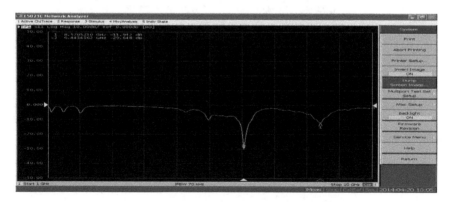

Fig. 8 Experimental return loss of fabricated frequency independent antenna

Table 9 Comparison of simulated and experimental results

Parameters	HFSS	Network analyzer
Start and stop frequencies	1–10 GHz	1–10 GHz
No. of bands	Double band	Double band
Resonant frequency	6.4 GHz, 9.3 GHz	6.4 GHz, 8.59 GHz

4 Conclusions

In this paper, the performance analysis and parametric study of frequency dependent and frequency independent microstrip patch antennas are presented. A simple idea of converting frequency dependent microstrip in to frequency independent patch by dimensional changes and spacing are discussed. Observing the comparative results of Simulated and experimental resonant frequencies, we can say that the frequency dependent antenna has single band and frequency independent antenna has double band characteristics. The proposed antennas used for ultra wideband applications as the operating frequency of both the antennas lies in the UWB (3.1–10.6 GHz) range. The deviations in practical results compared to simulated results are due to in-house manufacturing of proposed patch antenna structures and measurement environment.

References

1. C.A. Balanis, *Antenna Theory Analysis and Design*, 3rd edn. (Wiley, Hobokenm, 2005)
2. J.D. Kraus, *Antennas*, 2nd edn. (McGraw-Hill, New York, 1988)
3. M. Bemani, S. Nikmehr, A novel wide-band microstrip Yagi-Uda array antenna for WLAN applications. Prog. Electromagnet. Res. B **16**, 389–406 (2009)
4. R.K. Mishra, Transmission Line model for log periodic micrstrip antenna. in *Antennas and Propagation Society International Symposium* (APSURSI), pp. 168–169, IEEE 2013
5. R. Garg, P. Bhartia, I. Bahl, A. Ittipiboon, *Microstrip Antenna Design Handbook*. (Artech House Publications, Boston)
6. G.S.N. Raju, *Antennas and Wave Propagation* (Pearson Education (Singapore) Pvt. Ltd, New Delhi, 2005)
7. K.-L. Wong, *Compact and Broadband Microstrip Antennas*. (Wiley, New York, 2002)

Implementation of Gait Recognition for Surveillance Applications

K. Babulu, N. Balaji, M. Hema and A. Krishnachaitanya

Abstract There are various biometric measures that are used in industrial applications for identification of a human. They are signature verification, face recognition method, voice, iris recognition methods, and recognition using digital signatures. These existing human recognition methods have the following limitations of not being unique, low reliability, and could easily traceable by intruders. Gait is the walking style of a human. Gait can be recognized from a view-based approach. In this approach two different image features are required; they are the width of the outer contour of the silhouette and entire binary silhouette. Observation vector can be obtained from the image feature by modeling the frame to exemplar distance (FED) vector sequence with Hidden Markov Model (HMM) as it provides robustness to recognition. In this paper an effort is made for gait recognition useful in real time surveillance applications.

Keywords Gait · Feature vector · Frame to exemplar distance (FED) · Hidden markov model (HMM)

1 Introduction

In recent days automated human recognition is a major component of surveillance. An effective approach of identification is detecting physical characteristics of the person. The surveillance application based on biometric feature is a challenging

K. Babulu (✉) · N. Balaji · M. Hema · A. Krishnachaitanya
Department of ECE, JNTUK UCEV, Vizianagaram, A.P., India
e-mail: kapbbl@gmail.com

N. Balaji
e-mail: narayanamb@rediffmail.com

M. Hema
e-mail: hema_asrith@yahoo.co.in

A. Krishnachaitanya
e-mail: chaitanya.annepu@gmail.com

© Springer India 2016   91
S.C. Satapathy et al. (eds.), *Microelectronics, Electromagnetics
and Telecommunications*, Lecture Notes in Electrical Engineering 372,
DOI 10.1007/978-81-322-2728-1_9

task as it difficult to get the face and iris information from a far distance, also the captured image has less resolution. There are 24 different components to human gait, and that, if all the measurements are considered, gait is unique [1].

1.1 Hidden Markov Model (HMM)

Hidden Markov Model is a statistical model in which it uses a Markov process with hidden states. Hidden Markov Model (HMM) has a set of hidden states Q, an output alphabet (observations) O, transition probabilities A, output observation probabilities B, [2] and initial state probabilities Π. Usually the states, Q, and outputs, O, are understood, so an HMM [2] is said to be a triple, (A, B, Π).

The hidden states are defined by the following Eq. 2.

$$Q = \{q_i\},\ i = 1, 2, \ldots N \tag{1}$$

The transition probabilities are defined by Eq. 2.

$$A = \left\{ a_{ij} = P\left(\frac{q_{i(i=t+1)}}{q_{i(i=t)}} \right) \right\} \tag{2}$$

In the above equation P $(q_{i(i=t+1)} \mid q_{i(i=t)})$ is the conditional probability. Here A is the probability [3] that the next state is q_j given that the current state is q_i. Here Observations (symbols) [2] is defined by Eq. 3

$$O = \{o_k\},\ k = 1 \ldots M \tag{3}$$

The Emission probabilities are defined by Eq. 4

$$B = \left\{ b_{ik} = b_i(o_k) = P\left(\frac{o_k}{q_i} \right) \right\} \tag{4}$$

Here B is the probability that the output is o_k given that the current state is q_i. The initial state probabilities are defined by Eq. 5

$$\pi = \left\{ P_i = P\left(q_{i(t=0)} \right) \right\} \tag{5}$$

Figure 1 indicates hidden states and observation sequences with set of output probabilities b_{ik} and initial probabilities a_{ij}

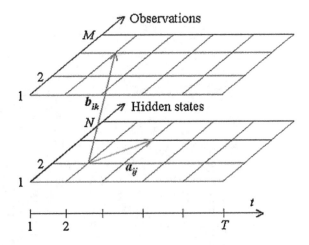

Fig. 1 Hidden Markov model

2 Recognition of Gait

The characteristics can be efficiently captured by extraction of gait features. The features should be chosen such a manner that they should be consistent and produce good discriminability across different humans. Silhouette is chosen as a good feature as it captures motion of most of the body parts [5]. After extracting the silhouette either the entire silhouette [3] or the outer contour of the silhouette can be selected based on the quality of the [4] silhouette. If the image data is of low quality and low resolution then the binarized silhouette is an efficient image feature.

2.1 HMM Based Human Recognition Using Their Gait

The different phases or stances of human can be identified from a gait cycle [6]. Figure 2 Shows the 5 different frames extracted from gait cycle of two humans.

The key differentiators for human gait recognition are the appearance of the stance and the transits across the stances during The better representation of a person is to extract N exemplars (or stances) from the image sequence which will minimize [1] the error in representation of person is given by following Eq. 6.

$$E = \{e_1 \ldots \ldots e_N\} \tag{6}$$

A human is better represented with minimum number of errors by extracting N exemplars from his image sequence. If we know the image sequence for an unknown person J = {j(1), ..., j(T)}, then the exemplars can be used for identifying a human as defined by the Eq. 7.

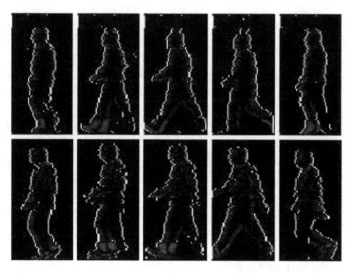

Fig. 2 stances for gait cycle of two humans

$$ID = \arg \min_{j} \sum_{t=1}^{T} \min d\big(y(t), e_n^j\big) \tag{7}$$

Here y(t) indicate the image sequence of a person j [3] for all time instances t, while n indicates the nth exemplar [3] of the jth individual. For a gait cycle [1], at the initial stage a frame is similar in its characteristics to the first [3] exemplar when compared to the other four. Further the frame will be similar to the second exemplar as compared [3] to the others and so on. The probabilistic dependence gives the information about the duration of a gait cycle [4] and the transition from one gait cycle exemplar to the other.

3 Logic and Technology Involved

Initially image [1] features for a person j $x^j = \{x^j(1), x^j(2), \ldots \ldots x^j(T)\}$ is extracted and then we build a model for the gait of person j in the data base and then use the output of the model which is an observation sequence, to identify this person from different persons in the database.

The gait cycle of each person is divided into N number of equal segments, and then we extract the image features for ith segment for all the gait cycles. Then for each segment the mean (centriods) of the features are calculated and they are called as the exemplar for that part. This process can be repeated for all the segments, then resultant exemplar set is the optimal exemplar [1] set $E = \{e_1^*, \ldots, e_N^*\}$. If x(t) is feature obtained from the silhouette at time instant t, then the distance of x(t) from the [3] corresponding exemplars $e_n \in E$ can be calculated. This computed distance is called as frame-to-exemplar distance (FED) vector. It is a one dimensional vector extracted from the two dimensional image to reduce complexity. For the jth person in the database FED vector can be obtained by following equation

$$f_j^{x^j} = d\left(X^j(t), e_n^j\right) \tag{8}$$

FED vector components are time varying with respective to different states. This temporal variation [3] in the FED vector components will results in transition across exemplars. In vision-based activity recognition, researchers have attempted a number of methods such as optical flow, Kalman filtering, Hidden Markov models, etc., under different modalities such as single camera, stereo, and infrared. In present work HMM is used which uses Markov process and has the advantage to model the changing statistical characteristics that are being observed to obtain the actual observations. In this model to calculate HMM parameters we assume exemplars as the states of the HMM and the FED [1] vector sequence is considered as observed process which is used as input for HMM [7].

The likelihood that the jth person will be matched by using the HMM parameters would be calculated by using the log probability by the following Eq. 9

$$P_j = \log\left(P\left(\frac{f_y}{\lambda_j}\right)\right) \tag{9}$$

We repeat the above process for all the persons in the database. If the probability of the unknown human with identity m is P_m, then Pm should be the largest among the database with probabilities of different persons. Then the human with probability P_m would be considered as the matched person with the human with probability P_m.

4 Implementation of Gait Using Java

The above process can be implemented using java, which is a platform independent language and can be used for any real time applications. The following is the process by which we recognized the humans from different databases.

4.1 Data Flow Diagram

The data flow diagram gives the preliminary steps used to create a system along with the inputs and outputs. Figure 3 shows the data flow diagram for identification of humans using gait.

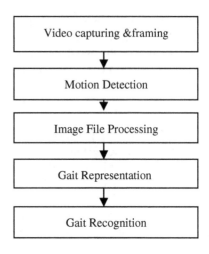

4.2 Video Capture and Framing

The test videos of different humans walking are captured and file operations are performed on the captured videos for extracting sequence of frames. On these captured videos, the frame grabber file operation is performed to extract sequence of frames. Here a JAVA class file called 'FFmpegFrameGrabber' is used for grabbing individual frames from the captured video file. For capturing the videos, the following assumptions are considered, they are, the camera is static and the human is moving and the other assumption is that multiple cameras can be used and at least any one camera captures the side view of the human [4] (Fig. 4).

4.3 Motion Detection

A motion detection algorithm called "background subtraction algorithm" [2] is applied to detect any moving object in the video and identify the object to classify as humans (Fig. 5).

For each frame I and for the reference image I_{ref}, the difference is classified as foreground and is defined by following Eq. 10

$$\left| I_{val} - I_{refval} \right| > Th \tag{10}$$

4.4 Image File Processing

In the Image file processing image operations such as reading and writing of the images, pre-processing algorithms like edge finding, binarizing, and thinning are performed. The Java built in classes File Input Stream and File Output Stream can be used to read and write the image files.

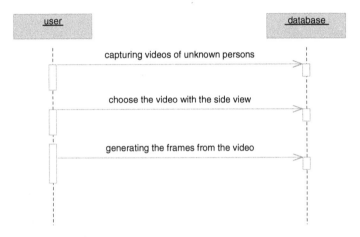

Fig. 4 Sequence diagram for video capturing and framing

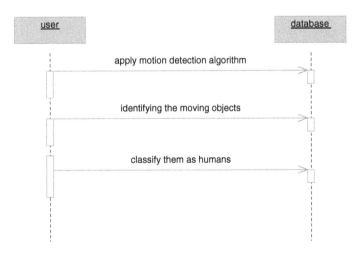

Fig. 5 Sequence diagram for motion detection

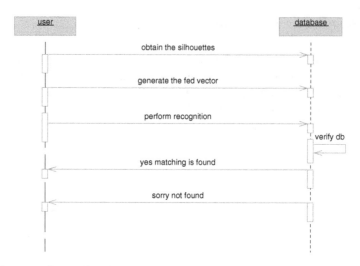

Fig. 6 Sequence diagram for gait recognition

4.5 Gait Representation and Recognition

Here the input image sequence is applied to the preprocessing algorithms as discussed in the image file processing step above. FED vectors are generated for all the elements in the image sequence and are trained with HMM to get the observation feaures. Then these features are compared with the stored features to find the best match. The Gait recognition process is explained through a sequence diagram in Fig. 6.

5 Experimental Results

When the system is executed, the output of initial user interface is obtained as shown Fig. 7. Here, the test path is given in the frame test path box and recognize button is pressed for further recognition process.

Once the test path is given, it generates FED vectors for each person present in the database as shown in the Fig. 8. After the generation of FED vectors, the silhouettes that are in the format of Portable Grey Map (PGM) present in the test folder are compared with the files present in the train folder and finally, if a person is matched then the matched person with that specific person identity number will be displayed.

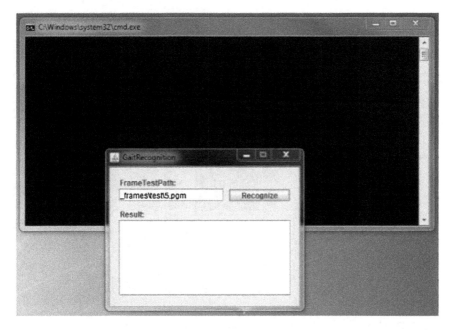

Fig. 7 Gait recognition window

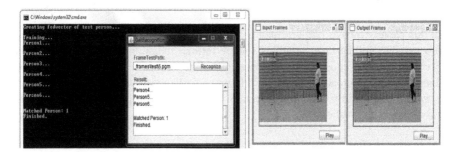

Fig. 8 Matched result

Once the person is recognized then the corresponding person's Input and Output video frames are displayed. An efficient Graphical User Interface is developed for Gait recognition as shown in the Fig. 9.

Fig. 9 Developed graphical user interface display for gait recognition

6   Conclusion

In this paper we have implemented an approach in which binarized silhouettes are processed to get the image features and are further trained with HMM to obtain the best observation sequence for recognition of an unknown human from the given database. This approach is tested on different gait databases and the implementation is working efficiently. The implementation is very much useful in all the real time surveillance applications.

References

1. A. Kale, A. Sundaresan, A.N. Rajagopalan, N.P. Cuntoor, A.K. Roy-Chowdhury, V. Kruger, R. Chellappa, Identification of humans using gait. Image Process. IEEE Trans. **13**(9), 1163–1173 (2004)
2. A. Elgammal, D. Harwood, L. Davis, Non-parametric model for background subtraction. in *Proceedings of the IEEE FRAME-RATE Workshop*, 1999
3. A. Kale, A. Sundaresan, A.K. Roychowdhury, R. Chellappa, Gait-based human identification from a monocular video sequence. in *Handbook on Pattern Recognition and Computer Vision*, 3rd edn. (World Scientific Publishing Company Pvt. Ltd., 2004)
4. R. Zhang, Ch. Vogler, D. MetaXas, Human gait recognition at sagittal plane. Image Vision Comput. **25**(3), 321–330,(2007)

5. L. Lee, W.E.L. Grimson, Gait analysis for recognition and classification. in *Proceedings of the Fifth IEEE International Conference on Automatic Face and Gesture Recognition*, May 2002
6. B. Fadaei, A. Behrad, Human Identification Using Motion Information and Digital Video Processing, in *7th Iranian Machine Vision and Image Processing (MVIP) 2011*, pp. 1–6, 16–17 Nov 2011. doi:10.1109/IranianMVIP.2011.6121603
7. L.R. Rabiner, A tutorial on hidden Markov models and selected applications in speech recognition. Proc. IEEE **77**, 257–285 (1989)

Comparative Analysis of HVS Based Robust Video Watermarking Scheme

Ch. Srinivasa Rao and V.S. Bharathi Devi

Abstract Digital watermarking is a technique of data hiding, which provide security of data. This paper presents a comparative analysis of robust video watermarking technique based on HVS (Human Visual system) quantization matrix. The host video is partitioned into frames and owner's identity watermark is embedded in quantized transform coefficients. Scrambled watermark are generated and embedded in each motionless scene of the host video. Experimental results show no perceptual difference between watermarked frame and original frame and it is robustness against a wide range attacks. PSNR (Peak Signal to Noise Ratio), NCC (Normalized Cross Correlation) are computed and compared with other recent video watermarking schemes. The proposed scheme offers high imperceptible and robust results in terms PSNR and NCC.

Keywords NCC · HVS · PSNR · Hadamard transform · Slant transform and discrete cosine transform (DCT)

1 Introduction

Digital video watermarking is a technique to hide the secret image (watermark image) into the each scene of the host video resulting "watermarked video". Digital watermarking is found to be a solution for digital copyright protection. Applications of video watermarking contain fingerprinting, broadcast monitoring, video authentication and copyright protection. Watermarking technique can be catego-

Ch. Srinivasa Rao (✉) · V.S. Bharathi Devi
Department of ECE, JNTUK-UCEV, Vizianagaram 535003
Andhra Pradesh, India
e-mail: chsrao.ece@jntukucev.ac.in

V.S. Bharathi Devi
e-mail: bharathi28velugula@gmail.com

© Springer India 2016   103
S.C. Satapathy et al. (eds.), *Microelectronics, Electromagnetics
and Telecommunications*, Lecture Notes in Electrical Engineering 372,
DOI 10.1007/978-81-322-2728-1_10

rized to in two ways they are visible and invisible watermarking. This paper is based on invisible watermarking. Two requirements of invisible watermarking are Imperceptibility and robustness. To maintain these two requirements optimally proposed video watermarking uses HVS mathematical model and Hadamard transform. Proposed video watermarking algorithm uses "Blind algorithm". This paper implemented in frequency domain based watermarking technique.

Digital image watermarking using DFT (Discrete Fourier transform) in phase and frequency domains are presented in [1]. J.R. Hernandez et al. presented watermarking using DCT in [2]. M.A. Suhail et al. in [3] watermarking is done using DCT-JPEG compression model. Watermarking based on DWT (Discrete wavelet transform) is presented in [4]. A novel block based watermarking is presented using SVD (Singular value decomposition) is presented in [5]. Human Visual system based base robust video watermarking is presented in [6]. H. Hartung et al. presented different approaches for video watermarking techniques in [7]. Robust Image watermarking based HVS in slant domain presented in [8].

The paper is organized as follows: Hadamard Transform is discussed in Sect. 2. Information about HVS weightage matrix is presented in Sect. 3. Proposed Watermark embedding and extraction processes are explained in Sect. 4. Experimental results with robustness check and comparative performance evaluation are discussed in Sect. 5.

2   Hadamard Transform

The proposed video watermarking algorithm uses Hadamard transform [9], it is sub-optimal orthogonal transform with maximum energy compaction. This transformation requires shorter processing time as it involves simpler integer manipulations. Integer transforms are very much essential in video compression. It has ease of hardware implementation than many common transform techniques. This transformation is computationally less expensive.

3   HVS Weighting Matrix for Hadamard Transform

Proposed quantization matrix for Hadamard kernel [10] and with viewing distance = 320 is given in the following Table 1.

Proposed HVS quantization matrix for watermark embedding and extraction process is presented in Table 1, based on this quantization matrix values proposed watermarking process is able to maintain its requirements.

Table 1 HVS quantization matrix

$$
\begin{bmatrix}
1.0000 & 0.6517 & 1.0000 & 0.9599 & 1.0000 & 0.7684 & 1.0000 & 0.8746 \\
0.4838 & 0.1836 & 0.3054 & 0.3168 & 0.6166 & 0.185 & 0.5786 & 0.2924 \\
0.9772 & 0.4735 & 0.8143 & 0.6439 & 0.9253 & 0.6723 & 0.9695 & 0.6398 \\
0.8792 & 0.5339 & 0.8280 & 0.6090 & 0.8266 & 0.3533 & 0.7619 & 0.4802 \\
1.0000 & 0.5806 & 0.9737 & 0.9475 & 1.0000 & 0.7581 & 1.0000 & 0.8025 \\
0.6101 & 0.3800 & 0.4450 & 0.5815 & 0.6206 & 0.2845 & 0.4619 & 0.5649 \\
1.0000 & 0.5743 & 0.8654 & 0.9146 & 1.0000 & 0.5965 & 0.9769 & 0.8105 \\
0.7470 & 0.2588 & 0.5404 & 0.5391 & 0.8507 & 0.3554 & 0.6050 & 0.4260
\end{bmatrix}
$$

4 Proposed Watermarking Scheme

Proposed watermarking process consists of two steps, watermark embedding and watermark extraction processes. Watermark embedding and extraction process steps are described below. Watermark embedding process detailed steps are given in Fig. 1 and watermark extraction process steps are given in Fig. 2.

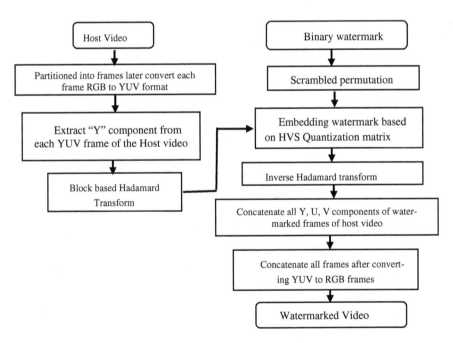

Fig. 1 Watermark embedding steps

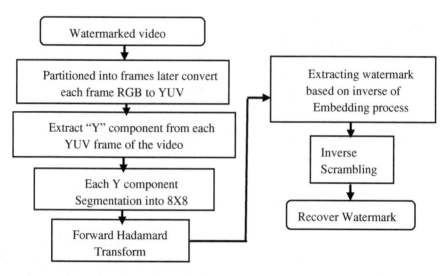

Fig. 2 Watermark extraction

5 Experimental Results

Proposed watermarking scheme is presented by using HVS based Hadamard transforms and compared the results with DCT and Slant transforms in terms PSNR and NCC values. Experimental results are verified by taking these as inputs-Host video-"Foreman.avi", embedding image in each scene is "javacup.gif", Embedding algorithm uses Threshold (T) value as 30 and HVS quantization step size (Q) as 16. By taking these parameters PSNR and NCC values are computed. Normally PSNR gives quality of watermarked image (Eq. 1).

$$PSNR = 20 \log_{10} \frac{255}{\sqrt{MSE}} \tag{1}$$

PSNR values are computed for Slant, DCT, and Hadamard kernels. By comparing the results proposed transform results are more superior to other two transforms. Robustness of the watermarking process is judged by using NCC value. Equation for NCC is given in Eq. 2. NCC is correlation between embedded and extracted images of the host video. NCC values are computed for all transforms and the results shown that proposed transform results are more robust than other transforms.

$$NCC = \frac{\sum_{i,j} w(i,j) . w^{\wedge}(i,j)}{\sum_{i,j} [w(i,j)]^2} \tag{2}$$

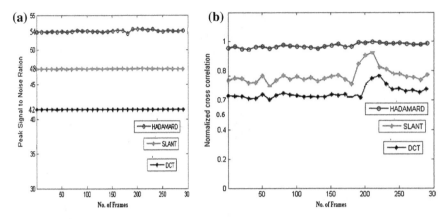

Fig. 3 **a** and **b** Comparative plot for PSNR and NCC versus no. of frames

The comparative plot drawn between PSNR values versus number of frames for host video is shown in Fig. 3a. It is observed that Propose method maintaining better image quality for the watermarked video. The comparative plot drawn between NCC values versus number of frames for host video is shown in Fig. 3b. It is evident that propose method more robust.

5.1 Robustness Check

The robustness of the proposed method is observed by applying various attacks on the watermarked frames of the watermarked video. Various attacks applied to the watermarked video are Noise attacks (Salt and Pepper, Gaussian noise), Cropping attacks (Central, Horizontal and vertical Cropping), Filtering attacks (Low pass, High pass and Median filtering), JPEG compression attack, Blurring attacks (Motion blur and Disk blur), Vertical Flipping attack, scaling attack and Rotation attack. Proposed watermarking scheme survive for all these attacks. In Table 2. Various attacks on the watermarked video with their PSNR and NCC values are furnished. Attacked watermarked scene and respective extracted scene are also included.

Different attacks applied on the watermarked which is prepared by using proposed watermarking algorithm. For all these attacks how much imperceptible and robust the algorithm will be observed by observing the attacked watermarked scene and extracted watermark of the watermarked video.

Table 2 Variour attacks and their respective PSNR and NCC values on watermarked video

S. no	Various attacks	Watermarked scene of a host video	Extracted logo
1	No attack		
2	Scaling		
3	Jpeg compression		
4	Median filtering		
5	Salt and pepper noise		
6	Smoothing filtering		
7	Central cropping		
8	Motion blur		
9	Disk blur		
10	Sharpening		
11	Vertical cropping		
12	Vertical flipping		

(continued)

Table 2 (continued)

S. no	Various attacks	Watermarked scene of a host video	Extracted logo
13	Horizontal cropping		
14	Gaussian noise		
15	Rotation		

5.2 Comparative Performance Evaluation

Comparative evaluation is done to evaluate performance quality metric using the equations given in Eqs. 1 and 2. The host video (foreman.avi) and watermark logo (javcup.jpg) are taken for watermarking technique, and is achieved by Slant, DCT and Hadamard transforms respectively. This is shown in Table 3.

Imperceptibility check for proposed method (Hadamard) is verified by calculating PSNR value of each scene of watermarked video and Robustness check for

Table 3 Comparative NCC values

PSNR				NCC		
Various attacks	DCT	Slant	Hadamard (proposed)	DCT	Slant	Hadamard (proposed)
No attack	36.6311	36.5544	54.4833	0.7841	0.7686	0.9544
Central cropping	32.0064	31.8874	33.6243	0.7858	0.7724	0.9605
Vertical cropping	10.7858	10.7851	10.7945	0.8271	0.8214	0.9661
Horizontal cropping	11.2855	11.2846	11.2954	0.8204	0.8158	0.9658
Vertical flipping	8.88674	8.8675	8.8745	0.7898	0.7742	0.9601
Low pass filtering	34.5629	34.1914	34.6230	0.8034	0.8050	0.9708
Sharpening	21.4742	22.2116	24.8953	0.7143	0.6962	0.7305
JPEG compression	40.5883	38.5481	46.9442	0.783	0.7947	0.9652
Median filtering	40.8188	40.9323	42.5819	0.8039	0.8099	0.9606
Gaussian noise	20.4868	20.4771	20.5692	0.6067	0.5881	0.0748
Motion blur	31.6208	31.6077	31.6323	0.8041	0.7988	0.9901
Salt and pepper	32.7922	32.3851	34.3053	0.7590	0.7540	0.8823
Rotation	4.9712	4.9710	4.9714	0.8337	0.8185	0.8957
Scaling	39.5976	38.1849	59.6735	0.7891	0.7703	0.9620
Disk blur	28.0090	28.0056	28.0145	0.8946	0.8525	0.9992

proposed method (Hadamard) is verified by calculating NCC values of each scene of Emebedded image and Extracted image. Compared the results with DCT, Slant transforms. Proposed method results are more optimal it is observed from Table 3.

Capacity, high perceptual quality and more robust against various attacks these can be understand by observing Table 3 by computing PSNR and NCC values for proposed and existing different transforms.

6 Conclusions

HVS based robust video watermarking technique by using Hadamard transform is presented in this paper. The proposed method (using Hadamard) results are compared with other transforms (Slant and Discrete Cosine Transforms). Proposed method is more robust and imperceptible and can be seen from PSNR and NCC values. Proposed method requires less computational time and also resilient to many attacks. This technique can be extended by using different optimization algorithms like Genetic Algorithm, Particle Swarm optimization Algorithm and Differential Evaluation Algorithms for improved performance.

References

1. J.J.K.O. Ruanaidh, W.J. Dowling, F.M. Boland, Phase watermarking of digital images, in *Proceedings of Conference on Image Processing*, vol. 3 (IEEE, Lausanne, Switzerland, 1996), pp. 239–242
2. J.R. Hernandez, M. Amado, F. Perez-Gonzalez, DCT-domain watermarking techniques for still images detector-performance analysis and a new structure. IEEE Trans. Image Process. **9** (1), 55–68 (2000)
3. M.A. Suhail, M.S. Obadiah, Digital watermarking based DCT & JPEG model. IEEE Trans. Instrum. Meas. **52**(5), 1640–1647 (2003)
4. M. Barni, F. Bartolini, A. Piva, Improved wavelet based watermarking through pixel-wise masking. IEEE Trans. Image Process. **10**(5), 783–91 (2001)
5. V. Aslantas, A.L. Dogan, S. Ozturk, DWT-SVD based image watermarking using particle swarm optimizer, in *IEEE International Conference on Multimedia and Expo*, IEEE (2008)
6. K. Meenakshi, Ch. Srinivasa Rao, K. Satya Prasad, A scene based video watermarking using slant transform, in *Department of E.C.E, JNTU, Kakinada, Department of E.C.E, University College of Engineering JNTUK*, Vizianagaram, A.P. India, 08 Oct 2014
7. H. Hartung, B. Girod, Watermarking of compressed and uncompressed video. Signal Process. **66**, 283–301 (1998)
8. K. Veeraswamy, B.C. Mohan, S.S. Kumar, HVS based robust image watermarking scheme using slant transform, in *Second International Conference on Digital Image Processing. International Society for Optics and Photonics* (2010)
9. K. Veeraswamy, S. Srinivaskumar, B.N. Chatterji, Designing quantization table for hadamard transform based on human visual system for image compression. ICGST-GVIP J. **7**(3), (2007)
10. S. Daly, Subroutine for the generation of a two dimensional human visual contrast sensitivity function, in *Tech. Rep Eastman Kodak* (1987)

Development of Microstrip Triple-Band Filter Using Hybrid Coupling Path Approach

**M. Manoj Prabhakar, M. Ganesh Madhan
and S. Piramasubramanian**

Abstract A multiple bandpass filter for 2.4, 3.6 and 5.2 GHz bands is designed using hybrid coupling paths approach and simulated. The hybrid coupling paths are implemented by coupled resonators structure. The filter is developed to provide different bandwidth at various bands. The microstrip filter is implemented in FR4 substrate and the simulation results are compared with experiment. The filter exhibits less than −10 dB return loss and low insertion loss in S_{11} and S_{21} characteristics respectively.

Keywords Triple band · Stepped impedance resonator (SIR) · Microstrip · Hybrid coupling · Bandpass filter

1 Introduction

Radio Frequency filters have become vital for modern wireless transceiver operations. Recent applications require wireless transceivers operating in multiple frequency bands. Today's wireless services operate at different frequencies and bandwidths. Hence multiband filters become essential in transceivers. Combining number of single band filters might be direct and simple, but large space requirement and additional combining circuits restricts this approach. An alternate technique utilizing the spurious frequency bands was reported by Makimoto and Yamashita [1]. They have implemented the filter using Stepped Impedance Resonator (SIR). Details of the resonance condition in SIR and their characteristics are also discussed

M. Manoj Prabhakar (✉) · M. Ganesh Madhan · S. Piramasubramanian
Department of Electronics Engineering, MIT Campus, Anna University,
Chennai, India
e-mail: mano_prabhak@yahoo.com

M. Ganesh Madhan
e-mail: mganesh@annauniv.edu

S. Piramasubramanian
e-mail: spsnanthan@gmail.com

© Springer India 2016   111
S.C. Satapathy et al. (eds.), *Microelectronics, Electromagnetics
and Telecommunications*, Lecture Notes in Electrical Engineering 372,
DOI 10.1007/978-81-322-2728-1_11

in Ref. [2]. Using the same approach, Chang et al. [3] have designed a dual band filter for 2.45 and 5.75 GHz bands. An ACSIR based dual band filter for UMTS and WiFi bands operating at 2 and 5.8 GHz, was also reported [4]. Further, Lin and Chu [5] have reported a multiple band pass filter operating at 1, 2.4 and 3.6 GHz, where SIR technique is adopted for the design. However, the SIR technique involves calculation of quality factors and coupling coefficients, which are cumbersome [6, 7]. A novel approach to overcome many of the short comings of the existing techniques was reported by Chuang [8], where different coupling paths were used for different bands, along with resonators. This approach leads to lesser computation time. A dual band pass filter design is reported in ref [9, 10]. A triple band pass filter for 900 MHz (GSM), 1.5 GHz (GPS) and 2.4 GHz (WLAN) is reported using trisection SIR scheme [11]. We consider this scheme for our design of a triple-band band pass filter for 2.4, 3.6 and 5.2 GHz bands. Two schemes viz., tapped feed line and coupled feed line are conventionally used in filter design. In the case of tapped feed line structure, additional resonators have to be implemented with appropriate coupling, to ensure the required quality factor. Hence, we resort to coupled line feed structure based implementation. The design involves evaluation of coupling coefficient with respect to coupling gap. Compared to the dual band design of Ref. [6], a similar structure with two additional resonators is required for the implementation. For the bands considered in the present study, resonators 4 and 5 are small compared with resonators 1 and 3, thereby necessitating a modification in the I/O coupling structure. We introduce another coupling line for I/O coupling structure to match the required Quality factor. With these modifications, the triple band filter is designed and analyzed using an EM software and implemented in a low cost FR4 substrate.

2   Filter Design

The hybrid coupling paths approach, allows different band signals to follow different paths, leading to better controllability of bandwidth [8]. Figure 1 shows the signal paths of the proposed filter. Five resonators are used to achieve the desired hybrid coupling paths. The 2.4 GHz band signals passes through resonators 1 and 3. The 3.6 GHz band passes through resonators 4 and 5 and 5.2 GHz travels through 1, 2 and 3.

Since both lower (2.4 GHz) and higher band (5.2 GHz) signals travels through resonators 1 and 3, they exhibits stepped impedance at respective central frequencies of the bands. A zero coupling at higher band is implemented by introducing quarter wavelength microstrip lines between first and third resonators. The higher band coupling between resonators 1 and 3 is avoided by using a transmission zero introduced by 90° coupled lines.

Resonator 2 does not allow lower band signal as it is non resonant. Further, it is characterized by uniform impedance and resonates for the higher band signal. The resonance conditions and impedance ratios are determined using the standard procedures reported for dual band filter [8]. Resonators 4 and 5 are also

Fig. 1 Coupling path for
a 3.6 GHz b 2.4 GHz band
c 5.2 GHz band

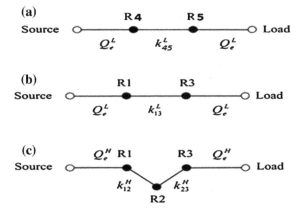

Fig. 2 Geometry of a SIR [2]

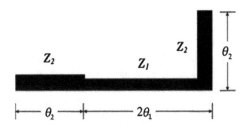

implemented as Stepped impedance resonators. As the resonators 4 and 5 are small compared with resonators 1 and 3, the I/O coupling structure has to be modified. This is realized by introducing another coupling line for I/O coupling structure to match the required Quality factor.

The resonance conditions for a single SIR as in Fig. 2 are determined by the following transcendental equations [2]

$$\tan \theta_1 = R_Z \cot \theta_2 \qquad \text{at} \quad f_0 \tag{1}$$

$$\cot a\theta_1 = -R_Z \cot a\theta_2 \qquad \text{at} \quad f_{s1} \tag{2}$$

where f_0 and f_{s1} represent the fundamental and the first spurious resonance frequencies. $a = f_{s1}/f_0$ denotes the frequency ratio and $R_Z = Z_2/Z_1$ is known as the impedance ratio. Terms θ_1 and θ_2 denote the electric lengths of the high and low impedance sections respectively. In this work, $a\theta_2 = 90^0$ generates a transmission zero at the higher band. The above conditions leads to $\theta = \theta_1 = \theta_2$ [2] and

$$R_Z = \left(\tan \left(\frac{\pi f_0}{2 f_{s1}} \right) \right) \tag{3}$$

Table 1 Specifications and parameters for triple band filter

Parameter	2.4 GHz band	3.6 GHz Band	5.2 GHz band
Central frequency, GHz	2.4	3.6	5.2
bandwidth, MHz	120	60	250
Filter order	2	2	3
Passband ripple, dB	0.5	0.5	0.5
Quality factor	28.058	84.174	33.20304
Coupling coefficient	0.0502	0.016734998	0.036335

The coupling coefficient between two adjacent resonators is controlled by the gap between them. Gaps g_{13}, g_{12} and g_{23} are applied for lower, higher bands respectively and given by

$$M_{i,i+1} = \frac{FBW}{\sqrt{g_i g_{i+1}}} \text{ for i} = 1 \text{ to m} - 1 \tag{4}$$

A coupled line feed structure is used for feeding the filter Table 1.

The required dimensions of the gaps between the resonators are calculated by coupling coefficient between the resonators, which is determined from the simulated S_{21} of the synchronous tuning circuit [8].

The external quality factor can be extracted by the simulated S_{11} of a singly loaded resonator. The two physical parameters to meet the required external quality factors of both bands simultaneously, are the coupling gap g_f and the coupling length L_f. The values of these two parameters are calculated numerically. External quality factor is given by [4]

$$Q_e = \frac{g_1}{FBW} \tag{5}$$

From (3) R_Z is found out as 0.78485. Considering Z_2 as 50 Ω, Z_1 is found as 63.705 Ω. The electric length θ is found out as 41.538°.

Figure 3, shows the structure for finding coupling coefficient between the resonators 1 and 3. In this structure, the ports are weakly coupled to the coupled

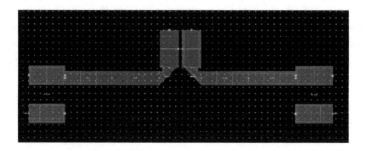

Fig. 3 Geometry of the filter [2]

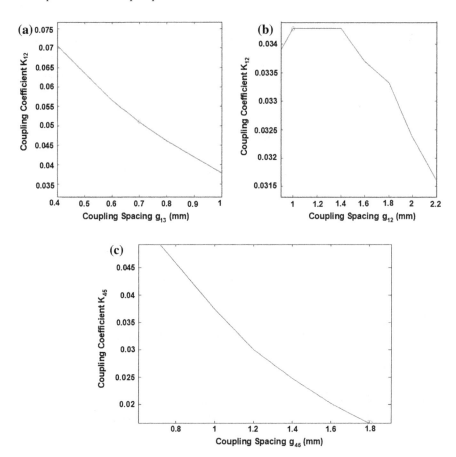

Fig. 4 Coupling coefficient variation with coupling spacing between the adjacent resonators **a** 2.4 GHz band **b** 5.2 GHz band **c** 3.6 GHz band

resonator structure. The structure is simulated, and from the S_{21} response, the coupling coefficient can be found as

$$k = \pm \frac{f_{p2}^2 - f_{p1}^2}{f_{p2}^2 + f_{p1}^2} \tag{6}$$

where f_{p1} and f_{p2} are the two peaks of the S_{21} response. Repeating this for different coupling spacing, the design curves are obtained and shown in the Fig. 4.

Figure 5 shows external quality factor variation coupling length L_f at different spacing, g_f, for the 2.4 and 5.2 GHz. Further, the two circles in the graphs, shows the required quality factors at two bands. The values of coupling spacing and coupling length at these points provide the data required for the final structure.

Fig. 5 External quality factor
variation with coupling length

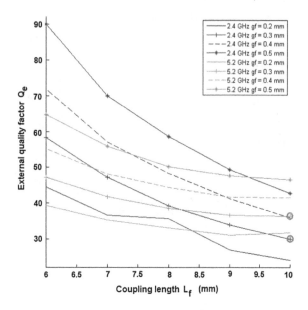

Figure 6 shows the quality factor for 3.6 GHz band versus coupling length for a coupling spacing of 0.2 mm. The coordinates corresponding to the circle, in the graph, denotes the required coupling length at I/O feeding structure for 3.6 GHz band.

Figure 7 illustrates final structure of the triple band microstrip filter, which is implemented in ADS software. Figure 8 depicts the photograph of the fabricated filter.

Fig. 6 External Quality
factor for 3.6 GHz band
versus coupling length for
coupling spacing s = 0.2 mm

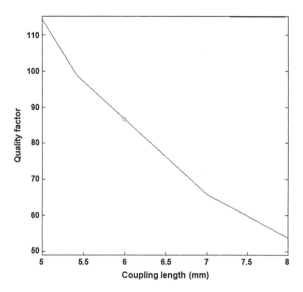

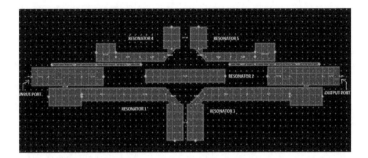

Fig. 7 Final layout of triple band filter

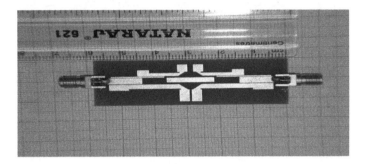

Fig. 8 Fabricated filter

Measurements are carried out on the fabricated filter using an Agilent 8722 Network Analyzer and the results are compared with the simulation. The S_{21} measurement plotted in Fig. 9 shows a stop band attenuation level of 30 dB between first two pass bands. However, the difference is only 15 dB, between the stop band and higher pass bands. In the case of S_{11} measurement (Fig. 10), the central band provides a return loss of −10 dB, whereas the other bands provide a return loss greater than −10 dB. The measured insertion loss is low in the lower band but increases at higher bands. The observed 3 dB bandwidths are 200, 120 and 310 MHz for 2.4, 3.6 and 5.2 GHz bands respectively. They closely match with the results obtained in ADS simulation. At higher frequency bands the fabricated filter shows increased attenuation. The deviations in the measured values with respect to simulation, are due to the fabrication tolerances and substrate imperfections.

Fig. 9 S_{21} response of the triple band filter

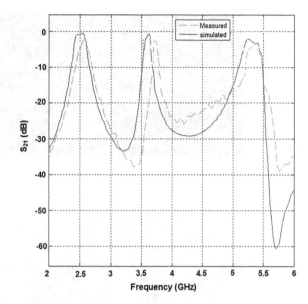

Fig. 10 S_{11} response of the triple band filter

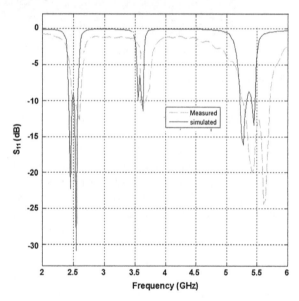

3 Conclusion

A multiple, band pass filter is designed in microstrip platform and simulated. The design is implemented in FR4 substrate which achieves the bandwidth controllability in the three bands. The simulation results are found to agree with the measurement.

References

1. M. Makimoto, S. Yamashita, Bandpass filters using parallel coupled stripline stepped impedance resonators. IEEE Trans. Microw. Theory Tech. (MTT) **28**(12), 1413–1417 (1980)
2. M. Makimoto, S. Yamashita, Microwave resonators and filters for wireless communication (Springer, 2001)
3. S.F. Chang, Y.H. Jeng, J.L. Chen, Dual-band step-impedance bandpass filter for multimode wireless LANs. Electron. Lett. **40**(1), 38–39 (2004)
4. A. Kamalaveni, M. Ganesh Madhan, Compact dual-band band pass filter using λ/2 ACSSIR. in *Proceedings of International Symposium on Antennas and Propagation, Cochin University of Science and Technology, Cochin*, pp. 363–366 (2014)
5. X.M. Lin, Q.X. Chu, Advanced triple-band bandpass filter using tri-section SIR. Electron. Lett. **44**(4), 295–296 (2008)
6. C.L. Hsu, J.T. Kuo, F.C. Hsu, Design of loop resonator filters with a dual passband response. in *Proceedings of Asia-Pacific Conference* (2005)
7. J.T. Kuo, T.H. Yeh, C.C. Yen, Design of microstrip bandpass filters with a dual-passband responses. IEEE Trans. Microw. Theory Tech. **4**(4), 1331–1337 (2005)
8. M.-L. Chuang, Dual-band microstrip coupled filters with hybrid coupling paths. IET Microw. Antennas Propag. **4**(7), 947–954 (2010)
9. J.S. Hong, M.J. Lancaster, *Microstrip Filters for RF/Microwave Applications* (Wiley, New York, 2001)
10. S. Sun, L. Zhu, Novel design of microstrip bandpass filters with a controllable dual passband responses: description and implementation. IEICE Trans. Electron. **2**(2), 197–202 (2006)
11. M. Manoj Prabhakar, M. Ganesh Madhan, Design and simulation of triple-band band pass filter for wireless communication applications. ICCNT, Coimbatore, 20180 (2012)

Security Issues in Cognitive Radio: A Review

Shriraghavan Madbushi, Rajeshree Raut and M.S.S. Rukmini

Abstract Electromagnetic spectrum is a scarce, important and a very useful resource. However, this resource has not been utilized effectively or in other words most of the spectrum remains vacant. Due to fixed or licensed spectrum allocation the spectrum remains vacant for a considerable amount of time. Cognitive radio, a wireless and an intelligent technology was proposed by Joseph Mitola to make efficient use of spectrum. The use of EM wave as a transmission medium makes security a major concern in Cognitive Radio as these are easily susceptible to attacks. Also, being a flexible wireless network technology it is susceptible to traditional threats as well as unique attacks which will have an adverse effects on its performance. This paper introduces the Cognitive Radio technology and mainly focuses on security issues related to it. Further, we present a layered classification of security issues of this exciting area of communication technology.

Keywords Cognitive radio · Spectrum · Software defined radio · Security issues

1 Introduction

Wireless Network technology has witnessed a remarkable growth during the last decade. In 1985 the FCC (Federal Communications Commission) issued a mandate defining several portions of the Electromagnetic spectrum as "license-exempt"

S. Madbushi (✉) · M.S.S. Rukmini
Department of ECE, Vignan University, Vadlamudi, Guntur Dist, A.P., India
e-mail: m.shriraghavan@gmail.com

M.S.S. Rukmini
e-mail: mssrukmini@yahoo.co.in

R. Raut
Department of ECE, RCEOM, Nagpur, Maharashtra, India
e-mail: raut.rajeshree@gmail.com

© Springer India 2016
S.C. Satapathy et al. (eds.), *Microelectronics, Electromagnetics and Telecommunications*, Lecture Notes in Electrical Engineering 372, DOI 10.1007/978-81-322-2728-1_12

[1, 2]. Some bands of the spectrum were allowed to operate without the need of license and this was called as the Industrial, Scientific and Medical (ISM) band. This declaration of FCC and the advent IEEE 802.11/a/b/g standards has brought a revolution in the wireless domain [1]. The ISM band, being license free, has thus become overcrowded resulting in increased interference and contention [1]. Although the ISM band is overcrowded, there are several licensed bands of spectrum which are being under-utilized. Licensed band here refers to band of spectrum which is allotted by FCC by the traditional and static allocation process of spectrum assignment serving a distinct service or several channels of distinct service [1]. The static allocation, however differs from country to country, it has been observed that the part of spectrum is not utilized efficiently. It was Joseph Mitola, who proposed a novel idea of the opportunistic use of spectrum which was under-utilized. He proposed an idea of using a novel device called Cognitive Radio (CR). Cognitive radio is the term coined by Mitola and Maguire in an article wherein they describe it as a radio that understands the context in which it finds itself and as a result can tailor the communication process in line with that understanding. The interconnection of Cognitive Radios (CRs) will form a Cognitive Radio network (CRN) [2]. Simon Haykin [3] defines Cognitive Radio as: An intelligent wireless communication system that is aware of its surrounding environments (i.e., outside world), and uses the methodology of understanding-by-building to learn from the environment and adapt its internal states to statistical variations in the incoming RF stimuli by making corresponding changes in certain operating parameters (e.g., transmit-power, carrier-frequency, and modulation strategy) in real-time. A cognitive radio is a device which has four broad inputs [4], namely, (i) an understanding of the environment in which it operates, (ii) an understanding of the communication requirement of the user(s), (iii) an understanding of the regulatory policies which apply to it and (iv) an understanding of its own capabilities. One of the major concerns in any Wireless Communication is that of security. This paper focuses mainly on the security issues related to the Cognitive Radio. The paper is further organized as follows. Section 2 describes the basics of Cognitive Radio. Section 3 discusses the security threats in detail. At the end the conclusions and motivation for future research are discussed.

2 Cognitive Radio Basics

Cognitive Radio in a way is a "smart radio". Generally there are two types of CRs [2]: Policy Radios and Learning Radios. Policy radios, by looking at the word "policy" we can conclude that, it is a radio that has to follow some predefined policies that will decide its behavior. Learning radios on the other hand have a learning engine which allows the radio to learn from its surroundings and can configure or re-configure its state. These radios make the use of a variety of Artificial Intelligence (AI) learning algorithms as well.

In CR scenario there will be two categories of users (i) primary users [PUs], holding a license of a particular portion of spectrum, and (ii) secondary users [SUs], also called cognitive users who do not hold the license but still, can use the portion of spectrum allotted to the PUs in an opportunistic manner. Here one thing has to be concentrated that, this should not interfere or hamper the performance the of the licensed user. The main objective of CR is to sense the spectrum and make its opportunistic use. The free spectrum portions are referred to as "white spaces" or "spectrum hole". Cognitive Radio technology is a Wireless Network Technology and is vulnerable to all traditional threats of Wireless Networks in addition to the unique threats introduced with the advent of this new area of Wireless Communication.

A Cognitive radio is based on what is called a Software Defined Radio (SDR). The software allows the radio to tune to different frequencies, power levels, modulation depending upon its learning and the environment in which it operates [15].

The Cognitive Radio is expected to perform the following four functions as follows [5]:

1. Spectrum Sensing: detection of "white spaces" or the portion of spectrum vacant for use. This must also ensure that no Primary User (licensed user) is not operating at the same time.
2. Spectrum Management: selection of the best spectrum hole for transmission.
3. Spectrum Sharing: sharing of spectrum with other potential users.
4. Spectrum Mobility: vacate the band when a licensed user is detected (spectrum handoff).

The functions mentioned above form a cognition cycle which forms the basis on which the Cognitive Radio operates. In his thesis Mitola [6] has described the cognitive cycle consisting of five states namely Observe, Orient, Plan, Decide and Act. The following figure, Fig. 1 shows the cognitive cycle described by Mitola.

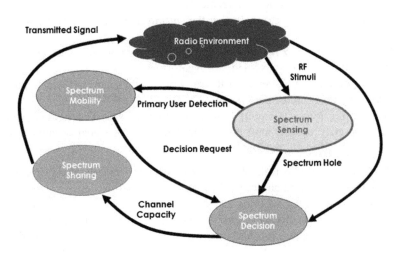

Fig. 1 The cognition cycle

Till now we have discussed basics of Cognitive Radio. The study of any Wireless Network would be incomplete unless we explore its topology. As far as the Cognitive Radio topology is concerned three different architectures have been discussed in [7]. These are as follows: (i) Infrastructure, (ii) Ad-Hoc and (iii) Mesh. Infrastructure Networks have Base Station (BS) also called as access points. A device can communicate with other devices within the vicinity through the base station. The communication between these devices can be routed via the base station. Ad- Hoc topology is formed by devices without the need of a base station These communicate with each other by establishing links between themselves making use of different communication protocols available. Mesh topology is a combination of the above two topologies.

3   Security Threats in Cognitive Radio

A "*security threat*" is described as a potential violation of security wherein it can be intentional like a deliberate attack or unintentional due to an internal failure or malfunctions [5].

An attack is considered strong if it involves a minimal number of adversaries performing minimal operations, but causing maximum damage or loss to the Primary Users and/or Cognitive Radio [8].

A detailed analysis of security requirements are done in [5] and they are defined as follows:

1. *Confidentiality*: confidentiality of the stored and communicated data must be ensured by the system.
2. *Robustness*: the system must resist against attacks and provide communication services as per service level agreements.
3. *Regulatory framework compliance*: the system must adhere to rules and regulations or policies.
4. *Controlled access to resources*: access to information or resources by unauthorized users must be avoided.
5. *Non-Repudiation*: non-denial of the responsibility for any of the activities performed by an entity of the system.
6. *System integrity*: system must guarantee the integrity of its system components
7. *Data integrity*: system must ensure the data, may be stored or communicated must not be illegally modified.

The Cognitive Radio technology must enforce the security triad of confidentiality, integrity and availability (CIA) [9].

We now discuss the security threats in Cognitive Radio with respect to the security requirements discussed earlier. The attacks generally follow a layered approach [7] and thus we also further categorize these threats according to different protocol layers they target. We will categorize these attacks as follows:

(i)   Physical layer attacks
(ii) Link layer attacks (also known as MAC attacks)
(iii) Network layer attacks
(iv) Transport layer attacks
(v)   Application layer attacks

In addition to above we also discuss cross layer attacks which are specifically targeting one particular layer but affect the performance of another layer.

As described in [2] the different classes of attacks which an attacker can construct is categorized as follows:

(i)   Dynamic Spectrum Access Attacks
(ii) Objective Function Attacks
(iii) Malicious Behavior attacks

3.1   Physical Layer Attacks

It is the bottom layer of the protocol stack and provides an interface to the transmission medium. It consists of a medium that makes two network devices communicate with each other such as cables, network cards etc. In Cognitive Radio the medium is atmosphere. The physical layer determines the bandwidth, channel capacity, bit rate etc. In Cognitive radio the spectrum is accessed as and when it is found unoccupied, this makes the physical layer complex. We now discuss some attacks specific to the physical layer.

3.1.1   Primary User Emulation Attack [PUEA]

A fundamental function of Cognitive Radio is to sense the spectrum which we refer to as spectrum sensing and the spectrum has to be shared in an opportunistic manner [16]. The secondary user has to vacate the currently used spectrum as and when it detects an incumbent (Primary User) signal to avoid the interference. This is referred to as the spectrum hand off.

For a fair spectrum sharing it is necessary that the CR must recognize the primary user signals. Nodes launching PUEA are of two types [1]:

- *Greedy nodes* which transmit fake primary user signals forcing all other users to vacate the band in order to acquire its exclusive use.
- *Malicious nodes* that copy primary user signals in order to cause Denial of Service (DoS) attacks. These nodes can cooperate and can transmit fake primary user signals in more than one band resulting in hopping of a CRN form band to band hampering its entire operation.

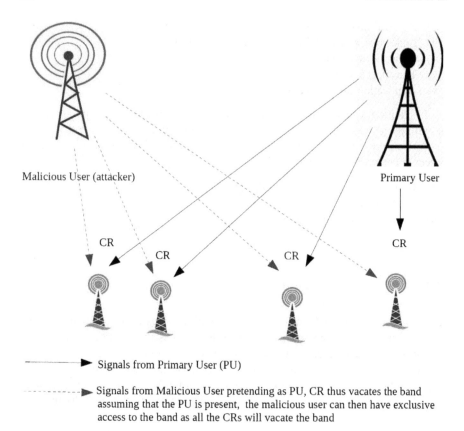

Signals from Primary User (PU)

Signals from Malicious User pretending as PU, CR thus vacates the band assuming that the PU is present, the malicious user can then have exclusive access to the band as all the CRs will vacate the band

Fig. 2 PUEA scenario

In PUEA [13], the attacker emulates the primary user signal to access the resources. The secondary user will be under the impression that a primary user is using the spectrum band and thus, it will vacate the band. Now, if the attacker's goal is to increase its share of spectrum then we refer this as *selfish* PUEA. This attack can be conducted simultaneously by two attackers by establishing a dedicated link between them [7].

If the attacker's goal is to prevent other legitimate users from using the spectrum then we refer this attack as a malicious PUEA.

The PUEA can target both types of cognitive radios Policy Radios and Learning Radios [2, 7] with different severity. The Fig. 2 shows the PUEA scenario [10].

3.1.2 Objective Function Attack

The radio parameters include center frequency, bandwidth, power, modulation type, coding rate, channel access protocol, encryption type and frame size [2, 7]. The cognitive engine manipulates these parameters to meet one or more objective functions. The objective function attacks are those in which algorithms that utilize the objective functions are attacked. Another name for objective function attacks is "*belief-manipulation attacks*" [9].

Intrusion Detection systems (IDS) are used for detecting Objective function attack. The IDS may be a mis-use based method or anomaly based method. The first method uses the signatures of attack. While the other makes use of the abnormal behavior of the system to detect an attack. Thus the attacks not known before hand can be detected by the later system. It is worth mentioning here that the IDS follows the FCC constraint of not modifying the Primary User system [17]. IDS operates in two phases, the profiling phase and the detection phase [10].

3.1.3   Jamming

Similar to the conventional wireless technology, the Cognitive Radio receiver requires a minimum signal-to-noise ratio (SNR) decode a digital signal. One of the traditional attack strategy is to reduce the SNR below a threshold value by transmitting noise over the channel this is also referred to as "*receiver jamming*" [11]. Now there may be a case wherein the performance of the receiver node is poor and this may or may not be due to jammer. Sometimes this performance may be poor due to natural causes, say network congestion, for example. Jamming is an attack that can be done in the physical and the MAC layers [7]. There are four types of jammers namely 1. Constant Jammer, 2. Deceptive Jammer, 3. Random Jammer, and 4. Reactive Jammer [7]. A Constant Jammer is the one that is continuously sending out data packets, while doing so it will not wait for channel to be idle and also it has no regard for other users on the channel. A Deceptive Jammer will send out packets continuously such that other users will switch to receive states and will remain in that state. A Random Jammer will take breaks between jamming signals and it may behave as a constant or a deceptive jammer during the jamming phase. Lastly the reactive jammer is the one that will sense the channel all the time. It starts transmitting the jamming signals whenever it detects a communication in the channel and thus it is difficult to detect a reactive jammer. "*Intentional Jamming*" is one of the most basic types of attacks [9] in which the attacker continuously and intentionally transmits data packets on a licensed band making it unusable for both the primary and secondary users. In Primary Receiver Jamming an attacker close to the primary user will send request for transmission from other secondary users. This will divert the traffic towards the primary user which will in turn create interference to the primary user.

3.1.4   Overlapping Secondary User

A geographical region may contain coexisting and overlapping secondary networks. An attacker in one network can transmit signals that may cause harm to the primary and secondary users of both networks. The malicious node is not in the control of the users of the network or under the direct control of the station. This is a direct attack on the capability of the cognitive radio network for spectrum sensing and sharing of both infrastructure and ad hoc based networks. This further results in the denial of service attack [9].

3.2 Link Layer Attacks

3.2.1 SSDF Attack (Byzantine Attack)

The 802.11 data link layer consists of two sub layers. These are LLC and MAC. MAC supports multiple users on a shared medium within the same network. The "*Spectrum Sensing Data Falsification*" (SSDF) attack is also called as "*Byzantine Attack*". The attacker in this attack is the legitimate user of the network and is called "*Byzantine*". This attack is for selfishly acquiring the spectrum availability and occupancy or with the goal of destructing the entire communication system. We analyse this attack from two different scenarios.

 i. distributed CRN
 ii. centralized CRN

In a distributed CRN the spectrum sensing information is based on the observations of the secondary user itself and on the basis of the observations shared by the secondary users. In other words the spectrum information is shared in collaboration between the secondary users. In a centralized CRN different secondary users sense the environment and the spectrum sensing information is send to a fusion center (FC). The FC is responsible for collecting the data and on the basis of this data the FC provides information regarding frequency bands. that are free or busy. The manipulation of fusion center data will prevent the user from accessing the vacant band or it may allow to use a band that is already in use resulting in interference.

This attack has more impact in a distributed CRN wherein false information can propagate quickly. However, in case of centralized CRN a smart FC can compare the data received by different CRs and identify which CR may be providing false information [14].

3.2.2 Control Channel Saturation DoS Attack (CCSD)

In a multi hop CRN a channel negotiation is necessary in a distributed manner. After channel negotiation the CRs can communicate with each other. Now, when many CRs want to communicate with each other the channel negotiation becomes a problem as the channel can support only a limited number of data channels. An attacker will therefore try to send MAC control frames thereby saturating the control channel which results in poor performance of the network with approximate zero throughput.

It has been further observed that this attack has more impact on multi hop CRNs as compared to centralized CRNs. This is because of the fact that the MAC control frames are authenticated by the base station and thus forging MAC frames is almost impossible.

3.2.3 Selfish Channel Negotiation (SCN)

A selfish channel in a multi hop CRN can refuse to forward any data for other hosts. This selfish behavior will result in conservation of energy and an increase in the throughput. Some other selfish host alters the MAC behavior of the other CR devices and thus claim the channel at the expense of other hosts. This scenario leads to the severe degradation of the throughput of the entire CRN.

3.2.4 Control Channel Jamming

Cognitive Radio users cooperate among themselves with the help of control channels. An attacker if attacks a control channel, known as control channel jamming (CCC) attack, the receivers are prevented from sensing the valid messages when an attacker injects a strong signal in the control channel. This scenario leads to the denial of service for the users in that particular network.

3.3 Network Layer Attacks

It has been found in literature that much research has been focussed on MAC and PHY layers. Network layer helps in routing of data packets from source node to destination node maintaining the quality of service. Routing in CR is a challenge due to spectrum handoff and dynamic spectrum sensing. As stated earlier Cognitive Radios are prone to attacks similar to classic Wireless networks. In addition to this CR is vulnerable to attacks which also plague the wireless sensor networks. Here we discuss two major attacks namely, Hello flood attack and Sink hole attack.

3.3.1 Hello Flood Attack

This attack has been investigated as an attack against a Wireless Sensor Network (WSN) and can be applied to a CR scenario as well due to its routing strategies. In this attack, an attacker first broadcasts a message to all the nodes in a network. This is an advertisement where an attacker offers a high quality link to a particular destination using high power to convince the node that the attacker is a neighbor.

As a lot of power is put, the strength of the received signal will be high and the node convinces itself that the attacker is its neighbor, although the attacker is far away from the node. In this way all nodes will forward data packets to the attacker assuming that the attacker is their neighbor. However, when the attack is detected the nodes will lose their data packets and find themselves alone with no neighbor nearby them. Some of the protocols used to exchange information between neighbors for maintaining topology may also be attacked.

3.3.2 Sinkhole Attack

Cognitive Radio uses multi hop routing similar to the Wireless Sensor Network (WSN). In this type of attack the attacker will be advertising itself as the best route to a specific destination thereby enticing the nodes neighboring the attacker. Once the attacker wins the trust of its neighbors, the neighbors themselves promote or advertise the attackers path as the best route. Now, the attacker has the capability to send the receives packets directly to the base station using high level power. Once the trust has been established that the attacker's route is the best route the attacker can begin other attacks for example, eavesdropping, selective forwarding attack by forwarding data packets from selected nodes, dropping received packets, modifying data packets etc.

3.3.3 Worm Hole Attack

It is closely related to the sink hole attack. A wormhole attack is generally per-petrated and administered by two malicious nodes. These nodes understate the distance between them by relaying the packets along an out-of-bound channel that is unavailable to other nodes [9]. In this attack the nodes are convinced by the attacker that they are only one or two hops away via the adversary. However, the nodes will be usually multiple hops from the base station. The attacker thus can receive the packets for forwarding or can capture the packets for eavesdropping. Another interesting scenario here is that the attacker can stop relaying the packets which would create the separation of the network. Network routing protocols thus have to be implemented and the attacker will be provided additional information which will help the attacker to perpetrate other potential attacks.

3.3.4 Sybil Attack

Some local entities perceive the other remote entities as informational abstractions (entities) without the physical knowledge of the remote entities. A sybil attacker will create a large number of pseudonymous identities so as to gain a large influence on the network. A system must exhibit the capability to ensure and identify that distinct identities refer to distinct entities. Generally an attacker will pair sybil attack with Byzantine attack or Primary User Emulation attack so that the entire decision making process gets effected.

3.3.5 Ripple Effect

The Cognitive Radio senses the spectrum and utilizes the portion of spectrum that is free. While doing so it performs spectrum handoff whenever it senses the Primary User (PU) so that interference is avoided. The ripple effect is similar to Byzantine attack or Primary User Emulation Attack [PUEA] [9]. The wrong channel information is shared so that the nodes change their channel. Here the attacker sends the false information hop by hop so that the network enters a state of confusion.

3.4 Transport Layer Attacks

3.4.1 Key Depletion

The protocols used for IEEE 802.11 are prone to key repetition attacks. The transport layer protocols establish cryptographic keys at the beginning of each transport layer session. Due to the generation of large number of keys there is a possibility that the sessions key that are generated will be repeated. This repetition of keys may result in the breaking of the cipher system.

3.5 Application Layer Attacks

3.5.1 Cognitive Radio Virus

This attack is due to the self propagating behavior of Cognitive radio [2]. In this attack a state that is *introduced* in a Cognitive Radio will cause a behavior that induces the same state in another Cognitive radio and so on. Thus, the said state will propagate through all the radios in that particular area. The self propagating behavior of radio thus can infect the entire network. These types of attacks can even spread between the radios that never had any protocol interaction.

3.6 Cross Layer Attacks

The cross layer attacks target the multiple layers and can affect the whole cognitive radio cycle. Attacks discussed in previous sections can be combined to form cross layer attacks. Generally the cross layer attacks target one layer but affect the performance of another layer.

3.6.1 Routing Information Jamming

In this attack the targeted node performs a spectrum handoff and it stops all ongoing communication, switches to a new spectrum band, identifies the neighbouring nodes. The targeted node will not be able to receive the routing information until the spectrum handoff is complete. This is referred to as *deafness* [9]. Till the completion of hand off process the targeted and its neighboring node will be using the stale route for communication We know that spectrum hand off is an important event in cognitive radio. There will be some delay while switching from one spectrum band to another spectrum band. This delay during transition is a matter of consideration. This delay allows the attacker to jam the routing information among the nodes. The nodes will therefore follow some stale routes and route the packets on an incorrect route. This attack can be made more severe if the targeted node performs spectrum hand off more frequently. This attack scenario has been discussed in detail in [12].

3.6.2 Small Back-Off Window

This attack is feasible against the cognitive radio networks using Carrier Sense Multiple Access with Collision Avoidance (CSMA/CA) protocol at the MAC layer [9]. The main motive of this attack is to gain more access to the channel by manipulating the contention protocol parameters (by choosing a very small back off, window) to obtain overall or more frequent access to the channel.

3.6.3 Lion Attack

This attack takes place at the physical layer or the link layer and targets the transport layer. The main motive behind this attack is to destroy the TCP connection. In this type of attack the attacker makes use of the Primary User Emulation Attack [PUEA]. As we know that the Secondary User [SU], whenever detects a primary signal has to perform spectrum handoff. However the TCP will not be aware of this handoff and will continue to send packets as usual without any acknowledgment. As the TCP receives no acknowledgment it considers that the segment is lost and re-transmit the segment resulting in delays and packet loss. This attack may be more severe if the attacker knows in advance and moves to the channel to which the secondary user will be hopping to. This scenario is similar to the Denial of Service (DoS) Attack.

3.6.4 Jelly Fish Attack

The jelly fish attack and the lion attack are related that they both target the TCP. This attack targets the transport layer, however, the attack is performed at the network layer. In this attack the attacker intentionally reorder the packets it receives and forwards. These re-transmissions severely degrades the performance. Further if the malicious node randomly delays the packets the TCP timers will be invalid resulting in network congestion [9]. The jelly fish also obeys the data plane and the control plane protocol rules making it difficult to distinguish between the attack and congested network [9].

3.7 Software Defined Radio (SDR) Security

SDR security is another area of critical importance. SDR security is important due to its capacity to reconfigure itself. SDR security falls in two main categories [1]:

 i. Software based protection, and
 ii. Hardware based protection.

Software based protection is to defend against the malicious software download, buggy software installation etc. Hardware protection involves the implementation of hardware components. These monitor several other parameters of the SDR [5].

4 Conclusions

It has been observed that several portions of the licensed spectrum are underutilized and on the other hand the user demand for spectrum usage is increasing. The solution proposed is in the form of a promising technology called as the *Cognitive Radio* technology. The Cognitive Radio technology being a wireless technology is vulnerable to various attacks including those which also plague the traditional Wireless Networks as well as Wireless Sensor Networks. This technology faces many security threats mainly due to its two interesting characteristics namely *cognitive capability* and *reconfigurability*. This paper discusses almost all types of attacks affecting the Cognitive Radio technology. Further, we have analyzed all these attacks from a layered approach point of view. The so called cross layer attacks and the Software Defined Radio (SDR) security threats can affect the overall functioning of the cognitive cycle. Being a review paper, only attacks affecting the Cognitive Radio, and no mitigation techniques of any of these attacks have been discussed in this paper. We further conclude that the security of primary user is of paramount importance, thus we emphasize more on security of primary user for our work in future.

References

1. A. Fragkiadakis, E. Tragos, I. Askoxylakis, A Survey on Security Threats and Detection Techniques in Cognitive Radio Networks. IEEE Commun. Surv. Tutorials **PP**(99), 1–18 (2012)
2. T. Clancy, N. Goergen, Security in cognitive radio networks: threats and mitigation. in *Third International Conference on Cognitive Radio Oriented Wireless Networks and Communications (CrownCom)* (2008), pp. 1–8
3. S. Haykin, Cognitive Radio: Brain-empowered wireless communications. IEEE J. Sel. Areas Commun. **23**, 201–220 (2005)
4. L.E. Doyle, Essentials of cognitive radio, Cambridge wireless essentials series (2009)
5. G. Baldini, T. Sturman, A.R. Biswas, Security aspects in software defined radio and cognitive radio networks : a survey and a way ahead. IEEE Commun. Surv. Tutorial **14** (2012)
6. J. Mitola, Cognitive radio: An integrated agent architecture for software defined radio, doctor of technology, Royal Institute of Technology (KTH), Stockholm, Sweden (2000)
7. W. El-Hajj, H. Safa, M. Guizani, Survey of security issues in cognitive radio networks
8. H. Kim, Privacy preserving security framework for cognitive radio networks, IETE Technical Review, vol. 30, Issue 2 (2013)
9. D. Hlavacek, J. Morris Chang, A layered approach to cognitive radio network security: a survey, computer network (2014)
10. M. Haghighat, S.M. Sadough, Cooperative spectrum sensing in cognitive radio networks under primary user emulation attacks. in *Sixth International Symposium on Telecommunications* (2012), pp. 148–151
11. A. Attar, H. Tang, A.V. Vasilakos, F. Richard Yu, V.C.M. Leung, A survey of security challenges in cognitive radio networks: solutions and future research directions
12. C.N. Mathur, K.P. Subbalakshmi, Security issues in cognitive radio networks: cognitive networks: towards self aware networks (2007)
13. R. Chen, J.M. Park, J.M. Reed, Defense against primary user emulation attacks. IEEE J. Sel. Areas Commun. **26**(1), 25–37 (2008)
14. A. Rawat, P. Anand, H. Chen, P. Varshney, Countering byzantine attacks in cognitive radio networks, acoustics speech and signal processing (ICASSP). in *IEEE International Conference* (2010), pp. 3098–3101
15. F.K. Jondral, Software-defined radio—basic and evolution to cognitive radio, EURASIP Journal on Wireless Communications and Networking (2005)
16. R. Chen, J.M. Park, Ensuring trustworthy spectrum sensing in Cognitive Radio Networks. in *IEEE Workshop on Networking Technologies for Software Defined Radio (SDR'06)* (2006), pp. 110–119
17. Y. Liu, P. Ning, H. Dai, Authenticating primary users' signals in cognitive radio networks via integrated cryptographic and wireless link signatures. in *Proceedings of IEEE Symposium Security and Privacy* (2010)

GPS C/A Code Multipath Error Estimation for Surveying Applications in Urban Canyon

Bharati Bidikar, G. Sasibhushana Rao, L. Ganesh
and M.N.V.S. Santosh Kumar

Abstract Global Positioning System (GPS) is satellite based navigation system implemented on the principle of trilateration, provides instantaneous 3D PVT (position, velocity and time) in the common reference system anywhere on or above the earth surface. But the positional accuracy of the GPS receiver is impaired by various errors which may be originating at the satellite, receiver or in the propagation path. These errors have assumed importance due to the high accuracy and precision requirements in number of applications like the static and kinematic surveying, altitude determination, CAT I aircrafts landing and missile guidance. In this paper, the error originating at the receiver due to multiple paths of the satellite transmitted radio frequency (RF) signal is estimated. Multipath phenomenon is prevalent particularly in urban canyons, which is the major error among other GPS error sources originating at the receiver. The algorithm proposed in this paper estimates the error using coarse/acquisition (C/A) code range, carrier phase range and Link1 (L1) and Link2 (L2) carrier frequencies. This algorithm avoids the complexity of the error estimation using conventional methods where sensitive parameters such as the geometry or the reflection coefficient of the nearby reflectors are considered. The error impact analysis presented in this paper will be useful in selecting the site for GPS receiving antenna where the reflection coefficients are hard to measure up to the required accuracy. Analysis of the change in intensity of this error with respect to elevation angle of the satellite will facilitate in selecting pseudoranges with least error. Error estimation and range modeling proposed in this paper will be a valuable aid in precise navigation, surveying and ground based geodetic studies.

B. Bidikar (✉) · G. Sasibhushana Rao
Department of Electronics and Communication Engineering,
AUCE(A), Visakhapatnam, India
e-mail: bharati.bidikar@gmail.com

L. Ganesh
Department of Electronics and Communication Engineering,
ANITS, Visakhapatnam, India

M.N.V.S. Santosh Kumar
Department of Electronics and Communication Engineering,
AITAM, Visakhapatnam, India

© Springer India 2016
S.C. Satapathy et al. (eds.), *Microelectronics, Electromagnetics
and Telecommunications*, Lecture Notes in Electrical Engineering 372,
DOI 10.1007/978-81-322-2728-1_13

Keywords GPS · Elevation · Satellite position · C/A code range · Carrier phase range · Multipath error · Navigation solution

1 Introduction

GPS applications are not limited only to precise positioning. GPS also finds its application in surveying, missile guidance, military and civil aviation [1]. However, the accuracy, availability, reliability and integrity of GPS navigation solution are impaired by various errors originating at the satellites like orbital errors, satellite clock errors [2]. Whereas the receiver clock errors, multipath errors, receiver noise and antenna phase centre variations are the errors originating at the receiver [3]. Also the propagation medium contributes, to the delays in the GPS signal propagation as it passes through the earth's atmosphere [4]. In addition to these errors, the accuracy of the navigation solution is also affected by geometric locations of the GPS satellites as viewed by the receiver. Hence error estimation and correction is of prime concern in precise navigation applications. In this paper, the error originating at the receiver i.e. multipath error is addressed.

Multipath affects both the C/A code range and carrier phase measurements [5]. But the C/A code range multipath is much larger than the carrier phase multipath. The multipath error originating at the receiver is very sensitive to the geometry and the reflection coefficients of the nearby reflectors which depend on the weather conditions [6]. These parameters limit the efficiency of the conventional multipath modeling methods. But in this paper an algorithm is proposed to calculate the multipath error on GPS L1 C/A code range measurement of L1 frequency using pseudorange, carrier phase, L1 and L2 carrier frequencies (f_{L1} = 1575.42 MHz and f_{L2} = 1227.60 MHz) [7]. The pseudorange measurements of visible satellite are corrected for the respective multipath error and the receiver position is calculated.

In this paper multipath errors for Satellite Vehicle Pseudorandom noise (SVPRN) 07, 23, 28 and 31 are estimated for the pseudorange and carrier phase measurements obtained for a geographical location in the Indian subcontinent for typical ephemerides collected on 11th March 2011. The proposed algorithm and the impact analysis done in this paper will also be a valuable aid for setting up the GPS receiver antenna.

2 Multipath Error Modelling

In precise positioning applications, multipath is major error source and impact needs to be calculated especially in urban canyon while setting up GPS receiving antenna [8]. This error is due to transmitted carrier signal taking multiple paths in addition to the line of sight signal path to antenna [9]. Multipath occurrence depends on the properties of the reflector, the antenna gain pattern and the type of correlator used in a receiver [10]. Multipath components or the reflected signal takes more time to reach the receiver than direct signal this interferes with the correlator in GPS receiver to precisely determine transit time of the signal [11]. This affects both C/A code and P code measurements on transmitted carrier signal [12]. In this paper, the multipath error is modeled considering the GPS observables and the carrier frequencies.

$$P_{L1} = \rho + I_{L1} + MP_{L1} \tag{1}$$

where,

P_{L1} Pseudorange on L1 frequency (m)
ρ Geometric Range (m)
I_{L1} Ionospheric delay on L1 frequency (m)
MP_{L1} Multipath error on P_{L1} (m)

$$\phi_{L1} = \rho - I_{L1} + \lambda_{L1} N_{L1} + m\phi_{L1} \tag{2}$$

$$\phi_{L2} = \rho - I_{L2} + \lambda_{L2} N_{L2} + m\phi_{L2} \tag{3}$$

where,

ϕ_{L1} Carrier phase measurement on L1 frequency (m)
ϕ_{L2} Carrier phase measurement on L2 frequency (m)
N_{L1}, N_{L2} Integer ambiguity on L1and L2 frequencies respectively
λ_{L1} Wavelength of L1 carrier frequency (m)
λ_{L2} Wavelength of L2 carrier frequency (m)
$m\phi_{L1}$ Multipath error on ϕ_{L1} (m)
$m\phi_{L2}$ Multipath error on ϕ_{L2} (m)

The multipath error on carrier phase measurements is negligible compared to error on pseudorange. Hence $m\phi_{L1}$ and $m\phi_{L2}$ are assumed as zero. The expression for MP_{L1} can be obtained by forming the appropriate linear combinations (Subtract Eq. (2) from Eq. (1)).

$$\begin{aligned} P_{L1} - \phi_{L1} &= 2I_{L1} - \lambda_{L1} N_{L1} + MP_{L1} \\ MP_{L1} - \lambda_{L1} N_{L1} &= P_{L1} - \phi_{L1} - 2I_{L1} \end{aligned} \tag{4}$$

To represent the above equation in ionospheric (I_1) delay term free form, the Eq. (3) is subtracted from Eq. (2) and rearranged as below,

$$\phi_{L1} - \phi_{L2} = I_{L2} - I_{L1} + \lambda_{L1} N_{L1} - \lambda_{L2} N_{L2} \tag{5}$$

In the above equation I_{L2} is substituted in terms of I_{L1} where, the ionospheric delays (I_{L1} and I_{L2}) are inter related to the respective carrier frequencies (f_{L1} and f_{L2}) as,

$$(f_{L1}/f_{L2})^2 = I_{L2}/I_{L1} \tag{6}$$

$$\phi_{L1} - \phi_{L2} = (f_{L1}/f_{L2})^2 \times I_{L1} - I_{L1} + \lambda_{L1}N_{L1} - \lambda_{L2}N_{L2}$$
$$\phi_{L1} - \phi_{L2} = ((f_{L1}/f_{L2})^2 - 1) \times I_{L1} + \lambda_{L1}N_{L1} - \lambda_{L2}N_{L2} \tag{7}$$

Simplifying the Eq. (7) we get,

$$I_{L1} = 1/((f_{L1}/f_{L2})^2 - 1) \times (\phi_{L1} - \phi_{L2}) + 1/((f_{L1}/f_{L2})^2 - 1) \times (\lambda_{L2}N_{L2} - \lambda_{L1}N_{L1}) \tag{8}$$

Substituting the above expression for I_1 in Eq. (4) we get,

$$MP_{L1} - \lambda_{L1}N_{L1} = P_{L1} - \phi_{L1} - 2/((f_{L1}/f_{L2})^2 - 1) \times (\phi_{L1} - \phi_{L2}) + 2/((f_{L1}/f_{L2})^2 - 1) \times (\lambda_{L2}N_{L2} - \lambda_{L1}N_{L1}) \tag{9}$$

Rearranging the terms in Eq. (9) we get,

$$MP_{L1} - (\lambda_{L1}N_{L1} - 2/((f_{L1}/f_{L2})^2 - 1) \times (\lambda_{L2}N_{L2} - \lambda_{L1}N_{L1}))$$
$$= P_{L1} - ((f_{L1}/f_{L2})^2 + 1)/((f_{L1}/f_{L2})^2 - 1) \times \phi_{L1} + 2/((f_{L1}/f_{L2})^2 - 1) \times \phi_{L2} \tag{10}$$

In above equation $(\lambda_{L1}N_{L1} - 2/((f_{L1}/f_{L2})^2 - 1) \times (\lambda_{L2}N_{L2} - \lambda_{L1}N_{L1}))$ is constant and expectation of MP_{L1} is assumed as zero. The impact of the multipath error and its variation with respect to elevation angle of the satellites for the entire duration of observation are analysed. This analysis will be helpful in kinematic applications where multipath signal become more arbitrary, particularly in aircraft navigation and missile guidance where the reflecting geometry and the environment around the receiving antenna changes relatively in random way [13].

3   Results and Discussion

Statistical analysis of the results shows that multipath error is too large to neglect. These errors are estimated for geographical location (x_u = 706970.90 m y_u = 6035941.02 m z_u = 1930009.58 m) in the Indian subcontinent for typical ephemerides collected on 11th March 2011 from the dual frequency GPS receiver located at Department of Electronics and Communication Engineering, Andhra

University College of Engineering, Visakhapatnam (Lat: 17.73°N/Long: 83.319°E), India. For the observation period of 24 h, analysis of the error which is supported by the relevant graphs and the tables are presented in this paper. During this observation period out of 32 satellites, minimum of 9 satellites were visible in each epoch. Though the error is computed and analyzed for all the visible satellites, in this paper the multipath error estimated for SV PRN07, 23, 28 and 31 are presented. Navigation solution for each epoch is calculated using the pseudoranges (multipath error corrected) of all visible satellites.

Table 1 illustrates the multipath error for four typical satellites. Similar results were also obtained for all the visible satellites. The table also details the error in receiver position distance from the surveyed location. Figure 1 shows the trajectories of the Satellites 07, 23, 28 and 31 with respect to elevation and azimuth angles. The subplots of Fig. 2 show the change in multipath error with respect to change in elevation angle. From the Fig. 2a–d it is observed that the for the elevation angle of less than 10° the multipath errors are 14.32, 18.79, 52.88 and 13.52 m respectively.

Table 1 Pseudorange multiipath error for satellites signal on L1 frequency

	C/A Code multipath error on L1 frequency (m)				Error in receiver position distance (m)
	SV PRN07	SV PRN23	SV PRN28	SV PRN31	
Min	7.362	14.11	40.21	9.136	−25.8
Max	14.32	18.79	52.88	13.52	31.49
Standard deviation	1.816	1.439	1.984	1.019	10.78

Fig. 1 Sky plot for the mentioned satellite orbits as viewed from GPS receiver located at Dept. of ECE, Andhra University (Lat: 17.73°N, Long: 83.31°E)

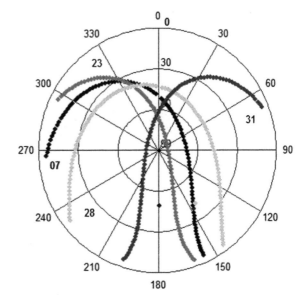

Fig. 2 **a**, **b**, **c** and **d** C/A code multipath error for respective satellites against the elevation angle for the observation period of 24 h

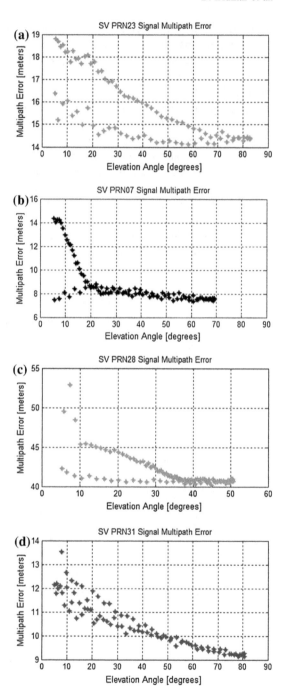

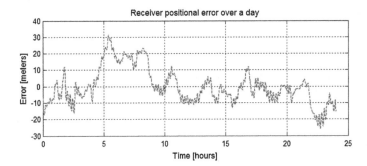

Fig. 3 Error in distance of GPS receiver position from surveyed position over the observation period of 24 h on 11th March 2011

The pseudoranges of all the visible satellites are corrected for estimated multipath error but due to the residual errors in range the GPS receiver position was not accurate. The inaccuracy in the distance from the surveyed location is estimated and 10.78 m of standard deviation was observed (as shown in Fig. 3).

4 Conclusions

The statistics and result analysis comprises of investigation of error magnitude variations over a period of 24 h. Signals transmitted from the satellites, visible at low elevation angles travel a longer path through the propagation medium than those at higher elevation angles, which are subjected to multiple reflections. From the results it is observed that the multipath error is maximum at the low elevation angle of satellite. The proposed algorithm to estimate multipath error is essential for all precise navigation applications. (e.g. CAT I/II aircraft landings, missile navigation) and especially in surveying application in urban canyon. The impact analysis done in this paper will also be a valuable aid in selecting a location of least multipath error to set up the GPS receiver antenna for surveying, aircraft navigation and tracking.

Acknowledgments The work undertaken in this paper is supported by ministry of Science and Technology, Department of Science and Technology (DST), New Delhi, India under woman scientist scheme (WOS-A). Vide sanction file No: SR/WOS-A/ET-04/2013.

References

1. D.A. Happel, Use of military GPS in a civil environment. In *Proceedings of the 59th Annual Meeting of the Institute of Navigation and CIGTF 22nd Guidance Test Symposium* (2003), Albuquerque, NM, June 2003, pp. 57–64
2. D. Sunehra, Estimation of prominent global positioning system measurement errors for gagan applications. Eur. Sci. J. May 2013 edition 9(15). ISSN:1857 – 7881 (Print) e – ISSN:1857–7431 68
3. K. Borre, G. Strang, *Linear Algebra Geodesy and GPS* (Wellesley-Cambridge Press, USA, 1997)
4. D.A. Bilal et al., Estimation propagation delays induced in GPS signals by some atmospheric constituents. IOSR J. Appl. Phys. (IOSR-JAP) **6**(5), 07–13. e-ISSN:2278-4861. Ver. I (Sep.–Oct. 2014)
5. H. Nahavandchi, et al., Correlation analysis of multipath effects in GPS-code and carrier phase observations. **42**(316), 193–206 (2010)
6. I.P. Shkarofsky et al., Multipath depolarization theory combining antenna with atmospheric and ground reflection effects. Annales des Télécommun. Janvier/Fevrier **36**(1–2), 83–88 (1981)
7. E.D. Kaplan, *Understanding GPS: Principles and Applications*, 2nd edn. (Artech House Publishers, Boston, USA, 2006)
8. P. Xie, et al. Measuring GNSS multipath distributions in urban canyon environments. IEEE Trans. Instrument. Measur. **64**(2) (2015)
9. P. Xie, M.G. Petovello, C. Basnayake, Multipath signal assessment in the high sensitivity receivers for vehicular applications. In: *Proceedings ION/GNSS*, Portland, OR, USA, Sept 2011
10. G.S. Rao, *Global Navigation Satellite Systems*, 1st edn. (McGraw-Hill, India, 2010)
11. M. Pratap, E. Per, *Global Positioning System: Signals, Measurements and Performance*, 2nd edn. (Ganga-Jamuna Press, New York, 2006)
12. L. Jingye, L. Lilong, GPS C/A code signal simulation based on MATLAB. In *2011 First International Conference Instrumentation, Measurement, Computer, Communication and Control*, pp. 4–6
13. B.W. Parkinson, J.R. Spilker, *Global Positioning System: Theory and Applications* (American Institute of Aeronautics and Astronautics, Washington DC, 1996)

QoS Discrepancy Impact (qdi) and Cohesion Between Services (cbs): QoS Metrics for Robust Service Composition

V. Sujatha, G. Appa Rao and T. Tharun

Abstract Service composition in service oriented architecture is an important activity. In regard to achieve the quality of service and secured activities from the web service compositions, they need to be verified about their impact towards fault proneness before deploying that service composition. Henceforth, here in this paper, we devised set of exploratory metrics, which enables to assess the services by multi objective QOS factors. These devised explorative measures reconnoitres the higher and lower ranges of the SCFI, which is from the earlier compositions that are notified as either fault inclined or hale. The experimental results explored from the empirical study indicating that the devised metrics are significant towards estimating the state of given service composition is fault tending or hale.

Keywords Web service compositions · Composition support · Service composition impact scale · Service descriptor impact scale · Web service composition fault proneness

1 Introduction

Service-Oriented Architecture (SOA) simplifies information technology related operational tasks by consumption of ready-to-use services [1]. Such SOA found to be realized currently in ecommerce domains such as B2B, B2C, C2B and C2C, in particular the web services are one that considered serving under this SOA.

V. Sujatha (✉)
GITAM School of Technology, GITAM University, Hyderabad, India
e-mail: varadi.sujatha@gmail.com

G. Appa Rao
GITAM Institute of Technology, GITAM University, Visakhapatnam, India
e-mail: apparao_999@yahoo.com

T. Tharun
GITAM University, Visakhapatnam, India

© Springer India 2016
S.C. Satapathy et al. (eds.), *Microelectronics, Electromagnetics and Telecommunications*, Lecture Notes in Electrical Engineering 372,
DOI 10.1007/978-81-322-2728-1_14

Web services are software components with native functionality that can be operable through web. Another important factor about this web services is that more than one service can be composed as one component by coupled together loosely. The standard WSDL is web service descriptive language that let the self-exploration of the web services towards their functionality and UDDI is the registry that lets the devised web services to register and available to required functionality [2].

Composition of web services is loosely interconnected set of Web service operations that acts as a single component, which offers solutions for divergent tasks of an operation. Since the task of composition is integrating divergent web services explored through different descriptors, it is the most fault prone activity. The functionality of service composition includes the activities such as (i) identify the tasks involved in a given business operation, (ii) trace related web services to fulfil the need of each task, (iii) couple these services by exploring the order of that services usage, which is based on the expected information flow, (iv) and resolve the given operation by ordering the responses of the web services that coupled loosely as one component.

In order to achieve quality of service and secure transactions in web service composition and usage, the impact of the composition should be estimated before deploying those loosely coupled web services as one component.

The Web service compositions used earlier that can be found in repositories and the services involved in those compositions helps to assess the impact of these web services towards fault proneness.

The current composition strategies [3–9] are error prone, since these State-of-the-art techniques are not mature enough to guarantee the fault free operations. However, finding these compositions as fault prone after deployment is functionally very expensive and not significant towards end level solutions, also may leads to serious vulnerable. Hence the process of estimating the composition scope towards fault proneness is mandatory.

2 Problem Identification

Let take the process of business to customer strategy as a model to explore the demand of QoS aware service composition. The set of tasks related to any B2C application are explored in Table 1.

These depicted interconnected tasks of the B2C strategy demands the services with max QoS objectives such as reliability, uptime, cost, response time, iterative scope, reputation and execution time, which is significant issue for research and many of the researchers delivered considerable solutions for QoS aware service composition in recent years. According to the service execution flow of the depicted B2C model are:

- Two services (card verification, payment clearance) are involved to fulfil the task 3,

Table 1 Tasks related to B2C application

Task id	Task	Description	Web services scope
1	Explores and walk through the site	Customer explores the menu to select choice items	Collecting available items details from associate Sellers
2	Choosing required items	Moving selected items to cart	No
3	Payment	Decide to pay	Card verification, payment transaction
4	Stock verification	Upon successful order, ordered elements are in stock or not to be checked	Stock statue verification
5	Pickup process to dispatch	If goods available, reserves for pickup dispatch	Reserving Items ordered for pickup, alert to courier about reserved items for dispatching
6	Dispatching	Confirm the courier to dispatch	Official confirmation of the item dispatched state
7	Invoice preparation and send	Once the items dispatched invoice should be prepared and send to customer through email	Invoice preparation and mail
8	Feedback collection from customer	Collect the experience of the customer about the deal	Prepare survey questioner and collect answers from customer

- The service to fulfil the task 5 is essentially need to wait for the stock verification process (task 4)
- The services related to task 6 and task 7 are independent, can be initiated in parallel
- The service related to task 8 is follower of the services related to task 6 and task 7

The observations above listed are not dependent of service QoS metrics, but they are the requirements raised according to the context of the service composition. The issue of selecting services for task 3 is sensitive since the verification and payment clearance are two individual services, but should perform together, which is further referred as the service connection constraint "togetherness". The service that opted for task 5 is dependent of service selected for task 4, until unless stock available, these two services are recursive and this constraint of service connection can referred further as "dependent". The services considered for task 6 and task 7 performs parallel, which we further referred as QoS connection constraint "parallel". The service selected for task is dependent of services selected for task 6 and task 7.

Another significant service connection constraint can depicted from the activity of task 4 and task 5. Since these two services are dependent and recursive, if the seller unable to arrange the items to be dispatched within the given time limit then the earlier functionality should revert and initiate refunding or assigning new seller, which is based on the customer choice. This constraint can be referred as "rollback" for further references.

Any of these service connection constraints can be simplified by achieving coherence of the connection between services. This coherence of the service connection is maximal and reliable, if both of the services are either from same provider or providers those are officially agreed to exchange support.

In our best of knowledge, this dimension of considering coherence between services towards QoS aware service composition was not considered in earlier research. Henceforth here in this paper we devised a new composition metric called Cohesion between services (*cbs*). The model devised here in this paper is defining a statistical assessment strategy that assesses the robustness of the composition under multiple QoS metrics and the cohesion between services.

3   Related Work

Service compositions with malfunctioned web services lead to form the highly fault prone compositions [10]. Henceforth the web service composition to serve as one component under SOA is complex and needs research domain attention to deliver effective strategies towards the QoS centric service. The model devised in [11] defined set of QoS factors to predict feasible services. Many of existing quality-aware service selection strategies aimed to select best service among multiple services available. The model devised in [9] considering the linear programming to find the linear combination of availability, successful execution rate, response time, execution cost and reputation, which is in regard to find the optimal service composition towards given business operation. The model devised in. [7, 12] is considering the temporal validity of the service factors. The authors in [13] modelled a mixed integer linear program that considers both local and global constraints.

The model devised in [8] is selecting services as a complex multi-choice multi-dimension rucksack problem that tends to define different quality levels to the services, which further taken into account towards service selection. All these solutions are depends strongly on the positive scores given by users to each parameter. However, it is not scalable to establish them in prospective order.

Though the QoS strategies defined are used in service composition the factor fault proneness of the service composition is usual. In regard to this a model devised in [14] explored a mechanism for fault proliferation and resurgence in dynamically connected service compositions. Dynamically coupled architecture outcomes in further complexness in need of fault proliferation between service groups of a composition accomplished by not depending on other service groups.

In a gist, it can be conclude that almost all of the benchmarking service quality assessment models are attribute specific, user rating specific or both. Hence importance of attributes is divergent from one composition requirement to other, and the user ratings are influenced by contextual factors, and another important factor is all of these bench mark models are assessing services based on their individual performance, but in practice the functionality of one service may influenced by the performance of other service. Henceforth here in this paper we devised a statistical approach that estimates the impact scale of service composition towards fault proneness, which is based on a devised metric called composition support of service compositions and service descriptors.

In contrast to all of the explored existing models, we devised a statistical assessment strategy in our earlier work called Web Service Composition Impact Scale towards Fault Proneness [15, 16].

4   QoS Discrepancy Impact (*qdi*) and Cohesion Between Services (*cbs*): Metrics for Service Composition

The Dataset opted is of 14 attributes (see Table 2) with values of type continues and categorical. The detailed exploration of these attributes given in our earlier article [16, 17]. The dataset opted is of the records, such that each record is of the 14 composition QOS representative attribute values. In regard to facilitate the attribute optimization process devised here in this paper, the values of the attributes in the given dataset should be numeric and categorical. Henceforth, initially we convert all continuous values to categorical.

Let us consider an application with set of m tasks and each task t can be fulfilled with any individual service among available services

$$S = \{st_1, st_2, \ldots, st_m \forall st_i = \{s_{i1}, s_{i2}, \ldots s_{ip}\}\}$$

The services in set $st_i = \{s_1, s_2, s_3, \ldots s_x\}$ are x number of similar services to resolve the task t_i of given application. Hence the solution to the given application is the composition of the services such that only on service among the x similar services of each task should be considered for composition. Thus, the objective of our proposal is which service should be selected from each set of x similar services.

The selected services toward service composition can influence the QoS. Hence, it is essential to pick optimal services. The meta-heuristic model proposed in this approach is based on the characteristics of services and their composition, which are described as follows:

- A service can be rated best in its independent performance. But might fail to deliver the some performance as a dependent service during composition.
- A service can be rated divergently with respect to its various QoS factors. As an example, a service s can be best with respect to uptime, but the service might be moderate in terms of cost, worst in the context of execution time.
- The importance of the QoS factors might vary from one composition requirement to other.

Table 2 Description of dataset attributes

Attribute ID	Attribute of complete record	Description	Value state of the attribute
1	Connotation	Services of same provider used for optimality	Ratio against expected
2	Cyclic	Number of services required to be cyclic	Services involved in composition with cyclic behavior
3	Dependent	No of services dependent of others	The count of service in composition dependent of other services
4	Parallel	No of services executes parallel	Count of services in composition with parallel execution
5	Repetitive	No of services invoked repeatedly due to failure	Count of services invoked repeatedly due to response failure
6	Uptime	Average of the services as composition uptime	Average of percentage of services uptime involved in composition
7	Services count	No of services in composition	Total number of services involved in composition
8	Diversity	No of services of divergent providers or environment	Services that are not of same provider or same environment
9	Roundtrip time	The completion time of the composition	Composition completion time
10	Cost	Composition cost	Total cost of the services as composition cost
11	Reliability	Response accuracy	Percentage of response accuracy
12	Response time	Composition response Time	Average response time of the services involved in composition
13	Versioning ratio	Composition versioning count	No of times composition changed due to change of services, removing existing or adding new services
14	Status	Indicates composition is fault inclined or hale	1 represents fault inclined, 0 represents hale

According to the characteristics of the services described, it is evident that the best ranked independent service is not always the optimal towards the composition. But at the same moment verification of the composition with all possible services of task is also not scalable and robust. The services under a composition that performed well under some prioritized QoS factors are always need not be the best fit for service composition under other prioritized QoS factors. In regard to this the said meta-heuristic model in its first stage, finds the fitness of the independent services, which is based on primary QoS factor opted. This process is labeled as local fitness evaluation of the services. Further services are ranked according to their fitness and will be used in the same order to finalize a service towards composition.

4.1 Input Data Format

Let a set of service level QoS metrics $M = \{m_1, m_2, m_3, m_4, \ldots m_{|M|}\}$ of each service in the given service set

$$S = \{st_1 = \{s_{11}, s_{12}, \ldots s_{1i}\}, st_2 = \{s_{21}, s_{22}, \ldots s_{2j}\}, \ldots st_m = \{s_{m1}, s_{m2}, \ldots s_{mp}\}\}.$$

Let $E = \{e_1, e_2, e_3, \ldots, e_n \forall [e_i : t_j \rightarrow t_{j+1}]\}$ be the set of n edges such that each edge connecting two tasks in composition sequence. Let $CT = \{[t_i, t_j, t_k, \ldots], [t_x, t_y, t_z, \ldots], \ldots\}$ be the set of task-sets such that the connections between tasks of each tasks-set is influenced by any of the connection QoS constraint called dependent, parallel, rollback or togetherness. This can be defined as, each tasks-set of CT is expecting cohesion between services. The term cohesion can be defined under our proposed model as the services used for these tasks should be from same provider or from the providers mutually agreed to support each other.

4.2 QoS Discrepancy Impact

If we consider a QoS metric m_{opt} as the prime metric for ranking the services, then the metrics of QoS of the services could be categorized as positive and negative. The positive e metrics are those which require higher values and the negative metrics are those which require optimal minimal values.

Henceforth the values of negative and positive metrics are normalized as if the metric m_k is positive then $v(m_k) = 1 - \frac{1}{m_k}$ or if negative then $v(m_k) = \frac{1}{m_k}$.

Next for each service set, based on the normalized values of the related services from maximum to minimum the services are given a ranking so that different metrics are given different ranking for each service. Further the QoS discrepancy is determined by applying this given ranking.

If we consider a rank set of a service $[s_j \exists s_j \in st_i \land st_i \in S]$ is $rs(s_j) = [r(m_1), r(m_2), \ldots r(m_n)]$, then for the service we can measure the QoS Discrepancy Impact qdi as below,

$$\mu(s_j) = \left(\frac{\sum_{i=1}^{|M_{s_j}|} r(m_i \exists m_i \in M_{s_j})}{|M_{s_j}|} \right) \tag{1}$$

In the above equation, $\mu(s_j)$ depicts the mean of the all the QoS metric ranks of the metrics M_{s_j}.

$$\sigma(s_j) = \sqrt{\frac{\sum_{k=1}^{|M_{s_j}|} \left(\mu(s_j) - \{r(m_k) \exists m_k \in M_{s_j}\}\right)^2}{|M_{s_j}|}} \tag{2}$$

In the above equation $\sigma(s_j)$ is the standard deviation of the QOS metric ranks given to a service s_j.

$$g(s_j) = \frac{\sum_{k=1}^{|M_{s_j}|} \left(\mu(s_j) - \{r(m_k) \exists m_k \in M_{s_j}\}\right)^3 \Big/ |M_{s_j}|}{(\sigma(s_j))^3} \tag{3}$$

$$g(s_j) = \sqrt{g(s_j) * g(s_j)} \tag{4}$$

In the above equation $g(s_j)$ represents the skewness [18, 19] related to the QOS metric ranks distributed for service s_j.

The value of skewness may be negative or positive, and close to 0 which denotes that all QoS metrics ranks are nearer, equal to 0 denotes all QoS metrics having same rank.

For only positive skewness values, we have to determine the square-root of the square of the resultant skewness.

As per ANOVA [19, 20, 21],

- The skewness if is less it denotes that the ranking is uniformly distributed that could be, best, moderate, or worst ranks and not a combination of the three types of ranks.
- The distributed ranks mean denotes the centrality of the distributed ranks.
- The deviation of these ranks with respect to each other is denoted by standard deviation.
- The less variance between skewness, mean and standard deviation represents the ranks distribution with less skewness, less deviation and moderately average of near ranks.

So the measurement of the fitness of the service s_j may be done as below,

$$\mu(g(s_j), \mu(s_j), \sigma(s_j)) = \frac{g(s_j) + \mu(s_j) + \sigma(s_j)}{3} \tag{5}$$

From this equation, the mean $\mu(g(s_j), \mu(s_j), \sigma(s_j))$ of the resulting skewness $(g(s_j))$, mean $(\mu(s_j))$ and standard deviation $(\sigma(s_j))$ of service s_j is measured as below,

$$\sigma^2(g(s_j), \mu(s_j), \sigma(s_j)) = \frac{\left[\begin{array}{l}(\mu(g(s_j), \mu(s_j), \sigma(s_j)) - g(s_j))^2 + \\ (\mu(g(s_j), \mu(s_j), \sigma(s_j)) - \mu(s_j))^2 + \\ (\mu(g(s_j), \mu(s_j), \sigma(s_j)) - \sigma(s_j))^2\end{array}\right]}{3} \tag{6}$$

$$\widetilde{qdi}(s_j) = \frac{1}{\sigma^2(g(s_j), \mu(s_j), \sigma(s_j)) + 1} \tag{7}$$

From this equation,

- $\widetilde{qdi}(s_j)$ denotes for a service s_j the inverse of the QoS discrepancy Impact.
- σ^2 or resultant variance for a value between 0 and 1 is normalized so that more variance results in more QoS discrepancy impact. Here the inverse results in lower qdi.
- For avoiding the error of divided by zero, we have added 1 to the variance.

4.3 Measuring Cohesion Between Services (cbs)

If we consider $C = \{c_1, c_2, c_3, \ldots, c_z\}$ as a set of all possible compositions that can be arranged,

For a composition c_i, the cohesion between services (cbs) represents the cohesion in the total number connections (connection created between services offered by the same provider or by the providers with mutual and official relationship) against the number of total connections that need cohesion (see Sect. 4.1). So the measurement of the cohesion between services $cbs(c_i)$ is as below,

$$cbs(c_i) = \left[\sqrt{\frac{(\mu(cbs(c_i), CBSR) - cbs(c_i))^2 + (\mu(cbs(c_i), CBSR) - CBSR)^2}{2}} + 1\right]^{-1} \tag{8}$$

In the above equation,

- $CBSR$ denotes the Cohesion count of the number of edges in total between tasks having the cohesion required for the service composition of the target application.
- $cbs(c_i)$ denotes for a composition c_i the cohesion between the services and its measurement is done with normalization of the standard deviation obtained from $cbs(c_i)$ and $CBSR$ to $0 \le cbs(c_i) \le 1$

- The resultant standard deviation is increased by 1 with which the error, divide by zero, is avoided.
- $\mu(cbs(c_i), CBSR)$ denotes the mean of the $cbs(c_i)$ and $CBSR$.

4.4 Measuring qdi of the Composition

Next for a composition c_i the overall composition fitness measurement can be done as below,

First for a services composition, the services are measured for the mean of fitness values as below,

$$\mu(c_i) = \frac{\sum_{j=1}^{|c_i|} \{\widetilde{qdi}(s_j) \forall s_j \in c_i\}}{|c_i|} \tag{9}$$

where $\mu(c_i)$ denotes the mean of the inverse of QoS discrepancy Impact related to the services comprising a composition c_i

Next for a service composition the services are measured for the standard deviation of the Inverse of the QoS discrepancy Impact as below,

$$\sigma(c_i) = \sqrt{\frac{\sum_{k=1}^{|c_i|} \left(\mu(c_i) - \{\widetilde{qdi}(s_k) \exists s_k \in c_i\}\right)^2}{|c_i|}} \tag{10}$$

Where $\sigma(c_i)$ denotes the standard deviation of the fitness distributed across the services comprising of a composition c_i

Then for a services composition the services are measured for the skewness of the $\widetilde{qdi}$ (inverse of the qdi)

$$g(c_i) = \frac{\sum_{k=1}^{|c_i|} \left(\mu(c_i) - \{\widetilde{qdi}(s_k) \exists s_k \in c_i\}\right)^3 \Big/ |c_i|}{(\sigma(c_i))^3} \tag{11}$$

$$g(c_i) = \sqrt{g(c_i) * g(c_i)} \tag{12}$$

Where $g(c_i)$ denotes the skewness seen across the fitness distributed over services comprising a composition c_i.

Next in a service composition the services, variance of the mean, standard deviation, and skewness of the fitness values are measured as below,

$$\sigma^2(g(c_i), \mu(c_i), \sigma(c_i)) = \frac{\left[\begin{array}{l} (\mu(g(c_i), \mu(c_i), \sigma(c_i)) - g(c_j))^2 \\ + (\mu(g(c_i), \mu(c_i), \sigma(c_i)) - \mu(c_j))^2 \\ + (\mu(g(c_i), \mu(c_i), \sigma(c_i)) - \sigma(c_j))^2 \end{array} \right]}{3} \tag{13}$$

where σ^2 denotes the variance between $g(c_i), \mu(c_i)$ and $\sigma(c_i)$

Finally the variance divides 1, resulting in the composition QoS Discrepancy Impact $qdi(c_i)$ (variance and qdi are proportionate), as below,

$$qdi(c_i) = \sigma^2(g(c_i), \mu(c_i), \sigma(c_i)) \tag{14}$$

$$\widetilde{qdi}(c_i) = [qdi(c_i) + 1]^{-1} \tag{15}$$

where the '$qdi(c_i)$' is increased by 1 for avoiding the error of divide by zero.

4.5 Ordering Resultant Compositions by Composition Aptness Value and Connotation Aptness Value

The resultant compositions are sequentially ordered by their values of inverse of QoS Discrepancy Impact $\widetilde{qdi}$ in the order of max to min. Then the best compositions i.e. "max best service compositions (*cbest*)" are selected. These *cbest* compositions are ordered considering their cohesion in between the services *cbs* in the order of max to min. Finally from the ordered *cbest* compositions the "final best compositions" (*fbest*) are selected.

5 Experiments and Results Exploration

The model devised is analysed for its performance with a dataset created to represent the tasks coupling requirements and the QoS factors different priority requirements. Table 3 explores the dataset applied in the performance assessment of the devised approach. The parameter of dependency scope of the services composition, requires the metric coupling between services (*cbs*) assessment. The significance of the divergent prioritization of the factors of QoS is highly relevant the service quality discrepancy impact (*qdi*) measurement. In service selection the

Table 3 The data used for experiments

Number of tasks	450
Range of tasks to be scheduled	70–250
Range of dependency scope	5–75

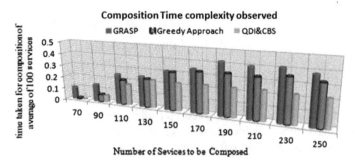

Fig. 1 Time complexity observed to compose an average of 100 services

metrics, *cbs*, *qdi* are applied in sequentially arranging the services. The services of a diverse set lying in range of 70–250 are used in the experiments. The expression language known as R is used in performing statistical analysis based on an explorative approach and the devised model performance analysis assessment is done using computational metrics known as, time complexity and time taken for task completion. The devised models scalability and robustness are also estimated with the additional performance assessment metric, optimal service selection for composition. In this context this model devised by us is contrasted with another two models called GRASP [22] and Greedy [5] which uses the strategies of assessment same as the model devised by us.

The results of the experiment performed show the proposed model is efficient in terms of scalability and robustness, according to the performance metrics, time complexity (see Fig. 1), service composition completion time (see Fig. 2) and the composed services task completion time (see Fig. 3). The inferences from the Fig. 1 show, in contrast to the GRASP and Greedy models our devised explorative statistical analysis model towards a ratio of 100 services the time factor involved is low and stable. As depicted in the Fig. 2 the devised model compared to the

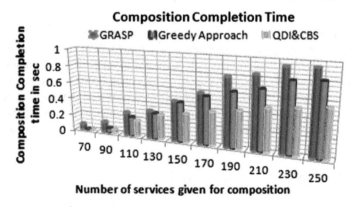

Fig. 2 Composition completion time observed

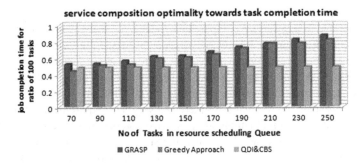

Fig. 3 Optimal resource utilization observed

remaining models of benchmark shows optimal scheduling completion time in the range of tasks from 70 to 250. The estimation of the selection of an optimal service for the composition is done towards task completion time considering a ratio of 100 tasks composition. In Fig. 3 the devised model for explorative statistical analysis is contrasted with GRASP and Greedy techniques for investigating scalable and robust factors.

We can infer a growth of 1.1 and 0.65 % respectively in Average Scheduling time complexity respectively with GRASP and Greedy techniques, in contrast to our devised QDI&CBS. A growth of 2 and 1.2 % in the Task Scheduling completion time respectively is seen with GRASP and Greedy techniques in contrast to QDI&CBS.

6   Conclusion

We have devised in this paper a meta-heuristic model for providing recommendations towards a QoS aware web service composition. In contrast to the remaining models of bench mark (as given in Sect. 2), the model devised is not limited to a single or just two QoS factors specifically. Our approach devised helps in selecting services based on a diverse range of QoS factors combination and combinations prioritization with the help of an important composition factor. The services of optimal requirements for a composition may be selected with the help of one of the devised metric, QoS Discrepancy Impact which overcomes the verification procedure for each available service towards the specified task applicability in a service composition. The services are assessed and ordered and the applicability of a service is determined based on the metric value associated with a service. Also the composition initiator involves those services only which have QoS Discrepancy Impact as per the given threshold values. This ability of the devised strategy considerably overcomes the difficulty associated with other prevailing models. The metric coupling between services is another metric which has significant role in reducing the complexity of the computing involved in finishing the task. The result achieved is possible with the strategy developed by us which uses the global fitness scale, with which in a service composition a selected service is assessed for its task compatibility based on the best QoS Discrepancy Impact. In case a service fails to qualify the fitness requirements then the next service associated with the task having best QoS Discrepancy Impact is verified for the service composition. In majority of the tasks, the service having best QoS Discrepancy Impact meets the global fitness scale requirements. So there is a very less requirement for verification of multiple services combinations in the service composition. The outcomes of the experiments are remarkable which show the significance of the devised approach. Also the scope of future research in various directions is shown possible with the experimental outcomes of the devised approaches. A possible direction for future research is the estimation of the correlation between QoS factors for the assessment of QoS Discrepancy Impact and in a similar way assessment of the correlation between services for determining the Coupling between Services. Another possible research topic that could be our next possible direction of research is applying fuzzy logic in the assessment of the QoS Discrepancy Impact metric.

References

1. C. Mao, J. Chen, X. Yu, An empirical study on meta-heuristic search-based web service composition. In: *2012 IEEE Ninth International Conference on e-Business Engineering (ICEBE)*, 2012, pp. 117–122
2. R. Anane, K.-M. Chao, Y. Li, Hybrid composition of web services and grid services. In: *Proceedings. The 2005 IEEE International Conference on e-Technology, e-Commerce and e-Service, EEE '05*, pp. 426–431 (2005)
3. J. El Hadad, M. Manouvrier, M. Rukoz, Tqos: Transactional and qos-aware selection algorithm for automatic web service composition. IEEE Trans. Services Comput. 3(1), 73–85 (2010)
4. Strunk, Qos-aware service composition: a survey. In: *2010 IEEE 8th European Conference on Web Services (ECOWS)*, pp. 67–74 (2010)
5. Y. Yu, H. Ma, M. Zhang, An adaptive genetic programming approach to qos-aware web services composition. In: *2013 IEEE Congress on Evolutionary Computation (CEC)*, pp. 1740–1747 (2013)
6. Z. Xiangbing, M. Hongjiang, M. Fang, An optimal approach to the qos-based wsmo web service composition using genetic algorithm. In: *Service-Oriented Computing—ICSOC 2012 Workshops*, vol. 7759, pp. 127–139 (Springer, Berlin, Heidelberg 2013)
7. D. Ardagna, B. Pernici, Adaptive service composition in flexible processes. IEEE Trans. Software Eng. 33(6), 369–384 (2007)
8. H. Zheng, W. Zhao, J. Yang, A. Bouguettaya, QoS analysis for web service compositions with complex structures. IEEE Trans. Services Comput. 1(99), 1939–1374 (2012). http://dx.doi.org/10.1109/TSC.2012.7
9. A. Strunk, QoS-aware service composition: a survey. In: *2010 IEEE 8th European Conference on Web Services (ECOWS)*, pp. 67–74 (2010)
10. Z. Zheng, Y. Zhang, M.R. Lyu, Investigating qos of real-world web services. IEEE Trans. Serv. Comput. 7(1), 32–39 (2014)
11. P.A. Bonatti, P. Festa, On optimal service selection. In: *WWW '05: 14th International Conference on World Wide Web*, pp. 530–538 (2005)

12. X. Zhao, B. Song, P. Huang, Z. Wen, J. Weng, Y. Fan, An improved discrete immune optimization algorithm based on pso for qos-driven web service composition. Appl. Soft Comput. **12**(8), 2208–2216 (2012)
13. D. Ardagna, B. Pernici, Global and local QoS guarantee in web service selection. In: *BPM Workshops*, pp. 32–46 (2005)
14. W. Li, H. Yan-xiang, Web service composition based on qos with chaos particle swarm optimization. In: *2010 6th International Conference on Wireless Communications Networking and Mobile Computing (WiCOM)*, pp. 1–4 (2010)
15. L. Xiangwei, Z. Yin, Web service composition with global constraint based on discrete particle swarm optimization. In: *Second Pacific-Asia Conference on Web Mining and Web-based Application*, 2009, pp. 183–186
16. Y. Zhang, Z. Zheng, M.R. Lyu, W.S. Pred: a time-aware personalized qos prediction framework for web services. In: *Proceedings of the 22th IEEE Symposium on Software Reliability Engineering (ISSRE 2011)*
17. W. Li, H. Yan-xiang, A web service composition algorithm based on global qos optimizing with mocaco. In: Algorithms and Architectures for Parallel Processing. Lecture Notes in Computer Science, vol. 6082. pp. 218–224 (Springer, Berlin, Heidelberg, 2010)
18. http://www.tc3.edu/instruct/sbrown/stat/shape.htm
19. K. Gottschalk, S. Graham, H. Kreger, J. Snell, Introduction to web services architecture. IBM Syst. J. **41**(2), 170–177 (2002)
20. http://www.statsoft.com/textbook/anova-manova
21. S. Varadi, G.A. Rao, Quality of service centric web service composition: assessing composition impact scale towards fault proneness. Global J. Comput. Sci. Technol. 14(9-C) (2014)
22. P. José Antonio, S. Segura, P. Fernandez, A. Ruiz-Cortés, QoS-aware web services composition using GRASP with Path Relinking. Expert Syst. Appl. **41**(9), 4211–4223 (2014)

Low Power and Optimized Ripple Carry Adder and Carry Select Adder Using MOD-GDI Technique

Pinninti Kishore, P.V. Sridevi and K. Babulu

Abstract Adders are the most important fundamental blocks of the digital systems which are used in a wide variety of applications. Among them the first basic adder is the Ripple Carry Adder (RCA) and the fastest adder is the Carry Select Adder (CSA). Though all these adders are existed, the speed adding using low power and optimized area is still a challenging issue. In this paper a new technique, Modified Gate Diffusion Input (MOD-GDI) is used to achieve an optimized design with less transistor count and low power dissipation. In the proposed work the 8-bit, 16-bit Ripple Carry Adder and the 8-bit, 16-bit Carry Select Adder have been designed using the CMOS and MOD-GDI techniques. The comparison of their power dissipations and overall transistor count is done and proved that low power and optimized designs are achieved through MOD-GDI technique. The simulation is carried out using Mentor Graphics tool with a supply voltage of 1.8 V at 90 nm technology.

Keywords Ripple carry adder · Carry select adder · Gate diffusion input (GDI) · Mod-GDI · Low power

1 Introduction

In the digital systems the adders play a vital role in computational circuits and other complex arithmetic circuits. The arithmetic function attracts a lot of researcher's [1] where they use them in many applications. Among the complex adders, the most

P. Kishore (✉)
Department of ECE, VNRVJIET, Hyderabad, India
e-mail: kishore_p@vnrvjiet.in

P.V. Sridevi
Department of ECE, Andhra University College of Engineering, Visakhapatnam, India
e-mail: pvs6_5@yahoo.co.in

K. Babulu
Department of ECE, University College of Engineering, JNTUK, Kakinada, India
e-mail: kapbbl@gmail.com

© Springer India 2016   159
S.C. Satapathy et al. (eds.), *Microelectronics, Electromagnetics and Telecommunications*, Lecture Notes in Electrical Engineering 372,
DOI 10.1007/978-81-322-2728-1_15

primary adder is the Ripple Carry Adder. The design of the Ripple Carry Adder (RCA) consists of multiple 1-bit full adders. Each full adder input of C_{in}, which is the C_{out} of the previous adder. As each carry bit ripples the next full adder, so it is called the Ripple Carry Adder. Since this adder has some disadvantages like the gate delay and the computational time is more, engineers had designed some adders with faster way to add the two binary numbers. One among those adders is the Carry Select Adder (CSA). In this adder it uses multiple pairs of RCA to generate the partial sum and carry by taking the carry input C_{in} and C_{out}. Then the final sum and the carry are selected by the Multipliers (MUX).

Modified Gate Diffusion Input (MOD-GDI) is a new technique which is used in low power circuits. This technique is originated from the Gate Diffusion Input (GDI) technique to overcome the demerits of GDI technique. Using this technique, the power dissipation and the overall transistor count is reduced. In this paper the implementation of the 8-bit and 16-bit RCA, 8-bit and 16-bit CSA is done using the MOD-GDI and CMOS technique and the comparison of power dissipation and the transistor count is tabulated.

2 Preliminaries

Modified Gate Diffusion Input (MOD-GDI) is a technique for the low power digital systems. Using this technique, the optimized schematics of the various digital circuits can be obtained. This technique w as adapted from the GDI technique [2], and so the structure of the MOD-GDI cell is similar to the GDI cell. Both the GDI [3] and MOD-GDI cells are three terminal. The only difference between these cells is that the substrate of PMOS is connected to the VDD and of the NMOS to the GND. Figure 1 shows the MOD-GDI cell with input terminals G, P and N.

Fig. 1 The basic Mod-GDI cell

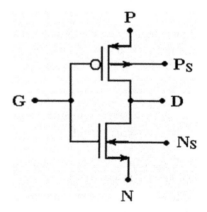

Table 1 Input configurations for different logic functions

N	NS	P	PS	G	D	Function
0	0	1	1	A	A^1	INVETER
B	0	0	1	A	AB	AND
A	0	B	1	A	A+B	OR
A^1	0	A	1	B	A^1B+AB^1	EX-OR (basic)
A	0	A^1	1	B	$AB+A^1B^1$	EX-NOR (basic)
0	0	B	B	A	A^1B	FUNCTION
C	0	B	1	A	A^1B+AC	MUX

Using this MOD-GDI cell the functions that can be achieved are tabulated in Table 1. To implement the RCA and the CSA, the gate functions needed are the Inverter gate EX-OR gate and the 2 × 1 MUX gate. The structures of these gates are shown in Figs. 2, 3 and 4 respectively. Both the RCA and the CSA uses the 1-bit Full Adder circuit. The 1-bit Full Adder circuit is designed using MOD-GDI technique is shown in the Fig. 5.

Fig. 2 Inverter using Mod-GDI technique

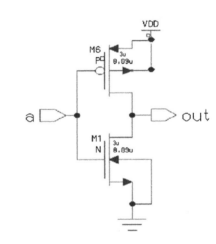

Fig. 3 Two input EX-OR gate using Mod-GDI technique

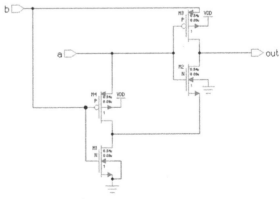

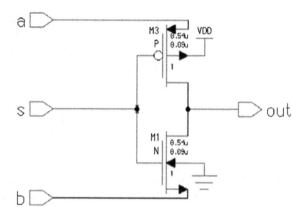

Fig. 4 2 × 1 MUX using Mod-GDI technique

The proposed 1-bit full adder consists of two EX-OR gates [4] and one 2 × 1 MUX gate which are connected as shown in the Fig. 5. It also consists of a feedback PMOS transistor which helps in achieving a full swing at the outputs of the Full adder.

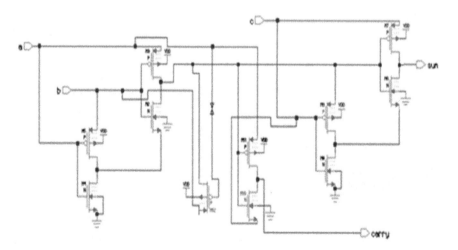

Fig. 5 1-bit full adder using Mod-GDI technique

3 Proposed RCA and CSA

3.1 Ripple Carry Adder

In a n-bit Ripple Carry Adder, there are n full adders connected in cascaded mode. Every 1-bit Full adder's carry output is given as input to the carry of the next stage Full adder. Every stage adder is a combinational circuit in which the output depends on the current inputs only. The C_{in}, carry input i.e. the LSB stage is always 0 (LOW). The construction of 2n-bit adder is done using two n-bit adders of RCA. Both modules can be simultaneously perform addition and then allow the carry propagation form one to the next module. In this paper the implementation of the 8-bit RCA's is done and followed by 16-bit RCA by cascading two 8-bit RCA. The cascading of 2n-bit RCA is shown in Fig. 6 [5].

The schematic of the 8-bit RCA using the MOD-GDI technique is shown in Fig. 7 followed by the Fig. 8 which shows the inputs given to 8-bit RCA for simulation process. In the Fig. 9 it shows the cascading of two 8-bit RCA to obtain 16-bit RCA and the inputs given to it is shown.

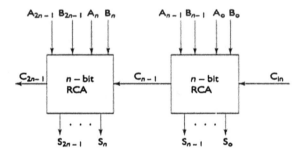

Fig. 6 Cascading of 2n-bit RCA

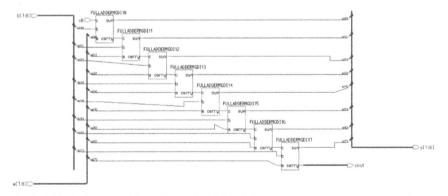

Fig. 7 8-bit ripple carry adder using MOD-GDI technique

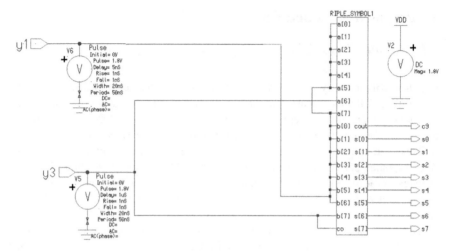

Fig. 8 Inputs given to 8-bit ripple carry adder

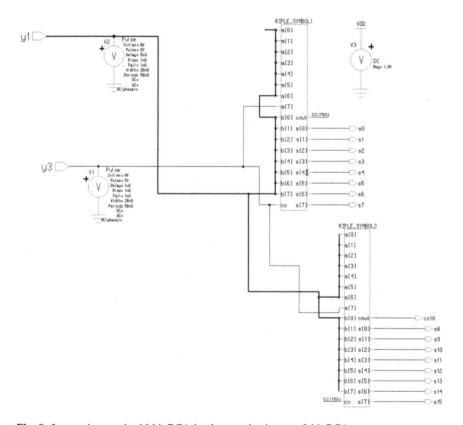

Fig. 9 Inputs given to the 16-bit RCA implemented using two 8-bit RCA

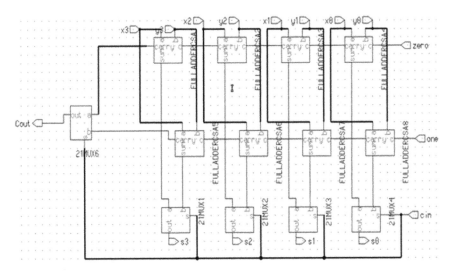

Fig. 10 Basic building block of CSA

3.2 Carry Select Adder (CSA)

Carry Select Adders use multiple narrow adders to create fast and wide adders. The CSA breaks the addition problem into smaller groups which helps for fast computation. It provides two separate group of adders for the upper words (MSB). A MUX is used to select the valid result. For example consider the addition of two n-bit numbers $a = a_0 \ldots a_{n-1}$ and $b = b_0 \ldots b_{n-1}$. The CSA splits the n-bits into (n/2)-bit sections and the higher group words $a_{n-1} \ldots a_{n/2}$ and $b_{n-1} \ldots b_{n/2}$ are added [6]. The carry out bit $c_{n/2}$ is produced by the sum of lower order words $a_{(n/2)-1} \ldots a_0$ and $b_{(n/2)-1} \ldots b_0$. The value of $c_{n/2}$ is then used to select the required result. The resultant can be designed in schematic form which is known as the basic building block of CSA is shown in Fig. 10. To implement a CSA the two important cells needed they are 1-bit full adder and the 2 × 1 MUX. These are implemented using MOD-GDI technique which is shown in Figs. 4 and 5. The 8-bit CSA and the 16-bit CSA schematics are shown in the Figs. 11 and 12 respectively.

4   Results and Analysis

4.1   Power Calculations

Basically in all the digital circuits the power consumption is occurred due to two main components: static power dissipation and dynamic power dissipation. The static power dissipation is occurred due to the sub threshold conduction of the

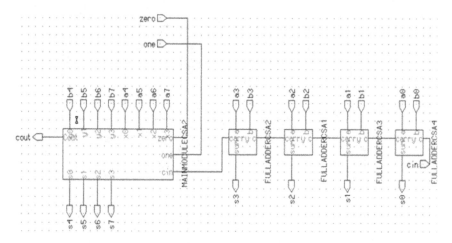

Fig. 11 8-bit carry select adder using MOD-GDI technique

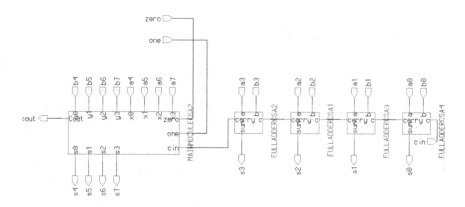

Fig. 12 16-bit carry select adder using MOD-GDI technique

Table 2 Comparison of total power dissipation and transistor count in CMOS and MOD-GDI techniques

Complex adders		CMOS		MOD-GDI	
		Transistor count	Total power dissipation (mw)	Transistor count	Total power dissipation (pw)
RCA	8-bit	308	0.66809	88	312.76
	16-bit	616	1.3363	176	625.51
CSA	8-bit	396	1.2177	142	501.72
	16-bit	878	2.9852	338	1192.4

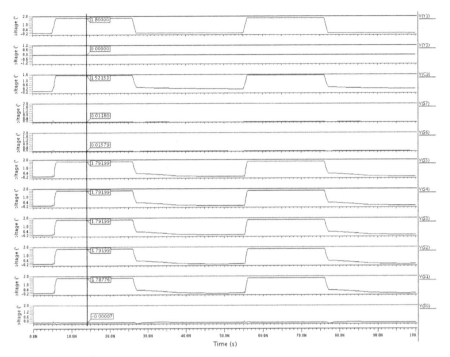

Fig. 13 Output waveforms of 8-bit RCA for the given example

transistors when they are off, also due to the leakage currents through the reverse biased diodes and the tunnelling current through gate oxide. The static power dissipation is very less when compared to the dynamic power for the digital circuit design. The dynamic power dissipation in the circuits is occurred due to the process of charging and discharging of the load capacitance during the switching process and also due to the short-circuit current flowing from the power supply to the ground during the transistor switching. In one complete cycle the current flows from supply to charge the load capacitance and then flows from the charged load capacitance to the ground at the time of discharge. The total power dissipation is given in Eq. (1).

$$P_{total} = P_{dynamic} + P_{static}$$
$$= \alpha CV^2F + I_{sub}V_{dd} \tag{1}$$

where,
A   switching activity
C   load capacitance
V_{dd} supply voltage
F   clock frequency
I_{sub} sub threshold leakage current

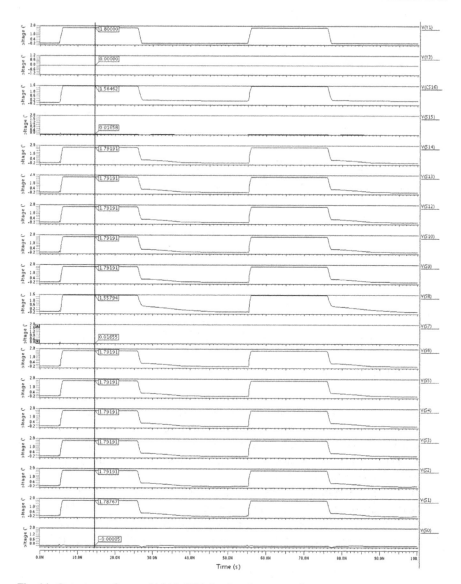

Fig. 14 Output waveforms of 16-bit RCA for the given example

The comparisons of Total power dissipation [7] and the overall transistor count using CMOS [8, 9] and MOD-GDI is taken for the 8-bit and 16-bit RCA and CSA which is tabulated in the Table 2.

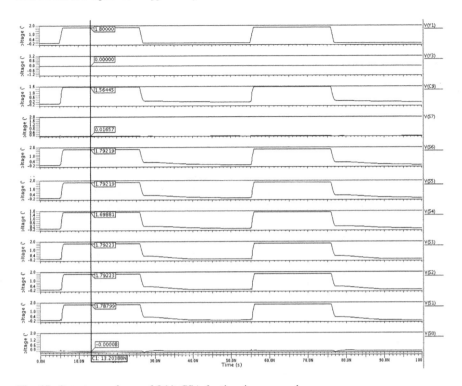

Fig. 15 Output waveforms of 8-bit CSA for the given example

4.2 Functionality Analysis of RCA and CSA

The Functionality of the RCA and CSA is tested by giving an example to it. The inputs given to the 8-bit RCA is a[7:0] = 10111111, b[7:0] = 01111111 and for and 16-bit RCA is a[15:0] = 1011111110111111, b[15:0] = 1111111111111111 as shown in Figs. 8 and 9. The expected output for the given inputs are S [7:0] = 100111110 and S[15:0] = 1011111110111110 respectively. Figures 13 and 14 shows the waveforms of 8-bit and 16-bit RCA which justifies the functionality by obtaining the expected outputs. Similarly the inputs given to 8-bit CSA is a [7:0] = 01111111, b[7:0] = 11111111 and 16-bit CSA is a[15:0] = 11111 11101111110, b[15:0] = 1111111111111111. The expected output for the given inputs are S[7:0] = 101111110 and S[15:0] = 11111111011111101 respectively. Figures 15 and 16 shows the waveforms of 8-bit and 16-bit CSA. The outputs S [7:0] and S[15:0] are observed by ignoring the first two inputs i.e. y(1) and y(3) in the all the wave forms.

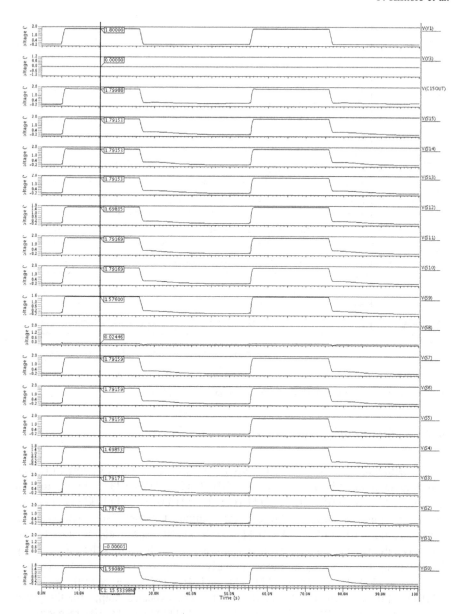

Fig. 16 Output waveforms of 16-bit CSA for the given example

5   Conclusion

The implementation of the 8-bit and 16-bit RCA, CSA were presented and their functionalities are verified. Total power dissipation and Transistor count of the proposed adders using MOD-GDI is shown in Table 2. Comparisons of CMOS based RCA and CSA with the MOD-GDI based RCA and CSA is done. From the Table 2 it is concluded that MOD-GDI technique can be the best technique than the CMOS for complex combinational circuits. The future research activities may include the integrate these complex adders in digital circuits like ALU. Processors and other combinational circuits.

References

1. O.J. Bedrij, Carry-select adder. IRE Trans. Eletron. Comput. Ec **11**(3), 340–344 (2005)
2. S. Ram, R.R. Ahamed, Comparison and analysis of combinational circuits using different logic styles. In: *IEEE–31661, 4th ICCCNT—2013*, Tiruchengode, India-(2013), pp. 157–162
3. A. Morgenshtein, A. Fish, I.A. Wagner, Gate-diffusion input (GDI)—a power efficient method for digital combinational circuits. IEEE Trans. VLSI **10**, 566–581 (2002)
4. S. Wairya, R.K. Nagaria, S. Tiwari, Comparative performance analysis of XOR/XNOR function based high-speed CMOS full adder circuits for low voltage vlsi design. Int. J. VLSI Design Commun. Syst. (VLSICS) **3**, 221–242 (2012)
5. B. Govindarajal, *Computer Architecture and Organization: Design Principles and Applications*, 8th edn, pp. 125–135 (Tata Mcgraw Hill Education Private Limited, 2008)
6. J.P. Uyemura, *Introduction to VLSI Circuits and Systems*, pp. 145–159 (Wiley-Indian Edition, 2009)
7. Y. Sunil Gavaskar Reddy, V.V.G.S. Rajendra Prasad, Power comparison of CMOS and adiabatic full adder circuits. Int. J. VLSI Design Commun. Syst. (VLSICS) **2**, 75–86 (2011)
8. D. Sinha, T. Sharma, K.G. Sharma, B.P. Singh, Design and analysis of low power 1-bit full adder cell. In: *2011 3rd International Conference on Electronics Computer Technology (ICECT)*, vol. 2, pp. 303–305 (2011)
9. K.-S. Yeo, K. Roy, *Low voltage, Low Power VLSI Sub Systems*, pp. 35–45 (Tata McGraw-Hill Edition, 2009)

Increasing the Lifetime of Leach Based Cooperative Wireless Sensor Networks

Simta Kumari, Sindhu Hak Gupta and R.K. Singh

Abstract Energy constrain is the major problem faced by WSN. In this paper LEACH protocol is analyzed. It is shown that distance; initial energy and election probability play a vital role in increasing the lifetime of sensor nodes. This paper implements LEACH in cooperative transmission and analyses its performance in comparison to SISO (Single Input Single Output). It is shown that for large network area cooperative transmission is more successful than SISO as it will consume less energy to transmit data over the large network area.

Keywords LEACH · Cooperative communication · WSN · V-MIMO

1 Introduction

Wireless sensor network is an upcoming research area in these days. WSNs consist of thousands of sensors nodes which are scattered randomly in harsh environment. These small sensor nodes have the capability to sense, process, and compute and transmit the data collected from particular event to base station [1]. Each single sensor node consists of microprocessor, power source, and low energy radios. Usually these sensor nodes are battery operated. Due to the limited battery resources which deplete at faster rate due to computation and communication operations. Low energy consumption is one of the key requirement of energy constrained WSNs. Multi Input Multi Output (MIMO) technology is very relevant

S. Kumari (✉) · S.H. Gupta
Amity University, Sec-125, Noida 201301, U.P., India
e-mail: simtarana16@gmail.com

S.H. Gupta
e-mail: shak@amity.edu

R.K. Singh
Technical University, Dehradun, Uttarakhand, India
e-mail: rksingh12kec@rediffmail.com

© Springer India 2016
S.C. Satapathy et al. (eds.), *Microelectronics, Electromagnetics and Telecommunications*, Lecture Notes in Electrical Engineering 372,
DOI 10.1007/978-81-322-2728-1_16

technology to increase the packet size. Practically it is very difficult to implement MIMO technology in wireless sensor network. This is because of the sensor node which has physical limitation that it can accommodate only one antenna practically for transmission. With the result it becomes difficult to implement MIMO, as it will be very difficult for a size wise constrained sensor node to support multiple transmitters and receivers [2]. Cooperative Communication helps to create virtual MIMO in a WSN. Cooperative Communication enables a wireless system to achieve diversity. It enables single antenna wireless sensor node into multiuser environment. This is made possible as sensor nodes share their antennas and achieve transmit diversity. The fundamental concept behind Cooperative Communication [3] was in the wake of information theoretic properties of the relay channel [4, 5]. Amplify and Forward, Decode and Forward and Coded Cooperative Communication [6] are the few Cooperative Protocols [7, 8]. Since we get all the benefits of MIMO from Cooperative Communication, we can term it as Virtual MIMO (V-MIMO).

LEACH is one of the basic clustering protocol which is firstly proposed by Heinzelman [9, 10]. This is Low Energy Adaptive Clustering Hierarchy. This protocol is based on probability and no of rounds. It depends on randomly selected cluster heads.

The operation of leach protocol consists of two phases:

1. Setup phase: In set up phase, clusters are organized and cluster heads are selected. Each node decides on the basis of probability whether or not become a cluster head for current round. This is made by the node n choosing a random number between 0 and 1. If any node which becomes clusters head once, it cannot become cluster head again for 'p' rounds. If number is less than threshold value t (n), the node become cluster head for current round.

The threshold is:

$$T(n) = \{p/1-p \times (r \times \mod 1/p\}, n \in G \tag{1}$$

Here
p   is the percentage of cluster head.
r   current round
G   set of nodes that are not cluster heads in the past 1/p rounds

2. Steady state phase: In this phase data is divided in frames and nodes send their data to cluster head. After some time new round begin and entered network renewed.

The main contribution of this paper is that:

It is shown that by increasing the initial energy and varying election probability of the node, network life time can be increased. This will be very beneficial factor in choosing a cluster head for future modification of LEACH algorithm.

It is shown that for long distance communication the energy consumed by cooperative communication is much less in comparison to energy consumed by non-cooperative transmission.

It is shown for same energy level the number of bits transmitted in cooperative transmission is much less in comparison to the number of bits transmitted in non-cooperative transmission.

2 System Model

The network model consists total 200 number of nodes which are deployed over a squared field network having area M × M in meters. Every sensor node has specific lifetime. During this lifetime, nodes utilize its energy for collecting, computing and processing and finally transmitting to base station. All sensor nodes are considered to be synchronized [11].

The simulation parameters are given in Table 1.

3 Energy Consumption Model

Energy consumption model in this paper is considered in [12]. The model consists of a source node, destination node and relay nodes. If only source transmits to destination it will be a non-cooperative SISO communication and if source transmits to destination via relay in addition to direct transmission it will be cooperative communication. In SISO (single input single output) non cooperative communication approach, sensors transmit their data to the cluster head and cluster head transmit all their aggregated data to the destination node without any cooperation.

Table 1 Parametres for simulation

Parameters	Values
Area	(100 * 100) m
Packet size	4000 bits
No of rounds	1500
Initial energy	0.5 nj/bit
p (probability)	0.05
Etx_elec	50 nj/bit
Erx_elec	50 nj/bit
Eamp	100 pj/bit/m^2
D	10 m
Mt	2
K (loss)	2
No of nodes	200

Energy required to transmit and receive b-bit data in distance d for single transmission is given by

$$E_{SISO-TX}(b, d) = bE_{transnitter} + bE_{amplifier}d^k \qquad (2)$$

$$E_{SISO-RX}(b) = bE_{receiver} \qquad (3)$$

In cooperative transmission, sensors nodes transmit their data to cluster head and cluster head further transmit to another cluster head to reach the sink cluster head is acting as relay. Data is transmitted to the sink with each other's cooperation.

We also consider M_t as the number of transmitting antennas, where as other parameters remain as same as in single transmission.

$$E_{COOPERATIVE-TX}(b, d) = M_t bE_{transmitter} + 1\ M_t E_{amplifier}d^k \qquad (4)$$

$$E_{COOPERATIVE-RX}(b) = M_t bE_{reciever} \qquad (5)$$

4   Experimental and Simulation Results

Figure 1 shows the simulation result of simple stable election protocol based on LEACH that if energy of node is 0.5 J, than node become dead after 700 no of rounds.

The formation of cluster head is done on the basis of probability and residual energy. Node which has highest residual energy will become the cluster head. Now all other nodes send their data to cluster head instead of direct sending to base station. Nodes become dead node after certain numbers of rounds. Now we observe that at which particular round the node becomes dead node.

Fig. 1 Number of rounds versus number of nodes

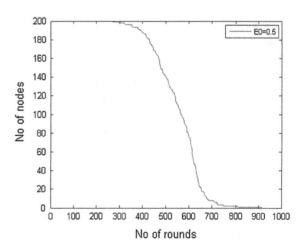

Figure 2 depicts the results of increasing initial energy will increase the lifetime of nodes. It is very evident from the graph that increase in initial energy results in direct increase in the number of rounds.

Table 2 gives a detailed insight into the effect of variation at initial energy on the number of rounds.

Figure 3 shows the graph between number of rounds and number of nodes on the basis of election probability. We see that number of alive node increases as the probability of the node to become cluster head decreases (Table 3).

The simulation result for single transmission is shown in Figs. 4 and 5. In the Fig. 4, we see that energy consumption of nodes increases as the distance of node increases. Here energy consumption of node is 5×10^{-8} J/bit when distance is around 1 m. As the distance increases, the energy consumption also increases and around 10 m distance, energy consumption of node increases to 6×10^{-8} J/bit (Fig. 6).

In Fig. 7, energy consumption of receiver increases as the no of bits increases. By using compression techniques no of bits can be reduced, which will save lot of energy. These results can be summarized from table given below:

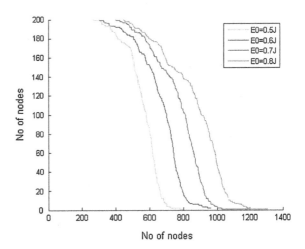

Fig. 2 Number of rounds versus number of nodes at different energy levels

Table 2 Effect of variation at probability on the number of rounds

Initial energy (J)	No of rounds
0.5	800
0.6	<900
0.7	1000
0.8	1200

Fig. 3 Effect of election probability on number of rounds

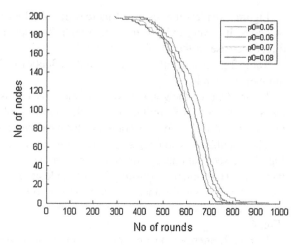

Table 3 Probability of cluster head node vs. number of rounds

Probability	No of rounds
0.05	<900
0.06	<800
0.07	800
0.08	<700

Fig. 4 Distance versus energy consumption for SISO in WSN

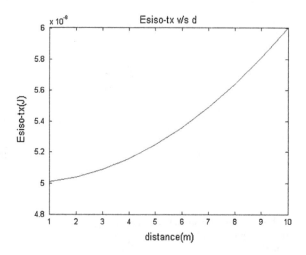

Table 4 depicts comparison between energy consumption for single transmission and cooperative transmission. We see that there is huge difference in the energy consumption. Cooperative transmission can save lot of energy.

Table 5 shows that if receiving node receives more number of bits than energy consumption also increase.

Fig. 5 Bits transmitted versus energy consumption for SISO IN WSN

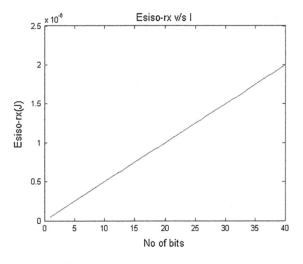

Fig. 6 Distance versus energy consumption for cooperative transmission

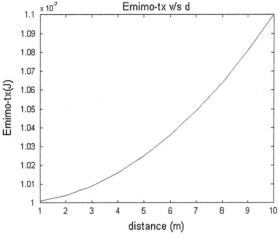

Fig. 7 Bits transmitted versus energy consumption for cooperative transmission

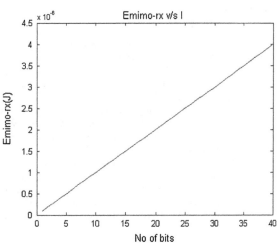

Table 4 Energy comparison of single and cooperative transmission

Distance (m)	Esiso_tx (J)	Emimo_tx (J)
1	5×10^{-8}	1×10^{-7}
3	5.1×10^{-8}	1.01×10^{-7}
7	5.5×10^{-8}	1.05×10^{-7}
8	5.6×10^{-8}	1.07×10^{-7}
10	6×10^{-8}	1.1×10^{-7}

Table 5 Energy consumption of SISO and MIMO vs number of bits

No of bits	Esiso_tx (J)	Emimo_rx (J)
10	0.5×10^{-6}	1×10^{-6}
20	1×10^{-6}	2×10^{-6}
30	1.5×10^{-6}	3×10^{-6}
40	2.5×10^{-6}	4×10^{-6}

5 Conclusion

The most important factor for Wireless Sensor Network is energy. In this paper LEACH and its implementation in cooperative WSN is analyzed. By critical analysis and simulation results it is shown that initial energy and election probability play a vital role in lifetime of a node. For future modification of LEACH these factors can be kept in view to choose cluster head, in order to manage energy efficiently. In further analysis, it is shown that implementation of LEACH protocol in cooperative WSN increases the energy efficiency of node as simulation results directly indicated that total energy consumption for the cooperative transmission gets decreased by approximately 14 % in comparison to non-cooperative SISO transmission when they are deployed over large network area.

References

1. W.B. Heinzelman, A.P. Chandrakasan, H. Balakrishnan, An application specific protocol architecture for wireless microsensor networks. IEEE Trans. Wireless Commun. **1**, 660–670 2002
2. T.M. Cover, A.E. Gamal, Capacity theorems for the relay channel. IEEE Trans. Inform. Theory **25**(5), 572–584 1979
3. Y. Jiang, Y. Yan, Y. Yang, Energy balanced clustering algorithm in large scope WSN based on cooperative transmission technology. J. Commun. **9**(8), 627–633 (2014)
4. A. Sendonaris, E. Erkip, B. Aazhang, User cooperation diversity Part I and Part II. IEEE Trans. Commun. **51**(11), 1927–1948 (2003)
5. J.N. Laneman, G.W. Wornell, D.N.C. Tse, An efficient protocol for realizing cooperative diversity in wireless networks. In: *Proceedings of IEEE ISIT*, Washington DC, p. 294 2001
6. X. Li, N. Li, L. Chen, Y. Shen, Z. Wang, Z. Zhu, An improved LEACH for clustering protocols in wireless sensor networks. In: *Proceedings International Conference Measuring Technology and Mechatronics Automation (ICMTMA 10)*, vol. 1, pp. 496–499 (2010)

7. A. Nosratinia, T.E. Hunter, Hedayat, Cooperative communication in wireless networks. Commun. Magazine IEEE **42**(10), 74–80 (2004)
8. H. Gou, Y. Yoo, An energy balancing LEACH algorithm for wireless sensor networks. In: *Proceedings International Conference Information Technology*, pp. 822–827 (2010)
9. S. Hak Gupta, R.K. Singh, S.N. Sharan, Performance evaluation of efficient cooperative MIMO system wireless communication. WSEAS Trans. Commun. 9–14 (2013)
10. G. Sachin, N. Choksi, M. Sarkar, K. Dasgupta, Comparative analysis of wireless sensor network motes. In: *2014 International Conference on Signal Processing and Integrated Networks (SPIN)*, pp. 426–431. IEEE (2014)
11. K. Rakesh, V. Pant, Energy prediction based heterogeneous mobile clustering protocol for wireless sensor networks. Int. J. Adv. Res. Eng. Appl. Sci. **3.8**, 35–42 (2014)
12. S. Cui, A.J. Goldsmith, A. Bahai, Energy-efficiency of MIMO and cooperative MIMO techniques in sensor networks. IEEE J. Selected Areas Commun. **22**(6), 1089–1098 (2004)

A U-Shaped Loop Antenna for In-Body Applications

Nageswara Rao Challa and S. Raghavan

Abstract This paper describes a flexible U shaped loop implantable antenna for in-body applications. The proposed antenna having a bandwidth which covers the Med Radio band (402–405 MHz) which is the standard band for Bio Implantable applications, ISM band 433.050–434.79 MHz. The antenna is made of polyamide substrate having a size of 3.3 mm radius and height of 13.5 mm. The simulated antenna is having a −10 dB Bandwidth of 158 MHz (335–493 MHz) resonates at 402 MHz having proper VSWR throughout the bandwidth. The Radiation pattern of this antenna is directional having a gain of −31.16 dB.

Keywords Biocompatible · ISM band · Med radio band · SAR · VSWR · MICS

1 Introduction

Implantable medical devices (IMDs) are the attractive scientific research area in the medical therapy and diagnosis [1, 2]. An important component of IMDs is the implantable antenna that is integrated with the IMD to communicate its telemetry with exterior medical equipment. The Medical Device Radio communications Service (Med Radio) band (402–405 MHz) is most commonly used for medical implant telemetry [3]. A healthcare provider can set up a wireless link between the implanted device (antenna) and base station to allow a high speed reliable system to monitor the health conditions of the patient in real time [4].

Implantable devices must be biocompatible. Whatever the material used as a substrate for the antenna, encapsulation must be done with a biocompatible material

N.R. Challa (✉) · S. Raghavan
ECE Department, National Institute of Technology, Tiruchirappalli
Tamil Nadu, India
e-mail: 308112051@nitt.edu

S. Raghavan
e-mail: raghavan@nitt.edu

© Springer India 2016
S.C. Satapathy et al. (eds.), *Microelectronics, Electromagnetics and Telecommunications*, Lecture Notes in Electrical Engineering 372, DOI 10.1007/978-81-322-2728-1_17

183

because if the encapsulating material is not biocompatible, its direct contact with the tissue may cause damage to the tissue. In the proposed antenna polyimide is used as encapsulation material. We have to prevent the undesirable short circuits in case of antennas used for the long-term implantation [5]. The design strategies are explained in [6]. There are so many biocompatible materials like zirconia (ε_r = 29; tan δ = 0.0002), PEEK (ε_r = 3.2; tan δ = 0.01), polyimide (ε_r = 3.5), ceramic alumina (ε_r = 9.8) etc. [7, 8]. Some more biocompatible materials are explained in [9].

2 Proposed Antenna

The proposed antenna is a modification of U shaped antenna mentioned in [10]. The configuration of the patch is shown in Fig. 1. The front and back views are shown in Figs. 2 and 3 respectively. The antenna is placed over an open cylindrical substrate made up of polyimide which has ε_r of 3.5 and thickness of 0.0254 mm. All the antenna patches are good conductors which can be excited over small power batteries. The patch is taken as 0.017 mm. The design is simulated using Computer Simulation Technology (CST) which analyses the antenna using the principle of Finite Difference Time Domain (FDTD). The parameters values are a = 5 mm, b = 11.5 mm, c = 2 mm, d = 6 mm, e = 0.5 mm.

The human body doesn't consists of a uniform dielectric constant throughout the body, because human body consists of different tissues like skin, fat, muscle, bone etc. These tissues have different thickness and different dielectric constants at different frequencies. Those specifications are mentioned in the Table 1 [7].

The antenna patch is having a length 12 mm and width is 20 mm. Later this antenna is wrapped over a hollow cylindrical substrate made of polyimide. The

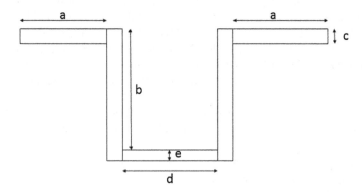

Fig. 1 Structure of the radiating element

Fig. 2 Front view

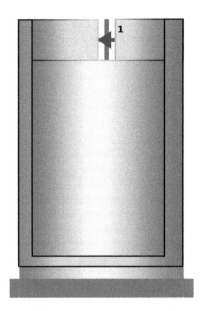

Fig. 3 Back view

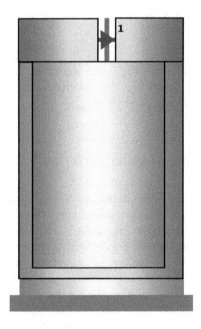

inner and outer radii of the substrate are 3.276 and 3.3 mm respectively. The height of the dielectric cylinder is 13.5 mm. The return loss curve is as shown in the Fig. 4 for free space conditions. It gives return loss of −18.2 dB and resonates at 7.6 GHz.

Table 1 Tissues and their relative permittivity and conductivity values

Tissue	Relative permittivity	Conductivity (S/m)
Muscle	57.129	0.79631
Fat	5.5798	0.41199
Skin	46.787	0.688

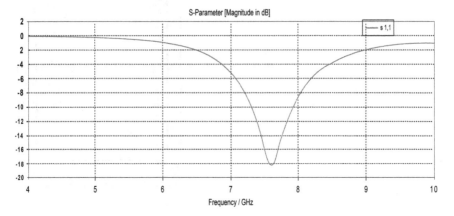

Fig. 4 Return loss curve for free space model

Encapsulation of the proposed antenna is taken with the polyimide cylinder of thickness 0.017 mm and height of 13.5 mm. Later the top and bottom parts of the cylinder are also covered by using polyimide material before placing it in the human body model.

3 Antenna Performance

First we place the antenna in the solid cylindrical muscle tissue because the result for the body model is very close to the results with muscle tissue. The muscle tissue dimensions are height is 110 mm and radius is 52 mm. Later we have to check the mechanical connections of the components with the muscle tissue properly. Later we add the fat and skin layers each of 4 mm thickness. Then we vary the dimensions of the parameters and position of the patch according to our required frequency using the parameter sweep. Some standard body models are explained in [10, 11]. The antenna is having enormous bandwidth. So, we can achieve more data rates. The antenna is surround by complex body model (i.e., Muscle, Fat, Skin models). This resonates in the Med Radio band (402–406 MHz) and ISM Band (433.05–434.79 MHz) where we are getting a proper VSWR. Simple model (muscle) giving a radiation efficiency of −36.47 dB with muscle model, whereas with complex model gives the radiation efficiency is −34.31 dB at 402 MHz. Gain value increases from −33.17 to −31.16 dB. This difference is due to the addition of

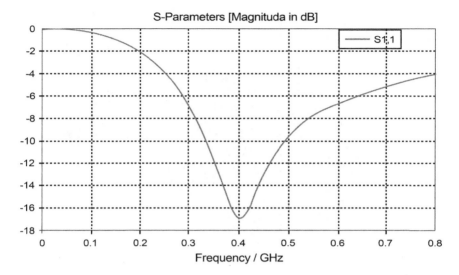

Fig. 5 Return loss curve for the proposed antenna

small Fat and Skin layers. Even though we added two different layers to the muscle model which are having different relative permittivity, conductivity and thickness values, there is only few Mega Hertz frequency shift which is due to the permittivity change from region to region. This frequency shift will not much affect the antenna performance. This model much related to the real time body model. The antenna is placed in the typical body model which consists of muscle, fat and skin. The antenna is surrounded by a cylindrical Muscle model having a thickness of 52 mm, Fat model having a thickness of 4 mm and Skin model having a thickness of 4 mm. All the three tissues are having height 110 mm. This configuration is having a return loss of −16.92 dB at 402 MHz frequency. This antenna model having a bandwidth of 158 MHz (335–493 MHz). The return loss curve is shown in Fig. 5. Radiation pattern at 402 MHz is shown in Fig. 6. The results of the proposed antenna at 402 and 433 MHz are compared in Table 2.

3.1 Specific Absorption Rate (SAR)

SAR indicates the rate of energy absorbed per unit mass of the human body when exposed to a RF electromagnetic field and also other forms of energy like ultrasound, and has units of watts per kilogram [12]. SAR is usually averaged either for the whole human body or over a sample volume of 1 or 10 g. Generally it should be as low as possible [12]. The antennas which are designed for brain applications should have low SAR values compared to that of body applications [13]. The measurement of SAR is officially done by using DASY SAR measurement system

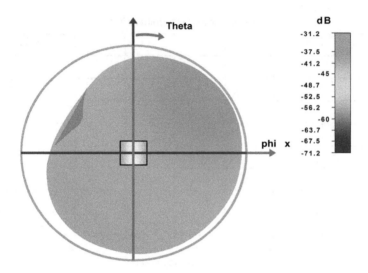

Fig. 6 Radiation pattern of the proposed antenna at 402 MHz

Table 2 The results of the proposed antenna at 402 and 433 MHz

Parameter	402 MHz	433 MHz
SAR	376 W/Kg	362 W/Kg
Radiation efficiency	34.31 dB	−33.19 dB
Maximum power	4.25 mW	4.419 mW
Gain	−31.16 dB	−30.1 dB

shown in [14]. The proposed antenna having SAR of 376 W/Kg for 1 W input power for 1-g SAR. So, we can give up to 4.25 mW for 1-g SAR. SAR distribution over the antenna at 402 MHz is shown in the Fig. 7. The comparison of results of proposed antenna with antenna in [10] are mentioned in Table 3.

In bio implantable antennas, antenna size is main criteria because we are embedding the antenna inside the body. When we try to design an antenna in ordinary procedures, the physical length is in the order of hundreds of millimeters for 402 MHz frequency. So, Instead of increasing the physical length we will increase the electrical length to reduce the antenna size. To reduce the antenna dimensions we go for the loop antennas and meander type antennas to reduce the antenna occupying area. The proposed antenna achieved 32 % reduction in the antenna size compared to the antenna in [10]. The proposed antenna covers two bands Med Radio band at 402–405 MHz, ISM bands at 433.050−434.79 MHz. This complicated model is more realistic and suitable for practical applications.

Fig. 7 SAR distribution over the antenna at 402 MHz

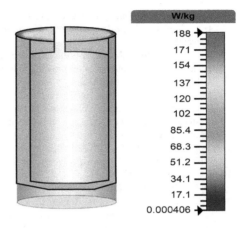

Table 3 Comparison of results of proposed antenna with the antenna in [10] at 402 MHz

Parameter	Previous antenna	Proposed antenna
SAR	500 W/Kg	376 W/Kg
Size	31.4×15 mm^2	20.724×12 mm^2
Maximum power	3.2 mW	4.25 mW
Gain	−28.95 dB	−31.2 dB
Return loss	−17.2 dB	−16.92 dB

4 Conclusions

A U-shaped loop directional antenna is designed for in-body applications with polyimide cylindrical substrate dimensions of 3.3 mm radius and 13.5 mm height and the patch dimensions are 20 mm width and 12 mm length with a size reduction of more than 30 % when compared to the antenna in [10] with a reduction of SAR by 25 % than the antenna in [10] with a bandwidth of 158 MHz with a realized gain of −31.2 dB resonates at 402 MHz. The proposed antenna also gives −30.1 dB at 433 MHz with SAR value 362 W/Kg for 1 W input power. We can give a maximum input power of 4.419 mW at 433 MHz.

References

1. T. Karacolak, A.Z. Hood, E. Topsakal, Design of a dual-band implantable antenna and development of skin mimicking gels for continuous glucose monitoring. IEEE Trans. Microw. Theory Tech. **56**(4), 1001–1008 (2008)
2. R. Warty, M.-R. Tofighi, U. Kawoos, A. Rosen, Characterization of implantable antennas for intracranial pressure monitoring: reflection by and transmission through a scalp phantom. IEEE Trans. Microw. Theory Tech. **56**(10), 2366–2376 (2008)
3. FCC, Washington, DC, USA, *Federal Communications Commission* (2012). http://www.fcc.gov

4. L. Huang, M. Ashoueil, F. Yazicioglu, *Ultra-low Power Sensor Design for Wireless Body Area Networks: Challenges, Potential Solutions, and Applications.* JDCTA 2009; 3 (September (3))
5. N. Vidal, S. Curto, J.M. Lopez Villegas, J. Sieiro, F.M. Ramos, Detuning study of implantable antennas inside the human body progress. Electromagn. Res. **124**, 265–283 (2012)
6. A.K. Skrivervik, F. Merli, Design strategies for implantable antennas. In: *Loughborough Antennas & Propagation Conference* (2011)
7. Calculation of the dielectric properties of body tissues. http://niremf.ifac.cnr.it/tissprop/
8. F. Merli, A.K. Skrivervik, Design and measurement considerations for implantable antennas for telemetry applications. In: *Proceedings of the 4th European Conference on Antennas and Propagation (EuCAP 2010)* (2010)
9. D.F. Williams, *The Williams Dictionary of Biomaterials* (Liverpool University Press, 1999)
10. R. Alrawashdeh, Y. Huang, P. Cao, A conformal u-shaped loop antenna for biomedical applications. In: *7th European Conference on Antennas and Propagation (EuCAP)*, 2013
11. M.L. Scarpelb, D. Kurup, H. Rogier, D. Vande Ginste, F. Axisa, J. Vanfleteren, W. Joseph, L. Martens, G. Venneeren, Design of an implantable slot dipole conformal, flexible antenna for biomedical applications. IEEE Trans. Antenn. Propag. **59**(10) (2011)
12. http://en.wikipedia.org/wiki
13. IEEE Standard for Safety Levels with Respect to Human Exposure to Radio Frequency Electromagnetic Fields, 3 kHz to 300 GHz, IEEE Standard C95.1-1999 (1999)
14. http://www.Antenna-theory.com

Design of NMEA2000 CAN Bus Integrated Network System and Its Test Bed: Setting Up the PLC System in Between Bridge–Bow Room Section on a Container Ship as a Backbone System

Jun-Ho Huh, Taehoon Koh and Kyungryong Seo

Abstract Following the recent changes in international agreements enforced by the International Maritime Organization (IMO), and considering the Korean environment where both the safety of old ships and the IoT technology have emerged as the major issues, even the old operational ships are required to establish the Controller Area Network (CAN Bus) system to be connected with a backbone network referred as the Ship Area Network (SAN) by installing communication cables to control and monitor shipboard electronic equipments when they are newly added to the ship's system. For existing operational ships, an extensive work may be needed for the establishment of the SAN system that uses unshielded twisted pair cables (UTP) depending on the location of the system to be installed as the most parts of the ship are made of steels. Such work could generate time, space and cost-related risks (e.g., large vessels require up to millions of Korean won per day as an anchorage charge). As the result of experiment, it was possible to install the PLC system within a few hours using existing power cables and placing a PLC modem at the spot on 180 m section (approximately 2/3 of ship's length). Also, with the proposed technology, costs were reduced as much as over 70 % compared to the UTP- and STP-based ship data communication system which requires additional cable installments. Therefore, we were able to confirm both the efficient communications and the cost-reducing effects and expect that this technology will become a foundation technology for the e-Navigation system using IoT technology, as well as for the various industrial settings.

Keywords NMEA2000 · PLC · Power line communication · IoT · Test bed

J.-H. Huh · T. Koh · K. Seo (✉)
Department of Computer Engineering, Pukyong National University
at Daeyeon, Busan, Republic of Korea
e-mail: krseo@pknu.ac.kr

J.-H. Huh · T. Koh
SUNCOM Co, Busan, Republic of Korea

© Springer India 2016
S.C. Satapathy et al. (eds.), *Microelectronics, Electromagnetics
and Telecommunications*, Lecture Notes in Electrical Engineering 372,
DOI 10.1007/978-81-322-2728-1_18

1 Introduction

Together with recent development of shipbuilding industry, marine equipment companies have been pursuing mutual growth also. Especially, the marine equipment industry continues to grow both quantitatively and qualitatively changing the industry's technical paradigm by shifting the concept of machine or electronic-based automation to the concept of IT-converged networks. It is expected that the regulations in various international agreements, which are enacted/revised by the International Maritime Organization (IMO), will be strengthened and as they've decided to fully implement e-Navigation system starting from 2018, a variety of studies and standardization plans are being proceeded actively for this task.

Meanwhile, infrastructures for core technologies such as the navigation system, geographic information system, communication network technology and its application system are required. The key technology among them is the maritime communications support technology and with this, an adequate set of data must be provided in accordance with the NMEA2000 protocol, which is a standardization plan for the serial data networking to meet increasing necessity of interfaces between an array of electronic equipments that have the direct effects on ship's navigation, when the application systems [e.g., Voyage Data Recorder (VDR) and Alarm Monitoring System (AMS)] essential for safe navigations request them to be transmitted. The ships' electronic equipments currently manufactured in Korea follow internationally agreed protocols and these equipments are operated in the form of integrated network across all levels of the ship thorough the ethernet-based Ship Area Network (SAN).

Following recent trend of building larger ships like container ships, the paradigm in infrastructure construction technology for ship's network is changing. Although NMEA2000 protocol stipulates the distance of 200 m and the speed of 250Kbps, the ship communication technology of the past could not overcome communication distance of 100 m for it used Ethernet-based UTP/STP cables. The problem has been settled with optical cables and wireless communication method. However, for those existing ships without adequate on-board communication infrastructure would have much obstacles (i.e., spatial, temporal and costly obstacles) when the e-Navigation comes into effect following the international agreements.

Thus, in this paper, we've designed a NMEA2000-based Ship Area Network (SAN) which has been converged with Power Line Communication (PLC) technology that utilizes existing power line on the ship to expeditiously respond to the changes in International agreements and to clear away obstacles, and carried out Test Bed experiment.

2   Related Research

2.1   Size of the Ship

The ship is a container ship with a length of approximately 399 m from stem to stern reaching about 3 times of a football field. Recently, Daewoo Shipbuilding & Marine Engineering has built a mega-container ship (19,224 TEU, the size of 4-football field combined) ordered from Swiss Mediterranean Shipping Company at Ok-po shipyard, Geoje island (Fig. 1).

2.2   Interior Structure of the Ship

As shown in Fig. 2, many processes are required to route communication cables when new or additional electronic equipments are to be installed to establish a connection with the intergrated SAN. Ship's bulkhead structures are made of steel plates so that it has an unfavorable environment. Due to recent demands of mobile device use within the ship, WiFi AP units have been installed but the problem is that because of the shielding function of steel bulkheads, they need to be installed at several places.

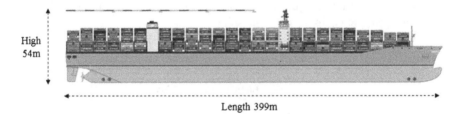

Fig. 1 Container Ship

Fig. 2 Ship's steel structure

2.3 Ship-PLC (Power Line Communication)

Ship-PLC is networking technology enabling the transfer of data through existing power lines at the ship [1–5]. It provides high-speed transfer and needs no extra cables Fig. 3 shows Ship-PLC. For the vessel which has been already wired entirely, the PLC can save costs and time. Our PLC unit provides up to 200 Mbps for data transfer. The PLC promises fast and efficient network on vessel deck. Definition of PLC The communication technology that transfers data on the high frequency signal, using carrier frequency on power lines as a medium [1].

As in Fig. 4, the Ship-PLC transmits data using the carrier frequency on the high-frequency power line.

However, ship operators are very conservative as long as frequency interference —which may influence ship's navigation safety—is concerned so that the frequency violations are strictly controlled (e.g., for maritime signaling bandwidth) by making notification of relevant laws. The Ship-PLC product used in the experiment has eliminated such problem in advance (Fig. 5).

2.4 Scalability of PLC

Since the PLC supports various communication methods like Ethernet- and Serial (RS-232/422/485)-based ones, its use can be extended to a variety of activities on the ship such as acquisition and analysis of all sorts of sensor information, signaling of dangers, status information monitoring through the connections with different types of electronic equipments on the ship.

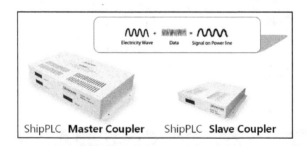

Fig. 3 Ship-PLC [1]

Fig. 4 Principle of PLC

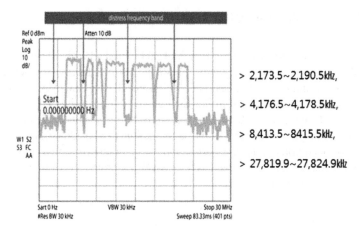

> 2,173.5~2,190.5kHz,

> 4,176.5~4,178.5kHz,

> 8,413.5~8415.5kHz,

> 27,819.9~27,824.9kHz

Fig. 5 Avoiding maritime distress signaling bandwidth

Thus, with this scalability in mind, we've verified the feasibility of the PLC-applied SAN by performing a Test Bed using shipboard CAN BUS-type electronic equipments that adapts the NMEA 2000 standard.

2.5 Example of NMEA 2000 Standard Application for a Ship

Figure 6 represents the NMEA 2000 standard CAN BUS. All the shipboard electronic equipments are built with the NMEA 0183/2000 standard CAN BUS in accordance with international specifications.

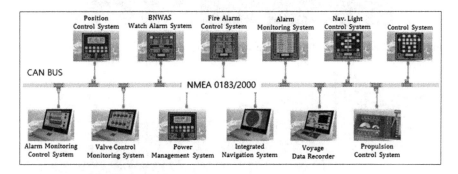

Fig. 6 NMEA standard CAN BUS

3 NMEA 2000 Integrated Network Using PLC

Using the container ship with a full length of 304 m and a height of 49 m (i.e., M/Y Hyundai Colombo owned by Hyundai Merchant Marine, Korea)—which can load 6,800+ 20-ft. containers—as a test bed, we've installed a PLC Master Coupler to the Emergency Breaker located in the Emergency Room positioned in the section between the Bridge and the Bow Room (a straight-line distance of 180 m) on the upper deck. Using the power line of the same section as a backbone, we connected a NMEA 2000 equipment to the serial port of a notebook and performed data transmission experiments for the CAN + SAN integrated network using a PLC modem and a PLC repeater (Fig. 7).

3.1 Goal of Final Test

We chose a spot furthest from the Bridge, where data collection and data transmission are carried out toward the land. The spot had the poorest environment for additional data cable installments and our premise was that the network had to overcome an extreme environment assuming that the NMEA 2000 standard CAN BUS-type electronic equipments will be installed at all levels of the ship (Fig. 8).

M/V HYUNDAI COLOMBO 6,800TEU / CONTAINER CARRIER

Fig. 7 Test Bed using a 6,800 TEU-class container ship

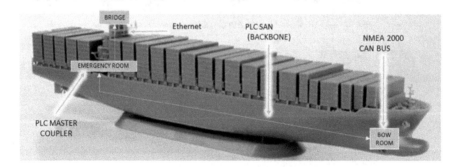

Fig. 8 PLC + NMEA-converged network

3.2 Outline of PLC (Power Line Communication)

Figure 9 shows 3-phase 4-wire Y-connection. The PLC technology was developed in US in 1924 and mostly used for the power-usage readings and home networks on land. As shown in the picture, it is comprised of 4 phases (R, S, T and N). And Fig. 10 is the power line structure used on a ship and unlike the land-use type, it is comprised with just 3 phases (R, S, and T), missing the N-phase [1].

3.3 Delta Connection Data Coupling

Figure 11 shows Problem associated with PLC. Network configuration with the Y-connection type can be possible regardless of phase changes in the same power line structure as all the connections are made with the same N-phase. However, the configuration with the Delta connection type will not be possible since there's no common phase like N such that the network cannot be comprised with the PCs on A, B and C decks. To solve above mentioned problem, we successfully established connections using the Delta connection by coupling R, S and T phases behind the switchboard where the power is supplied initially, as shown in Fig. 12.

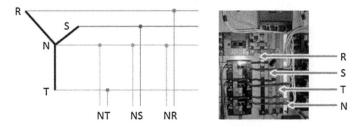

Fig. 9 3-phase 4-wire Y-connection

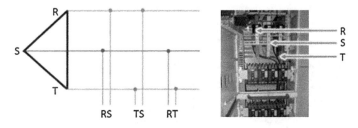

Fig. 10 3-phase 3-wire Delta connection

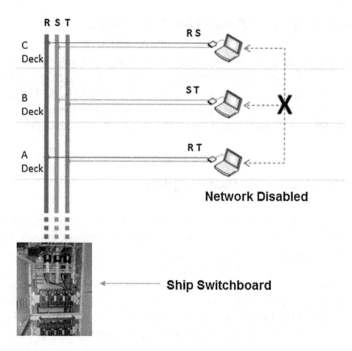

Fig. 11 Problem associated with PLC

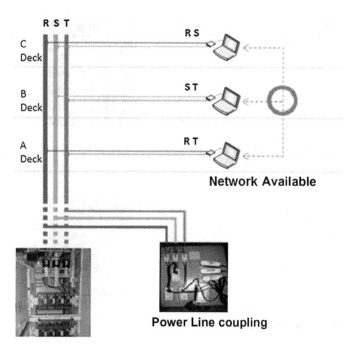

Fig. 12 A solution for the problem of network configuration with Delta connection

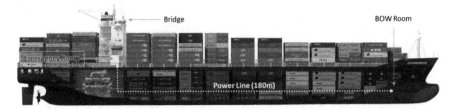

Fig. 13 Ship-PLC + NMEA test bed and communication testing section

3.4   Tests for PLC and NMEA Communication

As shown in Fig. 13, the section between the Bridge and the Bow Room where data collection and monitoring are carried out was chosen as the test bed and both the furthest distance and poor environment where additional cable installments are most difficult were considered. The test was conducted with existing power line without installing separate communication cables at the section.

Since the power lines furcate in each section to supply power, we applied the PLC repeater for each section (i.e., passage #1, #2, and #3), and by using the PLC we constructed a network between the Bridge and the Bow Room, followed by the test. Our goal was to apply the network to the cargo hall.

We first installed a PLC Master Coupler in the 200 V Emergency Breaker Panel of the Emergency Room located on the upper deck. Second, the PLC modem was installed in the Bow Room and third, a PLC Repeater Modem was installed at 100 m on the Passageway. Fourth, the PLC Repeater Modems were installed at 2/5 and 4/5 points of the Passageway. Fifth, the Passageway 220 V Emergency Breaker PLC Repeater was installed and finally, we confirmed the network establishment after installing the PLC Repeater on the Rotary Switch located at the entrance of Bow Room.

4   Performance Evaluation and Cost

Performance evaluation was repeated three times. In order to avoid 'speed down' problem due to introduction of noises, which is the most vulnerable factor for the PLC, we've conducted the test by minimizing the noise by installing the PLC Repeaters in respective sections. Our final goal was to reduce installment costs and maximize efficiency by using the least number of the PLC Repeaters.

The most vulnerable part of the PLC is the introduction of noise so that countermeasures must be studied to avoid any future communication disruptions. In this

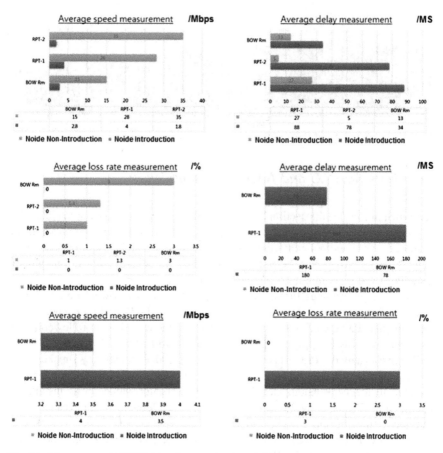

Fig. 14 Measurements of PLC tests due to noise introduction

paper, we've artificially introduced noises to see how the communication quality would deteriorate due to noises (Fig. 14).

Table 1 is the list of ships on which the PLC system has been installed and shows shipping companies, ship's names, installed sections and the number of modems supplied (EA). The total number of ships SUNCOM Co installed the PLC system from 2007 to 2015 was 39 and the average cost involved in the system construction was approx. USD 8,000 per ship. For the 39 ships, the distances for the system construction ranged from a minimum of 170–200 m maximum, showing total average distance of roughly 185 m. As of July of 2015, installing the PLC system will lead to about over 70 % cost reduction when compared to setting up internet network using UTP cables on a ship.

Table 1 List of ships on which the PLC system has been installed and additional information

No.	Shipping company	Name of ship	Installation date	Installed section	Number of supplied modems (EA)	
					Master coupler	Client model
1	STX Panocean	AUTO ATLAS	2008	ECR	1	2
2	STX Panocean	AUTO BANNER	2008	ECR	1	2
3	STX Panocean	ATLANTIC ADVENTURE	2011. 11	BRIDGE, CAPTAIN, CENG, SHIP'S OFFICE, ECR	1	5
4	STX Panocean	DK IMAN	2014. 10	BRIDGE, CAPTAIN, CENG, SHIP'S OFFICE, ECR	1	5
5	SIX Panocean	DK ITONIA	2014. 10	BRIDGE, CAPTAIN, CENG, SHIPS OFFICE, ECR	1	5
6	STX Panocean	EASTERN CARRIER	2008	BRIDGE, CAPTAIN, CENG, SHIPS OFFICE, ECR	1	5
7	STX Panocean	FRONTIER CARRIER	2009	BRIDGE, CAPTAIN, CENG, SHIPS OFFICE, ECR	1	5
8	STX Panocean	GOOD FIREND	2007	BRIDGE, CAPTAIN, CENG, SHIP'S OFFICE, ECR CO, 20, 30	1	8
9	STX Panocean	HAN JIN PIONEER	2011. 02	BRIDGE, CAPTAIN, CENG, SHIP'S OFFICE, ECR	1	5
10	STX Panocean	HARMONY CARRIER	2008	BRIDGE, CAPTAIN, CENG, SHIP'S OFFICE, ECR	1	5
11	STX Panocean	H5 ACASIA	2009	BRIDGE, CAPTAIN, CENG, SHIP'S OFFICE, ECR	1	5
12	STX Panocean	HS PIONEER	2008	BRIDGE, SHIP'S, OFFICE, ECR	1	3
13	STX Panocean	CUPID FEATHER	2008	BRIDGE, CAPTAIN, CENG, SHIP'S OFFICE, ECR	1	5
14	STX Panocean	MAGIC FORTE.	2008	BRIDGE, CAPTAIN, CENG, SHIP'S OFFICE, ECR	1	5
15	STX Panocean	MAGIC ORIENT	2008	BRIDGE, CAPTAIN, CENG, SHIP'S OFFICE, ECR	1	5

(continued)

Table 1 (continued)

No.	Shipping company	Name of ship	Installation date	Installed section	Number of supplied modems (EA)	
					Master coupler	Client model
16	STX Panocean	NEW DIAMOND	2008	ECR	1	2
17	SIX Panocean	LADY KADOORIE	2009	BRIDGE, CAPTAIN, CENG, SHIP'S OFFICE, ECR	1	5
I8	SIX Panocean	ORIENTAL HOPE	2008	BRIDGE, CAPTAIN, CENG, SHIP'S OFFICE, ECR	1	5
19	SIX Panocean	ORIENTAL FRONTIER	2007	BRIDGE, CAPTAIN, CENG, SHIP'S OFFICE, ECR	1	5
20	STX Panocean	ORIENTAL TREASURE	2009	BRIDGE, CAPTAIN, CENG, SHIP'S OFFICE, ECR	1	5
21	STX Panocean	OCEAN HOST	2007	BRIDGE, CAPTAIN, CENG, SHIP'S OFFICE, ECR	1	5
22	STX Panocean	OCEAN MASTER	2008	BRIDGE, CAPTAIN, CENG, SHIP'S OFFICE, ECR	1	5
23	STX Panocean	OCEAN UNIVERSE	2008	BRIDGE	1	2
24	STX Panocean	OCEAN ROYAL	2007	BRIDGE, CAPTAIN, CENG, SHIP'S OFFICE, ECR	1	5
25	STX Panocean	ORIENTAL FRONTIER	2008	BRIDGE, CAPTAIN, CENG, SHIP'S OFFICE, ECR	1	5
26	STX Panocean	PAN AMBmON	2008	ECR	1	2
27	STX Panocean	SEA OF FUTURE	2012	SHIP'S OFFICE, ECR	1	3
28	STX Panocean	SEA OF GRACIA	2012	ECR	1	2
29	STX Panocean	SEA OF HARVEST	2012	SHIP'S OFFICE, ECR	1	3
30	STX Panocean	SILVER CARRIER	2007	BRIDGE, CAPTAIN, CENG, SHIP'S OFFICE, ECR	1	5
31	STX Panocean	SJMOKOR HONGKONG	2010. 12		1	6

(continued)

Table 1 (continued)

No.	Shipping company	Name of ship	Installation date	Installed section	Number of supplied modems (EA)	
					Master coupler	Client model
				BRIDGE, CAPTAIN, CENG, SHIP'S OFFICE, ECR		
32	STX Panocean	STX A2AREA	2010. 06	BRIDGE, CAPTAIN, CENG, SHIP'S OFFICE, ECR	1	5
33	STX Panocean	STX FREESLA	2010. 12	ECR	1	2
34	STX Panocean	STX SINGAPORE	2008	BRIDGE, CAPTAIN, CENG, SHIP'S OFFICE, ECR	1	5
35	STX Panocean	VISION	2008	BRIDGE, CAPTAIN, CENG, SHIP'S OFFICE, ECR	1	5
36	MSC	MSC LONDON	2014. 07	BRIDGE, G DECK, F DECK	1	3
37	MSC	MSC NEWYORK	2014. 10	BRIDGE, G DECK, F DECK	1	3
38	MSC	MSC ISTANBUL	2015. 01	A DECK, B DECK, C DECK, D DECK, E DECK, F DECK	1	7
39	MSC	MSC AMSTERDAM	2015. 04	A DECK, B DECK, C DECK, D DECK, E DECK, F DECK	1	6

5   Conclusion

Purpose of carrying forward the policies (e.g., IMO, IEC, ISO and IALA, etc.) related to the e-Navigation and its standardization is to secure navigation and maritime safety through convergence of shipbuilding technology with Information technology. The technology enables collection of sensor data with various electronic equipments and machines and analysis of the data by sending them to the land side efficiently, during entire process of departure and entry. Should the e-Navigation policy become obligatory, existing ships will suffer overwhelming damages in costs following temporary suspension of ship operation, extra equipment purchase or replacement, and network cable installments to be certified.

In this paper, we've performed an actual test bed experiment on a ship to enable easy adaptation of NMEA 2000 protocol between the relevant parties using the PLC technology with 3-phase 3-wire Delta connection method applied within the ship without installing separate network cables. This technology can be applied to the old ships under the direct influence of changes made in international agreements or to the existing ships which are impossible to install additional network cables due to their features or types.

As the result of experiment, it was possible to install the PLC system within a few hours using existing power cables and placing a PLC modem at the spot on 180 m section (approximately 2/3 of ship's length). Also, with the proposed technology, costs were reduced as much as over 70 % compared to the UTP- and STP-based ship data communication system which requires additional cable installments.

Acknowledgments A part of the fundamental technology in this paper contains the Republic of Korea 'Patent No. 10-0942020' registered on Feb. 2010 after being submitted on April of 2008 related to the theory concerning 'The Ship-PLC Master Coupler Under the 3-Phase 3-Line Delta Connection Environment' by the author of this paper, SUNCOM Co., Ltd.

References

1. J.-H. Huh, T. Koh, K. Seo, NMEA2000 Ship Area Network (SAN) design and Test Bed using Power Line Communication (PLC) with the 3-Phase 3-Line Delta Connection Method, in *SERSC ASTL*, vol. 94 (Networking and Communication 2015) (2015), pp. 57–63
2. J.-H. Huh, K. Seo, PLC-Based Smart grid home network system design and implementation using OPNET simulation. J. Multimedia Inf. Syst. 113–120 (2014)
3. A. Akinnikawe, K.L. Butler-Purry, Investigation of broadband over power line channel capacity of shipboard power system cables for ship communication networks, in *2009 IEEE Power & Energy Society General Meeting*, Calgary (2009), pp. 1–9
4. S. Barmada, L. Bellanti, M. Raugi, M. Tucci, Analysis of power-line communication channels in ships. IEEE Trans. Veh. Technol. **59**(7), 3161–3170 (2010)
5. M. Antoniali, A.M. Tonello, M. Lenardon, A. Qualizza, Measurements and analysis of PLC channels in a cruise ship, in *2011 IEEE International Symposium on Power Line Communications and Its Applications*, Udine (2011), pp. 102–107

Inductively Powered Underground Wireless Communication System

Achyuta Gungi, Vikas Vippalapalli, K.A. Unnikrishna Menon and Balaji Hariharan

Abstract Underground Wireless Sensor Network (UWSN) is a newly emerging technology that is capable of replacing existing traditional wired connections. Research has proved that magnetic induction based communication performs better than electromagnetic wave communication, especially in the dynamic underground environment. Currently, data recorders are being used to collect the data from underground sensors. Data recorders have a few disadvantages such as the inability to produce data on time and the difficulty to deploy them. Many problems related to data recorders can be solved using UWSNs magnetic induction technique. Some of the main challenges are efficient power transfer and communication to underground sensors when distance between transmitter and receiver coils increases. This paper introduces a unique technique where both wireless power transfer and communication are achieved simultaneously for sensors which are present underground. A high frequency square pulse is used for inductive power transfer between transmitter and receiver coils. The same square pulse is pulse code modulated and is used for wireless communication between the transmitter and the receiver coils with a suitable data rate. Experimental results to prove the feasibility of this novel technique for concurrent wireless power transfer and communication make the paper different from its counter parts.

Keywords Inductive power · MI · UWSN · Relay coil · PWM

A. Gungi (✉) · K.A. Unnikrishna Menon · B. Hariharan
Amrita Vishwa Vidyapeetham, Kochi, Kerala, India
e-mail: achyuta.sun@gmail.com

V. Vippalapalli
G. Narayanamma Institute of Technology and Science (for Women), Hyderabad, India
e-mail: vikasdec23@gmail.com

© Springer India 2016
S.C. Satapathy et al. (eds.), *Microelectronics, Electromagnetics and Telecommunications*, Lecture Notes in Electrical Engineering 372,
DOI 10.1007/978-81-322-2728-1_19

1 Introduction

1.1 Overview

The concept of Wireless Sensor Network (WSN) emerged in late 2000s and has been exponentially growing. Sensor networks consist of multiple sensors which can communicate to each other and can send the sensed data to a sink node. Usually sensors are small in size and inexpensive, so they can be deployed in large numbers across any desired area. Since sensors are vastly used in various environments, some applications require random deployment of sensors across selected area, also power management is critical. Thus, choosing a proper power source for the sensors is important. Usually sensors work through battery power i.e. DC current. So amount of power consumed, energy saving, life time needs to be calculated and proper assumptions have to be considered for the sensor network to work properly. The number of sensors which die should not exceed the number of active sensors leading to loss of coverage or connectivity. Sensors should be deployed considering all the above aspects. For Underground sensor network most of the parameters to monitor as mentioned above and issues to be addressed changes because the channel is dynamic. Major criteria to be addressed in underground sensor network is communication to the channel, if the communication is done wirelessly then parameters to be considered will be power management, connectivity, and channel characteristics. If the distance between the transmitter and receiver coils exceeds a particular limit, then the efficiency goes down exponentially. This effect can be reduced by introducing relay coils in between transmitter and receiver coils.

1.2 Objective

The main objective of the paper is to present a technique based on strong, coupled, resonant magnetic induction to supply wireless power to underground sensor networks and also communicate with them. The real-time deployment results of the proposed technique are also presented so as to prove experimentally the feasibility of the proposed technique for wireless power supply and communication with the underground sensors.

1.3 Outline of Proposed Technique

A technique is proposed for continuous power supply and communication for underground wireless sensor networks using strong, coupled, resonant magnetic induction. The system uses a renewable or non-renewable energy source to sup- ply the continuous power. Source Coil is used in this system around which the magnetic field is generated and this magnetic field induces a voltage across the receiver coil, and this voltage can be used to supply for underground sensors. The system can be used in two ways i.e. it can be used to provide power continuously or periodically the system can be used to recharge the batteries present underground. The communication with underground sensors is achieved using pulse code modulation. The square pulse which is used for transfer of power is pulse code modulated with data that needs to be transmitted to the sensors deployed in the soil.

The related efforts to solve this problem have been explained in Sect. 2. A novel technique based on the principle of strong resonant coupled magnetic induction has been proposed for underground wireless power supply and communication to sensor networks in Sect. 3. Section 4 details the mathematical modeling of the proposed system. The analysis of real time deployment results are presented in Sect. 5. Finally the paper is concluded in Sect. 6.

2 Related Work

Wireless underground communication using magnetic induction for communication surpasses traditional wired based data collection. Currently data recorders are being used for collecting the data from underground sensors [1]. Data recorders have a few disadvantages such as the inability to produce data on time and difficulty in deploying them. UWSN with magnetic induction technique is advantageous as the channel is no longer air but soil with moisture content and rocks. Dynamic channel condition affects EM wave communication and it requires large antenna size which is not feasible underground [2]. Alternative technique to EM wave communication is MI waveguide technique [3]. The main challenge would be Efficient Power Transfer when the distance between the transmitter and the coils increase.

Presently wireless power transfer has been extensively used for many applications including Mobile phone charging, charging house hold appliances, and wireless charging of automobiles. This paper presents a unique technique of wireless power transfer to underground sensors using pulsating DC. As the efficiency in WPT degrades rapidly with distance, there is a need for a productive transfer of power. Usage of Resonant frequency for WPT is more advantageous than traditional MI waveguide technique [4]. Capacitive Power Transfer (CPT) is an alternative solution for Inductive WPT. In CPT, capacitors which are used for storing the electric Charge have high impedance at low frequencies. If high frequency is used for power transmission, capacitors provide low impedance and hence efficiency can be increased but the range would be limited to few centimeters [5]. Considering all the above challenges, WPT using resonant magnetic induction gives a better performance [6]. Witricity technology [7] explains how MI technique can be applied for transfer of wireless power and illustrates how resonators work in a closed environment. WPT for underground sensors using resonant MI technique has been implemented for shorter distances [8]. As the distance between transmitter and receiver coils increases efficiency goes down rapidly. This problem can be tackled by using repeater coils in between transmitter and receiver coils. Relay or repeater coils does not require any input power. They act as magnetic flux [9] repeaters once they are introduced into the transmitter's magnetic field.

The second major challenge would be communicating with underground sensors. As the EM wave communication in the soil has many disadvantages, due to dynamic soil conditions there is a need for an alternative communication technique [2]. An alternate technique for communication is using MI technique. As the power is transfered to the sensors using inductors which use resonant mutual induction principle, communication is also achieved using the same principle. This eliminates any extra hardware required for communication and also cost due to the communication transceivers.

When compared to a traditional EM wave communication, MI decays at higher rate. EM waves suffers underground due to dynamic channel condition, Large antenna height, and high path loss. Attenuation of the EM waves will be high due to the absorption of soil, water and rock. Also path loss depends on amount of water density, soil property and rock content. Considering the overall path loss, MI waveguide performs better than EM wave [2, 3]. MI waveguide has many advantages such as overall path loss is less, repeater coils which are used for flux relaying do not consume any power, and it surpasses traditional wired communication as it more pliable. The transmitter coil transmits MI waves at a particular frequency called resonant frequency. At resonant frequency impedance would be minimal as the reactance tends to zero, hence maximum power would be transferred [4].

2.1   Connectivity in Wireless Underground Sensor Networks (WUSN)

Underground the importance of Connectivity: Connectivity is considered to be the most important and crucial for network models like WUSNs. Compared to terrestrial wireless sensor networks connectivity analysis is more complicated because of path loss, communication through magnetic induction. When connectivity is considered in WUSNs various possibilities of communications are underground-to-aboveground channel (UG-AG), underground-to-underground channel (UG-UG) and aboveground-to-underground channel (AG-UG). Also the transmission ranges for all of these are quite different. So analysis of channel characteristics is necessary for proposing an efficient connectivity model.

Existing design model for connectivity: Considering the transmission ranges of the various communicating channels, UG-AG has higher transmitting range than UG-UG. UG-UG has very limited transmitting range when compared with the traditional wireless communication over the air. Most importantly AG-UG path loss is higher than UG-AG, because when air to ground path is taken, reflection of waves occurs on the surface.

Network Model: Considering WUSN is deployed in a region R. These sensors are distributed as poisson point process of constant spacial intensity lambda. There are n Fixed AG sinks uniformly distributed along the border of the region. As full connectivity is assured, Sensing region of the sensors is much larger than communication range. There are two phases in the functionality of this model, they are sensing phase and control phase. In sensing phase UG nodes sense and send data to AG nodes, while in control phase AG nodes send control messages to UG nodes. The latencies which are tolerable are considered to be Ts and Tc. The main difference between the connectivity analysis in the two phases is the latencies and transmission ranges.

Mobility Model: AG nodes can be carried by machinery, people or objects. This can be modeled as random walk model. This random walk has series of steps in the sensing region R. If the initial position is (x0, y0), their position after time t is (xt, yt), then it follows Gaussian distribution [2].

3 Proposed Technique

The proposed technique for a wireless power supply for underground sensor networks works on the principle of coupled, resonant, magnetic induction similar to that described in [9]. Magnetic induction was proposed solely for wireless communication in [2]. However in this proposed system, magnetic induction will not only be used for wireless communication but also for wireless power transfer (WPT), similar to that of power line communication in wired networks [10]. Magnetic induction technique appears ideal to support communication [2, 4] and power transfer in underground sensor networks, as magnetic fields remain much less affected by the unpredictable and changing nature of underground dynamic soil conditions, overtime, than other communication techniques. In order to achieve the most efficient power transfer a strong resonant coupling is required between the source and destination.

3.1 Designing of Coils

Coils are wound in such a way that it maximizes the magnetic field strength and reduces losses due to self inductance. Coils are wound using bifilar technique which reduces losses due to self inductance and maximizes power transfer. Transmitter, receiver and repeater coils are made identical so that impedance matching can be achieved which ensures maximum power transfer.

$$H = \frac{Ni}{L} \tag{1}$$

where H is Magnetic field strength, N is number of turns of coil, i is the current through the coil, L is the Length of the coil. Here as the number of turns in a coil increases, magnetic field strength increases. Magnetic field strength is inversely proportional to the diameter of the coil. So for maximum possible value of H, the ratio N/L should be as high as possible. This says that diameter of the coil should be as minimum as possible.

The inductance and capacitance can be determined and used for calculating resonant frequency theoretically. The inductance is given by [11].

$$L = \frac{10\pi\mu N^2 R^2}{9R + 10l} \qquad (2)$$

where L is the inductance, μ is the permeability of the medium, N is the number of turns of the coil, R is the radius of the col core used, l is the length of the coil. The capacitance of an helical coil with length l and number of turns N is given by [11].

$$C = \frac{4\mu\varepsilon r^2 N^2 k}{L \cos^2 \Psi} \qquad (3)$$

where C is the capacitance, μ is the permeability of the medium, r is the radius of the coil, k is proportionality constant, L is the length of the coil, ψ is the magnetic susceptibility of the coil.

Theoretical Resonant frequency: Resonance occurs at a particular Frequency when the reactance of the coil is equal to its resistance. At resonant frequency maximum power can be transferred, as reactance would be very low. Theoretically resonant frequency can be calculated by

$$F_r = \frac{1}{2\pi\sqrt{LC}} \qquad (4)$$

Here L is Inductance, C is capacitance, and Fr is resonant frequency of the coil. With the help of LCR meter, inductance and capacitance values of the coil are calculated. Measured L value was 500 µH and C value was 3.3 nF. After substituting these values for L and C in Eq. 3, the resonant frequency was found to be 246 kHz.

The system for concurrent wireless power supply and communication (as shown in Fig. 1) consists of different blocks namely: DC regulated power supply, PWM generator, gate driver circuit, H bridge circuit, communication signal block and transmitter coil.

4 Mathematical Modeling

4.1 Coils Used in the Proposed System

In the proposed system the coils used are shown in the Fig. 1. These coils are made up of 36 SWG (33 AWG) copper wires. 60 strands of copper wires are combined together and made as one conducting wire. This multi stranded wire is wound on a PVC(Poly vinyl chloride) tube with 6 cm diameter. Transmitting, relay and

Fig. 1 Block diagram of
proposes system

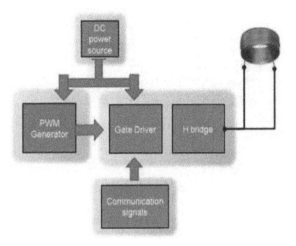

receiving coils are similar in the system in order to make the properties of both the
coils to be similar as much as possible.

4.2 Magnetic Field and Voltage Induced for the Coils Used in Proposed System

The equations for generated magnetic field and induced voltage for the coils used in
the present system can be derived from the basic equations of Biot-Savart's law and
Faraday's law. Due to multiple stranded coil the generated magnetic field on the
axis of the coil at a distance x is given as

$$B = \frac{\mu_0 I r^2 nN}{2(x^2 + r^2)^{3/2}} \tag{5}$$

We are using circular coils so the area of the coil is πr^2 for single turn and when
we consider for N turns and multiple strands then the voltage induced will be

$$V(t) = (B\omega \sin \omega t)nNA \cos \theta \tag{6}$$

5 Experimentation and Results

The experiments were designed for testing the feasibility of concurrent wireless
power supply and communication with underground sensor networks using the
principle of magnetic induction at resonance. The experiments were carried out in

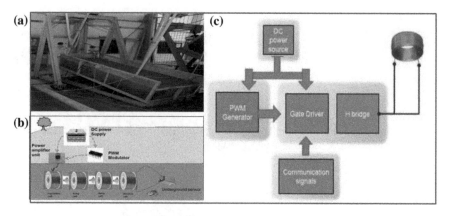

Fig. 2 Clock wise manner: **a** Test bed used to check the feasibility of a wireless power supply to underground sensor networks. **b** Wireless power transfer using resonant magnetic induction. **c** Block diagram of the proposed system

an artificially created test bed with soil. The image of the test bed which was used for conducting experiments is shown in Fig. 2a. The physical property of the soil (by volume) used is gravel −14 %, sand 61 %, silt 11 % and clay 14 %.

The coils used in these experiments were made up of copper and are hand wound. The coils were wound on a plastic core. An example view of the experimental setup is shown in Fig. 2b. The Fig. 2b shows the pictures taken by deploying the coils inside the test bed. The architecture block diagram is shown in Fig. 2c. Various blocks are DC power supply, Pulse Width Modulation (PWM) generator, gate driver, H Bridge, communication signals block, and coil (Fig. 3).

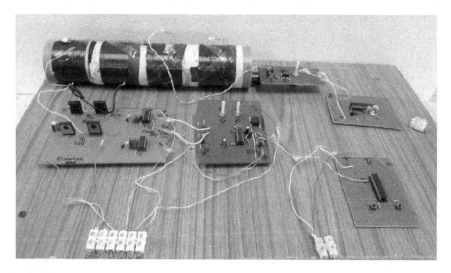

Fig. 3 Experimental setup

5.1 Experimental Setup

The experimental setup consists of PWM generator circuit, Power amplifier circuit, Communication transmitter block, communication receiver block, transmitter coil, receiver coil and relay coil. Communication transmitter block has ATMEGA 8L micro controller, which transmits a coded signal to the transmitter coil. Communication receiver coil which also has ATMEGA 8L micro controller receives the signal and decodes it.

5.2 Results

An input voltage of 50 V and current 0.01 A is given as input to the transmitter coil. The input power would be 500 mW. Figure 4 is plotted taking x axis as distance and y axis as output power. Underground testing is conducted with various moisture contents and by varying distances. The following experiment is conducted in soil type 1. Input power is fixed at 10 W, depth is fixed at 5 cm and at different moisture contents 0–5, 50, and 90 %. Figure 5a is plotted taking x axis as distance

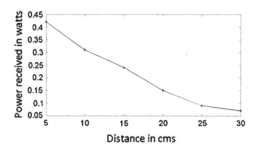

Fig. 4 Efficiency

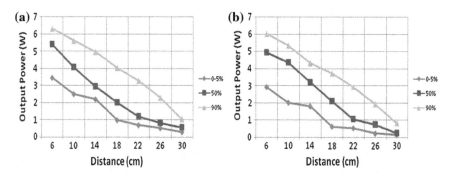

Fig. 5 Efficiency with varying moisture contents: Soil type-I **a** 5 cm depth, **b** 10 cm depth. Soil type-II **c** 5 cm depth, **d** 10 cm depth

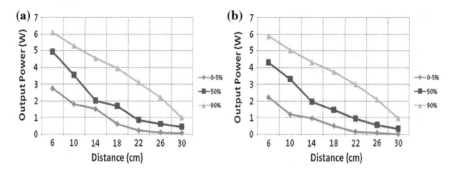

Fig. 6 Efficiency with varying moisture contents: Soil type-II. **a** 5 cm depth. **b** 10 cm depth

and y axis as output power. The following experiment is conducted in soil type 1. Input power is fixed at 10 W, depth is fixed at 10 cm and at different moisture contents 0–5, 50, and 90 %. Figure 5b is plotted taking x axis as distance and y axis as output power. The following experiment is conducted in soil type 2. Input power is fixed at 10 W, depth is fixed at 5 cm and at different moisture contents 0–5, 50, and 90 %. Figure 6a is plotted taking x axis as distance and y axis as output power. The following experiment is conducted in soil type 2. Input power is fixed at 10 W, depth is fixed at 10 cm and at different moisture contents 0–5, 50, and 90 %. Figure 6b is plotted taking x axis as distance and y axis as output power.

6 Conclusions

This paper presents the solution to the problems encountered in providing continuous power supply and communication for underground sensor networks thus enabling these networks to live much longer durations. This paper also describes the design of the coil required for using this technology. The experimental details and result analysis are also presented for various set of experiments conducted both in soil medium and air medium. It has been proved experimentally that this method is feasible for continuous power supply as well as communication with underground sensors.

7 Future Scope

Experiments can be conducted to check the variation of received power with respect to the depth at which the coils are deployed, for different moisture levels of the soil, for different soil types and also for different geometrical shapes of the coils. The performance of the system can also be further improved by including the impedance matching block. Reliability of the system depends on many aspects which could be refined. The performance of the system, by including relay coils in between the transmitter and receiver, and also using the coil arrays to make directional power transfer could also be investigated.

References

1. I.F. Akyildiz, E.P. Stuntebeck, Wireless underground sensor networks: research challenges. Ad Hoc Netw. **4**(6), 669–686 (2006)
2. Z. Sun, I.F. Akyildiz, Underground wireless communication using magnetic induction, in *IEEE International Conference Communications, ICC'09*
3. Z. Sun, et al., Magnetic induction communications for wireless underground sensor networks. IEEE Trans. Antennas Propag. (2010)
4. K. Kurs, A. Karalis, R. Moffatt, J.D. Joannopoulos, P. Fisher, M. Soljačić, Wireless power transfer via strongly coupled magnetic resonances. Sci. J. **317**, 83–86 (2007)
5. M. Kline, I. Izyumin, B. Boser, S. Sanders, Capacitive power transfer for contactless charging, in *2011 Twenty-Sixth Annual IEEE Applied Power Electronics Conference and Explosion*, 2011
6. T. Beh, Kato, et al., Wireless power transfer system via magnetic resonant coupling at fixed resonance frequency power transfer system based on impedance matching, EVS-25 Shenzhen, China, 2010
7. S.L. Ho, J. Wang, W.N. Fu, M. Sun, A comparative study between novel witricity and traditional inductive magnetic coupling in wireless charging. IEEE Trans. Magnet. (2011)
8. K.A. Unnikrishna Menon, V. Vikas, B. Hariharan, Wireless power transfer to underground sensors using resonant magnetic induction, in *WOCN*, 2013
9. S.F. Martin, S.H.B. Livi, J. Wang, The cancellation of magnetic flux. II. In a decaying active region. Aust. J. Phys CSIRO (1985)
10. C.P. Slichter, *Principles of Magnetic Resonance* (Springer, New York, 1990)
11. H.C Ferreira, H.M Grové, O. Hooijen, A.J. Han Vinck, *Powerline Communication* (Wiley Online Library, 2001)

Load Flow Analysis of Uncertain Power System Through Affine Arithmetic

Yoseph Mekonnen Abebe, Mallikarjuna Rao Pasumarthi and Gopichand Naik Mudavath

Abstract On this paper a novel load flow analysis using complex affine arithmetic (AA) based on Gauss-Seidel method for uncertain system is proposed. The Gauss-Seidel algorithm is used to find the uncertainty in each bus, which is the partial deviation value of the buses. The proposed algorithm is applied on an IEEE-14, 30 and 57 bus test systems. For comparison purpose a probabilistic load flow analysis based on Monte Carlo method is used. The proposed method is tested for different uncertainty level and in all the test cases AA based method is faster in convergence and gives slightly conservative bound than the probabilistic Monte Carlo approach.

Keywords Affine arithmetic · Conservative · Gauss-Seidel · Load flow analysis · Monte Carlo · Uncertainity

1 Introduction

The two mostly commonly known range analysis mechanisms are interval arithmetic (IA) and affine arithmetic (AA). IA was formalized by Moore in 1960 in order to estimate numerical errors while computing in a machine. It is based on the principle of representing a number in a closed interval with a lower and upper bound than approximating it in a floating point form. As a result, the basic principle of representing a real number is extended to deal with IA [1]. Though IA has been

Y.M. Abebe (✉) · M.R. Pasumarthi · G.N. Mudavath
Department of Electrical Engineering, College of Engineering (A), Andhra University,
Visakhapatnam 530 000, A.P., India
e-mail: yosephgod@gmail.com

M.R. Pasumarthi
e-mail: electricalprofessor@gmail.com

G.N. Mudavath
e-mail: gopi_905@yahoo.co.in

© Springer India 2016
S.C. Satapathy et al. (eds.), *Microelectronics, Electromagnetics
and Telecommunications*, Lecture Notes in Electrical Engineering 372,
DOI 10.1007/978-81-322-2728-1_20

playing a major role in range analysis, it suffers from dependency problems. As a result another mechanism which can address such a problem is needed and AA is invented.

Sources of error during manipulation may be external to the system like less precise and missing input data or a mathematical model which is uncertain by itself. Truncation and round off errors are categorized under internal errors. AA is first formalized and introduced in 1993 by Comba and Stolfi as a basic tool to deal with both internal and external sources of errors and other shortcomings of IA [2–4].

Though IA solves a lot of problem associated with fixed point calculation, its dependency problem is high. Dependency problem is simply defined as, having a different result for the same equation with different expressions. Since AA keeps the correlation between computed and input variables and exploit it at each operation, the dependency issue is not a problem. Due to all the aforementioned advantages AA is better than IA in dealing with uncertain system or function. Most application that has been done with IA can be done with AA ones the algorithm is formulated. AA mostly has been applied for computer graphics since its invention. Now a day's AA based reliable algorithm is also formulated for power flow analysis with uncertain data. A generalized Newton-Raphson method based load flow analysis using AA gives a conservative result in the worst case scenario than probabilistic Monte Carlo approaches [5, 6].

In [7] extension of AA is introduced to make it more applicable for different uses. The proposed extension is based on a general quadratic form. By using a branch and bound mechanism it is possible to find the bounds of a given function more reliably than before. The authors presented a new efficient inclusion function based on AA and they used Ichida-Fujii algorithm to solve global unconstrained minimization problem.

In wind and solar generation sources, which has a variable output in hourly basis, the uncertainty in load and generators is high. Voltage stability study should be applied in ordered to deliver reliable power to the utility. In [8] AA is applied to deal with such issue and the result is compared with a known Monte Carlo approach. The study is conducted on different bus test cases and it confirms the advantage of AA over Monte Carlo approach in reducing computational burden and getting a better results.

The effective use of genetic algorithm with AA helps to find the worst case circuit tolerance when the parameters are highly vernalable to uncertainty. Genetic algorithm is used to minimize underestimation error while AA is used to deal with over estimation problem. The joint use of genetic algorithm and AA provides a better result than traditional probabilistic methods for circuit tolerance analysis [9].

The problem in accuracy of the result in IA method, when there is linearization process, boosts the popularity of AA in power flow analysis. Because of its ability to consider all sorts of uncertainties, AA based algorithm provides an output which is more conservative than the probabilistic Monte Carlo approach [10].

Most of the load flow problem done in AA is based on Newton-Raphson method which has a lot of non affine operations resulting a high memory usage and computational burden [10].

In this paper a Gauss-Seidel analysis using complex AA is used to find the uncertainty values while the central values of voltage and angle are found using Newton-Raphson method. The proposed method is tested on an IEEE-14, 30 and 57 bus systems, and the result is compared with Monte Carlo method.

The remaining part of this paper is organized as follows: In Sect. 2, AA principle of operation is dealt. In Sect. 3 AA based Gauss-Seidel algorithm for power flow analysis is proposed. Section 4, is dedicated for result and discussion of the proposed method. At the end the conclusions is made in Sect. 5.

2 Generalized AA Principles

AA is simply an extension of interval mathematics. It is mathematically represented by (1).

$$\hat{x} = x_0 + \sum_{i=1}^{n} x_i \varepsilon_i. \tag{1}$$

The noise symbol 'ε' values lies between the interval $[-1, 1]$. The terms x_0 and x_i, are complex numbers and named central and partial deviation terms respectively. It is customary to covert IA to AA and vise versa. The interval to affine and affine to interval conversion is formulated in (2) and (3) respectively.

$$X = [a, b] \rightarrow \hat{x} = \frac{b+a}{2} + \frac{b-a}{2} \varepsilon_k \tag{2}$$

where 'a' and 'b' are the lower and upper bound values respectively. If the variables are not under the same condition, each conversion has different index value of k. In practical scenario, if two resistors are made from the same material with the same tolerance value they will share the same noise variables under the same room temperature assuming other conditions are under control [1].

$$\hat{x} = \frac{b+a}{2} + \frac{b-a}{2} \varepsilon_k \rightarrow X = x_0 + \sum_{1}^{n} x_i[-11] \tag{3}$$

Conversion from interval to affine form results one symbolic variable at a time. The result can be converted back to give the interval value. But, conversion from affine to interval is not reversible if the affine form has more than one symbolic variable. Since it loses its affine property there is no need of affine to interval conversion in the middle of the analysis. The conversion can be done at the end of the analysis to compare the result with other self validated mechanisms.

The following example shows the conversion of AA to IA and vice versa. Let $\hat{y} = 3 + 0.5\varepsilon_1 + 0.5\varepsilon_2$. Using (3) the interval value becomes $Y = [2, 4]$. Converting the interval result back to affine form using (2) gives $\hat{y} = 3 + 1\varepsilon_1$. The two affine forms are totally different confirming the irreversible nature of (3).

As in IA, all standard operations (addition, subtraction, multiplication and division) and others (square root, power, logarithm, trigonometry etc.) are redefined for affine forms. Operation in AA is divided into affine operations and non affine operations. Generally AA for both real and complex number is dealt in detain in [11].

2.1 Affine Operations

Those operations which do not approximate the result fall under this category. The two main affine operations are addition and subtraction. Let us take two complex affine forms. Where, both the central terms and the partial deviation terms are complex numbers. For (7) and (8) α is considered as a constant number.

$$\hat{x} = x_0 + \sum_{i=1}^{n} x_i \varepsilon_i \tag{4}$$

$$\hat{y} = y_0 + \sum_{i=1}^{n} y_i \varepsilon_i \tag{5}$$

Addition and subtraction is defined in (6). Equations (7) and (8) shows the use of constant number with affine function. They have the same form for both real and complex functions [11, 12].

$$\hat{x} \pm \hat{y} = (x_0 \pm y_0) + \sum_{i=1}^{n} (x_i \pm y_i) \varepsilon_i \tag{6}$$

$$a \pm \hat{y} = (a \pm y_0) + \sum_{i=1}^{n} y_i \varepsilon_i \tag{7}$$

$$a\hat{y} = ay_0 + \sum_{i=1}^{n} ay_i \varepsilon_i \tag{8}$$

Affine operations are simple and easy just like any numerical analysis with basic operations. There is no addition of any noises variable in using affine operations.

2.2 Non Affine Operations

Those operations which do approximate the results fall under this category. The most known of all non affine operation is multiplication. It is formulated in (9) for a complex number. Mathematical expressions containing square, square root, trigonometric, logarithmic operations and etc. are not simply found by applying simple affine rules. In order to deal with such function an affine approximation is mandatory [12]. So far a general formula for non affine operation is developed for multiplication which is defined in (9). The others are found using affine approximation mechanisms.

$$\hat{x}\hat{y} = (x_0 y_0) + \sum_{n=1}^{n} (x_0 y_i + x_i y_0)\varepsilon_i$$
$$+ \left(\sum_{i=1}^{n} x_i \sum_{i=1}^{n} y_i\right)\varepsilon_{n+1} \tag{9}$$

The main problem and difficulties in AA is selecting the best affine approximation. There are three distinct affine approximations. Affine approximations, that minimize the maximum absolute errors or minimize the range is a matter of choice. Namely they are called Chebyshev, min range and interval approximation. Chebyshev approximation is a well developed one amongst all with many non trivial results and with a vast literature done on it. For univariate function $\hat{y}$ the final approximated affine form is expressed by (10). On this paper Chebyshev approximation is used to get a precise result even though it is complicated than min range and interval approximation.

$$\hat{z} = \alpha\hat{y} + \xi + \delta\varepsilon_n \tag{10}$$

The detail of Chebyshev approximation and how to find the coefficients and constants of (10) can be found in detail in [12].

3 AA Based Load Flow Analysis

The Gauss Seidel algorithm is used to find the uncertainty value of the buses. The initial central voltage ($V_{i,0}$) and angle($\delta_{i,0}$) for the uncertain system is found using Newton-Raphson algorithm at the center of the load and generator power. Then from center voltage and angle the interval is formulated based on the percent of uncertainty in the demand and generator power by adding the percent of uncertainty for maximum value and by subtracting for minimum value to and from the center respectively. For generator and slack bus the voltage magnitude for upper and lower limit is the same.

$$V_{I,i} = [V_{i,\min}, V_{i,\max}]$$
$$\delta_{I,i} = [\delta_{i,\min}, \delta_{i,\max}] \tag{11}$$

From (11) a complex initial voltage can be formed and become:

$$v_I = [v_{i,\min}, v_{i,\max}] \tag{12}$$

Using interval to affine conversion formula in (2), the conversion of (12) to affine form yields (13).

$$\hat{V}_i = V_{i,0} + \frac{V_i}{2}(\varepsilon_{p,i} + \varepsilon_{q,i}) \tag{13}$$

where $\hat{V}_i$ the affine form of the voltage, $V_{i,0}$ is the central value found from ordinary load flow analysis without uncertainty, which can also directly found from (12) based on (2). The partial deviation, V_i, is found from (12) using (2) and divided equally for both active and reactive uncertainty. The terms $\varepsilon_{p,i}$ and $\varepsilon_{q,i}$ are symbolic variables for active and reactive power respectively. Similarly the maximum form of resultant real and reactive power in interval form is written as in (14). Both the real and reactive powers are the deference of the generator and load power and made to be interval form using the percent of uncertainty initially assumed.

$$
\begin{aligned}
P_{I,i} &= [P_{i,\min}, P_{i,\max}] \\
Q_{I,i} &= [Q_{i,\min}, Q_{i,\max}]
\end{aligned}
\tag{14}
$$

where $P_{I,i}$ and $Q_{I,i}$ are the interval form of real and reactive power at bus i. The terms $P_{i,\min}$ and $P_{i,\max}$ are the minimum and the maximum resultant real powers respectively. Similarly, $Q_{i,\min}$ and $Q_{i,\max}$ are the minimum and maximum resultant reactive powers respectively. The conversion of (14) to affine form using (2) yields (15) and (16).

$$
\begin{bmatrix} \hat{P}_1 \\ \hat{P}_2 \\ \vdots \\ \vdots \\ \hat{P}_n \end{bmatrix}
=
\begin{bmatrix} P_{1,0} \\ P_{2,0} \\ \vdots \\ \vdots \\ P_{n,0} \end{bmatrix}
+
\begin{bmatrix} P_{1,1} & & & & \\ & P_{2,2} & & & \\ & & \ddots & & \\ & & & \ddots & \\ & & & & P_{n,n} \end{bmatrix}
\begin{bmatrix} \varepsilon_{p,1} \\ \varepsilon_{p,2} \\ \vdots \\ \vdots \\ \varepsilon_{p,n} \end{bmatrix}
\tag{15}
$$

where $\hat{P}_1 \ldots \hat{P}_n$, $P_{n,0} \ldots P_{n,0}$ and $P_{1,1} \ldots P_{n,n}$ are the real power affine form, central terms and the partial deviations respectively. As described in affine voltage $\varepsilon_{p,1} \ldots \varepsilon_{p,n}$ are symbolic variables for active powers. On the same way the reactive power is developed from (14) using (2) and yields (16).

In both (15) and (16) most of the elements except the diagonals are zero. This is mainly due to each bus powers are considered independently and represented by a unique symbolic variable. Since the sensitivity to power change in bus voltage is different, the symbolic variables for each bus must be unique. This is one of the advantages of AA over IA. The latter do not consider such issue and difficult to differentiate two intervals of the same value, whether they are of different bus results or not, only by looking at the values.

$$
\begin{bmatrix} \hat{Q}_1 \\ \hat{Q}_2 \\ \vdots \\ \vdots \\ \hat{Q}_n \end{bmatrix} = \begin{bmatrix} Q_{1,0} \\ Q_{2,0} \\ \vdots \\ \vdots \\ Q_{n,0} \end{bmatrix} + \begin{bmatrix} Q_{1,1} & & & \\ & Q_{2,2} & & \\ & & \ddots & \\ & & & \ddots \\ & & & & Q_{n,n} \end{bmatrix} \begin{bmatrix} \varepsilon_{q,1} \\ \varepsilon_{q,2} \\ \vdots \\ \vdots \\ \varepsilon_{q,n} \end{bmatrix} \tag{16}
$$

where $\hat{Q}_1 \ldots \hat{Q}_n$, $Q_{n,0} \ldots Q_{n,0}$ and $Q_{1,1} \ldots Q_{n,n}$ are the reactive power affine form, central terms and the partial deviations respectively. The terms, $\varepsilon_{q,1} \ldots \varepsilon_{q,n}$ are symbolic variables for reactive powers. After the affine form of voltage and power is formulated the general Gauss-Seidel load flow analysis is used to find the uncertainty in the voltage.

$$
\hat{V}_{A,n} = \frac{1}{Y_{n,n}} \left[\frac{\hat{P}_n - j\hat{Q}_n}{\hat{V}_n^*} - \sum_{\substack{k=1 \\ k \neq n}}^{N} Y_{n,k} \hat{V}_k \right] \tag{17}
$$

where $\hat{V}_{A,n}$ is the affine form voltage result and $\hat{V}_n^*, \hat{V}_k$ are the conjugated affine voltage and normal affine voltage at nth and kth bus initially found from (13) respectively. The constants, $Y_{n,n}$ and $Y_{n,k}$ are the admittance at the nth bus and nth bus with kth bus respectively. Using the input in (13), (15), (16) and applying affine and non affine operation (1–10) on (17) yields (18). The central term is found from base power flow analysis at mid of the uncertain powers as mentioned earlier.

$$
\begin{bmatrix} \hat{V}_{A,1} \\ \hat{V}_{A,2} \\ \vdots \\ \vdots \\ \vdots \\ \vdots \\ \hat{V}_{A,n} \end{bmatrix} = \begin{bmatrix} V_{1,0} \\ V_{2,0} \\ \vdots \\ \vdots \\ \vdots \\ \vdots \\ V_{n,0} \end{bmatrix} + \begin{bmatrix} V_{1,1}^p & \cdots & V_{1,n}^p & V_{1,1}^q & \cdots & V_{1,n}^q \\ V_{2,1}^p & \cdots & V_{2,n}^p & V_{2,1}^q & \cdots & V_{2,n}^q \\ \vdots & \vdots & \vdots & \vdots & \vdots & \vdots \\ \vdots & \vdots & \vdots & \vdots & \vdots & \vdots \\ \vdots & \vdots & \vdots & \vdots & \vdots & \vdots \\ \vdots & \vdots & \vdots & \vdots & \vdots & \vdots \\ V_{n,1}^p & \cdots & V_{n,n}^p & V_{n,1}^q & \cdots & V_{n,n}^q \end{bmatrix} \begin{bmatrix} \varepsilon_{p,1} \\ \varepsilon_{p,2} \\ \vdots \\ \varepsilon_{p,n} \\ \varepsilon_{q,1} \\ \varepsilon_{q,2} \\ \vdots \\ \varepsilon_{q,n} \end{bmatrix} + \begin{bmatrix} V_{1,1} & \cdots & V_{1,k} \\ V_{2,1} & \cdots & V_{2,k} \\ \vdots & \vdots & \vdots \\ \vdots & \vdots & \vdots \\ \vdots & \vdots & \vdots \\ \vdots & \vdots & \vdots \\ V_{n,1} & \cdots & V_{n,k} \end{bmatrix} \begin{bmatrix} \varepsilon_{h,1} \\ \varepsilon_{h,2} \\ \vdots \\ \vdots \\ \vdots \\ \vdots \\ \varepsilon_{h,k} \end{bmatrix} \tag{18}
$$

where $\hat{V}_{A,n}$ is the affine form of the bus voltage as described before. The term $V_{n,0}$ is the central voltage value found initially applying base load flow analysis at the center of the active and reactive powers. The variables $V_{1,1}^p \ldots V_{n,n}^p$, $V_{1,1}^q \ldots V_{n,n}^q$ and

$V_{1,1} \ldots V_{n,k}$ are the partial deviations due to active power, reactive power and non affine operation respectively and found from (17) applying (1–16). Noise symbols, $\varepsilon_{p,n}$ $\varepsilon_{q,n}$ and $\varepsilon_{h,k}$ are symbolic variable for active power, reactive power and non affine operations respectively.

The next step is optimizing the error value within the limit. Due to inherent nature of affine approximation like multiplication in (9) optimization is mandatory [10]. This is mainly to limit the output voltage value within the maximum defined bound. In our case the bus voltage may become above the maximum limit if the partial deviations (error signals) are not optimized. For the optimization of error magnitude of voltage and angle within the limit, the magnitude and angle of error from (18) is taken separately. Mechanisms of solving optimization based power flow problems are intensively pointed in [6, 8, 10].

$$V_{e,n} = \sum_{j=1}^{n} |V_{n,j}^p| \varepsilon_{p,j} + \sum_{j=1}^{n} |V_{n,j}^q| \varepsilon_{q,j} + \sum_{j=1}^{k} |V_{nj}| \varepsilon_{hj} \tag{19}$$

$$\delta_{e,n} = \sum_{j=1}^{n} \delta_{n,j}^p \varepsilon_{p,j} + \sum_{j=1}^{n} \delta_{n,j}^q \varepsilon_{q,j} + \sum_{j=1}^{k} \delta_{n,j} \varepsilon_{h,j} \tag{20}$$

From (19) and (20) $V_{e,n}$ and $\delta_{e,n}$ are the voltage and angle partial deviation taken from the complex voltage in (18).

The maximum voltage limit in any power system network is a specified thing. As a result, the voltage error value in (20) must be within the limit between the center and the maximum limit of that bus at any percent of uncertainty in both directions. The main constraint for the optimization is then defined by (21).

$$t_n = \frac{\text{tol}}{2} (V_{lt,n} - V_{n,0}) \tag{21}$$

where $V_{lt,n}$ is the maximum defined voltage limit at each bus and '$V_{n,0}$' is the center voltage at any bus, and 'tol' represents percent of uncertainty diameter in which half of it represents the radius. In order for each voltage to be within the boundary the error value must be within the negative and positive values of t_n as written in (22). At any percent of uncertainty (21) guarantees the voltage at any bus is within the limit. The angle of bus voltage error can room with the wide tolerance value in both directions as defined in (23). This is mainly because of the limit of the voltage angle is in the whole circle and our initial guess for angle error is the radius of the tolerance in both direction.

$$
\begin{array}{c}
\text{min/max} \\
V_{e,n} \\
s.t \quad -t_n \leq V_{e,n} \leq t_n \\
-1 \leq \varepsilon_p \leq 1 \\
-1 \leq \varepsilon_q \leq 1 \\
-1 \leq \varepsilon_h \leq 1
\end{array} \tag{22}
$$

$$
\begin{array}{c}
\text{min/max} \\
\delta_{e,n} \\
s.t \quad \frac{-tol}{2} \leq \delta_{e,n} \leq \frac{tol}{2} \\
-1 \leq \varepsilon_p \leq 1 \\
-1 \leq \varepsilon_q \leq 1 \\
-1 \leq \varepsilon_h \leq 1
\end{array} \tag{23}
$$

The final voltage and angle at each bus become as in (24) and (25). The minimum and the maximum values in (24) and (25) are found from (22) and (23) respectively.

$$
V_n = V_{n,0} + [V_{e,n,\min} \ V_{e,n,\max}] \tag{24}
$$

$$
\delta_n = \delta_{n,0} + [\delta_{e,n,\min} \ \delta_{e,n,\max}] \tag{25}
$$

The value from (24) and (25) are used for the next iteration by applying interval to affine conversion in (3). Before next iteration starts checking the reactive power limit violation follow. If there is a violation, bus switching mechanism can be applied and the next iteration starts over. The iteration stops when convergence criterion is met.

4 Test Cases

An IEEE-14, 30 and 57 bus test cases are used in order to see the validity of the proposed method. Different percent uncertainty is applied. A traditional probabilistic power flow analysis for uncertain system based on Monte Carlo method is used for comparison. A Matlab programming environment is applied to analyze both methods and a multivariable linear programming solver is use to find the error values.

The Monte Carlo method is done by taking two thousand random variables of generator and demand power in the defined boundary. More than two thousand random variables are tested but do not have any significant change on the output results. As a result, after two thousand iterations the maximum and minimum value is selected by comparing the two thousand values of voltage and angle at each bus. The AA based algorithm proposed above converges in two iteration. In addition to

Fig. 1 IEEE-14 Bus Voltage
Magnitude of AA and Monte
Carlo

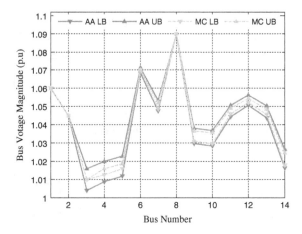

Fig. 2 IEEE-14 Bus Voltage
angle of AA and Monte Carlo

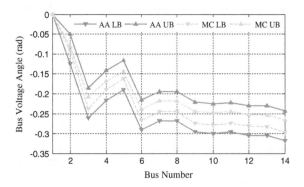

its fast convergence, its result as compared to Monte Carlo method is more conservative.

In all the figures, MC stands for Monte Carlo while AA stands for affine arithmetic respectively. The term LB and UB represents a lower and upper bound respectively. Figure 1 shows the voltage magnitude for IEEE-14 bus system for fifteen percent total uncertainty. Figure 2 shows the angle for the corresponding bus voltage. As seen from the figures the AA based output is slightly conservative than that of Monte Carlo based approach.

Figures 3 and 4 shows the voltage magnitude for different percent of uncertainty using the proposed AA approach and the classical Monte Carlo approach respectively for IEEE-14 bus system. Figures 3 and 4 indicates that when the percent of uncertainty become smaller and smaller for the generator and load power, the more the upper and lower bound become tight and marches to the center.

When the percent of uncertainty become zero the lower and the upper bound voltage become equal and gives the ordinary load flow result at the mid of the uncertain power or simply ordinary load flow analysis result without uncertainty.

Fig. 3 IEEE-14 Bus AA based Voltage for different percent of uncertainty

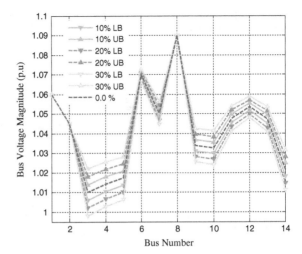

Fig. 4 IEEE-14 Bus Monte Carlo based Voltage for different percent of uncertainty

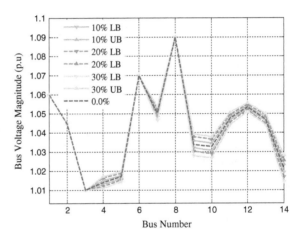

Figures 5 and 6 shows the voltage magnitude and angle for twenty five percent of uncertainty for IEEE-30 bus system respectively.

From Figs. 5 and 6 one can deduce that AA based approach has a conservative bound than Monte Carlo approach. This is mainly due to the inherent nature of AA to keep track of the correlation between variables to yield the worst case output.

Figures 7 and 8 shows the voltage magnitude for different percent of uncertainty using the proposed AA approach and the classical Monte Carlo approach respectively for IEEE-30 bus system. Whenever the percent of uncertainty decreases the voltage boundary marches to the center from both boundaries. When the percent of uncertainty is zero the load flow output become ordinary load flow result with single output at each bus confirming the accuracy of the proposed method.

Fig. 5 IEEE-30 Bus Voltage Magnitude of AA and Monte Carlo

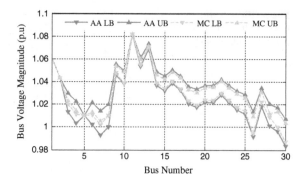

Fig. 6 IEEE-30 Bus Voltage angle of AA and Monte Carlo

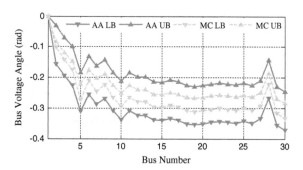

Fig. 7 IEEE-30 Bus AA based voltage magnitude for different percent of uncertainty

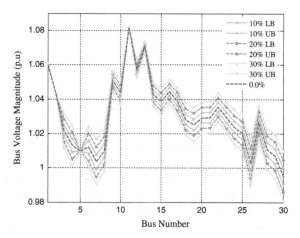

In all percent of uncertainty considered, AA is slightly conservative than Monte Carlo method. Finally when the percent of uncertainty is zero the two bounds become equal to the central voltage which can also be found using ordinary load flow analysis without uncertainty.

Fig. 8 IEEE-30 Bus Monte
Carlo based voltage value for
different percent of
uncertainty

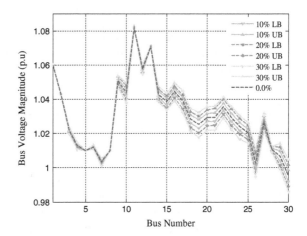

Table 1 Bus voltage uncertainty values for different percent of power uncertainty

Test systems	Power uncertainty tolerance (%)	Maximum bus voltage uncertainty magnitude (p.u)		Number of iterations
		AA	MC	AA
IEEE 14	±10	0.0040	0.0021	2
IEEE 30		0.0047	0.0026	2
IEEE 57		0.0095	0.0050	3
IEEE 14	±20	0.0080	0.0038	2
IEEE 30		0.0095	0.0074	2
IEEE 57		0.0190	0.0102	3
IEEE 14	±30	0.0120	0.0052	2
IEEE 30		0.0142	0.0081	2
IEEE 57		0.0285	0.0195	3

Table 1 shows the maximum bus voltage uncertainty magnitude for both AA and
MC methods for different power uncertainty. As mentioned earlier, two thousand
iterations are considered as the point of non varying output result for MC method.
From the table it can be seen that AA converges in two–three iterations depending
on the bus number.

5 Conclusion

In this paper a Gauss-seidel algorithm using complex AA is proposed to find the
uncertainty in bus voltage and angle while Newton-Raphson method is used to find
the center voltage. The proposed method is tested on an IEEE-14, 30 and 57 bus

test systems. A linear programming approach is applied to find the error value within the boundary. The test is performed for different percent of uncertainty on an IEEE-14, 30 and 57 bus systems. The proposed algorithm converges in two iterations for IEEE-14 and 30 and in three iterations for IEEE-57 bus systems. Such fast convergence makes it more advantageous than the two thousand iterations based Monte Carlo approach.

In all the study cases the results are compared with a known Monte Carlo probabilistic method. Beside fast convergence from the results, one can conclude that the proposed AA based algorithm is more conservative than Monte Carlo method. This is mainly due to the advantage of AA which considers input uncertainty, truncation, and round off errors. Moreover AA takes into consideration the correlations among variables in order to give a more conservative bound. In Monte Carlo approach the usual variation of the parameter is uniform and the assumption of mean value is zero. As a result of such consideration the worst case scenario in Monte Carlo approach is underestimated. The load flow results found from proposed method can help to control and forecast a huge power system in the worst case scenario more effectively than Monte Carlo results.

References

1. L.H. de Figueiredo, J. Stolfi, *Affine Arithmetic: Concepts and Applications* (Kluwer Academic Publisher Netherlands, 2004), vol. 37, pp. 147–158
2. J. Stolfi, L.H. de Figueiredo, An introduction to affine arithmetic. TEMA Tend. Mat. APL. Comput. **4**(3), 297–312 (2003)
3. W. Heindrich, Ph. Slusallek, H.-P. Seidel, Sampling procedural shaders using affine arithmetic. ACM Trans. Graph. **17**(3), 158–176 (1998)
4. F. Messine, Extensions of affine arithmetic: application to unconstrained global optimization. J. Univ. Comput. Sci. **8**, 992–1015 (2011)
5. A. Vaccaro, C. Cañizares, D. Villacci, An affine arithmetic-based methodology for reliable power flow analysis in the presence of data uncertainty. IEEE Trans. Power Syst. **25**(2), 624–632 (2010)
6. A. Dimitrovski, K. Tomsovic, A. Vaccaro, *Reliable Algorithms for Power Flow Analysis in the Presence of data Uncertainties*, Springer Series in Reliability Engineering, pp. 329–357 (2011)
7. F. Messine, A. Touhami, *A General Reliable Quadratic Form: An Extension of Affine Arithmetic* (Springer, New York, 2006), vol. 12, pp. 171–192
8. J. Muñoz, C. Cañizares, K. Bhattacharya, A. Vaccaro, Affine arithmetic based method for voltage stability assessment of power systems with intermittent generation sources, in *IREP Symposium-Bulk Power System Dynamics and Control—IX (IREP)*, Aug 2013, Rethymnon, Greece
9. N. Femia, G. Spangnuolo, True worst-case circuit tolerance analysis using genetic algorithms and AA. IEEE Trans. Circ. Syst.-I: Fundam. Theory Appl. **47**(9) (2000)
10. M. Pirnia, C.A. Cañizares, K. Bhattacharya, A. Vaccaro, An AA method to solve the stochastic power flow problem based on a mixed complementarily formulation, in *IEEE-PES General Meeting*, July 2012

11. G. Manson, Calculating frequency response function for uncertain systems using complex affine analysis. J. Sound Vibr. (2005)
12. O. Fryazinov, A. Pasko, P. Comninos, Fast Reliable Interrogation of Procedurally Defined Implicit Surfaces Using Extended Revised Affine Arithmetic, The National Center for Computer Animation, Bournemouth University, UK, July 2010

Advanced Parallel Structure Kalman Filter for Radar Applications

Seshagiri Prasad Teeparti, Chandra Bhushana Rao Kota,
Venkata Krishna Chaitanya Putrevu
and Koteswara Rao Sanagapallea

Abstract Normally in tracking applications, the target motion is usually modeled in Cartesian coordinates but, most sensors measure target parameters in polar coordinates. In this paper two contributions are considered in target tracking. One depends on position measurements and another one is on Doppler measurements. The position measurements are measured by taking the range and bearing (angle) of the target depending on the sensor location. Tracking the target Cartesian coordinates by using this range and bearing measurements is a nonlinear state estimation problem. To calculate the position measurements (range and angle), it is preferred to convert them to Cartesian coordinates by considering the linear form values. This is done, to avoid using nonlinear filters. This method is called as converted position measurement Kalman filter (CPMKF). In this paper another contribution is Doppler (range rate) measurement in target tracking systems. In this contribution the non-linear pseudo states are calculated. This method is called as Converted Doppler measurement Kalman filter (CDMKF). By considering these two methods a parallel filtering structure, called statically fused converted measurement Kalman filter (SF-CMKF) is proposed. The two methods are operated along with each other to construct the new state estimator SF-CMKF by a static estimator to obtain final state estimates.

Keywords Tracking · Kalman filter · Parallel filtering · Doppler measurement

S.P. Teeparti (✉) · C.B.R. Kota
Department of ECE, JNTUK-UCEV, Vizianagaram, A.P., India
e-mail: seshuprasad.teeparti@gmail.com

C.B.R. Kota
e-mail: cbraokota@yahoo.com

V.K.C. Putrevu
Department of ECE, GVP College of Engineering for Woman, Vizianagaram, A.P., India
e-mail: chaitanya_p_v_k@yahoo.co.in

K.R. Sanagapallea
Department of ECE, K L University, Vaddeswaram, Guntur, A.P., India
e-mail: rao.sk9@gmail.com

© Springer India 2016

233

S.C. Satapathy et al. (eds.), *Microelectronics, Electromagnetics and Telecommunications*, Lecture Notes in Electrical Engineering 372, DOI 10.1007/978-81-322-2728-1_21

1 Introduction

In much Doppler type of radars, the measurements are considered in the form of polar values, which gives the measurements like range, range rate (Doppler), and one or two angles of its position during moving. Then the Cartesian components errors in the converted measurements are correlated with each other is explored in [4, 5, 9, 11, 13, 14]. The other one is done by using extended Kalman filter (EKF) presented in [3, 6, 8, 10, 12, 13]. In this approach we have to consider the measurements of the target state estimation in a nonlinear fashion, which results the mixed coordinate filter [7, 8]. These measured terms results are considered to compare with the first two moment approximations which are presented here. The new converted measurement Kalman filter (CMKF) [14], is having estimation errors, which are compatible with the calculated covariance of the measured terms. The EKF is different from this method, because it is consistent only for small errors. So that the CMKF is having the correct covariance, it processes all the target measurements with a gain, which is nearly optimal and gives smaller errors compared with the EKF [3]. In the moderately accurate sensors, the EKF performs very poorly in tracking the target at long range for RMS azimuth error of 1.5° or more [10]. But the CMKF [12] is consistent for 10° RMS azimuth error also.

In this paper to rectify these shortcomings, a new method is proposed. In the proposed method, the use of the nonlinear recursive filtering methods is avoided during the processing of Doppler measurements [6]. In the first one, a pseudo state vector is considered, in which the existing converted Doppler measurements of the target are linear functions and they are constructed. These pseudo state vectors consist of the converted Doppler measurements and its derivatives [7, 8]. The pseudo state equations are derived from the measurements and proven to be linear in two commonly used target motion models. One model is the constant velocity (CV) and the other one is constant acceleration (CA) models. By using these converted Doppler measurement Kalman filter (CDMKF), is proposed to estimate the pseudo states [7]. This is also used to filter the noise in the converted Doppler measurements Kalman filter. Finally, the CDMKF is combined with the CPMKF [13, 14] to construct a new filter which gives a new state estimator called as statically fused converted measurement Kalman filters (SF-CMKF).

2 Problem Description

2.1 System Formulation

In Cartesian coordinates target's parameters are considered by depending on the conversion measurements of the target from polar coordinates to Cartesian. It is modeled as

$$X(k+1) = \Phi(k)X(k) + \Gamma(k)V(k) \tag{1}$$

where $X(k) = [x(k), y(k), \dot{x}(k), \dot{y}(k)]^T$, here $X(k) = R^n$ is the state vector consisting of target's position components and corresponding target's velocity components along x and y directions, respectively, at every time step k. If a moving target is considered, the state vector can be taken by other components such as acceleration. Here, $\Phi(k) \in R^{n \times n}$ is the target's state transition matrix, $v(k)$ is zero-mean Gaussian random process noise with covariance $Q(k)$, and $\Gamma(k)$ is noise gain matrix [2].

If we considered a 2D Doppler radar, which is assumed to report measurements of moving targets in polar coordinates, including range, range rate (Doppler) and angle presented in [3, 13, 14]. The measurement equation can be expressed as

$$\begin{aligned} z(k) &= [r_m(k), \theta_m(k), \dot{r}(k)]^T \\ &= h[X(k)] + w(k) = [r(k, \theta(k), \dot{r}(k))]^T + w(k) \end{aligned} \tag{2}$$

where

$$r(k) = \sqrt{x^2(k) + y^2(k)} \tag{3}$$

$$\theta(k) = \tan^{-1}[y(k)/x(k)] \tag{4}$$

$$\dot{r}(k) = [x(k)\dot{x}(k) + y(k)\dot{y}(k)]/\sqrt{x^2(k) + y^2(k)} \tag{5}$$

$$w(k) = \left[\tilde{r}(k), \tilde{\theta}(k), \tilde{\dot{r}}(k) \right] \tag{6}$$

Normally the measured range and bearing of the target [14] are considered by taking the true range r and bearing θ as

$$r_m = r + \tilde{r}, \quad \theta_m = \theta + \tilde{\theta} \tag{7}$$

2.2 Measurement Conversion Equations

The errors like range $\tilde{r}$ and bearing $\tilde{\theta}$ are taken to get independent with zero mean and standard deviations presented in [1]. These polar measurements are converted into Cartesian coordinate measurements by using the following conversion techniques [14]

$$x_m = r_m \cos\theta_m; \quad y_m = r_m \sin\theta_m; \tag{8}$$

The errors can be found by expanding these terms

$$
\begin{aligned}
x_m &= x + \tilde{x} = (r + \tilde{r})\cos(\theta + \tilde{\theta}); \\
y_m &= y + \tilde{y} = (r + \tilde{r})\sin(\theta + \tilde{\theta})
\end{aligned}
\tag{9}
$$

The mean error of Cartesian positions becomes

$$
\mu_t(r, \theta) = \begin{bmatrix} E[\tilde{x}|r, \theta] \\ E[\tilde{y}|r, \theta] \end{bmatrix} = \begin{bmatrix} r\cos\theta(e^{-\frac{\sigma_\theta^2}{2}} - 1) \\ r\sin\theta(e^{-\frac{\sigma_\theta^2}{2}} - 1) \end{bmatrix}
\tag{10}
$$

After some algebraic manipulation of the measurements, the elements of the targets converted measurement covariance [14] are given by

$$
\begin{aligned}
R_t^{11} = \mathrm{var}(\tilde{x}|r, \theta) &= r^2 e^{-\sigma_\theta^2}[cos^2\theta(\cosh(\sigma_\theta^2) - 1) + \sin^2\theta\sinh(\sigma_\theta^2)] \\
&+ \sigma_r^2 e^{-\sigma_\theta^2}[cos^2\theta\cosh\sigma_\theta^2 + \sin^2\theta\sinh(\sigma_\theta^2)]
\end{aligned}
\tag{11a}
$$

$$
\begin{aligned}
R_t^{22} = \mathrm{var}(\tilde{y}|r, \theta) &= r^2 e^{-\sigma_\theta^2}[\sin^2\theta(\cosh(\sigma_\theta^2) - 1) + \cos^2\theta\sinh(\sigma_\theta^2)] \\
&+ \sigma_r^2 e^{-\sigma_\theta^2}[\sin^2\theta\cosh\sigma_\theta^2 + \cos^2\theta\sinh(\sigma_\theta^2)]
\end{aligned}
\tag{11b}
$$

$$
R_t^{12} = \mathrm{var}(\tilde{x}, \tilde{y}|r, \theta) = \sin\theta\cos\theta e^{-2\sigma_\theta^2}[\sigma_r^2 + r^2(1 - e^{\sigma_\theta^2})]
\tag{11c}
$$

Equations (10) and (11a, 11b, 11c) are the expressions of the bias and covariance of the targets converted measurements. The converted measurements of the target have a significant bias for long range and large bearing error. The true bias and covariance of the measured parameter values depend on the true range and bearing. They are denoted as with elements (10) and with elements (11a, 11b, 11c), respectively.

A conversion of the Doppler measurements is also made in this paper to yield the converted Doppler measurements as [7]:

$$
\eta_c(k) = r_m(k)\dot{r}_m(k) = \eta(k) + \tilde{\eta}(k)
\tag{12}
$$

where $\eta(k)$ is the converted Doppler (i.e., the product of range and range rate), given by

$$
\eta(k) = x(k)\dot{x}(k) + y(k)\dot{y}(k)
\tag{13}
$$

and $\tilde{\eta}(k)$ is the error in the converted Doppler Measurement $\eta_c(k)$, [10].

The use of the zero-mean expressions of measured one cannot be taken for the bias of the target at long ranges with the bearing error and the covariance approximation (13) is poor. The expressions in (10) and (11a, 11b, 11c) cannot be used because; the fact is that they are conditioned on the target's true values of range and bearing [14]. But these are not available in practice. The results become useful, when the expected values of target true moments are evaluated with the measured position. The expected bias and covariance are examined as [5]:

$$E\left[\mu_t(r,\theta)\big|r_{m,\theta_m}\right] = \mu_a \tag{14}$$

$$E\left[R_t(r,\theta)\big|r_{m,\theta_m}\right] = R_a \tag{15}$$

Then the expected value of the target's true bias is considered with elements of its classical approximations and target's true covariance is considered with elements of the same, which are conditioned on the measured position. These are called as the target's average true bias and target's average true covariance. Expanding the expected bias and covariance using (1) and applying the trigonometric identities gives the mean (14) as [5]

$$\mu_a = \begin{bmatrix} r_m \cos\theta_m(e^{-\sigma_\theta^2} - e^{-\sigma_\theta^2/2}) \\ r_m \sin\theta_m(e^{-\sigma_\theta^2} - e^{-\sigma_\theta^2/2}) \end{bmatrix} \tag{16}$$

and the covariance as [5]:

$$R_a^{11} = r_m^2 e^{-2\sigma_\theta^2}[\cos^2\theta_m(\cosh 2\sigma_\theta^2 - \cosh\sigma_\theta^2) + \sin^2\theta_m(\sinh 2\sigma_\theta^2 - \sinh\sigma_\theta^2)]$$
$$+ \sigma_r^2 e^{-2\sigma_\theta^2}[\cos^2\theta_m(2\cosh 2\sigma_\theta^2 - \cosh\sigma_\theta^2)$$
$$+ \sin^2\theta_m(2\sinh 2 + \frac{2}{\theta} - \sinh\sigma_\theta^2)] \tag{17a}$$

$$R_a^{22} = r_m^2 e^{-2\sigma_\theta^2}[\sin^2\theta_m(\cosh 2\sigma_\theta^2 - \cosh\sigma_\theta^2) + \cos^2\theta_m(\sinh 2\sigma_\theta^2 - \sinh\sigma_\theta^2)]$$
$$+ \sigma_r^2 e^{-2\sigma_\theta^2}[\sin^2\theta_m(2\cosh 2\sigma_\theta^2 - \cosh\sigma_\theta^2)$$
$$+ \cos^2\theta_m(2\sinh 2\sigma_\theta^2 - \sinh\sigma_\theta^2)] \tag{17b}$$

$$R_a^{12} = \sin\theta_m \cos\theta_m e^{-4\sigma_\theta^2}[\sigma_r^2 + (r_m^2 + \sigma_r^2)(1 - e^{\sigma_\theta^2})] \tag{17c}$$

Note that the average covariance (17a, 17b, 17c) is larger compared to the covariance (11a, 11b, 11c), which is conditioned on the exact position; it gives the additional errors by evaluating it by the measured position [7]. This is difficult in showing the consistency later. The bias and increase in the covariance, is always significant for long ranges and also for large bearing errors. So the new polar-to-Cartesian conversion [8], is an unbiased consistent conversion [8], with the correction of the average bias which is taken, instead of (7), given by

$$z^c = \begin{bmatrix} x_m^c \\ y_m^c \end{bmatrix} = \begin{bmatrix} r_m \cos\theta_m \\ r_m \sin\theta_m \end{bmatrix} - \mu_a \tag{18}$$

where the elements of μ_a are taken from (16) and the average covariance of the converted measurements is Ra with elements (17a, 17b, 17c).

Similarly, one can get the bias and variance of the converted Doppler measurements as [8]

$$\mu_n(k) = \rho \sigma_r \sigma_{\dot{r}} \tag{19}$$

$$R^{\eta\eta}(k) = r_m^2(k)\sigma_{\dot{r}}^2 + \sigma_r^2 \dot{r}_m^2(k) + 3(1+\rho^2)\sigma_r^2\sigma_{\dot{r}}^2 + 2r_m(k)\dot{r}_m(k)\rho\sigma_r\sigma_{\dot{r}} \tag{20}$$

The debiased converted position measurements are given as

$$z_c^\eta(k) = \eta_c(k) - \mu^\eta(k) \tag{21}$$

The covariance between the converted position measurements and the converted Doppler measurements can be given as [8]

$$R^{p\eta}(k) = \begin{bmatrix} R^{x\eta}(k) \\ R^{y\eta}(k) \end{bmatrix} = \begin{bmatrix} \left[\sigma_r^2 \dot{r}_m(k) + r_m(k)\rho\sigma_r\sigma_{\dot{r}}\right] \cos\theta_m(k) e^{-\sigma_\theta^2} \\ \left[\sigma_r^2 \dot{r}_m(k) + r_m(k)\rho\sigma_r\sigma_{\dot{r}}\right] \sin\theta_m(k) e^{-\sigma_\theta^2} \end{bmatrix} \tag{22}$$

2.3 Converted Doppler Kalman Filter

Normally, nonlinear filtering methods like the EKF and UKF are taken to deal with Doppler measurements. Here, the nonlinear problem is solved by building a pseudo state vector, which is having a linear relationship with the already existing converted measurements of the target and by deriving the particular linear filtering equations. This shows the results of the CDMKF. In this particular section, the two commonly used target motion models are, one is nearly constant velocity (NCV) and the other one is nearly constant acceleration (NCA) models [2], are evaluated. Now, the pseudo state equations of derivatives are considered by taking the second order and third order derivatives are as zero [2].

$$\begin{bmatrix} \ddot{x}(k) \\ \ddot{y}(k) \end{bmatrix} = \begin{bmatrix} 0 \\ 0 \end{bmatrix} \tag{23}$$

for CV model and

$$\begin{bmatrix} \dddot{x}(k) \\ \dddot{y}(k) \end{bmatrix} = \begin{bmatrix} 0 \\ 0 \end{bmatrix} \tag{24}$$

for CA model.

The pseudo state vector of the dynamic system for a CV model and CA model is considered as [9, 10]

$$\eta(k) = \begin{bmatrix} \eta(k) \\ \dot{\eta}(k) \end{bmatrix} = c[X(k)] = \begin{bmatrix} x(k)\dot{x}(k) + y(k)\dot{y}(k) \\ \dot{x}^2(k) + \dot{y}^2(k) \end{bmatrix} \tag{25}$$

$$\eta(k) = \begin{bmatrix} \eta(k) \\ \dot{\eta}(k) \\ \ddot{\eta}(k) \\ \dddot{\eta}(k) \end{bmatrix} = c[X(k) = \begin{bmatrix} x(k)\dot{x}(k) + y(k)\dot{y}(k) \\ \dot{x}^2(k) + \dot{y}^2(k) + x(k)\ddot{x}(k) + y(k)\ddot{y}(k) \\ 3\dot{x}(k)\ddot{x}(k) + 3\dot{y}(k)\ddot{y}(k) \\ 3\ddot{x}^2(k) + 3\ddot{y}^2(k) \end{bmatrix} \tag{26}$$

The derivatives of the NCV and the NCA in Cartesian coordinates are considered by using zero-mean white noise [2]. Then the NCV and NCA are expressed as follows

$$\begin{bmatrix} x(k+1) \\ y(k+1) \\ \dot{x}(k+1) \\ \dot{y}(k+1) \end{bmatrix} = \begin{bmatrix} 1 & 0 & T & 0 \\ 0 & 1 & 0 & T \\ 0 & 0 & 1 & 0 \\ 0 & 0 & 0 & 1 \end{bmatrix} \begin{bmatrix} x(k) \\ y(k) \\ \dot{x}(k) \\ \dot{y}(k) \end{bmatrix} + \begin{bmatrix} T^2/2 & 0 \\ 0 & T^2/2 \\ T & 0 \\ 0 & T \end{bmatrix} \begin{bmatrix} v_x(k) \\ v_y(k) \end{bmatrix} \tag{27}$$

$$\begin{bmatrix} x(k+1) \\ y(k+1) \\ \dot{x}(k+1) \\ \dot{y}(k+1) \\ \ddot{x}(k+1) \\ \ddot{y}(k+1) \end{bmatrix} = \begin{bmatrix} 1 & 0 & T & 0 & T^2/2 & 0 \\ 0 & 1 & 0 & T & 0 & T^2/20 \\ 0 & 0 & 1 & 0 & T & 0 \\ 0 & 0 & 0 & 1 & 0 & T \\ 0 & 0 & 0 & 0 & 1 & 0 \\ 0 & 0 & 0 & 0 & 0 & 1 \end{bmatrix} \begin{bmatrix} x(k) \\ y(k) \\ \dot{x}(k) \\ \dot{y}(k) \\ \ddot{x}(k) \\ \ddot{y}(k) \end{bmatrix} + \begin{bmatrix} T^2/2 & 0 \\ 0 & T^2/2 \\ T & 0 \\ 0 & T \\ 1 & 0 \\ 0 & 1 \end{bmatrix} \begin{bmatrix} v_x(k) \\ v_y(k) \end{bmatrix} \tag{28}$$

Now by considering the mean of the squared noise as a known input, the state equation can be written as

$$\eta(k+1) = \Phi_\eta \eta(k) + Gu(k) + \Gamma_x v_x(k) + \Gamma_s v_s(k) \tag{29}$$

where for NCV model

$$\Phi_\eta = \begin{bmatrix} 1 & T \\ 0 & 1 \end{bmatrix}, \quad G = \Gamma_s = \begin{bmatrix} T^3/2 & 3T/2 \\ T^2 & T^2 \end{bmatrix}, \quad u(k) = E\left(\begin{bmatrix} v_x^2(k) \\ v_y^2(k) \end{bmatrix}\right) = \begin{bmatrix} q \\ q \end{bmatrix}$$

$$\Gamma_x(k) = \begin{bmatrix} T & 3T^2/2 \\ 0 & 2T \end{bmatrix}, \quad v_x(k) = X_\Gamma v(k) = \begin{bmatrix} x(k) & y(k) \\ \dot{x}(k) & \dot{y}(k) \end{bmatrix} \begin{bmatrix} v_x(k) \\ v_y(k) \end{bmatrix}$$

$$v_s(k) = \begin{bmatrix} v_x^2(k) - q \\ v_y^2(k) - q \end{bmatrix}$$

for NCA model

$$
\phi_\eta = \begin{bmatrix} 1 & T & T^2/2 & T^3/6 \\ 0 & 1 & T & T^2/2 \\ 0 & 0 & 1 & T \\ 0 & 0 & 0 & 3 \end{bmatrix}, \quad G = \Gamma_s = \begin{bmatrix} T^3/2 & T^3/2 \\ 3T^2/2 & 3T^2/2 \\ 3T & 3T \\ 3 & 3 \end{bmatrix}, \quad u(k) = E\left(\begin{bmatrix} v_x^2(k) \\ v_y^2(k) \end{bmatrix} \right) = \begin{bmatrix} q \\ q \end{bmatrix}
$$

$$
\Gamma_x = \begin{bmatrix} T & 3T^2/2 & T^3 \\ 1 & 3T & 3T^2 \\ 0 & 3 & 6T \\ 0 & 0 & 6 \end{bmatrix}, \quad v_x(k) = X_\Gamma v(k) = \begin{bmatrix} x(k) & y(k) \\ \dot{x}(k) & \dot{y}(k) \end{bmatrix} \begin{bmatrix} v_x(k) \\ v_y(k) \end{bmatrix},
$$

$$
v_s(k) = \begin{bmatrix} v_x^2(k) - q \\ v_y^2(k) - q \end{bmatrix}
$$

3 Statically Fused Converted Measurement Kalman Filters

3.1 Filtering Structure

The CDMKF provides a new method to exploit Doppler measurements. But the resulting pseudo states from the CDMKF are quadratic [7, 8], not linear, in Cartesian states. Additional processing is needed to extract the final target states from the pseudo states. The Cartesian states can be provided by the CPMKF, which is used along with the CDMKF, leading to a new tracking filtering approach, the SF-CMKF [13].

Figure 1 illustrates the structure of the SF-CMKF. The original sensor measurements (i.e., range, Doppler, and angle) are divided into two parts to be processed separately by two linear filters first.

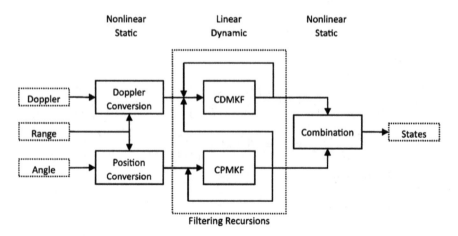

Fig. 1 Filtering of SF-CMKF

The prior mean of the state to be estimated is

$$\bar{x}(k+1) = E[x(k+1)/\hat{x}_p(k+1,k+1)] \tag{30}$$

Debiased converted measurement is

$$z(k+1) = \eta(k+1) - \tilde{\eta}(k+1,k+1) \tag{31}$$

The covariance between the states to be estimated and the measurement is

$$P_{XZ} = E\left[(x - \bar{x})(z - \bar{z})^T\right] \tag{32}$$

The covariance of the measurement is

$$P_{ZZ} = E\left[(z - \bar{z})(z - \bar{z})^T\right] \tag{33}$$

The static nonlinear estimation equation is obtained as

$$\hat{X} = \hat{x}_p + P_{xz}(P_{zz})^{-1}(\hat{\eta} - \bar{z}) \tag{34}$$

4 Simulation Results

Considering the target starting and moving with two trajectories which gives the effectiveness of the CDMKF and CPMKF methods in the forms nearly constant velocity trajectory and also nearly constant acceleration trajectory which are starting at (10, 10 km) and the target moves with a speed of 10 m/s heading to 60°. In the second scenario the acceleration is of 0.2 m²/2. The process noise of the target is assumed to be zero-mean white Gaussian noise with standard deviation 0.001 m/s². The sensor is located at origin (0, 0 km) and the sampling interval is T = 1 s. The standard deviations of target's range, azimuth, and Doppler measurements are taken as, $\sigma_r = 50$ m, $\sigma_\theta = 2.5°$, $\sigma_{\dot{r}} = 0.1$ m/s. The correlation coefficient of the target between range and bearing is taken as $\rho = 0$. Simulations are performed here, over 200 time steps with the 50 Monte Carlo experiments.

The motion of the targets and the root mean squared (RMS) errors of the following methods are considered and are shown in Figs. 2, 3, 4, and 5. The results are shown with Root Mean Square (RMS) error for the NCV and NCA trajectories, respectively. The effectiveness of the SF-CMKF as tracking filter is illustrated by comparing the performance of this method with that of the sequential nonlinear filtering method based on the sequential extended Kalman filter (SEKF) and the sequential filtering approach with the sequential unscented Kalman filter (UKF). The three tracking filters are having approximately the same RMSEs.

Fig. 2 RMS position error in NCV scenario

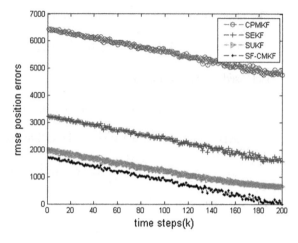

Fig. 3 RMS velocity error in NCV scenario

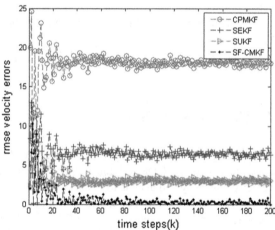

Fig. 4 RMS position error in NCA scenario

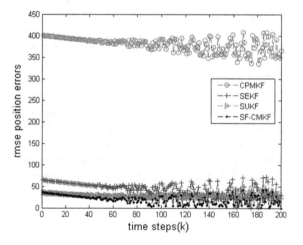

Fig. 5 RMS velocity error in NCA scenario

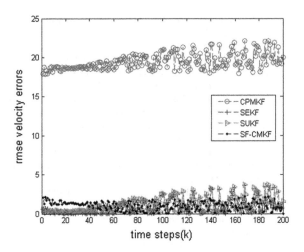

5 Conclusion

In this paper, the use of nonlinear recursive filtering approaches is avoided while processing the Doppler measurements. A linear filter, the converted Doppler measurement Kalman filter (CDMKF), is proposed to estimate the pseudo states and filter the noise in the converted Doppler measurements.

CDMKF can be used to operate along with the CPMKF to construct a new state estimator, statically fused converted measurement Kalman filter (SF-CMKF). Cartesian state and pseudo state estimates are produced by CPMKF and CDMKF, respectively, and are then combined by a static estimator to obtain final state estimates. The non-linearity of the pseudo states is quadratic and is handled by expanding the pseudo states up to the second term around the estimated states of the CPMKF.

References

1. Y. Bar-shalom, X.R. Li, T. Kirubarajan, *Estimation with Applications to Tracking and Navigation: Theory, Algorithms, and software* (Wiley, New York, 2001)
2. Y. Bar-Shalom, P.K. Willett, X. Tian, *Tacking and Data Fusion: A Handbook of Algorithms* (YBS Publishing, Storrs, CT, 2011)
3. D.F. Bizup, D.R. Brown, The over-extended Kalman filter—Don't use it, in *Proceedings of the 6th International Conference on Information Fusion*, Cairns, Queensland, Australia, July
4. S.V. Bordonaro, P. Willet, Y. Bar-Shalom, Tracking with converted position and Doppler measurements, in *Proceedings of SPIE Conference on Signal and Data Processing of Small Targets*, 2011, pp. 81370D-1-4
5. S.V. Bordonaro, P. Willet, Y. Bar-Shalom, Unbiased tracking with converted measurements, in *Proceedings of 2012 IEEE Radar Conference*, pp. 741–745

6. Y.P. Dai, et al., A target tracking algorithm with range rate under the color measurement environment, *Proceedings of the 38th SICE Annual Conference*, Morioka, Japan, July 1999, pp. 1145–1148

7. Z. Duan, C. Han, X.R. Li, Sequential nonlinear tracking filter considered with range-rate measurements in spherical coordinates, in *Proceedings of the 7th International Conference held on International Conference on Information Fusion*, 2004, pp. 599–605

8. Z. Duan, C. Han, Radar target tracking with range rate measurements in polar coordinates. J. Syst. Simul. **16**(12), 2860–2863 (2004). (in Chinese)

9. Z. Duan, C. Han, X. Li, R Comments on unbiased converted position and Doppler measurements for tracking. IEEE Trans. Aerosp. Electron. Syst. **40**(4), 1374–1377 (2004)

10. Z. Duan, et al., Sequential unscented Kalman filter for radar target tracking with range rate measurements, in *Proceedings of the 8th International Conference held on Information Fusion*, 2005, pp. 130–137

11. S.J. Julier, J.K. Uhlmann, A consistent, debiased method for converting between polar and Cartesian coordinate systems, in *Proceedings of the 1997 SPIE Conference on Acquisition, Tracking, and Pointing XI*, vol. 3086

12. D. Lerro, Y. Bar-shalon, Unbiased kalman filter using converted measurements versus EKF. IEEE Trans. Aerosp. Electron. Syst. **29**(3), 1015–1022 (1993)

13. W. Mei, Y. Bar-Shalom, Unbiased Kalman filter using converted measurements: revisit, in *Proceedings taken from SPIE Conference held on Signal and Data Processing in Small Targets*, 2009, pp. 7445–38

13. G. Zhou et al., Statically fused Converted Position and Doppler measurement Kalman filters. IEEE Trans. Aerosp. Electron. Syst. **50**(1), 300–318 (2014)

Performance of Fusion Algorithm for Active Sonar Target Detection in Underwater Acoustic Reverberation Environment

Cheepurupalli Ch. Naidu and E.S. Stalin

Abstract Classically, automatic detection of targets in active sonar system is addressed by a matched filter processing followed by a constant False Alarm Rate (CFAR) thresholding method. Even though, various CFAR techniques viz. CA CFAR, GO CFAR and SO CFAR etc. are available in literature, none of them alone is sufficient to eliminate the false echoes. In certain applications, such as active sonar where the probability of false alarm p_{fa} requirements are very stringent, the performance of CFAR alone cannot be used as detection criteria. Further, the choice of a particular CFAR algorithm is also a complex task, as the non-homogenous nature of the acoustic medium is difficult to predict. In this paper, a fusion algorithm is proposed for active sonar application where, in addition to CFAR technique, a support vector machines (SVM) based classification algorithm is also used to eliminate the false echoes. The performance of the algorithm is verified using practically measured data.

Keywords CA-CFAR · GO-CFAR · SO-CFAR and SVM

1 Introduction

Primary goal of a sonar receiver designer is to detect the presence of signal echo of an actual target which in most cases is buried in noise as well as reverberation. If we are assuming that the noise is additive white and reverberation is negligibly small, a replica correlator gives the optimum performance [1]. But, in scenarios where

C.Ch. Naidu (✉)
NSTL, Visakhapatnam, India
e-mail: challamnaidu@gmail.com

E.S. Stalin
NPOL, Kochin, India
e-mail: stalines@gmail.com

© Springer India 2016 245
S.C. Satapathy et al. (eds.), *Microelectronics, Electromagnetics and Telecommunications*, Lecture Notes in Electrical Engineering 372,
DOI 10.1007/978-81-322-2728-1_22

automatic detection of target is required, a thresholding scheme is to be incorporated to identify the actual echo. Since, the noise power is an ever varying quantity, a fixed threshold detection scheme will not control the false alarm rate. One widely accepted scheme is to fix the threshold adaptively by locally estimating the noise statistics. In these schemes, the threshold is computed on a cell by cell basis using the estimated noise power from nearby reference range cells. A CFAR is ensured in this scheme if the assumed statistics are not changed drastically [2]. Amongst all the CFAR schemes discussed in literature, Cell average (CA) CFAR is by far the most widely used CFAR scheme. The CA-CFAR processor is the optimum CFAR processor (maximizes detection probability) in a homogeneous background when the reference cells contain independent and identically distributed noise [2–7]. But the scheme produces excessive false alarms when reference cells are at reverberation edges. Hansen [8, 9] proposed Greatest Off (GO) CFAR to overcome the excessive false echoes at reverberation edges. But, both the above schemes fail in multi target scenario. Trunk [10] proposed Smallest Off (SO) CFAR to avoid false echoes in multi target scenario. Choosing a particular scheme depends on the apriori knowledge about the channel condition. But practically, it is a very difficult task to estimate channel conditions apriori. Moreover, there will be severe detection losses and high false alarms if the schemes are wrongly chosen [11]. Hence, none of the CFAR algorithms alone sufficient to control the false alarm rate as the channel conditions are difficult to predict practically.

In recent years, SVM based classification techniques are found to be significantly useful for a wide range of real-time applications, including handwritten digit recognition [12], object recognition [13], speaker identification [13], face detection in images [14], text categorization [15] etc. Taking these applications, a two class SVM is considered as the best classification algorithm. SVM maps the data into a high dimensional feature space, where it is linearly separable. Here, we use SVM classifier to differentiate the actual target echo from the false echoes due to reverberation and noise.

Section 2 briefly describes the threshold needed to achieve the CA-CFAR. Section 3 summarizes SVM classification for sample data with simulation. Section 4, describes the proposed fusion algorithm. Section 5 shows the experimental results of CA-CFAR detection, SVM classification and fusion algorithm. Section 6 concludes the paper.

2 CFAR Processing

Constant false alarm rate (CFAR) processors are useful for automatic detection of sonar/radar targets in background for which all parameters in the statistical distribution are not known. In CFAR systems, target decision is commonly performed using the sliding window technique. The data available in the reference window is fed to algorithm for the calculation of the decision threshold. The procedure for the computation of threshold is nearly same in all CFAR systems. The first step is to measure the mean noise plus reverberation power level. The second step is to scale estimated background power. The scaling factor is chosen based on the required false alarm rate. The resulting power level is directly used as the threshold.

The output of the matched filter over a number of range bins is collected. For each of the range cell under test (CUT), $n/2$ range cells preceding and succeeding CUT is considered as reference cells for background estimation. The cells immediately preceding and succeeding CUT are called masking cells and are avoided for background estimation to eliminate the effect of target echoes.

In CA CFAR processor the average value of the reference cells is taken as the estimated background power. The computation of the scaling factor to find the threshold is thoroughly discussed in [11]. As given in [11], for CA CFAR processor the relation between probability of false alarm p_{fa}, probability of detection p_d obtained by

$$p_d = \left[1 + \frac{T}{(1+S)}\right] - N,$$

where,

$$T = (p_{fa})^{-\frac{1}{N}} - 1$$

and N is the size of reference window, S is the SNR (Signal to Noise Ratio) and T is the fixed scale factor used for achieving a desired p_{fa} for a given N, when the total noise of the background is homogeneous.

3   SVMs for Linearly Separable Classes

Support Vector Machines are considered is a robust algorithm for classification [16]. The SVM concept is a classification algorithm based on statistical learning theory (SLT) and the main objective of this concept is to classify the data based on the different classes which are linearly separable by the decision boundary [17]. SVM usually has two stages, which are called learning stage and a forward (prediction) stage. In the learning stage the machine sees the training data sets and learns a rule to be able to separate data in two groups/classes according to data sets. In forward (prediction) stage the machine is asked to predict labels of new and unseen data sets.

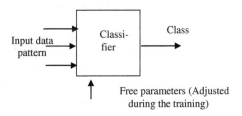

Given a set of training samples, $\{(p_j, q_j), j = 1, 2, 3 \ldots M\}$, where $p_i \in R^n$ and $q_j \in \{0, 1\}$ and q_j is two class classification method which belongs to class 0 and class 1. Assume that the classes are linearly separable then there exists $w \in R^n$, where w is weighting function and $b \in R$, b is the bias [18].

$$
\begin{aligned}
w^T p_j + b \succ 0, \forall_i \text{ such that } q_j = 1; \\
w^T p_j + b \prec 0, \forall_i \text{ such that } q_j = 0;
\end{aligned}
\tag{1}
$$

For a linear separable data set, there exists at least one linear classifier defined by the pair (w, b) which correctly classifies all training data sets. This linear classifier is represented by $H(w^T p_h + b) = 0$, where H indicates the hyper plane, if $H > 0$ then training data sets belongs to the class1 otherwise data sets belongs to the class0.

There are many algorithms available in literature to find out the separating hyper plane. In SVM, the separating hyper plane is computed by finding maximally separated support vectors from the data sets [17]. The effectiveness of the SVM based classification for different applications can be seen in [19–22]. The effectiveness of the SVM for low SNR sonar case is discussed in [23].

4   The Fusion Algorithm

The matched filter output contains the target information spread over a number of range bins. The maximum number of range bins over which the echo is spread can be estimated from the length of the expected target. In CFAR processing, in order to decide whether a range cell contains a target, the matched filter output is compared with the estimated background. For the estimation of background, for a cell under test, the nearby cells are not considered to avoid the effect of target. Hence, in this process, the pattern of the matched filter output when an echo is present is completely neglected.

In fact, the range cells nearby the cell under test, are rich in information, and hence can be used to classify the received signal as echo or reverberation. If we have a large collection of experimental data, it is possible to train a classification algorithm to find the weighting vector. Two class linear SVM method can be used for separating the valid data sets and invalid data sets. Here, linear SVM is chosen as the classification algorithm because of its successfulness in classifying similar kind of data. All the data sets which are considered for training contains either range cells having targets embedded in noise or reverberation signals along with noise. The target echo signals embedded in reverberation are avoided in order to hold the linear separable condition. The CFAR algorithm can effectively detect targets embedded in noise, but it fails in non homogenous conditions such as multiple targets and reverberation. On the other hand, the classification algorithm performs better in conditions where CFAR fails, but gives a less superior performance to CFAR in homogenous conditions. In the fusion algorithm, the goodness of both

CFAR as well as classification algorithm is combined. Here, the decision to declare a target is taken if and only if the range cell is passed in both CFAR and classification algorithm.

5 Experimental Results

A sequence of experiment in mono-static condition is conducted to verify the effectiveness of the algorithm. Different pings of Linearly Modulated Signals (LFM) are transmitted from a moving platform against a moving target approximately 80 m long. Transmission pings of duration 50 and 100 ms are used

Table 1 Performance comparison of SVM, CA-CFAR and fusion algorithm

	Detection of valid target (%)	False detection of invalid target (%)
SVM	60.60	4.5
CFAR	59.09	3.03
Fusion	51.51	0

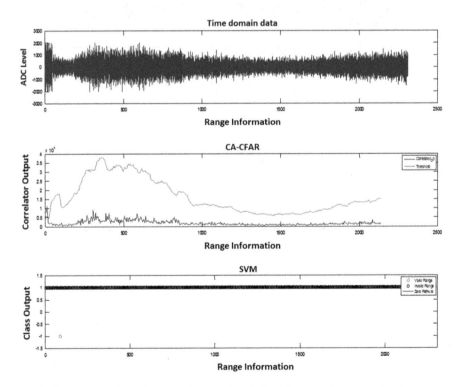

Fig. 1 Performance of SVM and CA-CFAR where in both the cases target is not declared

250   C.Ch. Naidu and E.S. Stalin

throughout the experiment. Different geometries of transmitter and target are also considered to get different aspect combinations. All the data collected during the trial is recorded. The recorded data is passed through a matched filter to get the row data for processing. The range cells are manually labeled as Class1 (actual target echo) and Class2 (reverberation signal). This is not a complex task since geometry of the transmitter target position is known apriori.

From the data sets, 100 Class1 data sets and 100 Class2 data sets are chosen randomly as the training set for SVM. Using the training set the weight vector and bias are computed. The weight vector and bias is used to verify the performance on the remaining data set. The same test data set is used to verify the performance of CFAR algorithm. Further, the results of fusion method is also computed and tabulated. Table 1 indicates the performance comparison of SVM, CA-CFAR and Fusion algorithms. Different scenarios to illustrate the performance of CFAR, SVM and fusion algorithm are shown in Figs. 1, 2, 3 and 4.

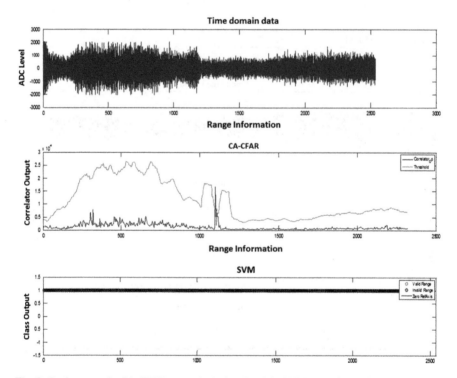

Fig. 2 Performance in CA-CFAR target is declared and in SVM target is not declared

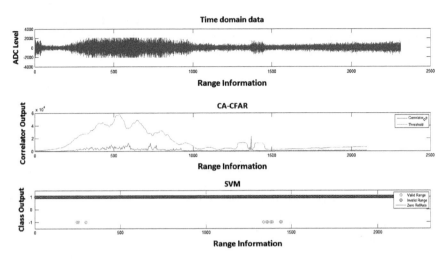

Fig. 3 Performance in both CA-CFAR and SVM target is declared

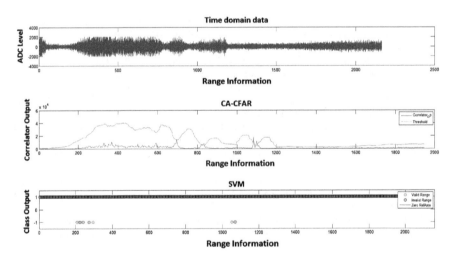

Fig. 4 Performance in CA-CFAR three targets are declared where as in SVM only one target is declared

6 Conclusion

In this paper a fusion algorithm based on classical CFAR processing and SVM classification method is adopted for discriminating between the valid target echoes and false echoes due to reverberation. The CFAR method performs better than the SVM method in homogeneous case whereas its performance diminishes in non-homogeneous condition. The validity of the echoes in CFAR processing is further verified by an SVM method and it is seen that the fusion algorithm outperforms CFAR processing in all the cases.

References

1. T.H. Glisson, C.I. Black, A.P. Sage, On sonar signal analysis. IEEE Trans. Aerosp. Electron. Syst. **AES**-6, 37–50 (1970)
2. B.O. Steenson, Detection performance of a mean-level threshold. IEEE Trans. Aerosp. Electron. Syst. **AES**-4, 529–534 (1968)
3. G.M. Dillard, Mean-level detection of nonfluctuating signals. IEEE Trans. Aerosp. Electron. Syst. **AES**-10, 795–799 (1974)
4. R.L. Mitchell, J.F. Walker, Recursive methods for computing detection probabilities. IEEE Trans. Aerosp. Electron. Syst. **AES**-7, 671–676 (1971)
5. R. Nitzberg, Analysis of the arithmetic mean CFAR normalizer for fluctuating targets. IEEE Trans. Aerosp. Electron. Syst. **AES**-14, 44–47 (1978)
6. J.D. Moore, N.B. Lawrence, Comparison of two CFAR methods used with square law detection of Swirling I targets, in *Proceedings of the IEEE International Radar Conference*, pp. 403–409 (1980)
7. M. Weiss, Analysis of some modified cell-averaging CFAR processors, in multiple-target situations. IEEE Trans. Aerosp. Electron. Syst. **AES**-18, 102–113 (1982)
8. V.G. Hansen, Constant false alarm rate processing in search radars, in *Proceedings of the IEEE 1973 International Radar Conference*, London, pp. 325–332 (1973)
9. V.G. Hansen, J.H. Sawyers, Detestability loss due to greatest of selection in a cell averaging CFAR. IEEE Trans. Aerosp. Electron. Syst. **AES**-16, 115–118 (1980)
10. G.V. Trunk, Range resolution of targets using automatic detectors. IEEE Trans. Aerosp. Electron. Syst. **AES**-14, 750–755 (1978)
11. P.P. Gandhi, S.A. Kassam, Analysis of CFAR processors in non-homogeneous Background. IEEE Trans. Aerosp. Electron. Syst. **AES**-24, 427–455 (1988)
12. M. Pontil, A. Verri, Support vector machines for 3-D object recognition. IEEE Trans. Pattern Anal. Mach. Intel. **20**, 637–646 (1998)
13. V. Wan, W.M. Campbell, Support vector machines for speaker verification and identification, in *Proceedings of IEEE Workshop Neural Networks for Signal Processing*, Sydney, Australia, Dec 2000, pp. 775–784
14. E. Osuna, R. Freund, F. Girosi, Training support vector machines: application to face detection, in *Proceedings of Computer Vision and Pattern Recognition*, Puerto Rico, 1997, pp. 130–136
15. T. Joachims, Transductive inference for text classification using support vector machines, in *Presented at the International Conference on Machine Learning*, Slovenia, June 1999
16. C.J. Burges, A tutorial on support vector machines for pattern recognition. Knowl. Discov. Data Mining **2**, 121–167 (1998)
17. T. Evgenuiu, M. Pontil, Statistical Learning Theory: A Primer (1998)
18. B. Scholkopf, S. Kah-Kay, C.J. Burges, F. Girosi, P. Niyogi, T. Poggio, V. Vapnik, Comparing support vector machines with Gaussian kernels to radial basis function classifiers. IEEE Trans. Signal Process. **45**, 2758–2765 (1997)
19. M. Ning, S.C. Chin, False alarm reduction by LS-SVM for manmade object detection from side scan sonar images. Senior Member IEEE, Member IEEE
20. D. He, H. Leung, CFAR Intrusion Detection Method Based on Support Vector Machine Prediction

21. Department of Electrical and Computer Engineering University of Calgary 2500 University Drive, NW Calgary, Alberta, Canada, T2N IN4
22. S.S. Salankar, B.M. Patre, SVM Based Model as an Optimal Classifier for the Classification of Sonar Signals
23. S. Jeong, S.-W. Ban, S. Choi, D. Lee, M. Lee, Surface Ship-Wake Detection Using Active Sonar and One-Class Support Vector Machine Sungmoon

Investigation of Optimum Phase Sequence for Reduction of PAPR Using SLM in OFDM System

Srinu Pyla, K. Padma Raju and N. Balasubrahmanyam

Abstract Orthogonal Frequency Division Multiplexing (OFDM) is able to mitigate the detrimental effects of multipath fading but, OFDM signal suffers from high peak to average power ratio (PAPR). Increase in PAPR, decrease the power amplifier efficiency otherwise leads to in-band distortion and out of band radiation due to signal clipping and spectral broadening respectively. There are many techniques to reduce PAPR and Selective Level Mapping (SLM) is a particularly promising technique. In this technique, array multiplication of data sequence with phase sequence reduces PAPR. The combination with the minimum PAPR is considered for transmission. There are many phase sequences for reduction of PAPR such as Riemann, Rudin Shapiro, Chaotic, Chu, Pseudorandom, Hadamard and Novel phase sequences. The selection of phase sequence is very crucial and it depends on PAPR improvement and simplicity in the recovery of signal. In this paper PAPR improvement for various phase sequences and digital modulation schemes is compared and a new phase sequence is developed.

Keywords OFDM · PAPR · Selective level mapping · Chaotic sequence · Riemann sequence · Chu sequence · Pseudorandom sequence · Rudin–Shapiro sequence · Modified sequence

1 Introduction

Orthogonal Frequency Division Multiplexing (OFDM) is a technique in which high rate serial data is converted into low rate parallel data and increases symbol duration. International standards using OFDM in wireless LAN, WiMAX, Mobile broadband

S. Pyla (✉) · N. Balasubrahmanyam
Department of ECE, GVP College of Engineering, Vishakhapatnam, India
e-mail: srinupyla@gvpce.ac.in

N. Balasubrahmanyam
e-mail: gvpcoe_ece_hod@gvpce.ac.in

K. Padma Raju
Department of ECE, J N T University, Kakinada, India
e-mail: kpr.ece.kkd@jntukakinada.edu.in

© Springer India 2016

255

S.C. Satapathy et al. (eds.), *Microelectronics, Electromagnetics and Telecommunications*, Lecture Notes in Electrical Engineering 372, DOI 10.1007/978-81-322-2728-1_23

wireless access (MBWA) and Broadcasting Radio Access Network (BRAN) committees [1, 2]. OFDM provides multipath fading and impulse noise free transmission and minimizes complexity of equalizers (single tap equalizer is sufficient). Hardware complexity of OFDM implementation is greatly reduced by using Fast Fourier Transform (FFT). OFDM suffers from Inter carrier interference (ICI) due to frequency offsets and increase in Peak to Average Power Ratio (PAPR) due to coherent addition of subcarriers [1, 3]. Increase in PAPR causes in band distortion and out of band radiation due to clipping and spectral broadening respectively [4].

There are many techniques to reduce PAPR such as amplitude clipping, recursive clipping and filtering (RCF), coding, tone reservation (TR), tone injection (TI), active constellation extension (ACE), Partial Transmit Sequence (PTS), Selective Level Mapping (SLM) and interleaving. SLM and PTS reduce PAPR significantly without loss of any information [1–3], but in PTS latency increases with number of sub blocks. In this paper, SLM is considered for reduction of PAPR using various phase sequences. The PAPR improvement changes with phase sequences. Hence the selection of phase sequence is very important for better improvement of PAPR. This paper investigated for optimum phase sequences for better PAPR performance improvement. Firstly, the improvement of PAPR using SLM technique and pseudorandom phase sequence is carried out. Later the phase sequences namely Riemann, Hadamard, Chaotic, modified Chu, Rudin Shapiro, Novel and Modified phase sequences are replaced and the results are compared. Riemann, chaotic, modified Chu, Hadamard and Novel phase sequences are well described in [5–10] respectively.

2 PAPR Problem in OFDM System

OFDM System

Let total information bits (M) are grouped into N symbols (X_k, k = 0, 1, ...N − 1) and each symbol is modulated on one of a set of orthogonal subcarriers The total bandwidth of system W is divided into N sub bands, then the individual subcarrier bandwidth is W/N and it is less than the coherence bandwidth B_c. Hence the frequency selective fading channel is converted into flat fading channel (W/N < B_c), and OFDM symbol duration is increased by N times. OFDM signal can be defined as

$$x(t) = \frac{1}{\sqrt{N}} \sum_{K=0}^{N-1} X_k e^{j2\pi f_k t}, \quad 0 \leq t \leq NT \tag{1}$$

where $j = \sqrt{-1}$. As input data streams are orthogonal, the real and imaginary parts of x(t) are uncorrelated and according to the central limit theorem, the distribution of these parts approach Gaussian distribution for large N with zero mean and variance $\sigma^2 = E\left[|Re\{x(t)\}|^2 + |Im\{x(t)\}|^2\right]/2$, where E[x] is the average value of x. the Probability Density Function (PDF) of OFDM signal is

$$P_r\{x(t)\} = \frac{1}{\sqrt{2\pi\sigma^2}} e^{-\frac{|x(t)|^2}{2\sigma^2}} \tag{2}$$

Original OFDM signal has Rayleigh nature, uniform phase and its PDF is

$$P_r(r) = 2re^{-r^2} \tag{3}$$

where 'r' indicates OFDM signal level.

PAPR Problem

In OFDM, the peak power increases with number of subcarriers and much larger compare to average power due to coherent addition of subcarriers. This high PAPR moves signal into non-linear region of amplifier and degrades its efficiency. The PAPR of OFDM signals x(t) is

$$PAPR[x(t)] = \frac{P_{peak}}{P_{average}} = 10 \log_{10} \frac{\max[|X(n)|^2]}{E[|X_n|^2]} \tag{4}$$

where E[] denotes expected value. If this ratio is beyond threshold level, signal distortion due to clipping and radiation due to spectral broadening will occur. For large N, Rayleigh distributed OFDM signal peak will increase with non zero probability and its probability exceeds a threshold

$$P_0 = \frac{\sigma_0^2}{\sigma_n^2}. \tag{5}$$

$$P(PAPR \geq P_0) = 1 - (1 - e^{-P_0})^N \tag{6}$$

Relation Between PAPR and Subcarriers

Let there are N Gaussian independent and identically distributed random Variables $x_n; 0 \leq n \leq N - 1$ with zero mean and unit power. Average power $E_n = (x[n])^2$ is

$$E\left[\{\frac{1}{\sqrt{N}}(|X_0 + X_1 + X_2 + \cdots + X_{N-1}|)\}^2\right] \tag{7}$$

$$\frac{1}{N}E(|X_0 + X_1 + X_2 + \cdots + X_{N-1}|)^2 = \frac{E|x_0|^2}{N} + \frac{E|x_1|^2}{N} + \cdots + \frac{E|x_{N-1}|^2}{N} = 1$$

Value increases with number of coherent subcarrier

$$\max\left[\frac{1}{\sqrt{N}}(|X_0 + X_1 + X_2 + \cdots + X_{N-1}|)\right]^2 = \left[\left|\frac{N}{\sqrt{N}}\right|\right]^2 = \left|\sqrt{N}\right|^2 = N$$

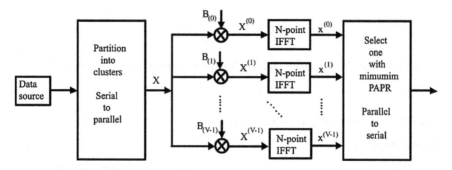

Fig. 1 Block diagram of OFDM transmitter with SLM

Therefore maximum PAPR of OFDM with N subcarriers is N.

Selective Level Mapping

In SLM, multiple data blocks of same information is generated and array multiplied with various phase vectors as shown in the Fig. 1. The sequence with minimum PAPR is considered for transmission. Every data block is array multiplied by V different phase sequences, each of length N, $V(m) = [v_{m,0}, v_{m,1}, \ldots, v_{m,N-1}]^T$, m = 1, 2, ..., M, resulting in M new (modified) data blocks.

The new block for the mth sequence is

$$X(m) = [X_{0V_{m,1}}, X_{1V_{m,2}}, \ldots X_{N-1v_{m,N-1}}]^T \tag{8}$$

where m = 1, 2, ..., M.

Among new data blocks X (m), m = 1, 2... M, the one with the minimum PAPR is considered for transmission. The amount of PAPR performance improvement depends on number of duplicate candidates (M) and type of phase sequence. The reduction of PAPR is proportional to M, but increase in M increases overhead information in terms of index length (log_2M) as side information. Hence M should not be large. Therefore selection of phase sequence play vital role in improving PAPR performance.

3 PAPR Mitigating Phase Sequences

Riemann Sequence

Riemann matrix (**R**) can be defined as

$$R(p, q) = \begin{array}{ll} p - 1 & \text{if p divides q} \\ -1 & \text{otherwise} \end{array} \tag{9}$$

Here first row and column of the Riemann matrix have to remove to obtain required matrix.

Chaotic Sequence

In this approach concentric circle constellation (CCC) based mapping is considered to obtain the data symbols. This sequence exhibits random behaviour. The M-ary chaotic sequence $C_n \in \{0, 1, 2,...M - 1\}$, $0 \leq n \leq N - 1$ and C_n is given by

$$C_n = \left[\frac{M_{y_{n+1}}}{2}\right] + \frac{M}{2}, \quad y_{n+1} = f(y_n) = 1 - \alpha y_n^2,$$

$$\alpha \in [1.4015, 1.99], \quad y_n \in (-1, 1) \tag{10}$$

C_r, the rth element of first phase sequence becomes $P_r^1 = \exp(j2\pi C_r/M)$.

Rudin Shapiro Sequence

Every element of this sequence is either $+1$ or -1. The nth term of the sequence, b_n, is

$$a_n = \sum \varepsilon_i \varepsilon_{i+1} \quad \text{and} \quad b_n = (-1)^{a_n} \tag{11}$$

where ε_i indicates the digits of n (binary form). Therefore a_n gives the number of occurrences of the sub-string 11 in the binary expansion of n.

$b_n = +1$ for a_n is even and $b_n = -1$ for a_n is odd. This sequence can be generated by

$$b_{2n} = b_n \quad \text{and} \quad b_{2n+1} = (-1)^n b_n \tag{12}$$

Chu Sequence

In this scheme, the phase sequence is defined as $B^{(u)} = \left[B_0^{(u)}, B_1^{(u)}, \ldots\ldots B_{N-1}^{(u)}\right]$.

The $b_k^{(u)}$ is the kth element of a phase sequence. It is defined as

$$b_k^{(u)} = \begin{cases} \exp\left(i\frac{2\pi}{N}\left[\frac{uk^2}{2}\right]\right), & N \text{ even} \\[2mm] \exp\left(i\frac{2\pi}{N}\left[\frac{uk(k+1)}{N}\right]\right), & N \text{ odd} \end{cases} \tag{13}$$

Pseudo Random Sequence

This sequence can be generated randomly in terms of exponential sequence. The phase sequence is defined as

$$p(u) = \exp\left(\frac{\pi i}{n}\right) \tag{14}$$

where n is a random angle value to generate the exponential phase sequence.

Modified Sequence

In the proposed approach, the matrix elements obtained by the multiplication of the normalised matrices is considered as phase rotation sequence. The first matrix is defined as

$$A(p, q) = 0 \text{ if } k = 0$$
$$= -1 \text{ if } k < 0$$
$$= 1 \text{ if } k > 0 \text{ where } k = p - q.$$

The second matrix is defined as

$$B(p, q) = 1 \text{ if } p \text{ divides } q$$
$$= -1 \text{ else}$$

The final matrix (phase elements) is obtained by multiplying the above matrices.

$$C(i,j) = A(i,j) * B(i,j) \tag{15}$$

4 Digital Modulation Schemes

Digital modulation schemes are of binary and M-ary. If the bandwidth of the channel is not sufficient for desired data transmission, M-ary digital modulation schemes are preferred. In general OFDM systems are designed for high data rate applications, so M-ary signaling schemes especially M-ary PSK and QAM are used. Detailed description about M-ary PSK and QAM performance are given in [11].

5 Results and Conclusions

The probability of the OFDM Signal PAPR exceeds a threshold value γ is

$$P_r\{PAPR[x(n)] > \gamma\} = 1 - (1 - e^{-\gamma})^N$$

Here the PAPR improvement is verified by considering 4096 bits for various phase sequences. From the results it is observed that PAPR reduction improvement is better for Riemann sequence and also it is observed that improvement increases with order of modulation. Increase in order improves the spectral efficiency. Modified phase sequence performance is better over other sequences at higher order modulation schemes (Figs. 2, 3, 4, 5, 6, 7 and 8).

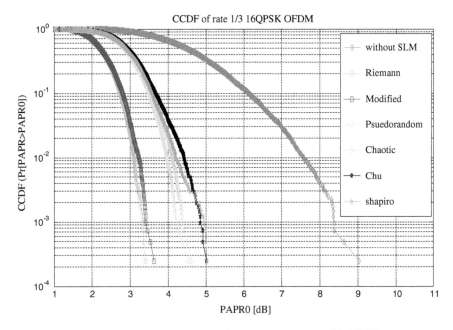

Fig. 2 Comparision of PAPR reduction for various phase sequences with 16PSK

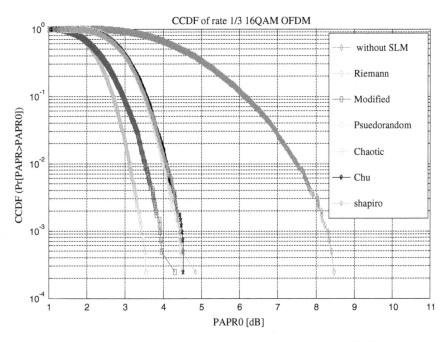

Fig. 3 Comparision of PAPR reduction for various phase sequences with 16QAM

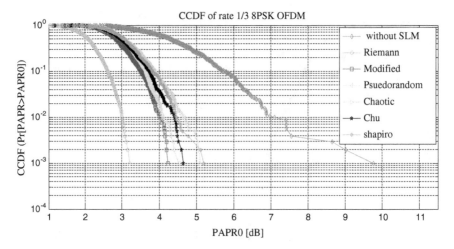

Fig. 4 Comparision of PAPR reduction for various phase sequences with 8PSK

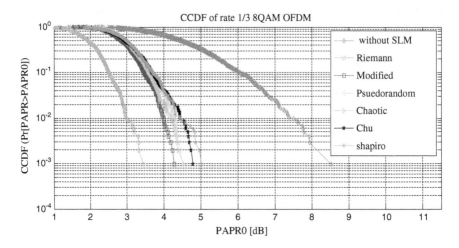

Fig. 5 Comparision of PAPR reduction for various phase sequences with 8QAM

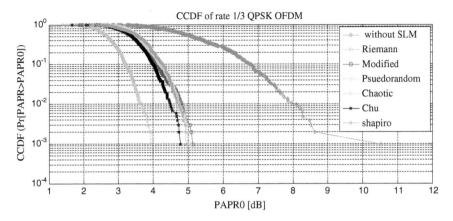

Fig. 6 Comparision of PAPR reduction for various phase sequences with QPSK

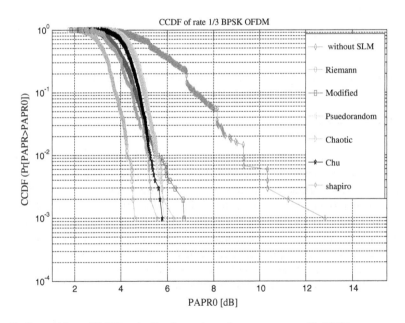

Fig. 7 Comparision of PAPR reduction for various phase sequences with BPSK

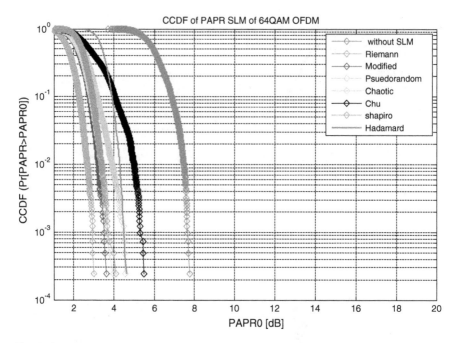

Fig. 8 Comparision of PAPR reduction for various phase sequences with 64QAM

6 Conclusions

In this paper, the main problem of OFDM system, high PAPR and its significance is discussed. SLM scheme is implemented for reduction of PAPR and the reduction performance for various PAPR minimization phase sequences is compared. For higher order modulation schemes, the PAPR performance improvement is better. The PAPR reduction improvement using Modified phase sequence is far better for higher order modulation schemes compare to other phase sequences and competent with Riemann phase sequence. This sequence is less complex and easier for implementation and transmission of side information along with desired data is not required. Hence it is more bandwidth efficient.

References

1. S.H. Han, J.H. Lee, An overview: peak-to-average power ratio reduction techniques for multicarrier transmission. IEEE Wirel. Commun. **12**(2), 56–65 (2005)
2. V. Vijayarangan, R. Sukanesh, An overview of techniques for reducing peak to average power ratio and its selection criteria for orthogonal frequency division multiplexing radio systems. J. Theor. Appl. Inf. Technol. **5**(1), 25–36 (2009)

3. T. Jiang, W. Yiyan, An overview: peak – to - average power ratio reduction techniques for OFDM Signals. IEEE Trans. Broadcast. **54**(2), 257–268 (2008)
4. P. Sharma, S. Verma, PAPR reduction of OFDM signals using selective mapping with turbo codes. Int. J. Wirel. Mob. Netw. (IJWMN) **3**(4), 217–223 (2011)
5. P.M.Z. Goyoro, I.J. Moumouni, S. Abouty, SLM using Riemann sequence combined with DCT transform for PAPR reduction in OFDM communication systems. World Acad. Sci. Eng. Technol. **6**(4), 304–309 (2012)
6. M. Chandwani, A. Singhal, N. VishnuKanth, V. Chakka, A low complexity SLM technique for PAPR reduction in OFDM using Riemann sequence and thresholding of power amplifier, in INDICON-2009-*IEEE Conference* (2009), pp. 445–448
7. A. Geol, M. Agarwal, P. Gupta Poddar, M-ary chaotic sequence based SLM-OFDM system for PAPR reduction without side-information. Int. J. Comput. Commun. Eng. **6**(8), 677–682 (2012)
8. P.K. Sharma, C. Sharma, PAPR reduction using SLM technique with modified Chu sequences in OFDM system. MIT Int. J. Electron. Commun. Eng. **2**(1), 23–26 (2012)
9. L. Ning, M. Yang, Z. Wang, Q. Guo, A novel SLM method for PAPR reduction of OFDM system, in *VTC Spring-7th IEEE Conference* (2012)
10. N. VishnuKanth, V. Chakka, A. Jain, SLM based PAPR reduction of OFDM signal using new phase sequence. Electron. Lett. **45**(24), 1231–32 2008. (IET Publications, IEEE)
11. W. Tomasi, Electronic communications systems fundamentals through advanced, in *Fifth Edition Pearson Education Publications* (2000)

Optimization of Peak to Average Power Ratio Reduction Using Novel Code for OFDM Systems

R. Chandrasekhar, M. Kamaraju, K. Rushendra Babu and B. Ajay Kumar

Abstract Communication is one of the important aspects of our day to day life. The field of communication has been increasing rapidly in order to fulfill the requirements of the human needs. In the first generation communication, information was transmitted in analog domain and later digital domain to overcome disadvantages existed in the analog. For better transmission, single carrier modulation has been replaced by multicarrier modulation. Orthogonal Frequency Division Multiplexing (OFDM) is a form of multicarrier modulation technique that transmits information bearing signals through multiple carriers. The carriers are different frequencies and orthogonal to each other. Orthogonality in OFDM reduces bandwidth usage and frequency selective fading due to multipath. However, the main drawback of OFDM system is its high peak-to-average power ratio (PAPR). Increasing Peak power is a significant problem in OFDM system because it degrades the power amplifier performance. Above the threshold range they become non linear resulting in signal distortion. Several techniques are found in literature to reduce the PAPR viz. Amplitude clipping and filtering, Selected level mapping (SLM), Partial transmit sequence (PTS), and Interleaving. In this paper, a Novel code is proposed to reduce PAPR. Simulations are performed by using MATLAB tool and results are analyzed for different systems (Uncoded, Hamming Code, and Novel Code). The results demonstrate that Novel Code is an efficient technique that reduces PAPR when compared to Hamming code and uncoded system.

R. Chandrasekhar (✉) · M. Kamaraju · K. Rushendra Babu · B. Ajay Kumar
Department of ECE, Gudlavalleru Engineering College (A),
Gudlavalleru 521356, Andhra Pradesh, India
e-mail: chandrasekharece403@gmail.com

M. Kamaraju
e-mail: madduraju@yahoo.com

K. Rushendra Babu
e-mail: rushendrababu.k@gmail.com

B. Ajay Kumar
e-mail: aiay.bhupati@gmail.com

© Springer India 2016
S.C. Satapathy et al. (eds.), *Microelectronics, Electromagnetics and Telecommunications*, Lecture Notes in Electrical Engineering 372,
DOI 10.1007/978-81-322-2728-1_24

Keywords Orthogonal frequency division multiplexing · Hamming code novel code · Peak-to-average power ratio (PAPR) · Inter symbol interface

1 Introduction

Orthogonal Frequency Division Multiplexing is one of the key techniques for fourth generation (4G) wireless communication. It transmits information bearing signals through multiple carriers. Orthogonality in OFDM reduces bandwidth usage and frequency selective fading due to multipath. Multiple carrier system increases symbol duration and also decreases the relative amount of dispersion. In OFDM Inter-symbol interference (ISI) is eliminated by providing cyclic prefix or guard time interval to every OFDM symbol [1]. Orthogonal frequency division multiplexing is mainly used in Digital Audio Broadcasting (DAB), Digital Video Broadcasting (DVB), European Wireless LAN Standard-Hyper LAN/2, High bit rate Digital subscriber line (HDSL at 1.6 Mbps), Asymmetric Digital Subscriber Line (ADSL up to 6 Mbps), IEEE 802.11a and 802.11g wireless LANs [2]. Some of the drawbacks faced by OFDM are Inter carrier Interference and Peak to Average Power Ratio (PAPR). PAPR occurs due to the addition of all orthogonal signals at a maximum point, which causes the nonlinear effects in power amplifier. The non-linearity in power amplifiers leads to spectral spreading, changes in the signal constellation and in band, and out band interferences to the signals. Hence distortion less transmission is possible only when the power amplifier consist a back off, which is equal to Peak to Average Power Ratio (PAPR). The efficiency of Power amplifier decreases because of the larger values of back off. Therefore, the main intention of this paper is to reduce PAPR [3–7].

In OFDM, Peak-to-Average Ratio (PAPR) is defined as "The ratio of the maximum instantaneous power to the average power"

$$PAPR = \frac{\max |x(t)|^2}{E\left[|x(t)|^2\right]} \tag{1}$$

Several techniques have been projected in the literature to reduce the PAPR. Amplitude Clipping and Filtering, Coding, Partial Transmit Sequence, Selective Level Mapping, Tone Reservation, Interleaving, Tone Injection and Active Constellation Extension Technique [1, 5, 6, 8–10].

The remainder of this article is organized as follows. Section 3 elucidates the Novel Code Algorithm. Section 4 describes the Novel Code Design. Section 5 presents simulation results and analysis. Finally, Sect. 6 gives conclusions.

2 Hamming Code

For the block of k message bits (n − k), parity bits or check bits are added. Hence the total bits at the output of channel encoder are n. Such codes are called (n, k) block codes [11, 12]. Hamming codes are (n, k). Linear Block Codes must satisfy the following conditions:

(1) Number of Check bits q ≥ 3
(2) Block Length $n = 2^p - 1$
(3) Number of message bits k = n − q
(4) Minimum distance $d_{min} = 3$ (Fig. 1).

2.1 (7, 4) Hamming Code Encoder

The following Fig. 2 shows the Hamming code encoder. The lowest register contains check bits C1, C2, and C3. These bits are obtained from the message bits by mod-2 additions. These additions are performed by the following equations, $C1 = (M1 \oplus M2 \oplus M3)C2 = (M1 \oplus M2 \oplus M4)C3 = (M1 \oplus M3 \oplus M4)$. First the Switch "S" is connected to message register then all message bits are transmitted, next switch "S" is connected to check bit register the combination of message and check bits gives hamming code word length "7" (4 Bits message+3 Bits Check Bits) for a transmitted message [13–15].

2.2 (7, 4) Hamming Code Decoder

In every (n, k) linear block code there exists a parity check matrix H, which is equivalent to H = [PT: Iq] q × n, through which H^T is calculated using the formula. Suppose the transmitted code vector is X and corresponding received code vector is represented by Y, then $YH^T = (0\ 0...0)$, if X = Y that means there is no error in the

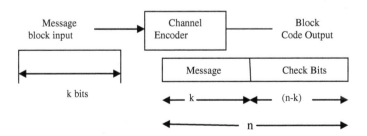

Fig. 1 Functional block diagram of block coder

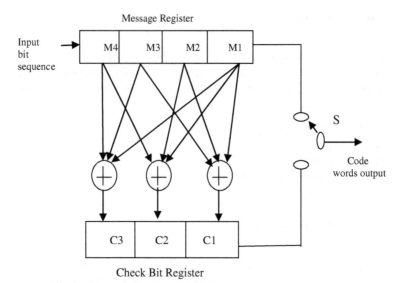

Fig. 2 (7, 4) Hamming code encoder

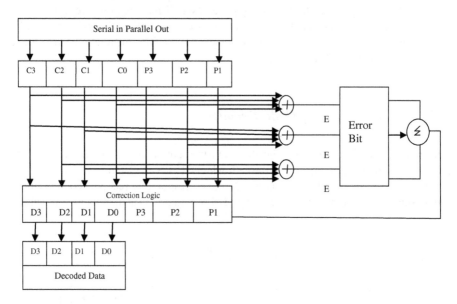

Fig. 3 (7, 4) hamming code decoder

received codeword. YH^T = Non-Zero, if $X \neq Y$ that means there is an error in the received code word. The non-zero output of the product YH^T is called Syndrome and it is used to detect the errors in Y. $S = YH^T$ this is shown in Fig. 3.

3 Novel Code Algorithm

In this paper, a new technique is proposed to reduce PAPR in OFDM. Figure 4
shows the transmitter section of proposed system in that four bit message is converted into eight bit code word by adding four parity bits (or) redundant bits. The
resultant eight bit code word is EX-OR with a bias vector which gives least Peak to
Average Power Ratio. The Resultant code word along with BPSK modulation
effectively reduces the Peak to Average Power Ratio (PAPR). Figure 5 shows the
receiver section of proposed system in which the received codeword is demodulated
by using BPSK demodulator, thus the demodulated codeword is EX-OR by same

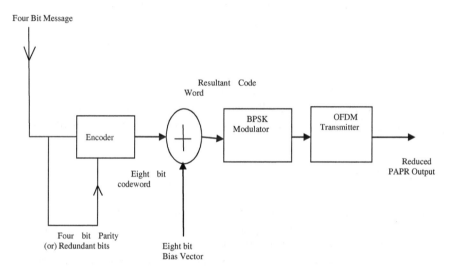

Fig. 4 Transmitter section of novel code system

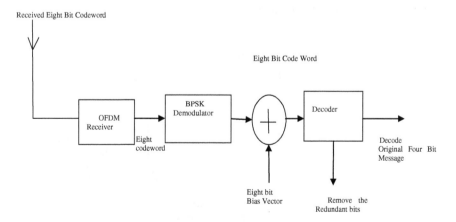

Fig. 5 Receiver section of novel code system

bias vector which is used at transmitter section. Finally the resultant codeword is transmitted to decoder which gives a four bit original message.

3.1 Encoding

(1) Encode the message word into 8 bit codeword using generator matrix which indirectly helps in determination of least power codeword's from set of possible codeword's.
(2) XOR 8 bit codeword with bias vector to yield resultant codeword.
(3) Resultant code words are applied to BPSK modulation.

3.2 Decoding

(1) Apply the BPSK demodulation to the received code words.
(2) XOR the received vector with bias vector which used at the transmitter.
(3) Decode the resultant word and applied FEC to obtain the transmitted message.

4 Novel Code Design

Basically NOVEL CODE is a linear block code. In this four bit message words are encoded into eight bit codeword by appending four parity bits to message words as shown in Fig. 3. The generated parity bits have some relation with message bits, and the resultant code words ex-or with a bias vector. In general, the representation of linear block code is (n, k). In that the 'n' represents the number of bits in code word whereas 'k' represents number of bits in message word. Here the total numbers of check bits (Parity bits) are n − k (Fig. 6).

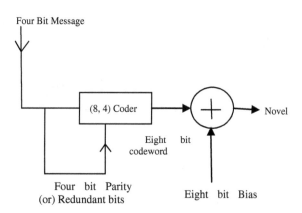

Fig. 6 Block diagram of novel code

To design the Novel code, consider all possible combinations of 8 bit code words. That means a total of $2^8 = 256$ code words. From all these possible code words, identify 16 code words based on following two conditions:

(1) Minimum Hamming Distance.
(2) Low PAPR of code words.
(3) In this Novel coding scheme, the minimum Hamming distance is chosen as three for detection and correction of single bit errors.

4.1 Properties of Novel Codes

(1) Novel code is a Systematic code.
(2) If any two code words present in the set are operated with modulo-2 addition, the resultant codeword is also present in that set.
(3) If any two codes add in the generator matrix the resultant code should not in the generator matrix.
(4) The minimum hamming distance = Minimum weight of the code. $d_{min} = w_{min}$

5 Results and Analysis

Simulations are performed by using MATLAB tool. To evaluate the performance of OFDM System, complementary cumulative distribution function (CCDF) curves are used. CCDF curve gives information about the percentage of time the signal spent at or above the level which defines the probability for that particular power level. Figure 7 shows the plot for the CCDF with PAPR for Different Systems.

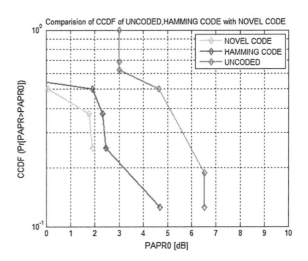

Fig. 7 Comparison of complementary cumulative distribution function of uncoded, hamming code novel code

From the figure it can be analyzed that, in uncoded system, maximum peak to average power ratio (PAPR) is 6.569 dB and it is caused by code words 00000000, 11111111. In hamming code system, maximum peak to average power ratio (PAPR) are 4.456 dB and it is caused by code words 00000011, 11111100. In NOVEL code system, maximum peak to average power ratio (PAPR) is 1.973 dB and it is caused by codewords 00001010, 00010001, 11101110, and 11110101.

6 Conclusions

Orthogonal Frequency Division Multiplexing is a bandwidth and power efficient system; it reduces the frequency selective fading by its orthogonality principle. In this paper a Novel Code technique is proposed to reduce PAPR. From the analysis of three different coding techniques it is observed that the Novel code is an efficient technique to reduce PAPR (1.973 dB) when compared to Hamming Code (4.456 dB) and uncoded system (6.569 dB).

References

1. Y. Su, D.X.M. Liu, X. Jiang, Combined selective mapping and extended hamming codes for PAPR reduction in OFDM systems, in *4th IEEE International Conference on Information Science and Technology (ICIST)* (2014), pp. 582–585
2. C. Rajasekhar, D. Srinivasa Rao, V. Yaswanth Raghava, D. Hanith, PAPR reduction performance in OFDM systems using channel coding techniques, in *International Conference on Electronics and Communication Systems (ICECS)* (2014), pp. 1–5. ISBN:978-1-4799-2321-2
3. T. Jianga, Y. Wu, An overview: peak-to-average power ratio reduction techniques for OFDM signals. IEEE Trans. Broadcas. **54**(2), (2008)
4. Y. Rahmatallah, S. Mohan, Peak-to-average power ratio reduction in OFDM systems: a survey and taxonomy. IEEE Commun. Surv. Tutorials **15**(4), (Fourth Quarter 2013)
5. S.H. Han, J.H. Lee, An overview of peak-to average power ratio reduction techniques for multicarrier transmission. IEEE Wirel. Commun. **12**(2), 56–65 (2005)
6. S.H. Han, J.H. Lee, An overview of peak-to-average power ratio reduction techniques for multicarrier transmission. Modulation Coding Signal Process. Wirel. Commun. IEEE Wirel. Commun. (2005)
7. T. Jiang, Y. Wu, An overview: peak-to-average power ratio reduction techniques for OFDM signals. IEEE Trans. Broadcast. **54**(2), (2008)
8. J.-C. Chen, Partial transmit sequences for peak-to-average power ratio reduction of OFDM signals with the cross-entropy method. IEEE Signal Process. Lett. **16**(6), (2009)
9. Y. Wu, W.Y. Zou, Orthogonal frequency division multiplexing: a multi-carrier modulation scheme. IEEE Trans. Consum. Electron. **41**(3), 392–399 (1995)
10. D. Wulich, Reduction of peak to mean ratio of multicarrier modulation using cyclic coding. IEE Electron. Lett. **32**(29), 432–433 (1996)
11. S. Fragicomo, C. Matrakidis, J.J. OReilly, Multicarrier transmission peak-to-average power reduction using simple block code. IEEE Electron. Lett. **34**(14), 953–954 (1998)

12. T. Jiang, G.X. Zhu, OFDM peak-to-average power ratio reduction by complement block coding scheme and its modified version, in *The 60th IEEE Vehicular Technology Conference 2004 Fall*, Los Angeles, USA, Sept. 2004, pp. 448–451
13. T. Jiang, W.D. Xiang, P.C. Richardson, J.H. Guo, G.X. Zhu, PAPR reduction of OFDM signals using partial transmit sequences with low computational complexity. IEEE Trans. Broadcast. **53**(3), 719–724 (2007)
14. S. Litsyn, G. Wunder, Generalized bounds on the crest-Factor distribution of OFDM signals with applications to code design. IEEE Trans. Inform. Theory **52**(3), 992–1006 (2006)
15. J.A. Davis, J. Jedwab, Peak-to-mean power control in OFDM, Golay complementary sequences, and reed-muller codes. IEEE Trans. Inform. Theory **45**, 2397–2417 (1999)

Scattering of SODAR Signal Through Rough Circular Bodies

M. Hareesh Babu, M. Bala Naga Bhushanamu, D.S.S.N. Raju,
B. Benarji and M. Purnachandra Rao

Abstract Scattering of SODAR signal from a thin rough circular bodies is studied using an integral equation method. The relation between Incident and scattered fields governs by the transition matrix (T-matrix). T matrix has determined using an Integral representation of the Helmholtz equation. The prototype rough surfaces are modeled by a superposition of a number of the sinusoidal surface that are randomly translated and rotated with respect to each other. Numerical results for the scattered fields are presented in the case of an incident plane wave only. For scattered wave are going to discussed the further paper. The conclusion has made here that the amplitude of the backscattered wave is not only dependent on parameters like signal frequency, incident angle and surface characteristics like rms height and correlation function. It's also concluded that the exact geometry of rough objects is also important for scattering properties, which Implies that a statistical approach to the problem is of limited values. All these mathematical algorithms and plots have developed in the Matlab scientific software.

Keywords Acoustic · Circular · Deterministic · Gaussian · Helmholtz · Numerical · Rough · Scattering · Sodar · T-Matrix

M. Hareesh Babu (✉) · M. Bala Naga Bhushanamu · D.S.S.N. Raju · M. Purnachandra Rao
Department of Systems Design, Andhra University, Visakhapatnam, India
e-mail: hareesh.makesu@gmail.com

M. Bala Naga Bhushanamu
e-mail: balanagabhusanamu@gmail.com

D.S.S.N. Raju
e-mail: trijeth@gmail.com

M. Purnachandra Rao
e-mail: raomp17@gmail.com

B. Benarji
Department of ECE, Andhra University, Visakhapatnam, India
e-mail: benarjiec@gmail.com

© Springer India 2016
S.C. Satapathy et al. (eds.), *Microelectronics, Electromagnetics
and Telecommunications*, Lecture Notes in Electrical Engineering 372,
DOI 10.1007/978-81-322-2728-1_25

277

1 Introduction

A SODAR signal is an acoustic wave with 2 kH frequency propagates through a turbulent medium which interacts with various types of bodies and aerosols in the medium [1]. Internal momentum fluctuations are induced by eddy motions, which cause changes in the internal pressure and radiate away as a scattered acoustic wave. These type of interactions is also referred to as local scattering or internal scattering. Internal and thermal scattering effects are to direct some of the incident acoustic energy, away from the actual propagation direction. Temperature fluctuations in the lower atmosphere and local change in the acoustic index also some extend responsible for scattering. However, there is no reduction in the total transmitted acoustic power but that power is diffused over a wide area. Therefore the received Intensity along the actual direction has diminished. There is a no way for the receiver to distinguish between this effect and a bona-fide absorption loss. Scattering of acoustic wave (Sodar signal) from a smooth circular body is a classical problem, which has been considered by a large number of authors throughout the year. More recently a review of the literature is given by Sleator [2] and studied by Kristensson and waterman [3] using the null field approach. Bostrom and Peterson [4] has solved the problem of scattering from the circular body in the interface between two fluids. However, there is no such things as a smooth surface in reality. All real surfaces are more or less rough and there has immense volume of literature on scattering from rough surface. Despite the large amount of research efforts that has been devoted to this subject there are still many unsolved problems. Acoustic scattering from a rough circular bodies does not seem to have received much attention so far. Rough surfaces can be divided into two categories. These are 1. Random Rough surface and 2. Deterministic surfaces. 1. Random rough surface have the advantage of being realistic and having absence of an analytical expression for the surface complicates the matter substantially. Numerical simulations based on approximate models like Kirchhoff's theory are computationally intensive. Unfortunately, the result does not tell us anything about the scattering properties of a particular object belonging to the circular scattered body. 2. Deterministic surfaces are very tractable analytically. The drawback is the lack of resemblance to real surfaces. Here, in this paper has attempted both advantages and approaches. The rough surface is modeled by a superposition of a number of deterministic surfaces, which are randomly displaced with regard to each other. In this manner we have developed a prototype circular body surface which has a very simple shape from a mathematical point of view. However, this prototype is still is sufficiently irregular to serve as a realistic circular bodies.

2 Mathmatical Algorithm Evalution

Here in this paper, we have consider a thin rough circular disk as a rough circular body of radius "a" immersed in a non-viscous fluid of infinite extent and an incident acoustic wave i.e., Sodar signal. We have used cylindrical coordinates (ρ, φ, z) to form the equation of the disk i.e., rough cylindrical body, is

$$z = \in a\eta(\rho, \varphi) \quad \rho < a \tag{1}$$

where is a small parameter? Here we also assume time harmonic condition of angular frequency ω and suppress a factor $\exp(-i\omega t)$ throughout the velocity potential is denoted by $u^1(r)$ for $z > \in a\eta(\rho, \varphi)$ and $u^2(r)$ for $z < \in a\eta(\rho, \varphi)$. Now thee wave equation has reduces to the Helmholtz equation.

$$\nabla^2 u^n + k^2 u^n = 0, \quad n = 1, 2 \tag{2}$$

where $k = \omega/c$ is the wave number and c is the speed of sound. In this paper we assumed that the disk i.e., circular body, is sound hard we have the following boundary conditions.

$$u^1 = u^2, z = \in a\eta(\rho, \varphi), \rho < a \tag{3}$$

In the next step we expanded u^n in a power series in the small parameter $\in$

$$u^n = u_0^n + \in u_1^n + O(\in^2) \tag{4}$$

By inserting into Eqs. (2) and (3) and expanding the boundary conditions in Taylor series $z = 0$ we obtain the following differential equations with pertinent boundary conditions at Order $\in^0$ and Order $\in^1$. Here we are not stating the total equation.

For Order $\in^0$

$$\nabla^2 u^n + k^2 u_0^n = 0, \quad n = 1, 2 \tag{5}$$

For Order $\in^1$

$$\nabla^2 u^n + k^2 u^n u_1^n = 0, \quad n = 1, 2 \tag{6}$$

To solve the Eqs. (5) and (6) has employed the integral equation method by Kernk and Schmidt [5]. Further the scattered fields are expanded in cylindrical partial waves involving a Fourier series in φ and Hankel transform in ρ. Now the function $\eta(\rho, \varphi)$ has been expanded in a Fourier series as

$$\eta(\rho, \varphi) = \sum_n g_n(\rho) e^{in\varphi} \tag{7}$$

2.1 Transition Matrix

The T matrix is used to express the relation between the incident field u^i and the scattered field u^s. The basic idea is to expand both fields in conveniently chosen systems of basic functions and determine a relation between the coefficients. For the incident field we use the expansion in regular spherical waves. The relation between the coefficients a_{lm} and f_{lm} of the incident and scattered fields, respectively, is governed by the T matrix as

$$f_{lm} = \sum_{l',m'} T_{lml'm'} \cdot a_{l'm'} \tag{8}$$

Here, T matrix is a very useful concept since the scattered fields can be computed in very simple manner for any incident field once the T matrix has been determined. The starting point is the following well-known integral representation of the scattered field is

$$u^s(r') = \int_s [u^1(r) - u^2(r)] \frac{\partial G}{\partial n} dS \tag{9}$$

where $G(r, r') = \frac{e^{ik|r-r'|}}{4\pi|r-r'|}$ is the free-space Green's function and the integration is carried out over the "upper" surface of the disk i.e., rough circular body. In next step we expand u^s and the T matrix in power series in terms of the small parameter $\in$ Thus it has shown that analytically that $T_{lml'm'}^{(0)}$ as well as t $T_{lml'm'}^{(1)}$ satisfy the "symmetry condition"

$$T_{lml'm'} = (-1)^{m+m'} T_{l',-m'l,-m} \tag{10}$$

which states that the system is invariant under a change of direction of time.

3   Modeling of the Rough Surface

To model a rough surface we have taken a superposition of finite number of corrugated surfaces which are randomly rotated and translated with respect to each other, i.e.,

Fig. 1 Backscattered far-field amplitude form a circular body A. Azimuthal angle (φ) for ka = 5. Angle of incident (θ) = 20°. ∈ = 0 for (*solid curve*) and 0.02 for (*dotted curve*)

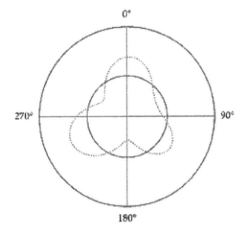

$$\eta(\rho, \varphi) = \sum_{i=1}^{N} 2\omega_i \sin((\rho \cos(\varphi - \phi i) - Xia)/\lambda_i) x \sin((\rho \sin(\varphi - \phi i) - Yia)/\lambda_i)$$

(11)

Here X_i, Y_i, and ϕ are randomly generated numbers such that $0 \le X_i \le 1$, $0 \le Y_i \le 1$, and $0 \le \phi_i \le \pi$. The λ_i's are typical length scales and the weight factor ω_i are normalized according to $\sum w_i^2 = 1$.

Here we have taken three typical surfaces, which will be referred to as surfaces A, B, and C are shown in Figs. 1, 2 and 3. For all three surfaces N = 11. In Fig. 1 λ values are evenly distributed in the interval $0.1a \le \lambda_i \le 0.3a$. For Figs. 2 and 3 λ values are evenly distributed so that $0.05a \le \lambda_i \le 0.15a$. In all three cases the weight factor ω_i are all equal. In order to better reveal the roughness the value $\in = 0.05$ has been chosen when plotting Figs. 1, 2 and 3. The value of X_i, Y_i, and ϕ are given below (Tables 1, 2 and 3).

Table 1 Randomly generated data for surface A

λ_i	X_i	Y_i	ϕ_i
0.10	0.356	0.501	2.936
0.12	0.279	0.507	1.017
0.14	0.624	0.690	2.006
0.16	0.670	0.280	1.300
0.18	0.876	0.564	2.628
0.20	0.285	0.603	2.617
0.22	0.611	0.782	3.107
0.24	0.719	0.156	1.936
0.26	0.118	0.921	0.561
0.28	0.847	0.795	2.045
0.30	0.090	0.936	2.870

Table 2 Randomly generated data for surface B

λ_i	X_i	Y_i	ϕ_i
0.05	0.428	0.960	1.904
0.06	0.743	0.333	0.797
0.07	0.607	0.962	1.113
0.08	0.716	0.809	1.920
0.09	0.027	0.540	2.216
0.10	0.371	0.105	0.088
0.11	0.317	0.955	0.656
0.12	0.937	0.614	1.642
0.13	0.079	0.523	1.214
0.14	0.152	0.962	0.748
0.15	0.350	0.299	0.764

Table 3 Randomly generated data for surface C

λ_i	X_i	Y_i	ϕ_i
0.05	0.638	0.472	2.860
0.06	0.594	0.803	0.351
0.07	0.382	0.748	1.038
0.08	0.342	0.146	1.388
0.09	0.651	0.270	1.898
0.10	0.147	0.886	0.069
0.11	0.013	0.166	1.250
0.12	0.725	0.334	0.642
0.13	0.782	0.902	1.349

Fig. 2 Backscattered far-field amplitude form a circular body A. Azimuthal angle (φ) for ka = 5. Angle of incident (θ) = 30°. $\in$ = 0 for (*solid curve*) and 0.02 for (*dotted curve*)

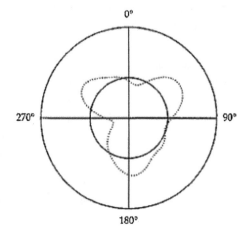

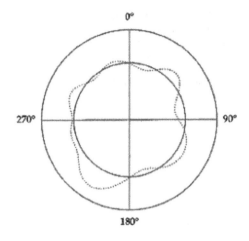

Fig. 3 Backscattered far-field amplitude form a circular body A. Azimuthal angle (φ) for ka = 5. Angle of incident (θ) = 45°. ∈ = 0 for (*solid curve*) and 0.02 for (*dotted curve*)

4 Simulation and Results

Here, the numerical example has chosen to calculate the far-field amplitude of the back scattered field for an incident plane wave of unit amplitude. The three scatterers A, B, and C have been utilized this purpose only. These results are presented as a functions of the azimuthal angle (φ) for various values of the angle of incident (θ) (Figs. 4, 5, 6, 7, 8, 9, 10, 11, 12, 13, 14, 15, 16, 17, 18 and 19).

Similarly, we have calculated and compared the λ_i versus X_i, Y_i and ϕ_i for circular body B and C also. Observations through graphs and required changes has been carried out further journal.

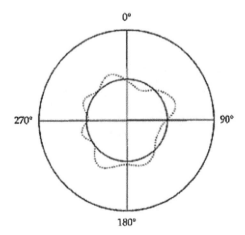

Fig. 4 Backscattered far-field amplitude form a circular body A. Azimuthal angle (φ) for ka = 5. Angle of incident (θ) = 60°. ∈ = 0 for (*solid curve*) and 0.02 for (*dotted curve*)

Fig. 5 Backscattered far-field amplitude form a circular body A. Azimuthal angle (φ) for ka = 10. Angle of incident (θ) = 20°. $\in$ = 0 for (*solid curve*) and 0.02 for (*dotted curve*)

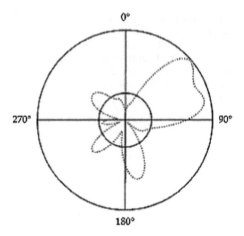

Fig. 6 Backscattered far-field amplitude form a circular body A. Azimuthal angle (φ) for ka = 10. Angle of incident (θ) = 30°. $\in$ = 0 for (*solid curve*) and 0.02 for (*dotted curve*)

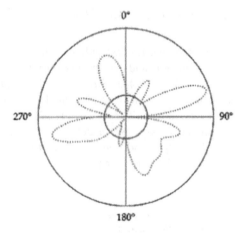

Fig. 7 Backscattered far-field amplitude form a circular body B. Azimuthal angle (φ) for ka = 10. Angle of incident (θ) = 45°. $\in$ = 0 for (*solid curve*) and 0.02 for (*dotted curve*)

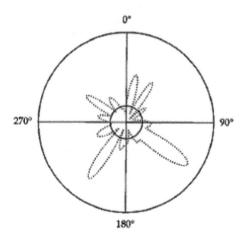

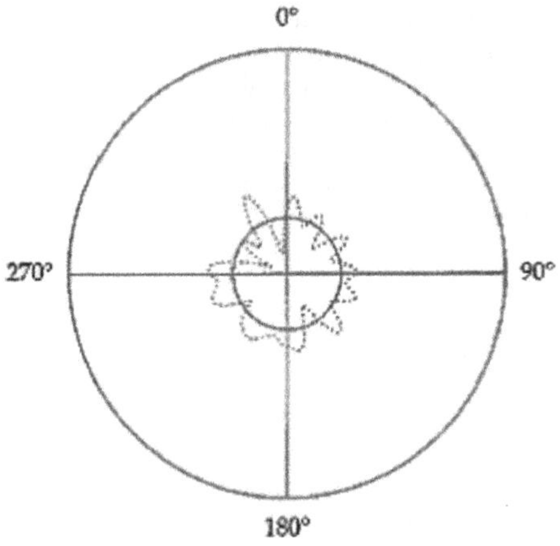

Fig. 8 Backscattered far-field amplitude form a circular body B. Azimuthal angle (φ) for ka = 10. Angle of Incident (θ) = 60°. $\in$ = 0 for (*solid curve*) and 0.02 for (*dotted curve*)

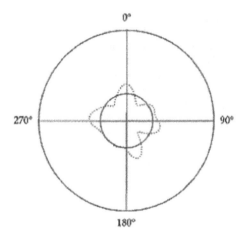

Fig. 9 Backscattered far-field amplitude form a circular body B. Azimuthal angle (φ) for ka = 10. Angle of incident (θ) = 20°. $\in$ = 0 for (*solid curve*) and 0.02 for (*dotted curve*)

Fig. 10 Backscattered far-field amplitude form a circular body B. Azimuthal angle (φ) for ka = 10. Angle of incident (θ) = 30°. $\in$ = 0 for (*solid curve*) and 0.02 for (*dotted curve*)

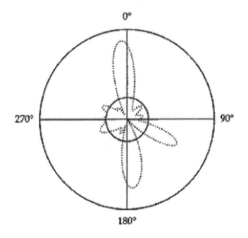

Fig. 11 Backscattered
far-field amplitude form a
circular body B. Azimuthal
angle (φ) for ka = 10. Angle
of incident (θ) = 45°. ∈ = 0
for (*solid curve*) and 0.02 for
(*dotted curve*)

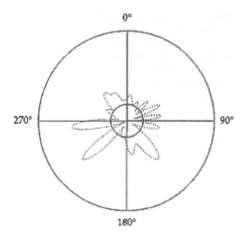

Fig. 12 Backscattered
far-field amplitude form a
circular body B. Azimuthal
angle (φ) for ka = 10. Angle
of incident (θ) = 60°. ∈ = 0
for (*solid curve*) and 0.02 for
(*dotted curve*)

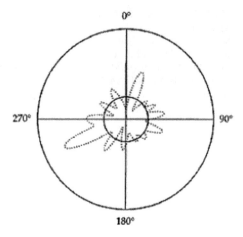

Fig. 13 Backscattered
far-field amplitude form a
circular body C. Azimuthal
angle (φ) for ka = 10. Angle
of incident (θ) = 20°. ∈ = 0
for (*solid curve*) and 0.02 for
(*dotted curve*)

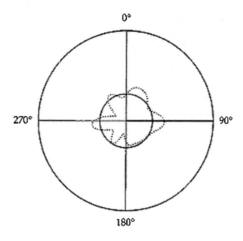

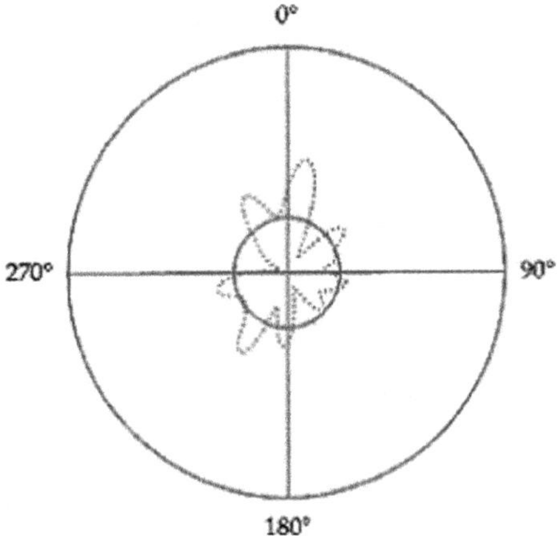

Fig. 14 Backscattered far-field amplitude form a circular body C. Azimuthal angle (φ) for ka = 10. Angle of incident (θ) = 30°. ∈ = 0 for (*solid curve*) and 0.02 for (*dotted curve*)

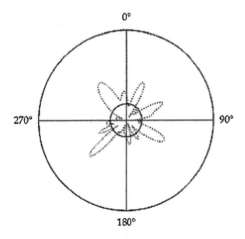

Fig. 15 Backscattered far-field amplitude form a circular body C. Azimuthal angle (φ) for ka = 10. Angle of incident (θ) = 45°. ∈ = 0 for (*solid curve*) and 0.02 for (*dotted curve*)

Fig. 16 Backscattered far-field amplitude form a circular body C. Azimuthal angle (φ) for ka = 10. Angle of incident (θ) = 60°. ∈ = 0 for (*solid curve*) and 0.02 for (*dotted curve*)

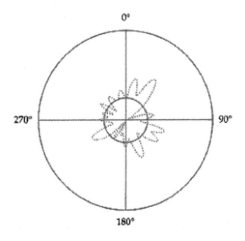

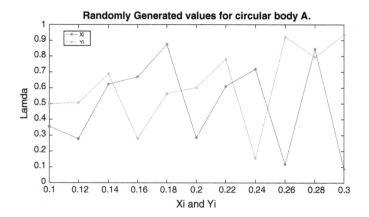

Fig. 17 Randomly generated values for circular body A

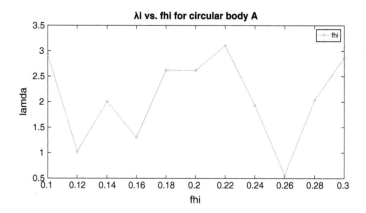

Fig. 18 λ_i versus ϕ_i for circular body A

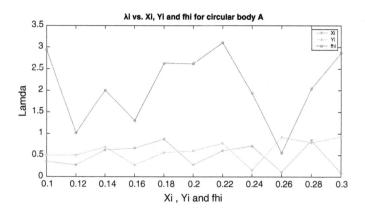

Fig. 19 λ_i versus X_i, Y_i and ϕ_i for circular body A

5 Conclusion

SODAR (acoustic) signal scattering from a rough circular body has studied using an integral equation method. The prototype of the experiment has carried out with a circular disk and developed mathematical equations later these equations has applied to circular bodies. The prototype modeled rough surface is a function of superposition of a number of sinusoidally doubly corrugated surfaces that are randomly translated and rotated with respect to each other. Here we concluded that two different surfaces with identical statistical properties show a completely different scattering behavior and circular disks/bodies with different statistical properties may give the result that are having nearby values, just as similar. In other words, statistical techniques would not lead to any result of practical values as long as the scattering from one specific body is of interest. This is a result that should be examined more extensively. Finally, we concluded that scattering of the SODAR signal will not depend on the statistical properties of the circular bodies lies in the aerosols in the lower atmosphere.

References

1. M. Hareesh Babu, A. Sarvani, D.S.S.N. Raju, Scattering of Sodar Signal by Turbulence in Homogenous, Isotopic and other Mediums, vol. 2, Issue 4, ISSN:2347-2693 (2014, April)
2. A. Bostrom, L. Peterson, Scattering of acoustic waves by a circular disk in the interface between two fluids. J. Acoust. Soc. Am. **90**, 3338–3343 (1991)
3. A.G. Voronovich, *Wave Scattering from Rough Surfaces*. (Springer, Berlin, 1994)
4. E.I Thorsos, D.R. Jackson, J. Acoust. Soc. Am. **86**, 261–277 (1989) (The validity of the perturbation approximation for rough scattering using a Gaussian roughness spectrum, 1989)
5. F.B. Sleator, J. Acoust. Soc. Am. **86**, (1969) (The disc in electromagnetic and acoustic scattering by simple shapes, ed. by J.J. Bowman, T.B.A. Senior, P.L.E. Uslenghi (North–Holland, Amesterdam, 1969))
6. G. Kristensson, P.C. Waterman, J. Acoust. Soc. Am. **72**, 1612–1625 (1982) (The T matrix for acoustic and electromagnetic scattering by circular disks, 1982)
7. J.A. Ogilvy, *Theory of Wave Scattering from Random Rough Surfaces*. (Hilger, Bristol, 1991)
8. S. Krenk, H. Schmidt, Elastic wave scattering by a circular crack. Phillos. Trans. R. Soc. London Ser. A **308**, 167–198 (1982)
9. U. Ingard, A review of the influence of meteorological conditions on sound propagation. J. Acoust. Soc. Amer. **25**(3), 405–411 (1953)
10. U. Ingard, S.K. Oleson, Measurements of Sound Attenuation in the Atmosphere, AFCRL-TR-60-431, U.S. Air Force, 4 Nov 1960
11. V.O. Knudsen, The absorption of sound in air, in oxygen, and in nitrogen-effects of humidity and temperature. J. Acoustic. Soc. Amer. **5**(2), 112–121 (1933)
12. M.A. Kallistratova, V.I. Tatarskii, Accounting for wind turbulence in the calculation of sound scattering in the atmosphere. Soviet Phys. Acoust. **6**(4), 503–505 (1961)

13. M.A. Kallistratova, FTD-TT-63–447, U.S. Air Force, (Available from DDC as AD 412 821.), Experimental Investigation of Sound Wave Scattering in the Atmosphere (1963)
14. M. Hareesh Babu, M. Bala Naga Bhushanamu, M. Purnachandra Rao, Design and simulation of graphical user interface for SODAR system using VC++. Int. J. Innovative Res. Dev. IJIRD 2(12), ISSN(O): 2278–0211 (2013)
15. H.J. Sabine, V.J. Raelson, M.D. Burkhard, Part III, U.S. Air Force. Sound Propagation near the Earth's Surface as Influenced by Weather Conditions. WADC Tech. Rep. 57–353 (1961)

Mitigation of Fault in 5 Bus System with Switch Type Voltage Controlled FCL

P. Sridhar, V. Purna Chandra Rao and B.P. Singh

Abstract In modern power distribution system increasing the growth of electrical power demand is destined results in a corresponded maximization of the short circuits in the power system network. So many imperative methodologies are concerned and evaluated to establish satisfactory operation, in that Fault Current Limiter (FCL) suggests the best way to minimize the short circuited stresses and may bourn the electromagnetic stress over associated devices as well as reduce the rated capacity of circuit breakers for fault initiated transients from effecting spurious trips. This analysis illustrates the utilization of proposed FCL for analyzing power system dynamics and mitigation of multiple fault conditions at 5 buses can have gallant power quality concerns. This paper highlights the integration of a single switch with voltage controlled three phased FCL is planned for 5 bus system by using Matlab/Simulink.

Keywords Fault current limiter (FCL) · Point of common coupling (PCC) · Power quality (PQ) · Semiconductor switch · Current controller and voltage sag

1 Introduction

Expeditious development of the power system network causes the various faults of the system increased critically. Levels of fault currents in many situations have much overtaken the withstand capability of the existed power system apparatus. As

P. Sridhar (✉)
Department of E.E.E, Institute of Aeronautical Engineering, Hyderabad, India
e-mail: sridharp35@gmail.com

V. Purna Chandra Rao
Lords Institute of Engineering and Technology, Hyderabad, India

B.P. Singh
BHEL R&D, Director R&D, St. Martins Engineering College, Secunderabad, India

© Springer India 2016
S.C. Satapathy et al. (eds.), *Microelectronics, Electromagnetics and Telecommunications*, Lecture Notes in Electrical Engineering 372, DOI 10.1007/978-81-322-2728-1_26

entails to this situation; stability, reliability, security of power system dynamics will be negatively deviated observed in [1]. Moreover, restrict the fault current and voltage dips at PCC of the power system network to a safe level can critically minimize the risk of failure to the power system apparatus because of huge current flows through the system. Due to that, there is no capture to fault currents limiting methodology ought to a hotspot of fault protection technology may limit the fault current to a mean level [2, 3].

The virtual common (PQ) power quality issues today are harmonic distortions, low power factor, voltage sags. Voltage sags/dips are the most crucial power quality concern, caused by a fault within the customers' facility side or a utility system with a high commercial loss, severe disturbances from end user equipments. The utilization of power semiconductor technology (FACTS) has introduced for enhancing power system dynamics with versatile advanced control methods in transmission/distribution networks [4]. The crucial demerits of FACTS devices are more expensive to provide smooth outcome, vast devices, more complex to implement. Implementation of this simple technology preferred by authors in this work as, utilization of fault current limiting (FCL) scheme in the electric power system network is never restricted to minimize the short circuit current amplitudes. Performance evaluation of power system transient stability improvement with a reliable operation, maximization of the power transfer capability of the power system network while using FCL operating under voltage controlled mode. Several topologies are introduced in literatures [5–7]. Moreover, a single switch voltage controlled FCL has ideal characteristics as listed as follows;

- Zero/Low Impedance in the normal operating region.
- Less/No power loss in normal operation and large impedance region into faulty sections.
- Fast appearance of impedance in a faulty section.
- Quick recovery after the fault suppression.
- Should play load impedance role during fault region for better operative performances.

So as to confine the fault current levels near to pre-fault regions and maintain voltage as a constant at the point of common coupling (PCC). From the optimal point of view, by engaging line current under fault section as well as before fault section will experienced effective restriction of fault current stress may control the power electronic switch in FCL topology for enhancing voltage sag during fault conditions and attaining power should be virtue [8]. The voltage constrained apparatus is obliged to be familiarized in power system network for climbing to its good prospective worth. This will evade from the updated switchgear units throughout the framework of conveying customers are more required for empowering legitimate coordination of compensation scheme over formal protection schemes.

As Fig. 1 depicts the schematic diagram of FCL operating under voltage controlled mode in a power system network, in that main source comes from the grid and engaged to 33 kV to 440 V step-down transformer for highly sensitive loads are connected at end-user level and fault should be occurred at nearby second load bus

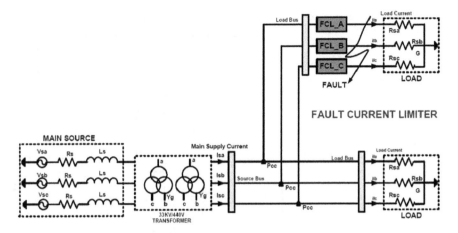

Fig. 1 Schematic diagram of fault current limiter operating under current controlling device in three phase power system network

and FCL also interfaced premises of sensible load. For an excellent dependable power supply, the (FCL) fault current limiter is accompanied into fundamental parts of power system control and operations. This paper categorizes the analysis of voltage controlled fault current limiter should be validated under several fault conditions such as line to ground (LG) fault, double line to ground (LLG) fault, triple line to ground (LLLG) fault using Matlab/Simulink tool and results are conferred.

2 Proposed FCL Configuration and Its Operation

As Fig. 2 depicts the schematic diagram of the proposed FCL topology this consists of two following parts. Presence of these four diodes in a bridge manner as a coordination of small limiting DC reactor (Ldc) with respect to internal (Rdc) resistor with an IGBT switch (Vswf) with a supportive diode (D5) acts as primary part. A shunt impedance branch supports as a limiter for minimization of fault currents which involves a series resistor (Rsh) with a shunt inductor (Lsh) they form (Rsh + iωLsh) as a secondary part. Some researchers introduce several types of FCL structures of fault envision and eradication [9–12] in that utilized more number of active devices and more complicated control unit with operational delay because of higher impedance value to excavate the fault current between the occurrence of fault period and the switch turn-off period instead of one switch device with unique control unit methodology have superior features compare to classical schemes.

This essential estimation of Ldc has a respectable voltage drop on the FCL as well as misfortunes the power loss in the shunt impedance part and the presence of DC reactor loss when operated during general condition. By employing the power

Fig. 2 Schematic diagram of intended FCL topology

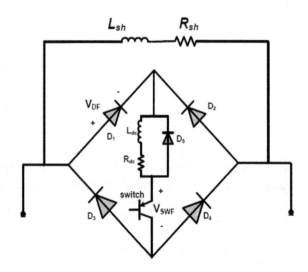

electronic active switch within the intended structure have quick operation, highest feasibility to pick a little esteem for an L_{dc} to fence off serious di/dt at a herd of fault event. However use of self controlled active device rather than SCRs in the proposed structure in practice and prompts more expansible cost [13–17]. Due to power loss perspective intended FCL has the typical loss on the diode bridge topology and active switch with a R_{dc}. The suitable diode of bridge topology is ON at the half period of a cycle, whilesame way the FCL loss compo the active switch is turned ON continuously. In this nent in general operation could be ensured as;

$$P_{loss} = P_R + P_D + P_{SW}$$
$$= R_{dc} + 4V_{DF}I_{ave} + V_{SWF}I_{dc}$$

where

I_{dc} dc side current which is same as peak of line current;
V_{DF} diode forward voltage drops;
V_{SWF} Switch forward voltage drop;
I_{ave} averaged diodes current in each cycle that should be equal to I_{peak}/π

3 5 Bus System Power System Network

A bus is a node at which one or many lines, one or many loads and generators are connected. In a power system each node or bus is associated with 4 quantities, such as magnitude of voltage, phage angle of voltage, active or true power and reactive power in load flow problem two out of these 4 quantities are specified and remaining 2 are required to be determined through the solution of equation. The

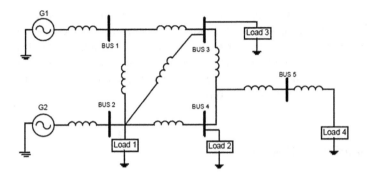

Fig. 3 5 Buses power system network without fault

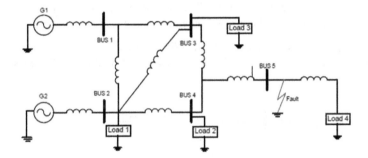

Fig. 4 5 Bus power system network with fault

quantities in the proposed 5 bus power system network would get effect due to the faults. To reduce the effects of the fault on the power system network Fault Current Limiter (FCL) has been used. Analyzing the fault in the buses is classified into 3 categories.

(i) 5 Bus power system network without fault. (ii) 5 Bus power system network with fault.

(iii) 5 Bus power system network with fault and FCL (Figs. 3 and 4).

In the above, network configuration the fault time has been taken from 0.5 to 1.5 s. At fault occurrence time the voltages behavior at the proposed 5 buses has been analyzed.

The 5 bus system is proposed with the configured FCL which limits the faults by passing the faults current from high impedance path of the FCL.

Figure 6 depicts the generation of switching states by voltage sensing scheme for the proposed FCL. In this normal operating region of the power system, the power-electronic switch should be operated as ON condition. Hence, Ldc is to be charged to peak value of line current and act as a short circuit mode. Utilizing the power electronic device and small DC reactor causes a negligible voltage drop on the FCL circuit. When a fault occurs PCC voltage goes to drops and sensors detects that dropped voltage and compare to reference p.u voltage value with relational operator generates switching pulse to turns-off the power electronic switch. At this mode, the free-wheeling diode Df conducts and supports the free-wheeling path for discharging the DC reactor. During the bridge non-conducts, the fault current passes through the providing shunt impedance part of the FCL. Imperatively, presence of large impedance in power system network to prevent the fault condition and maintain PCC voltage as a constant.

4 Simulation Results

The power circuit topology in Fig. 5 is used for simulation in the various fault condition to validate the proposed FCL with the voltage controlled device. The fault starts at 0.5 s and continues to 1.5 s. The simulation results for the proposed FCL operating under fault conditions as framed as below cases (Figs. 6).

Case 1 Bus voltages without Fault and Without FCL

The voltages at Buses 1 to 5 without fault and Without FCL are shown in Figs. 7, 8, 9, 10 and 11.

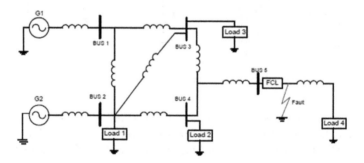

Fig. 5 5 Bus power system network with fault and FCL

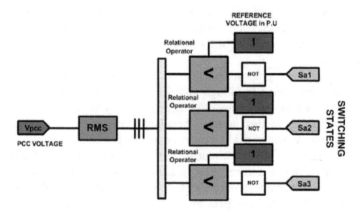

Fig. 6 Generation of switching states by using voltage sensing scheme

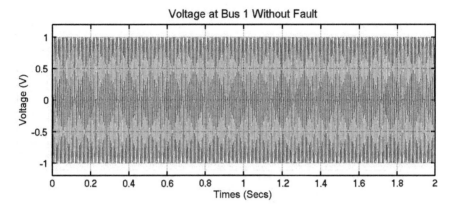

Fig. 7 Three phase voltage with constant 1 pu magnitude at bus 1, without fault condition

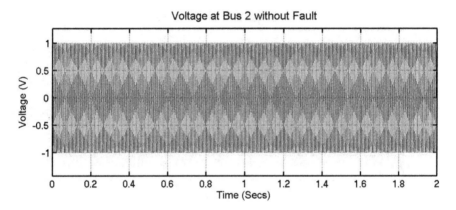

Fig. 8 Three phase voltage with constant 1 pu magnitude at bus 2, without fault condition

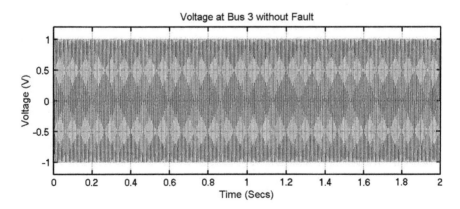

Fig. 9 Three phase voltage with constant 1 pu magnitude at bus 3, without fault condition

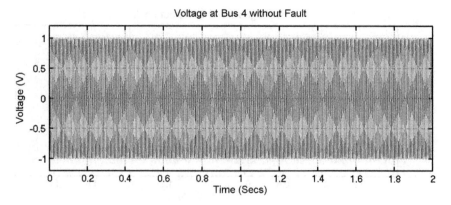

Fig. 10 Three phase voltage with constant 1 pu magnitude at bus 4, without fault condition

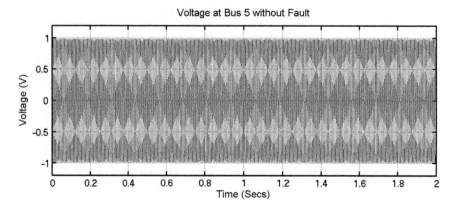

Fig. 11 Three phase voltage with constant 1 pu magnitude at bus 5, without fault condition

Case 2 The fault starts at 0.5 s and continues to 1.5 s. The simulation results for the proposed operating under fault conditions without FCL as framed as below

The voltages at Buses 1 to 5 with Fault and Without FCL are shown Figs. 12, 13, 14, 15, 16 and 17.

Case 3 The fault starts at 0.5 s and continues to 1.5 s. The simulation results for the proposed operating under fault conditions with FCL as framed as below cases (Figs. 18, 19, 20, 21, 22 and 23)

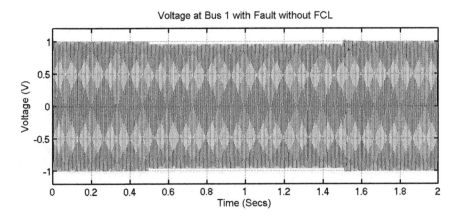

Fig. 12 Three phase voltage with 0.95 pu magnitude variation at bus 1, during fault condition from 0.5 to 1.5 s

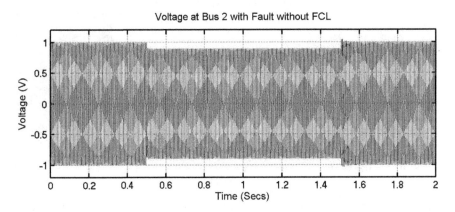

Fig. 13 Three phase voltage with 0.92 pu magnitude variation at bus 2, during fault condition from 0.5 to 1.5 s

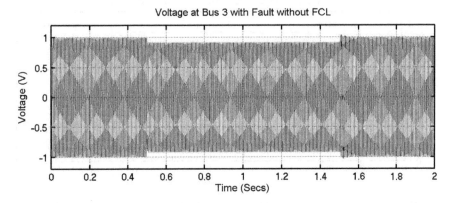

Fig. 14 Three phase voltage with 0.9 pu magnitude variation at bus 3, during fault condition from 0.5 to 1.5 s

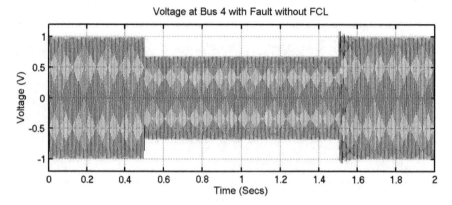

Fig. 15 Three phase voltage with 0.68 pu magnitude variation at bus 4, during fault condition from 0.5 to 1.5 s

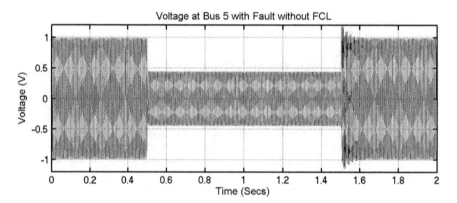

Fig. 16 Three phase voltage with 0.45 pu magnitude variation at bus 5, during fault condition from 0.5 to 1.5 s

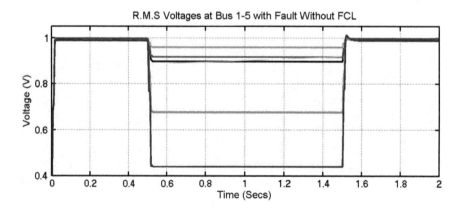

Fig. 17 R.M.S Values of the voltages of the buses 1–5 with Fault duration from from 0.5 to 1.5 s and Without FCL from these wave forms it is identified that fault severity is high at bus 5

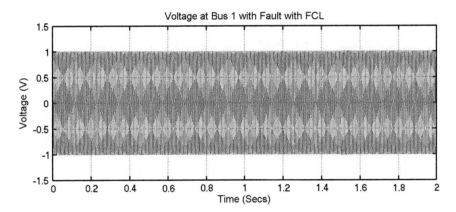

Fig. 18 Voltage at bus 1 is 0.99 pu even at fault condition due to the presence of FCL

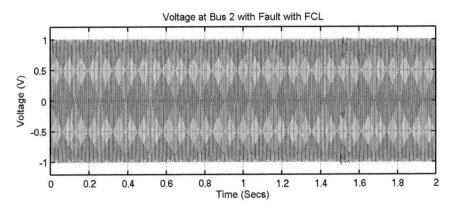

Fig. 19 Voltage at bus 2 is 0.92 pu even at fault condition due to the presence of FCL

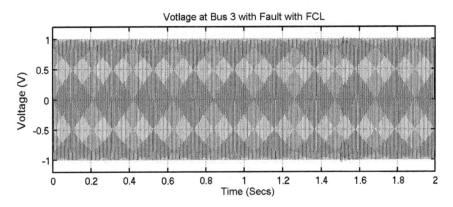

Fig. 20 Voltage at bus 3 is 0.91 pu even at fault condition due to the presence of FCL

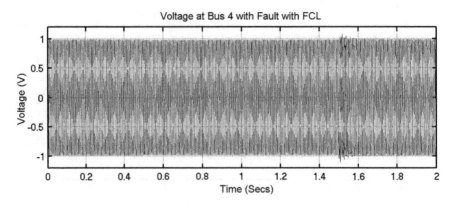

Fig. 21 Voltage at bus 4 is 0.982 pu even at fault condition due to the presence of FCL

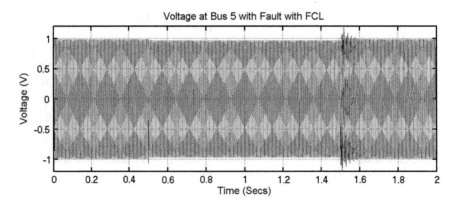

Fig. 22 Voltage at bus 5 is 0.987 pu even at fault condition due to the presence of FCL

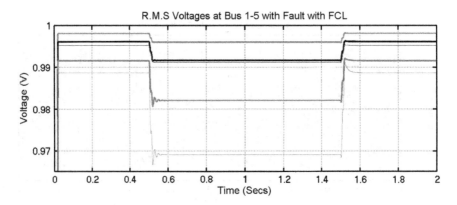

Fig. 23 R.M.S voltages at bus 1–5 with a very less variation in voltage magnitude at fault condition from 0.5 to 1.5 s due to the presence of FCL

5 Conclusions

This paper highlights the design and analysis of fault current limiter (FCL) using powerful computer simulations. The basic importance of FCL in the power system network is to operate as a circuit breaking element, limiting the fault current, advocate the PCC voltage as a constant. FCL is incorporated for the resolution of short circuit level in power distribution network because of low commercial option compared to formal methods to diminish this type of fault issues. In spite of limit the fault current, FCL offers optimal merits to power supply industries, technically more perfect and economically also more perfect. Proposed FCL is tested under a 33 kV/440 V distribution feeder fed from a three phase power system network. Simulation results show the detection of fault current and activate the control action; it may send the firing signal to IGBT switch to divert the fault current through limiting reactor by using voltage controlled scheme. Comparison of fault current, PCC voltage, shunt impedance values, DC reactor currents between the system without FCL and system with FCL for various types of multiple fault conditions and the operation of proposed FCL topology is validated and confirmed with the simplified topology based on test system results are attained from a simulation analysis.

References

1. H.-R. Kim, S.-E. Yang, S.-D. Yu, H. Kim, W.-S. Kim, K. Park, O.-B. Hyun, B.-M. Yang, J. Sim, Y.-G. Kim, Installation and testing of SFCLs. IEEE Trans. Appl. Supercond. **22**(3), (2012)
2. M. Firouzi, G.B. Gharehpetian, M. Pishvaie, Proposed new structure for fault current limiting and power quality improving functions. in *International Conference on Renewable Energies and Power Quality (ICREPQ'10)* Granada (Spain), 23rd to 25th March, 2010
3. N. Ertugrul, A.M. Gargoom, W.L. Soong, Automatic classification and characterization of power quality events. IEEE Trans. Power Del. **23**(4), 2417–2425 (2008)
4. S.M. Blair, C.D. Booth, N.K. Singh, G.M. Burt, Analysis of energy dissipation in resistive superconducting fault-current limiters for optimal power system performance. IEEE Trans. Appl. Supercond. **21**(4), (2011)
5. M. Brenna, R. Faranda, E. Tironi, A new proposal for power quality and custom power improvement: Open UPQC. IEEE Trans. Power Del. **24**(4), 2107–2116 (2009)
6. W.M. Fei, Y. Zhang, Z. Lü, Novel bridge-type FCL based on self turnoff devices for three-phase power systems. IEEE Trans. Power Del. **23**(4), 2068–2078 (2008)
7. E. Babaei, M.F. Kangarlu, M. Sabahi, Mitigation of voltage disturbances using dynamic voltage restorer based on direct converters. IEEE Trans. Power Del. **25**(4), 2676–2683 (2010)
8. M. Moradlou, H.R. Karshenas, Design strategy for optimum ratingselection of interline DVR. IEEE Trans. Power Del. **26**(1), 242–249 (2011)
9. S. Quaia, F. Tosato, Reducing voltage sags through fault current limitation. IEEE Trans. Power Del. **16**(1), 12–17 (2001)
10. L. Chen, Y. Tang, Z. Li, L. Ren, J. Shi, S. Cheng, Current limiting characteristics of a novel flux-coupling type superconducting fault current limiter. IEEE Trans. Appl. Supercond. **20**(3), 1143–1146 (2010)

11. Y. Cai, S. Okuda, T. Odake, T. Yagai, M. Tsuda, T. Hamajima, Study on three-phase superconducting fault current limiter. IEEE Trans. Appl. Supercond. **20**(3), 1127–1130 (2010)
12. G.T. Son, H.-J. Lee, S.-Y. Lee, J.-W. Park, A study on the direct stability analysis of multi-machine power system with resistive SFCL. IEEE Trans. Appl. Supercond. **22**(3), (2012)
13. J.-S. Kim, S.-H. Lim, J.-C. Kim, Study on application method of superconducting fault current limiter for protection coordination of protective devices in a power distribution system. IEEE Trans. Appl. Supercond. **22**(3), (2012)
14. J. Miret, M. Castilla, A. Camacho, L. Garci de Vicuna, J. Matas, Control scheme for photovoltaic three-phase inverters to minimize peak currents during unbalanced grid-voltage sags. IEEE Trans. Power Electron. **27**, 4262–4271 (2012)
15. A. Abramovitz, K.M. Smedley, Survey of solid-state fault current limiters. IEEE Trans. Power Electron. **27**, 2770–2782 (2012)
16. M. Fotuhi-Firuzabad, F. Aminifar, I. Rahmati, Reliability study of HV substations equipped with the fault current limiter. IEEE Trans. Power Del. **27**, 610–617 (2012)
17. J. Sheng, Z. Jin, B. Lin, L. Ying, L. Yao, J. Zhang, Y. Li, Z. Hong, Electrical-thermal coupled finite element model of high temperature superconductor for resistive type fault current limiter. IEEE Trans. Appl. Supercond. **22**, 5602004 (2012)

Energy Efficiency in Cognitive Radio Network: Green Technology Towards Next Generation Networks

Seetaiah Kilaru, Y.V. Narayana and R. Gandhiraja

Abstract Energy efficiency of mobile network is always a challenging task. From the past decade, it is observable that the users who are using multimedia services are increasing in a rapid way. These multimedia applications require higher data rates. High data rates will consume more energy of mobile network, which results poor energy efficiency. To meet higher data rates and to achieve energy efficiency, Cognitive Mobile Network with small cell model was explained in this paper. Dynamics of the power grid also have significant impact on mobile networks, hence smart grid implementation was proposed instead of traditional power grid. Most of the existed studies on cognitive mobile network focused on spectrum sensing only. This paper focused on the cognitive radio network implementation by considering spectrum sensing and smart grid environment. An iterative algorithm was proposed to attain equilibrium condition to the problem. Interference management and energy efficient power allocation were achieved with the introduction of smart grid. Simulation results proved that optimum power allocation and energy efficiency are possible with the introduction of smart grid in the cognitive network.

Keywords Cognitive radio network · Green technology · Energy efficiency · Smart grid

S. Kilaru (✉) · Y.V. Narayana
Department of Electronics & Communication Engineering,
Tirumala Engineering College, Narasaraopet, Andhra Pradesh, India
e-mail: dr.seetaiah@gmail.com

Y.V. Narayana
e-mail: narayana_yv@yahoo.com

R. Gandhiraja
Department of Computer Science Engineering,
Tirumala Engineering College, Narasaraopet, Andhra Pradesh, India

© Springer India 2016
S.C. Satapathy et al. (eds.), *Microelectronics, Electromagnetics and Telecommunications*, Lecture Notes in Electrical Engineering 372,
DOI 10.1007/978-81-322-2728-1_27

1 Introduction

The issue of higher data rate is directly related with higher power consumption. If the device is providing high data rates will use more power when compared to traditional device. From the network provider point of view, they are spending almost half of their total amount only on power expenses. High data rates are demanding high energy which leads to high power consumption. The power consumption has a considerable impact on environment due to CO_2 emission. All these issues led to think about the energy efficiency of mobile networks.

Conversion of homogeneous networks into Heterogeneous Network (HETNET) [1] is one of the techniques which can increase the energy efficiency. Macro cell can provide the large coverage area, but they are not providing high data rates for multimedia applications (particularly at cell edges). To meet these constraints, micro cells were introduced in macro cell region to increase data rates for multimedia applications [1]. Micro cell covers small coverage region, hence they require less power [2]. They can provide higher data rates in the vicinity of microcell. Microcell base stations are energy efficient, as they are using the small power level. If we increase number of microcells, then addressing the handoff problem is always a problematic task [3]. Improper handoff rates will degrade the performance of the network. Hence, there should be a joint deployment of macro cell with different micro cell will solve energy efficiency problem with proper handoff rates [4].

The concept of Cognitive Radio Network (CRN) is introduced to address the problems of Spectral Efficiency (SE). With CRN, it is possible to improve SE. CRN will collect information about current spectrum usage and find the unused frequencies [5]. After getting the list of all unused frequencies, it will use them in proper way. There is huge data rate requirement for mobile networks to handle multimedia communications [1, 4]. Scarcity of the spectrum is one of the reason for this problem [6], hence we should use spectrum in an efficient way.

There are many studies which can address energy efficiency in CRN and HETNET. Seetaiah et al. [1] presented solution to maximize CRN efficiency. Lasaulce et al. proposed power control mechanism in wireless networks. Kilaru et al. [6] studied cooperative sensing schedule algorithm with the aid of partially observable Markov decision process. Treust et al. proposed distributed power control mechanism in CRN [6]. All these studies cannot consider the power grid, which can provide power to all communication networks. Power blackout is burning issue in all traditional power grids [7]. Optimized electricity transmission and distribution will convert conventional grid to smart grid. With the help of intelligence control algorithms and network technologies, it is possible to reduce the power usage. In this paper, we explained the green cognitive mobile networks with small cells.

2 Design Strategies

Cognitive radio environment can be implemented with different micro cells which are power by a smart grid.

2.1 CRN with Micro Cell

Let us assume that the entire region covered by Macro cell Base station and several number of small cells. This set up will convert homogeneous structure into heterogeneous network [1]. These microcells connected to macro cell via broadband connection. Each microcell has defined capacity and serves for limited number of users. Macro cell should aware the spectrum usage of microcell [2, 8, 9]. This can be done by cognitive radio technology. With this technology, micro cell also monitor surrounding radio spectrum and access the channels in intelligent way. The cellular system under this condition will operate on time slotted principle. In every time slot, the spectrum resource licensed to macro base station is divided into number of sub channels. The communication between macro cell and its users can be done through OFDMA [7, 9].

2.2 Electricity Consumption Design

The number of information bits that are transmitted per unit transmit energy cycle is normally taken as a measurement for energy efficiency. For a proper tradeoff between energy efficiency and data rate, we should not only consider the transmitted energy but also other ways of energy consumption [4, 7]. Broadly, we can divide the total power consumption at base station into 3 types.

1. Static power (independent of bandwidth of the base station and also active antennas. It will take the power consumption of cooling system etc.)
2. Dynamic power (it depends on active antennas and bandwidth)
3. Conversation power (which defines the total power over a defined or desired amplifier efficiency).

2.3 Smart Grid Implementation with Real Time Pricing (RTP) and Demand Side Management (DSM)

Dynamic electric consumption is possible by DSM. It was designed by utility companies to satisfy the needs of defined customers. With this mechanism, we can decrease emission of dangerous gases and also to reduce electricity bills [2]. RTP is one of the efficient dynamic pricing protocols. In this protocol, pricing mechanism depends on cost of energy supply.

3 Problem Formulation

By considering the macro cell, microcell and available retailers, we can divide the smart grid process into 3 stages as shown in Fig. 1.

From Fig. 1, it is clear that stage I has retailers and offer services to both micro and macro cell with real time electricity price x_r. Now in stage II, Macro cell decides to choose the particular retailer which depends on amount of electricity consumed i.e. q_m. Based on q_m, energy efficient power allocation p_m was calculated. In stage II, macro cell is leader and it will offer interference price to microcell i.e. y. in stage III, based on the inputs of retailer's electricity price and interference price y, microcell will select the best retailer.

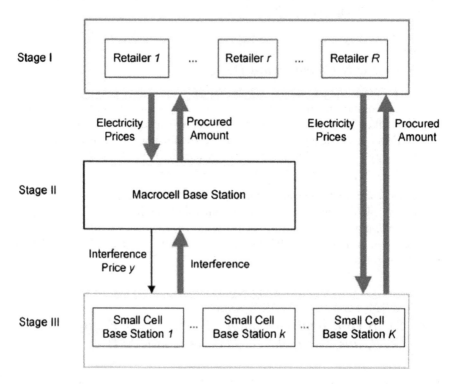

Fig. 1 Three stage process

In this process, each stage strategy is affecting next strategy. The following analysis will explain the optimization of defined stages in such a way that the one stage decision should not affect other stages in huge manner. We can't reduce this effect completely, but we can reduce it in considerable way.

4 Problem Solution

The Equilibrium between three stages is one of the finest solutions to reduce the effect. Hence we can discuss the power allocation strategy at every stage.

4.1 Power Allocation for Microcell

Microcell should choose possible retailer from the information of retailer price and also on the interference price from macro cell [1, 6, 10]. Let us assume that a microcell k with its net utility function μ_k. Let p_k is the transmitted power of microcell. From [9], we can define net utility function as

$$\frac{\partial^2 \mu_k}{\partial p_k^2} = -\frac{Wh_k^4}{\left(\sigma_k^2 + p_k h_k^2\right)ln2} < 0 \tag{1}$$

where W is the transmission bandwidth, h_k is channel gain and is a Gaussian noise. From the above equation we can derive optimum power allocation of microcell and can be obtained as

$$p_k = \frac{W}{\mu_k\left(\sum_{r=1}^{R} x_r B_{kv} s_{rk}\right) + \lambda_k g_{km}^2)ln2} - \frac{\sigma_k^2}{h_k^2} \tag{2}$$

where x_r is the real time electricity price offered by a retailer, B_{kv} is the power allocation efficiency of micro cell and s_{rk} is the parameter which denotes procurement of electricity from r or not, if this value is 1, it is taking from retailer r and if it zero it is not taking from r. λk is weight factor of micro cell and g_{km}^2 is the gain between macrocell user and microcell.

4.2 Power Allocation for Macro Cell

Macro cell can procure electricity from one of the retailer and adjust its power strategy according to individual micro cell requirements and offer interference price to micro cells [3, 9]. Hence, the net utility factor in the case of macro cell completely depends on μ_m and p_m. From the decomposition theory, we can divide this power allocation problem into two divisions. They are

1. To maximize μ_m, first we should fix the interference price and then calculate p_m.
2. Find the final interference price in optimal configuration.

$$I_m = \sigma_m^2 + \sum_{k=1}^{K} g_{km}^2 \left[\frac{W}{\mu_k (B_{kv} x_{rk}) + \lambda_{ky} g_{km}^2) ln2} - \frac{\sigma_k^2}{h_k^2} \right] \qquad (3)$$

From the above equation, optimum power allocation of Macro cell can be obtained as

$$\frac{\partial \mu_m}{\partial p_m} = \frac{W h_m^2}{(I_m + p_m h_m^2) ln2} - \alpha \left(\sum_{r=1}^{R} x_r B_{mv} s_{rm} \right) = 0 \qquad (4)$$

where, B_{mv} is the power allocation efficiency of macro cell and s_{rm} is the parameter which denotes procurement of electricity from r or not, if this value is 1, it is taking from retailer r and if it zero it is not taking from r with respect to macro cell. If macro cell selects the lowest retailer price value, and then s_{rm} should equal to 1. Hence p_m can be written as

$$p_k = \left[\frac{W}{(B_{mv} x_{rm}) \propto ln2} - \frac{I_m}{h_k^2} \right] \qquad (5)$$

4.3 Equilibrium Algorithm

To get equilibrium condition between retailer, macro cell and microcell, simple iterative algorithm was proposed.

Step1: Assign electricity price value x_r foe all retailers
Step 2: Macro cell should decide the interference price y
Step 3: Macro cell should select the particular retailer and also about how much to procure
Step 4: Microcell has to perform power allocation with respect to energy efficiency.
Step 5: All retailers has to update their prices by following the condition

$$x_r(t) = \beta_r(x_{-r}[t-1])$$

where, β_r is the best response function of retailer r, x_{-r} is the offered price of other retailers and t − 1 represents the iteration level.
Step 6: Repeat the steps 1–5 until to get the following condition

$$\frac{\|x(t) - x(t-1)\|}{\|x(t-1)\|} \le \varepsilon$$

Here, ε should be very small value, considerably 1–2 %.

Step 7: Stop the process of iteration.

5   Simulation Setup and Results

Let us assume the defined conditions such as

Radius of Macro cell = 1000 m

Radius of Microcell = 20 m

Penetration loss = 20 dB for exterior and 10 dB for interior

Path loss between microcell user and its macro cell base station =

15.3 + 37.6 log (distance between macro cell base station and microcell user + penetration loss of exterior wall)

Path loss between microcell base station and its user = 46.86 + 20 log (distance between microcell and its user + penetration loss of interior wall)

W = 5

Static power = 88 W

Dynamic power = 75 W

Power allocation efficiency = 0.34

$\mu_k = \lambda_k = 1$

$\alpha = 0.5$ and $\beta = 10$

Fig. 2 Effect of α on energy efficiency

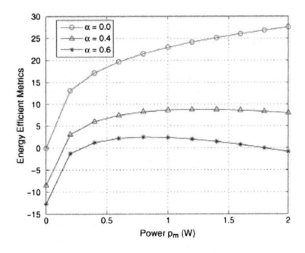

Fig. 3 Total electricity cost
with retailer for existed and
proposed scheme

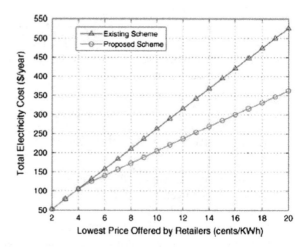

Each base station has 50 channels.

Figure 2 shows the effect of α on energy efficiency metric.

As the α increases, the energy efficiency will decrease.

Figure 3 shows the tradeoff between total electricity cost and prices offered by retailers for existed and proposed schemes.

6 Conclusion

In this paper, cognitive radio network implementation with deployment of microcell was discussed. Energy efficient power allocation possibilities were discussed by adjusting the amount of electricity of macro and micro cells. It was also shown that, interference price problem and optimum power allocation was possible with the introduction of smart grid. To achieve equilibrium condition to the problem, an iterative algorithm was proposed. Finally, it was shown that the proposed method was much efficient with respect to energy efficiency and total electrical cost.

References

1. S. Kilaru, Y. Ashiwini Prasad, K. Sai Kiran, N.V. Sarath Chandra, Design and analysis of heterogenious networks. Int. J. Appl. Eng. Res. **9**(17), 4201–4208
2. D.P. Palomar, M. Chiang, A tutorial on decomposition methods for network utility maximization. IEEE J. Sel. Areas Commun. **24**(8), 1439–1451 (2006)
3. D. Fudenberg, J. Tirole, *Game Theory* (MIT Press, Cambridge, 1993)
4. S. Bu, F.R. Yu, Y. Cai, P. Liu, When the smart grid meets energy efficient communications: Green wireless cellular networks powered by the smart grid. IEEE Trans. Wireless Commun. **11**(8), 3014–3024 (2012)

5. H. Zhang, A. Gladisch, M. Pickavet, Z. Tao, W. Mohr, Energy efficiency in communications. IEEE Commun. Magazine **48**(8), 48–49 (2010)
6. S. Kilaru, K. Harikishore, T. Sravani, C.L. Anvesh, T. Balaji, Review and analysis of promising technologies with respect to fifth generation networks. in *Networks and Soft Computing (ICNSC), 2014 First International Conference on*, pp. 248–251, IEEE (2014, August)
7. Y. Narahari, D. Garg, R. Narayanam, H. Prakash, *Game Theoretic Problems in Network Economics and Mechanism Design Solutions* (Springer, London, 2009)
8. S. Kilaru, A. Gali, Improving quality of service of femto cell using optimum location identification. IJCNIS **7**(10), 35–41 (2015). doi:10.5815/ijcnis.2015.10.04
9. J. Hoadley, P. Maveddat, Enabling small cell deployment with HetNet. IEEE Wirel. Commun. **19**(2), 4–5 (2012)
10. S. Bu, F.R. Yu, Y. Cai, P. Liu, When the smart grid meets energy efficient communications: green wireless cellular networks powered by the smart grid. IEEE Trans. Wirel. Commun. **11**(8), 3014–3024 (2012)

Performance Evaluation of Rectangular Enclosure for Any Arbitrary Polarization Angle

G. Kameswari and P.V.Y. Jayasree

Abstract This paper intends to develop an analytical formulation for the shielding effectiveness of a rectangular enclosure with rectangular aperture having multiple holes of identical and distinct sizes with arbitrary polarization angle. A single mode normal incident transverse electric wave containing arbitrary polarization is decomposed into two orthogonal components in such a way that the electric field is situated perpendicular to both the length and the breadth sides of the aperture. Analysis of each individual component as per TE_{10} mode is made by implementing Transmission Line Model and then combined to obtain the se. It is important to consider the mutual admittance between the aperture holes for accurate estimation of the SE. The formula under consideration is further extended to include the suitable admittance which takes the mutual intertwining between the aperture holes into account to signify the group of aperture holes. Simulation also has been performed to verify the SE of rectangular enclosure with identical and distinct sizes of aperture holes. It is evident from the simulation results that the SE is at its best when mutual admittance is considered and at the worst when mutual admittance is ignored. It can also be observed that the SE is comparatively more for aperture holes of identical size than aperture holes of distinct sizes.

Keywords Transmission line model · Transverse electric field · The se · Arbitrary polarization · Aperture with identical and distinct sizes of holes · Mutual admittance

G. Kameswari (✉) · P.V.Y. Jayasree
Department of ECE, GIT, GITAM University, Visakhapatnam, A.P., India
e-mail: kameshwarreephd@gmail.com

P.V.Y. Jayasree
e-mail: y_pappu@rediffmail.com

© Springer India 2016
S.C. Satapathy et al. (eds.), *Microelectronics, Electromagnetics and Telecommunications*, Lecture Notes in Electrical Engineering 372,
DOI 10.1007/978-81-322-2728-1_28

1 Introduction

Electromagnetic shielding effectively reduces emissions or enhances the immunity factors of electronic apparatus [1, 2]. But the shielding effect is largely determined by the aperture located in the wall of the enclosure. Aperture in the wall at times becomes indispensable for free flow of air and dissipation of heat. Thus the SE of the enclosure with aperture is a hot topic for research. The SE can be found by several numerical methods or analytical approach.

Several studies have focused on this concept of the SE of enclosure with apertures. A focus on the SE of enclosure with center aperture and TE_{10} mode, using transmission line method can be found in [3–7]. A further analysis can be observed from [8] including the high order mode and enclosure-loss limited to centre aperture with only one perpendicular polarization of the happening electromagnetic field. Supplementary observations on off-center aperture and high order mode and enclosure-loss can be obtained from [9, 10]. The consequences of normal incident electromagnetic wave having an arbitrary angle of polarization on the electric SE of a rectangular enclosure with aperture, is evident from the studies carried out in [11]. The effect of mutual admittance on SE of rectangular enclosure with multiple number of apertures for normal incident electromagnetic wave is evaluated in [12]. But all the previous studies ignored the vital aspect of mutual admittance between arrays of holes in an aperture when electromagnetic wave incidence has any arbitrary polarization angle. The validity of the studies becomes questionable if the apertures are too close. These restrictions have enlarged the scope for further research in order to extend the model for wider applications. An analytical approach of the transmission line provides a comparatively less expensive alternative to the numerical techniques [13–18], which saves considerable resources of computing.

It is important to consider the mutual admittance between the apertures for accurate estimation of the se. The formula under consideration is further extended to include the suitable admittance which takes the mutual coupling between the arrays of holes into account to signify the group of holes. The designed formula deals with the consequences of normal incident electromagnetic wave, having an arbitrary angle of polarization on the electric SE of a rectangular enclosure with aperture with and without considering the mutual admittance between the holes. Hence, the decomposition of the electric field takes place splitting into two orthogonal components. Analysis of each individual component as per TE_{10} wave mode is made and then combined to obtain the se. This paper also studies rectangular enclosure with rectangular aperture of multiple holes of distinct sizes.

2 Analysis of Rectangular Enclosure with Apertures

According to the transmission line theory, the enclosure with aperture can be modeled as in Fig. 1. The location of the test, P, is situated in the middle portion of the enclosure. The normal irradiation of the electromagnetic wave is towards the

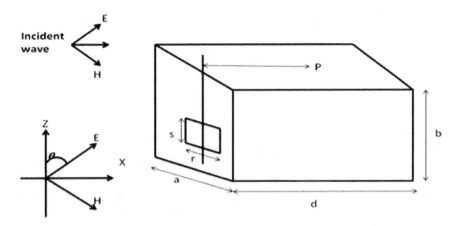

Fig. 1 Rectangular enclosure with rectangular aperture

front face of the cavity having an aperture of multiple holes of identical and distinct sizes shown in Figs. 2 and 3 respectively. The voltage V_0 and the impedance $Z_0 = 377$ are used to represent the radiating source. The representation of the enclosure is made by loaded lines of transmissions, whose characteristic impedance

Fig. 2 Identical size multiple rectangles partially perforating the wall of an enclosure

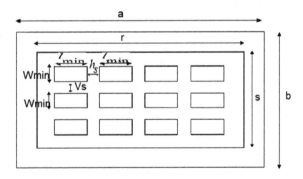

Fig. 3 Distinct sizes multiple rectangles partially perforating the wall of an enclosure

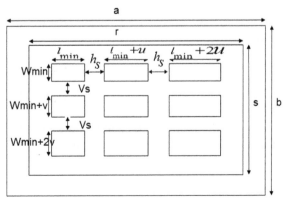

and propagation constants are Z_{TE} and γ respectively, determined by waveguide formulas.

2.1 SE Without Mutual Admittance

The decomposition of incident wave into two orthogonal components may be attributed to any polarization angle θ, the electric strength of which is $E \cos \theta$ and $E \sin \theta$ respectively. Consequentially, we can obtain the source voltage $V_o \cos \theta$ and $V_o \sin \theta$ in the corresponding short end circuit.

Let us consider the condition where the electric field stays perpendicular to length side of the aperture. In such cases, the impedance of the aperture can be represented as

$$A_{zl} = \frac{1}{2}\frac{r}{a}jS_z \tan\frac{\beta r}{2}$$
$$\text{Where } \beta = \frac{2\pi}{\lambda_0} \tag{1}$$

According to [19, 20], the characteristic impedance of the transmission line can be given by

$$S_z = 120\pi^2 \left[\ln\left(2\frac{1+\sqrt[4]{1-h^2}}{1-\sqrt[4]{1-h^2}} \right) \right]^{-1} \tag{2}$$

where $h = w_e/b$, $\phi = \frac{4\pi}{t}$ and the effective width w_e is

$$w_e = s - \frac{5}{\phi}[1 + \ln(\phi s)] \tag{3}$$

where 't' is the thickness of the enclosure wall, 'r' is the length and 's' is width of the aperture.

Now from the perspective of incidence wave, the equivalent source voltage and impedance are

$$V_{al} = V_0 \cos\theta A_{zl}/(Z_0 + A_{zl}) \tag{4}$$

$$Z_{al} = Z_0 A_{zl}/(Z_0 + A_{zl}) \tag{5}$$

Due to the existence of multiple modes such as $TE_{10}, TE_{20}, TE_{m0}$ etc., the characteristic impedance and propagation constants are

$$Z_{TEl} = Z_0/\sqrt{1 - (m\lambda_0/\lambda_c)^2} \tag{6}$$

$$\gamma_l = \beta\sqrt{1 - (m\lambda_0/\lambda_c)^2} \tag{7}$$

From the perspective of P, the corresponding source voltage V_{bl}, corresponding source impedance Z_{bl} and load impedance Z_{cl} are

$$V_{bl} = \frac{V_{al}}{\cos(\gamma_l p) + jg_l \sin(\gamma_l p)}$$
$$\text{Here } g_l = \frac{Z_{al}}{Z_{TEl}} \tag{8}$$

$$Z_{bl} = Z_{TEl}\frac{g_l + j \tan(\gamma_l p)}{1 + jg_l \tan(\gamma_l p)} \tag{9}$$

$$Z_{cl} = jZ_{TEl} \tan \gamma_l(d - p) \tag{10}$$

For TE_{m0} mode, the total voltage at P is $V_{tl} = \sum_l V_l$ where $V_l = V_{bl}Z_{cl}/Z_{bl} + Z_{cl}$.

Let us now consider another condition where the direction of the electric field remains perpendicular to the breadth side of the aperture. In this condition we can investigate it in the identical way as in the previous condition, except with small modifications of the parameters $V_o \cos\theta$, r and a in the above formulation into $V_o \sin\theta$, s and b respectively.

As a result, for TE_{m0} mode, the total voltage at P is $V_{ts} = \sum_s V_s$ where $V_s = V_{bs}Z_{cs}/Z_{bs} + Z_{cs}$.

The obtained end results regarding the two conditions with the sum voltage at P can be represented as

$$V_{total} = \sqrt{V_{tl}^2 + V_{ts}^2} \tag{11}$$

In the absence of shielding enclosure, the voltage at P is $V_o/2$. Consequently, SE is

$$SE = -20 \log_{10}|(2V_{total}/V_0)| \tag{12}$$

2.2   SE with Mutual Admittance

It is important to consider the mutual admittance between the apertures for accurate estimation of shielding.

The normalized shunt admittance for aperture of multiple holes of identical sizes displayed in Fig. 2, is given by [11].

$$\frac{Y_{array}}{Y_o} = -j\frac{3h_s v_s \lambda_0}{\pi d_1^3} + j\frac{288}{\pi \lambda_0 d_1^2} \cdot \left[\sum_{\substack{m=0 \\ m \neq odd}}^{\infty} \sum_{\substack{n=0 \\ n \neq odd}}^{\infty} (\varepsilon_m n^2/v_s^2 + \varepsilon_n m^2/h_s^2)J_1^2(X)\right] \quad (13)$$

Here the free-space wavelength and intrinsic admittance are respectively indicated by λ_0 and Y_0. v_s and h_s are the vertical and horizontal separations between the holes and $d_1 = 0.636(l_{min} + w_{min})$ for small rectangular holes. Here 'l_{min}' constitutes length of small rectangular hole and 'w_{min}' stands for width of small rectangular hole. In case of h_s, v_s, d_1 being much less than the wave length the second term in (13) can be ignored. The impedance $Z_{array} = 1/Y_{array}$ models the multiple small rectangles connecting the free space with the waveguide. Figure 2 shows same size multiple rectangles partially perforating the wall of an enclosure and its wall impedance Z_{array}^1 is a fraction of Z_{array}. According to impedance ratio concept

$$Z_{array}^1 = Z_{array} \times \left(\frac{r \times s}{a \times b}\right) \quad (14)$$

where length r and width s of the array are

$$r = (h_s/2) + (m_1 - 1)h_s + (h_s/2) \quad (15)$$

$$s = (v_s/2) + (n_1 - 1)v_s + (v_s/2) \quad (16)$$

Here, m_1 and n_1 constitute the number of multiple small holes in each row and column of the array respectively. We can get the SE by substituting (14) in (4) proceeding with the same process.

The normalized shunt admittance for aperture of multiple holes of distinct sizes as shown in Fig. 3, is given by

$$\frac{Y_{array}}{Y_o} = -j\frac{3h_s v_s \lambda_0}{\pi d_1^3} + j\frac{288}{\pi \lambda_0 d_1^2} \cdot \left[\sum_{\substack{m=0 \\ m \neq odd}}^{\infty} \sum_{\substack{n=0 \\ n \neq odd}}^{\infty} (\varepsilon_m n^2/v_s^2 + \varepsilon_n m^2/h_s^2)J_1^2(X)\right] \quad (17)$$

Here the free-space wavelength and intrinsic admittance are respectively indicated by λ_0 and Y_0. v_s and h_s are the vertical and horizontal separations between the holes and $d_1 = 0.636(l_{min} + u + w_{min} + v)$ for small rectangular holes. Here 'l_{min}' constitutes minimum length of small rectangular hole and 'w_{min}' stands for minimum width of small rectangular hole. 'u' is increase in length from hole to hole and 'v' is increase in width from hole to hole. In case of h_s, v_s, d_1 being much less than the wave length the second term in (17) can be ignored. The impedance $Z_{array} = 1/Y_{array}$

models the multiple small rectangles connecting the free space with the waveguide. Figure 3 shows distinct size multiple rectangles partially perforating the wall of an enclosure and its wall impedance Z^1_{array} is a fraction of Z_{array}.

According to impedance ratio concept

$$Z^1_{array} = Z_{array} \times \left(\frac{r \times s}{a \times b}\right) \tag{18}$$

Here, it can be perceived that the rectangular apertures are in geometric progression corresponding to the dimensions. Then, the resultant length and width of the rectangular aperture can be calculated as

$$r = m_1\left(l_{min} + u\right) + (m_1 - 1)h_s \tag{19}$$

$$s = n_1\left(w_{min} + v\right) + (n_1 - 1)v_s \tag{20}$$

Here, m_1 and n_1 constitutes the number of multiple small holes in each row and column of the array respectively. We can get the SE by substituting (18) in (4) proceeding with the same process.

3 Results

In simulation to the design, it is observed with the consideration of a rectangular enclosure of size 400 mm × 160 mm × 400 mm (a × b × d) with a 100 mm × 10 mm (r × s) aperture and a copper wall thickness of the enclosure as 1.5 mm, the

Fig. 4 SE for distinct polarization angles $\theta_1 = 30°$, $\theta_2 = 45°$

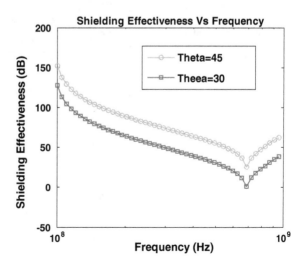

Fig. 5 Comparison of SE in electric field being perpendicular to the length and width of the aperture at $\theta = 30°$

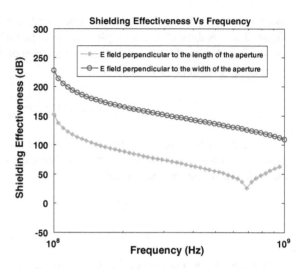

following results are generated. Figure 4 shows calculated SE at distinct polarization angles θ by using TLM. It can be observed that SE increases as θ increases.

Figure 5 shows the comparison of the SE in electric field being perpendicular to the length and the breadth of the aperture. Enhanced SE can be obtained when the electric field is at a 90° angle to the breadth of the aperture as the θ increases compared to the electric field, when it is at a 90° angle to the length of the aperture.

Figure 6 shows the comparison of the SE when incident wave of arbitrary polarization with and without mutual admittance between the aperture holes. Approximately an increase of 60 dB in SE can be obtained, when it is calculated by

Fig. 6 Comparison of SE when the incident wave of arbitrary polarization with and without mutual admittance between the aperture holes

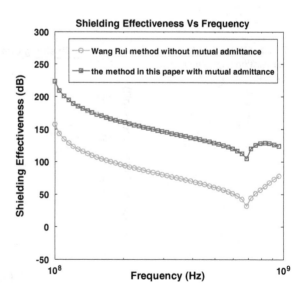

including the mutual admittance between the apertures compared to the SE, ignoring the mutual admittance between the apertures.

Figure 7 shows the comparison of the SE of rectangular enclosure with rectangular aperture of identical and distinct sizes of holes. The SE is considerably higher when the aperture has multiple holes of identical size than when the aperture has multiple holes of distinct sizes, since the total number of holes becomes reduced when distinct sizes are used than when identical size is considered. The theory of transmission line predicts that the SE increases corresponding to the enhanced number of holes by keeping the aperture area constant.

Figure 8 shows the result with array of 3 × 3 and 6 × 6 identical and distinct sizes of aperture holes by changing the aperture area. It can be observed that the SE

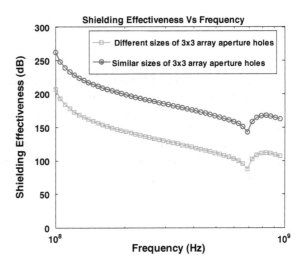

Fig. 7 Rectangular enclosure with rectangular aperture of identical and distinct sizes of holes

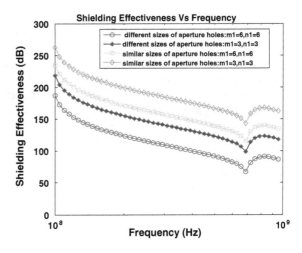

Fig. 8 SE of rectangular aperture having increased number of holes by changing the area of the aperture

Fig. 9 Calculated SE at
distinct 'u' and 'v' values

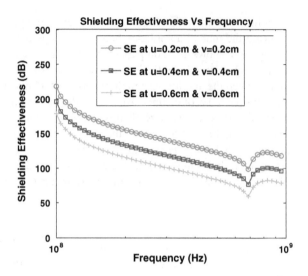

decreases if the number of holes in the aperture increases by changing the aperture area in both the cases of identical and distinct sizes of holes.

Figure 9 shows the calculated SE at distinct 'u' and 'v' values. SE can be calculated for 3 × 3 array having nine holes of distinct sizes by changing the values of u and v as 0.2, 0.4 and 0.6 cm. It can be observed that the SE is considerably higher when the aperture has less increase in both length and width from hole to hole than when the aperture has more increase in both length and width from hole to hole.

4 Conclusion

The resultant derivation is in accordance with the study and design analysis of the SE of the enclosure with an aperture analyzed by incident wave with any angle of polarization, with and without taking the mutual admittance between the apertures. Transmission Line Model is employed to analyze each orthogonal component of the incident wave. Then, they are combined into a formulation between the SE and the polarization angle θ. The final results are successfully obtained and verified through MATLAB R2009a simulation tool. It can be observed that an improved SE can be obtained when the electric field is at a 90° angle to the breadth of the aperture as the θ increases in comparison to, when the electric field is at a 90° angle to the length of the aperture. It can be concluded that considering the mutual admittance between the apertures for an accurate estimation of the shielding is essential, because the SE can be approximately increased by 60 dB due to this condition. Considering the aperture with multiple holes of identical sizes can enhance the SE considerably as has been established through the simulation than with variation in sizes. An analytical approach of the transmission line provides a comparatively less expensive alternative to the other existing techniques (FDTD, FEM, MOM and Hybrid) which saves considerable resources of computing. This analytical method has significant advantage in terms of solution time.

References

1. Y. Gao, *Shielding and grounding* (Beijing University of Posts and Telecommunications, China, 2004)
2. S. Celozzi, R. Araneo, G. Lovat, *Electromagnetic Shielding* (Wiley, 2008)
3. I. Belokour, J. Lovetri, S. Kashyap, Se estimation of enclosures with apertures. In: *IEEE Electromagnetic Compatibility Symposium Proceedings*, pp. 855–860 (2000)
4. P. Argus, P. Fischer, A. Konrad, et al., Efficient modeling of apertures in thin conducting screens by the TLM method. In: *IEEE International Symposium on Electromagnetic Compatibility*, **1**, 101–106 (2000)
5. T. Konefal, Improved aperture model for shielding prediction. In: *IEEE International Symposium on Electromagnetic Compatibility*, vol. 1 (2007)
6. P. Dehkhoda, A. Tavakoli, R. Moini, An efficient se calculation (a rectangular enclosure with numerous apertures). In: *IEEE International Symposium on Electromagnetic Compatibility* (2007)
7. P. Dehkhoda, A. Tavakoli, R. Moini, An efficient and reliable se evaluation of a rectangular enclosure with numerous apertures. IEEE Trans. Electromagn. Compat. **50**, 208–212 (2008)
8. I. Belokour, J. Lovetri, S. Kashyap, A higher order mode transmission line model of the se of enclosures with apertures. In: *IEEE International Symposium on Electromagnetic Compatibility*, vol. 2 (2001)
9. F. Ahamad Po'ad, Z. Mohd Jenu Mohd, C. Christopoulos, et al., Analytical and experimental study of the se of a metallic enclosure with off-centered apertures. In: *17th International Zurich Symposium on Electromagnetic Compatibility* (2006)
10. D. Shi, Y. Shen, Y. Gao, *3 High-order mode transmission line model of enclosure with off-center aperture* (Beijing University of Posts and Telecommunications, Beijing, China, 2007)
11. W. Rui, Y. Gao, D. Shi, et al., Application of the transmission line method to the se with incident wave of arbitrary polarization. Beijing University of Posts and Telecommunications (Beijing, China, 2009)
12. X. Zhang, Y. Zhou, R. Ma, et al., Se of rectangular enclosure with higher order modes of transmission line model. CEEM 2012/shang'hai (2012)
13. S. Benhassine, L. Pichon, W. Tabbara, An efficient finite-element time-domain method for the analysis of the coupling between wave and shielded enclosure. IEEE Trans. Electromagn. Compat. **38**(2), 709–712 (2002)
14. G. Cerri, R. De Leo, V.M. Primiani, Theoretical and experimental evaluation of the electromagnetic radiation from apertures in shielded enclosures. IEEE Trans. Electromagn. Compat. **34**, 423–432 (1992)
15. W. Ulrich, B. Achim, High frequency characteristics of modular metric, high density connector systems. In: *IEEE International Conference*, pp. 2481–2585 (2007)

16. C. Lu, B. Shanker, Generalized finite element method for vector electromagnetic problems. IEEE Trans. **55** (2007)

17. M. Li, J. Nuebel, L. James, et al., EMI from airflow aperture arrays in shielding enclosure-experiments, FDTD and MOM modeling. IEEE Trans. Electromagn. Compat. **42** (2000)

18. S. Tharf, G.I. Costache, A hybrid finite element-analytical solutions for in-homogeneously filled shielding enclosures. IEEE Trans. Electromagn. Compat. **36**, 380–385 (1994)

19. K.C. Gupta, R. Garg, I.J. Bahl, Micro strip lines and slot lines, vol. 7 (Norwood, MA, Artech House, 1979)

20. D.M. Pozar, Microwave engineering (Wiley, 2005)

PAPR Reduction in SFBC OFDM System-MCMA Approach

M. Vijayalaxmi and S. Narayana Reddy

Abstract In high speed data transmission, Orthogonal Frequency Division Multiplexing (OFDM) is a spectrally proficient method over multipath fading channel. This OFDM signal suffers from high Peak-to-average power ratio (PAPR) and Bit Error Rate (BER) (Van Nee and de Wild, Vehic. Technol. Conf. 3:2072–2076, 1998) [1]. MIMO OFDM configuration step-up the capacity of the system and diversity gain on frequency selective and time variant channels. Preference is on SFBC-MIMO-OFDM to operate fading channel much more effectively. Constant Modulus Algorithm (CMA) is highly proficient technique for alleviating high PAPR. But the main demerit associated with CMA is slow convergence rate, extensive steady state mean square error (SSMSE), and phase blind nature. We propose an enhanced modified constant modulus algorithm (MCMA) where the step size is carefully assessed to maintain a balancing between convergence rate and final accuracy. Simulations demonstrates the efficiency of proposed MCMA.

Keywords MIMO · OFDM · CMA · CCDF · SFBC. etc.

1 Introduction

In real world applications of CMA algorithm, a key parameter is the step size, which is adopted to arrange the update of equalizer weights. Computation of equalizer weights by using CMA formula for various blind adaptive equalizer algorithms. With increasing in the step size the rate of convergence will be rapid with CMA but only drawback is raise in SSMSE and it will be vice versa with lowered step size [2].

M. Vijayalaxmi (✉)
Department of ECE, S.K.I.T, Srikalahasti, A.P., India
e-mail: Mvlpr3@gmail.com

S. Narayana Reddy
Department of ECE, S.V.U, Tirupati, A.P., India
e-mail: snreddysvu@yahoo.com

© Springer India 2016
S.C. Satapathy et al. (eds.), *Microelectronics, Electromagnetics and Telecommunications*, Lecture Notes in Electrical Engineering 372,
DOI 10.1007/978-81-322-2728-1_29

327

So, the step size is the critical factor to maintain equilibrium between convergence rate and SSMSE. An innovative technique adopted to modify the step size for MCMA by balancing absolute difference error and iteration number. Jones [3] presented one of such method by adopting the channel output signal vector energy $||x(k)||^2$ with controlled the step size. Chahed et al. [4] arrange step size by taking a time varying step size specification which relies on squared Euclidian norm of the channel output vector and the equalizer output. Xiong et al. [5] adopted the lag (1) error autocorrelation function between e(k) and e(k − 1). Where, e(k) is the blind equalization output error. Zhao [6], proposed that the variable step size of CMA algorithm can be controlled by relative difference among present and past MSE. Banovi′c [7] projected this method of adjusting step size based on the size of the output radius of equalizer. Liyiet al. [8] proposed a method for adjusting the step-size criterion based on a nonlinear function of instantaneous error. Zhang [9], presented an affine projection blind equalization CMA algorithm found on quantized evaluation errors and volatile step-size. The method advocated by Xue [10] based on the location of the received signal lying on the constellation plane for adjustment of step size. MCMA has the advantage of correction of both phase rotation and spinning phase error using both modules and phase of the equalizer output. In the proposed algorithm, initially a larger step size is consider to gain the convergence and then the step size is steadily reduced to a smaller value to attain steady state mean square error (SSMSE). Blind equalization is the method for adaptive equalization which brings about a channel beyond the assistance of usual training sequence. In MCMA the crucial factor is the step size which should be carefully assessed to maintain a balancing among convergence rate and final accuracy. This paper confer a novel technique to alter the step size of MCMA based on absolute difference error and iteration number. Using complementary cumulative distribution function (CCDF) the relative results of MCMA algorithm are compared with CMA.

2 MIMO OFDM System Model

In high-data rate wireless communication, OFDM can be adopted to transform into number of parallel flat MIMO channels from frequency-selective MIMO channel. Due to this even in multipath fading environment also high data rate hefty and stable transmission can be acquired by mitigating the intricacy of the receiver. MIMO-OFDM systems attain coding gain and diversity gain by space-time coding more over the OFDM system can be accomplished with simple structure. Hence MIMO-OFDM system has become a vital role for 4G mobile communication systems. The basic structure of PAPR reduction system model is shown in Fig. 1. The SFBC contrubute two blocks each of the length N, for OFDM at the transmitter. The input blocks are encoded to achieve space frequency diversity, which are transmitted by the first and second transmit antenna respectively. The original signal is effectively recovered by feeding the incoming signals into a detector at the receiving end.

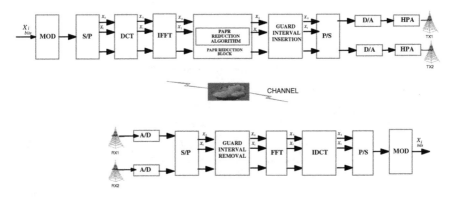

Fig. 1 PAPR reduction system model in MIMO-OFDM SYSTEMS

3 Space Frequency Block Codes (SFBC)

In this section, SFBC technique is enforced into the OFDM system. Simply, 2Tx/1Rx antenna configurations are contemplate to correlate the system behavior of MIMO OFDM system. Here, we advice the conventional MIMO SFBC-OFDM structure with 2Tx/1Rx antenna. The combination of OFDM with SFBC allowed the creation of codes known as SFBC-OFDM offering low decoding complexity and bandwidth efficiency [2].

Alamouti's scheme with 2Tx-1Rx antenna: Alamouti introduced a very transparent scheme of SFBC by allowing transmission from two antennas with the unique data rate as on each single antenna. Alamouti's algorithm uses space and frequency domain to encode data with increase in the functional behavior of the system is illustrated in Fig. 2 [11]. Thus, the Alamouti's code accomplish diversity as it transmits two symbols in two frequency slots.

In the first frequency slot, symbols s_0 and s_1 are transmitted by the transmitting antennas Tx1 and Tx2. During the next frequency slot, symbols $-s_1^*$ and s_0^* are sent. Furthermore, it is supposed that the channel, which has transmission coefficients h_1 and h_2 remains constant and frequency is flat over the two consecutive time steps. The received vector R is formed by stacking two consecutively received data samples in frequency, i.e.

$$R = Sh + w \tag{1}$$

$$where\ R = [r_0, r_1]^T,\ S = \begin{pmatrix} s_0 & s_1 \\ -s_1^* & s_0^* \end{pmatrix} \tag{2}$$

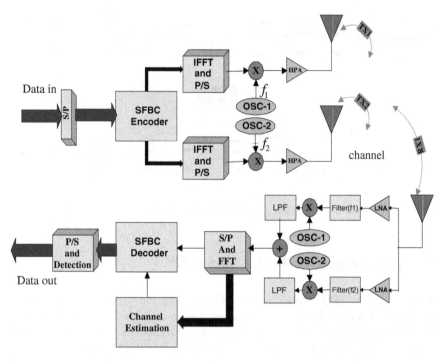

Fig. 2 2 × 1 SFBC-OFDM Transceiver

$h = [h_1, h_2]^T$ is the complex channel vector, $w = [w_1, w_2]^T$ is the noise at the receiver and S defines the SFBC.

The plot in Fig. 3 shows the BER comparative performance of Alamouti's SFBC-OFDM with different channels. The 2Tx and 1Rx SFBC-OFDM with Rayleigh channel shows outperformance than standard OFDM system.

Fig. 3 BER performance of
2 × 1 Alamouti SFBC OFDM

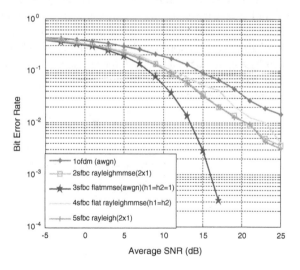

The basic concept of OFDM is to sub-divide the information bit stream into a large number of bit streams, each with low individual bit rates, which are then carried on individual orthogonal subcarriers. This transmission technique is especially suited for mitigating the effect of the multipath fading channel. SFBC-OFDM has very high frequency spectrum efficiency. Since in the OFDM system, sub-carriers are orthogonal to each other, channel spectrum overlapping is allowed, which can utilize limited spectrum resources maximally. It adequately mitigates BER.

4 PAPR

High peaks in OFDM system can be denoted as PAPR. In some literatures, it is also treated as PAR. It is usually defined as:

$$\text{PAPR} = \frac{P_{peak}}{P_{average}} = 10 \log_{10} \frac{\max\left[|x_n|^2\right]}{E\left[\left[|x_n|^2\right]\right]}$$

for continuous signals

$$\chi_n = \frac{\max_{t \in [n, n + T_s]} |x(t)|^2}{\int_n^{n+t_s} |x^2(t)| dt} \tag{3}$$

For sampled signals

$$\chi_n = \frac{\max_k |x_n[k]|^2}{E\left\{|x_n[k]|^2\right\}} \tag{4}$$

where P_{peak} represents peak output power, and $P_{average}$ is the average output power. $E[\bullet]$ denotes the prospective value, x_n represents the OFDM signals which are symbols of X_k. Mathematically, x_n is expressed as:

$$x_n = \frac{1}{\sqrt{n}} \sum_{k=0}^{N-1} X_k W_N^{nk} \tag{5}$$

High PAPR Power amplifiers are employed to access the required power level for transmission in MIMO OFDM systems. To achieve the peak power efficiency, high power amplifiers (HPAs) are functioned at or near the saturation region which causes to distortion and experience inter-modulation products between various subcarriers. A large number of high frequency carrier signals are adopted in MIMO-OFDM system for effective conveying of large number of narrow-band input signals. Large number of narrow band signals combined to form a multi-carrier signal in time domain. This gives the peak value of the signal higher than average value leads to reduction in the power efficiency of the system. To mitigate the high PAPR, techniques like Discrete Fourier Transform (DFT) Pulse Shaping Selected Mapping (SLM) and Partial Transmit Sequence (PTS) are used [12].

When CCDF, $PAPR_0$ threshold is exceed, PAPR value of OFDM signal can be expressed as:

$$PAPR(X(n) = p_{r}(PAPR(X(n)) > PAPR_{0})) \tag{6}$$

Depending on the independent data block N, SISO OFDM PAPR-CCDF represented as:

$$p = p_r(PAPR(X(n)) > PAPR_0) = 1 - (1 - e^{-PAPR_0})^N \tag{7}$$

If this equation is composed of MIMO-OFDM system, PAPR value on the ith transmitted antenna is,

$$PAPR_i \, p_r(PAPR_{MIMO-OFDM} > PAPR_0) = 1 - (1 - e^{-PAPR_0})^{M_i N} \tag{8}$$

Based on above it can be infer that MIMO-OFDM system has better PAPR performance.

5 Constant Modulus Algorithm (CMA)

CMA is a well known algorithm used to develop a new PAPR reduction approach for MIMO-OFDM systems. This combines two ideas: (1) The linear combination of time domain signals (consisting of several subcarriers) with the help of precoding weights, transparent to the receiver; (2) The mitigation in PAPR is due to minimization of modulus variations of the resulting signal, with proper designing of precoding weights. In CMA technique, the input to the channel is a constant magnitude modulated signal at each and every time instant. Any deviation in the incomming signal amplitude at the receiver from the fixed magnitude is considered as distortion due to channel. This distortion is primarily because of band-limiting or multi-path effects in the channel. Both the effects appearing in ISI and leads to distortion in the incoming signal at the receiver. The error e(n) is then computed by assuming the nearest valid amplitude level of the modulated signal as the desired value [2]. The CMA algorithm is exploited for blind equalization of signals that have a constant modulus. The error is different and defined by [1]:

$$\varepsilon(n) = (1 - |y(n)|^2)y^*(n) \tag{9}$$

Which is known as Godard's algorithm. The error function for a derived version of CMA is given by

$$\varepsilon(n) = \frac{y(n)}{|y(n)|} - y(n) \tag{10}$$

The Beam formed Data Matrix X is given by

$$X = W^H D \tag{11}$$

where W and D is a block-diagonal matrix with guard bands. The IFFT of the beam formed data matrix is given by

$$Y = XF^H = W^H DF^H \tag{12}$$

where F^H denotes the IFFT matrix
The resulting MIMO OFDM transmit matrix is

$$S = W^H \Omega DF^H \tag{13}$$

where Ω is a diagonal precoding matrix (unimodular).
By applying Kronecker products, we can rewrite S as

$$S = A\omega \tag{14}$$

where ω is the vecdiag (Ω). The cost function is provided by $\min_\omega J(\omega)$, where $J(\omega) = E\left[|S|^2 - \alpha^2\right]$ and α is the average transmit power.
The weight update equation for MCMA is given by,

$$\omega^{i+1} = \omega^{k+1}\theta|\omega^{k+1}| \tag{15}$$

where $\omega^{i+1} = \omega^k - \mu\Delta J(\omega)^i$, μ is the step size and θ is the point wise division.
The simulations in Fig. 4. Shows that the CMA attains a PAPR reduction of up to 5.5 dB for 50 iterations. Figure 5 shows the appearance of interferer signals with an angle of -10 and $-40°$, where both interferers are rejected. The signal to be

Fig. 4 PAPR performance of
SFBC OFDM system using
CMA method

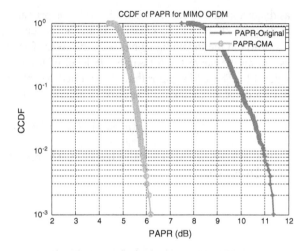

Fig. 5 The interferer signals
arrival with CMA method

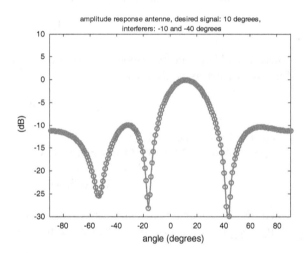

received appears at an angle of 10°. The instantaneous amplitude of the array output
is sufficient and no synchronization requirement is the backbone of the CMA
method. Because of this functional behavior, the CMA algorithm is simple and
reliable.

6 Modified Constant Modulus Algorithm (MCMA)

The MCMA for PAPR mitigation was derived from the well known
Algorithm CMA. Consider the Transceiver model as shown in Fig. 2. Here an
OFDM block contains N number of subcarriers. These Subcarriers comprises of

Nus useful subcarriers with two guard bands. Again these useful subcarriers are further divided into M resource blocks each consisting of Nrb = Nus/M subcarriers. Here pilot subcarriers act as training symbols. Data in the resource blocks is converted to space time domain using an IFFT. Below is a basic algorithm for an SFBC OFDM based on the smart-gradient method [4].

(1) Starting with the input bit stream, break it up into blocks of size N for parallel transmission, where there are N sub channels. Consider the transmission of p OFDM blocks for each antenna. Perform BPSK/QAM modulation to obtain the frequency-domain symbols, X_k.

(2) Adopt transmit diversity by using Space-Frequency Block Coding (SFBC) and determine the permissible extensions for each sub channel.

(3) Consider the IFFT for each antenna's signal, $x^0[n, l]_t$ (n is the sub channel index, l is the antenna index, and t is the transmission period index). Set $i = 0$.

(4) Clip any value $|x^i[n, l]_t| \geq A$ in magnitude for some clip level A to get

$$\bar{x}[n, l]_t = \begin{cases} x^i[n, l]_t, & |x^i[n, l]_t| \leq A \\ Ae^{j\theta[n,l]_t}, & |x^i[n, l]_t| \geq A \end{cases} \tag{16}$$

Where $x^i[n]_t = |x^i[n, l]_t| e^{j\theta[n,l]_t}$

(5) Compute the clipped signal portion

$$c_{clip}[n, l]_t = \bar{x}[n, l]_t - x^i[n, l]_t \tag{17}$$

(6) Employ an FFT to each antenna's clipped signal to obtain $c_{clip}[n, l]_t$

(7) Preserve only the components of $c_{clip}[n, l]_t$ which are admissible expansion directions for the given sub channel constellations and set all remaining components to zero.

(8) Implement the SFBC compulsion (i.e., project onto S_B) by considering the maximum ACE extension and enforcing it on each and every block according to the code matrix.

(9) Adopt an IFFT to get $c[n, l]_t$.

(10) Derive a step size $\mu[l]_t$ based on the smart gradient criterion.

$$x^{i+1}[n, l]_t = x^i[n, l]_t + \mu[l]_t c[n, l]_t \tag{18}$$

If a tollerable PAPR or an end iteration count has not been reached, update $i = i + 1$ and return to Step 4. Else, stop PAPR reduction. The computational complexity of MCMA is O (MTNL log(NL)) based on the oversampling of IFFT/FFT operations. The algorithm merges adequately after 3–4 iterations. This algorithm can be applied to SFBC and it effectively improves the PAPR performance and Bit Error Rate performance for SFBC-OFDM system.

7   Simulation Results and Discussion

We assume the channel coefficients are always flat Rayleigh on different implementations of space-frequency block codes. The simulation plot in Fig. 3 demonstrates the Bit Error Rate performance of 2 × 1 Alamouti SFBC OFDM. Figure 4
shows PAPR achievement of SFBC OFDM with CMA method. For BPSK modulation the CMA attains a PAPR reduction of up to 5.5 dB. Figure 5 shows the
amplitude response of the adaptive array. The plot in Fig. 6, illustrates the rela-
tiveness of CMA and MCMA with BPSK. The Bit Error Rate (BER) behaviour of
the MCMA method with BPSK modulation is approximately 3–4 dB better than
CMA method. For BPSK modulation the proposed MCMA is about 1.2 dB superior
CMA. From Fig. 6, the Modified CMA is analogous to a Rayleigh fading channel
based on its BER performance.

Thus the same error correcting codes used for a fading channel can be applied
here. This inspires the use of MCMA technique. Figure 7 shows the CCDF
Vs PAPR curve for a MIMO OFDM system with QPSK modulation scheme is
found to be 5 dB.

We achieved a PAPR reduction of about 5 dB using CMA and 6.6 dB using
MCMA. Hence, from Table 1, it proves that the PAPR reduction efficiency of
MCMA is superior than CMA. We also analyzed the convergence property of
MCMA and its PAPR reduction performance. We got a PAPR reduction of about
6.6 dB using MCMA. The performance of PAPR reduction in MCMA is better than
that of CMA is shown in the Table 1.

Fig. 6 BER performance of
the proposed MCMA in time
flat, frequency selective
Rayleigh fading (BPSK)

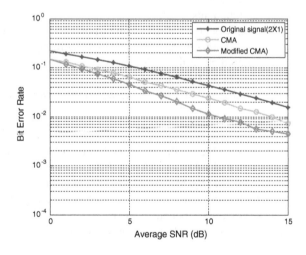

Fig. 7 PAPR performance of SFBC MIMO-OFDM signal

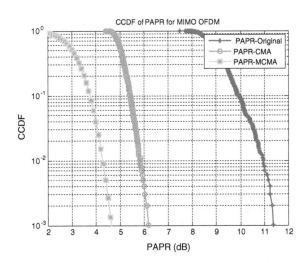

Table 1 Comparison of PAPR reduction in original, CMA and MCMA

Parameter	Original (SFBC-OFDM)	CMA	MCMA
PAPR	11.2	6.2	4.6
PAPR reduction in dB	–	5	6.6

8 Conclusion

In this paper, the fundamental concepts of a MIMO SFBC OFDM system with significant design and performance characteristics are studied. The basic MIMO SFBC OFDM system with Rayleigh channel shows outperformance than standard OFDM system. Further, the BER and PAPR performance analysis of MIMO-OFDM system has been covered in this paper. We present an enhanced modified constant modulus algorithm (MCMA) where the step size is carefully assessed to maintain a balancing between convergence rate and final accuracy is outperforms than the Constant Modulus Algorithm (CMA) is highly proficient technique for alleviating high PAPR. Simulation results illustrates that the signals of proposed scheme attains the diversity of SFBC without the complexity of number of antennas at the source.

References

1. R. Van Nee, A. de Wild, Reducing the peak-to average power ratio of OFDM. Vehic. Technol. Conf. **3**, 2072–2076 (1998)
2. S. Khademi, Student Member, IEEE, A.-J. van der Veen, Fellow, IEEE, Constant modulus algorithm for peak-to-average power ratio (PAPR) reduction in MIMO OFDM/A. IEEE Signal Process. Lett. **20**(5), 531–534

3. D.L. Jones, A normalized constant modulus algorithm. In: *IEEE Conference Record of the Twenty-Ninth Asilomar Conference on Signals, Systems and Computers Pacific Grove* (USA), vol. 1, pp. 694–697 (1995)
4. I. Chahed, J. Belzile, A.B. Kouki, Blind decision feedback equalizer based on high order MCM. In: *Canadian Conference on Electrical and Computer Engineering,* vol. 4, pp. 2111–2114. Niagara Falls (Canada) (2004)
5. Z. Xiong, L. Linsheng, Z. Dongfeng, A new adaptive step-size blind equalization algorithm based on autocorrelation of error signal. In: *7th International Conference on Signal Processing.* Bejing, China (2004)
6. B. Zhao, J. Zhao, L. Zhang, A variable step size constant modulus blind equalization algorithm based on the variation of MSE. J. Taiyuan University of Technol. **36**(4) (2005)
7. K. Banovic, A. Esam, A novel radius-adjusted approach for blind adaptive equalizer. IEEE Signal Process. Lett. **13**(1), 8 (2006)
8. Z. Lilyi, C. Lei, S. Yunshan, Variable stepsize CMA blind equalization based on nonlinear function of error signal. In: International Conference on Communications and Mobile Computing. Kunming (China), vol. 1, pp. 396–399 (2009)
9. M. Zhang, A novel blind equalization algorithm based on affine projection and quantization estimation errors. IEEE 978-1-4244-5668 (2009)
10. W. Xue, A variable step size algorithm for blind equalization of QAM signals. In: PIERS Proceedings, Cambridge, USA, July 5, 8 (2010)
11. N. Sreekanth, M.N. Giriprasad, Effect Of TO and CFO on OFDM and SIR analysis and interference cancellation in MIMO OFDM systems. Int. J. Comput. Technol. Electron. Eng. (IJCTEE) **2**(3), 159–171 (2012). ISSN 2249-6343
12. S.H. Han, J.H. Lee, An overview of peak-to-average power ratio reduction techniques for multicarrier transmission. IEEE Wireless Commun. Mag. **12**, 56–65 (2005)

Reduction of Mutual Coupling Effect in Adaptive Arrays

A.V.L. Narayana Rao, N. Balaankaiah and Dharma Raj Cheruku

Abstract This paper deals with the problem of mutual coupling in circular and elliptical antenna arrays. Input impedance method is used to compensate the Mutual coupling among the array elements. Improved LMS method is used to optimize the weights to generate proper beam form. Directive elements are used in the array. Convergence speed, directivity, SNR and resolution of angle are discussed and compared for circular and elliptical geometrical arrays.

Keywords Antenna arrays · Elliptical (oval) arrays · Mutual coupling · Compensation methods · Beam forming

1 Introduction

Increased demand for multimedia applications in wireless communication requires more efficient signal processing. Wide application of wireless communication technology in daily life demands efficient and reliable signal transmission. Smart antenna is one of the alternatives for achieving this. Smart antenna can increase directivity, user capacity, battery life and source separation. Smart antennas are popularly used in radar systems, wireless communications and in vehicle collision avoidance applications [1]. Antenna array is an integral part of smart antenna

A.V.L. Narayana Rao (✉)
Department of ECE, SSNCET, Ongole, A.P., India
e-mail: narayanaavlrao@gmail.com

N. Balaankaiah
CSIR, Bangalore, India
e-mail: bala.nunna8@gmail.com

D.R. Cheruku
Department of ECE, GIT, GITAM University, Visakhapatnam, A.P., India
e-mail: dharmarajc@yahoo.com

© Springer India 2016   339
S.C. Satapathy et al. (eds.), *Microelectronics, Electromagnetics and Telecommunications*, Lecture Notes in Electrical Engineering 372,
DOI 10.1007/978-81-322-2728-1_30

system. In Smart antenna system, array of elements are designed to achieve required Beam form for a given application.

A systematic arrangement of group of radiating elements is called an array. Size of the array and aperture proportionally increases with gain of the antenna [2]. Two methods to increase the aperture size of the array is to increase the number of elements and/or inter element spacing of the array. Increasing the number of elements increases the area leading to computational complexity. However, increasing inter element spacing of the array increases grating lobes. Due to grating lobes the power efficiency of the system is reduced [3]. A compromise between these two parameters can be achieved by reorienting the geometry of the array and also number of elements is the array.

In general, the linear array is simple in structure compared to other geometries. It is limited to estimation of the elevation angle (1D) of incoming signal only. In addition, the linear array can only estimate the range of azimuth angle limited by $1800(-900 < \theta < 900)$ [4]. Hence the accuracy of Direction of Arrival (DOA) estimation is greatly affected when the incoming signal is within the range of end-fire array $(700|\theta| < 900)$.

Signal processing algorithms of most antenna arrays assume that the signals and the elements are independent of each other. Such assumptions will not be true if the inter elemental space is small [5]. The entire transmitted signal is equal to the sum and reradiated signal from the elements. Similarly, induced current on the antenna element reradiates electromagnetic field which would be received by the other neighboring elements in the array. Such mutual coupling effect usually considered as a defect which degrades the performance of the array [6].

2   Conventional Mutual Impedance Method

A circuit theory approach for reducing or compensating mutual coupling effect is discussed in this work. This method is called conventional mutual impedance method (CMI) [7]. Common mutual coupling can be easily measured directly or from s-parameters.

The relation between the terminal voltage and current can be given by

$$V_1 = I_1 Z_{K,1} + I_1 Z_{K,2} + \ldots + I_i Z_{k,i} + \ldots + I_K Z_{K,K} + \ldots I_N Z_{K,N} + V_{OK} \quad (1)$$

Z_L being impedance, relation between voltage and current is given by

$$I_1^t = -\frac{v_i^t}{Z_L} \quad (2)$$

The relationship between the open-circuit voltages and terminal voltages can be written as

$$
\begin{bmatrix}
1+\frac{Z_{11}}{Z_L} & \frac{Z_{12}}{Z_L} & \frac{Z_{1N}}{Z_L} \\
\frac{Z_{21}}{Z_L} & !+\frac{Z22}{ZL} & \frac{Z_{22}}{Z_L} \\
.... & .... & .... \\
\frac{Z_{N1}}{Z_L} & \frac{Z_{N2}}{Z_L} & !+\frac{Z_{NN}}{Z_L}
\end{bmatrix}_{NXN}
\begin{bmatrix}
V1 \\ V2 \\ .... \\ VN
\end{bmatrix}_{1XN}
=
\begin{bmatrix}
V01 \\ V02 \\ .... \\ VN
\end{bmatrix}_{1XN}
\tag{3}
$$

3 Geometry and Array Factor

3.1 Circular Array

In circular array assume that N equally spaced isotropic elements are placed on X–Y plane along a circular ring. Let the radius is given by a. Circular array is able to scan 360° azimuthally. The geometry of N-element circular array antenna is shown in Fig. 1. The array factor is given by the Eq. (4)

$$
AF(\theta, \phi) = \sum_{N-1}^{N} I_n \, e^{j[ka \sin\theta \cos(\phi-\phi_n)+\alpha_n]}
\tag{4}
$$

where,

I_n = amplitude of excitation. α_n = phase of the nth element, θ = elevation angle from z axis. The radius of the array increases the directivity of uniform circular array and tends to N.

3.2 Elliptical Antenna Array

The geometry of the elliptical antenna array with origin as center is shown in Fig. 2 [8].

Fig. 1 Geometry of circular antenna in XY plane

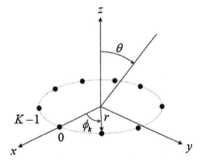

Fig. 2 Geometry of ellipse in XY-plane

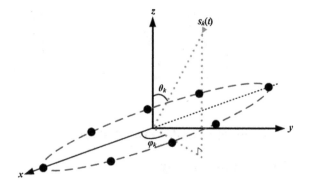

The array factor is given by Eq. (5).

$$AF(\theta, \phi) = \sum_{n-1}^{N} I_n \exp(j[k \sin(\theta)(a \cos((\Phi_n)) \cos(\phi) + b \sin(\Phi_n) \sin(\phi)) + \alpha_n])$$

(5)

where $K = 2\Pi/\lambda$, $\Phi_n = 2\Pi(n-1)/N$, $e = \sqrt{1 - b^2/a^2}$.

In = amplitude of excitation: α_n = phase of the nth element: θ = elevation angle from z axis: $\emptyset_n$ = Azimuth angle measured from x axis for nth element.

a, b = semi major and minor axises respectively: e = eccentricity of elliptical array and is 0.5

$$\alpha_n = k \sin(\theta_0)(a \cos(\emptyset_0) + b \sin(\emptyset_n) \sin(\phi_0))$$
$$\theta_0 = 90°, \emptyset_0 = 0°.$$

(6)

N = no of elements is 8 or 10.

4   Results

4.1   Circular Array

The objective is to analyze the response of elliptical antenna array. The source and noise are 90° and 180° azimuth directions respectively. A modified LMS algorithm optimizes the weight factor for DOA estimation. Mutual coupling is minimized by the CMI method.

Example 1: 8 Dipole Element Circular Array

In this example, an 8-element Circular array is optimized. The best amplitude value determined by the optimized technique is shown in Fig. 3. From Fig. 4 it can be noted that the output noise power varies with the number of iterations below 50.

Fig. 3 Circular array with CMIM

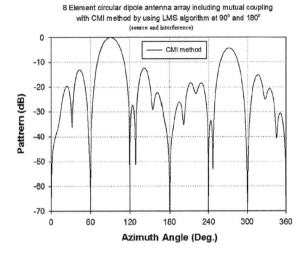

Fig. 4 Iterations versus output noise in circular array with CMIM

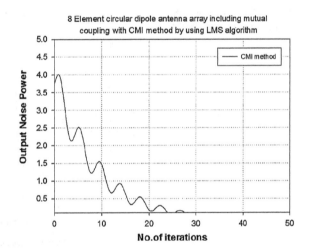

Example 2: 10 Dipole Element Circular Array

From Fig. 5 it is to be observed that the output noise power is very low when the number of iterations below 40. Figure 6 presents the comparison between the patterns generated with CMI and without CMI method for Circular array.

4.2 Elliptical Antenna Array

The response of elliptical antenna array is also analyzed. The source and noise are assumed to 90° and 150° azimuth directions respectively. The weight factor is

Fig. 5 Iterations versus SNR
in circular array with CMIM

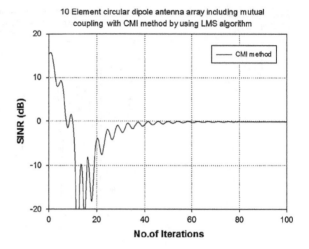

Fig. 6 Circular array with
CMIM and without CMIM

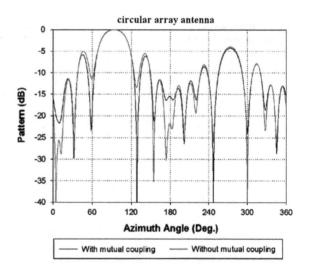

optimized using modified LMS algorithm for DOA estimation. Mutual coupling is
minimized by adopting CMI method.

Example 3: 8 Dipole Element Elliptical Array

In this example, an 8-element elliptical array is considered and optimized. The best
amplitude value is determined by the optimized technique given in Fig. 7. From
Fig. 8, it is to be noted that the output noise power is very low as number of
iterations increases above 8.

Fig. 7 Elliptical (oval) array
with CMIM

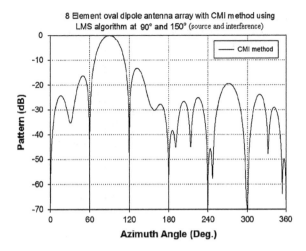

Fig. 8 Iterations versus
output noise in elliptical
(oval) array using CMIM

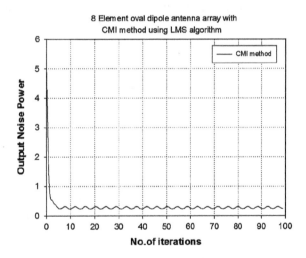

Example 4: 10 Dipole Elements Elliptical Array

From Fig. 9 it can be concluded that the signal to noise power ratio is very low
when the number of iterations are below 20. This response is better than circular
array. Figure 10 gives the comparison between performance of elliptical array with
and without CMI.

Fig. 9 Iterations versus
output SNR in elliptical array
using CMIM

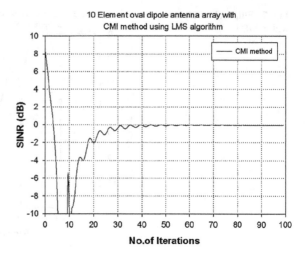

Fig. 10 Elliptical array with
and without CMIM

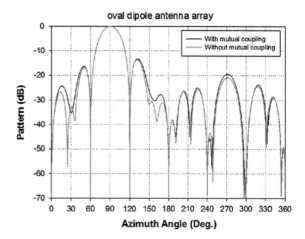

5   Conclusion

Mutual coupling compensation method CMI is implemented with circular and
elliptical antenna arrays. 8-element and 10 element arrays are considered and
analyzed. Output power is optimized for direction of interest. However elliptical
array has better performance than circular array particularly in side lobe suppression
and directivity. Since the dipole elements are used in the array, this antenna array is
suitable for many wireless communication applications.

References

1. G.T. Okamoto, *Smart antenna systems and wireless LANs*, 1st edn. (Kluwer Academic, Santa Clara, California, 2002)
2. R.L. Haupt, *Antenna Arrays: A Computational Approach* (Wiley, New Jersey, 2010)
3. C.A. Balanis, *Modern Antenna Handbook* (Wiley, New Jersey, 2008), p. 135
4. P. Ioannidis, C.A. Balanis, Uniform circular arrays for smart antennas. IEEE Antennas Propag. Mag. **47**, 192–206 (2005)
5. V. Ammula, Dual excited planar circular array antennas for direction agile applications. In: *42nd Southeastern Symposium on System Theory*, pp. 138–142 March 2010
6. H. Hema Singh, L. Sneha, R.M. Jha, Mutual coupling in phased arrays: a review. Int. J. Antenn. Propag. (2013)
7. J.Y.-L. Chou, *An investigation on the impact of antenna array geometry on beam forming user capacity, master of science (Engineering)* (Queen's University, Kingston, Ontario, 2002)
8. N. Ahmidi, A. Neyestanak, R. Dawes, Elliptical array antenna design based on particle swarm method using fuzzy decision rules. In: *24th Biennial Symposium on Communications*, pp. 352–355. Kingston, ON (2008)

Parameter Estimation of Solid Cylindrical Electromagnetic Actuator Using Radial Basis Function Neural Networks

V.V. Kondaiah, Jagu S. Rao and V.V. Subba Rao

Abstract The electro-magnetic actuator presents a solution for most of the technical problems of the traditional mechanical bearings since it ensures the total levitation of a body in space eliminating any mechanical contact between the stator and the levitated body. In practice there is lot of difference between theoretical force and actual force developed between the stator and rotor of an actuator and it varies with air gap and current. This difference is mainly due to different losses in the system. In the present work a correction factor is introduced to account for different losses. A radial basis function neural network (RBFNN) has been implemented to estimate the correction factor and validated with experimental values. The RBF network has been used to estimate the actuator parameters namely force, position stiffness and current stiffness. The RBF predicted values have been validated with the experimental values.

Keywords Magnetic actuator · Parameter estimation · Neural networks · Correction factor

1 Introduction

Solid cylindrical magnetic actuator works on the principle of electromagnetic levitation. It consists of an electromagnet assembly, a set of power amplifiers which supply current to the electromagnets, a controller, and gap sensors with associated

V.V. Kondaiah
Department of Mechanical Engineering, Tirumala Engineering College,
Narasaraopeta, Guntur, Andhra Pradesh, India
e-mail: vvkondaiah@gmail.com

J.S. Rao (✉)
Department of Mechanical Engineering, RGUKT, Nuzvid, Andhra Pradesh, India
e-mail: jagusrao@rgukt.in

V.V. Subba Rao
Department of Mechanical Engineering, JNTUK, Kakinada, Andhra Pradesh, India
e-mail: rao703@yahoo.com

© Springer India 2016
S.C. Satapathy et al. (eds.), *Microelectronics, Electromagnetics
and Telecommunications*, Lecture Notes in Electrical Engineering 372,
DOI 10.1007/978-81-322-2728-1_31

electronics to provide the feedback required to control the position of the rotor (floater) within the gap. The magnetic actuators could be used as magnetic bearings which has wide range of applications such as in flywheels, turbo generators and high speed pumps etc.

Magnetic actuator consists of a stator and a rotor. The stator controls the axial position of a rotor. In the design of magnetic actuator the parameter identification between the force and the air gap and current is a key factor. A theoretical model is used for preliminary designs. However there will be a significant difference between the theoretical force and the actual force developed due to fringing, leakage, variation in material properties, misalignments, vibrations, and temperature effects etc. In practice these losses are accounted by introducing corresponding loss factors in the model [1–3]. These corrected models would be further used in finding system parameters for control applications.

Allaire et al. [1] presented the design of a prototype of single acting magnetic actuator for a high load-to-weight ratio and an expression is used to find leakage correction factor. David et al. [2] explained the leakage, fringing and eddy current effects in the design of magnetic bearings and correction is calculated using FEM. Groom and Bloodgood [3] proposed a model by adding the loss and leakage factors to ideal models with and without bias permanent magnets. Subsequently Bekinal et al. [4, 5] experimented on permanent magnet thrust bearing and compared the theoretical and practical force generated in their test setup Aenis et al. [6] used AMB as diagnosis tool and taken constant correction factors. Gunzberg and Buse [7] used AMB for measurement of force in the centrifugal pump and taken constant geometric correction factor. Marshal et al. [8, 9] used constant correction factor named as derating factor, calculated with initial readings, in multi point technique. Minihan et al. [10] considered constant fringe factor in control of magnetic thrust bearing. Rao and Tiwari [11] implemented multi-objective genetic algorithms (MOGAs) for the optimization of active magnetic thrust bearings (AMTB) with pure electro magnets considering constant loss factors. From [12–18], artificial intelligent techniques had been used for control of active magnetic bearing systems however the system identification was done online.

In the above mentioned literature constant loss factors with different names have been used to account for losses in magnetic actuators. But no literature was reported the variation of loss (or correction) factor as a function of air gap and current input. A better estimation of actuator system parameters in offline would result in reducing the effort of system identification online. In the present work the correction factor has been modeled as a function of air gap and current using RBF neural networks and validated with experimental results. The actuator parameters namely position stiffness and current stiffness are also modeled using RBF predicted correction factor and compared with experimental values.

This article is organized as follows. Section 2 mathematical modeling of the system is carried out. In Sect. 3 the correction factor, force parameters are modeled using RBFNN. The description of test setup and testing procedure is presented in Sect. 4. The discussion of the results and conclusions are provided in Sects. 5 and 6 respectively.

2   Mathematical Modeling of Solid Cylindrical Magnetic Actuator

0(a) shows C-type electromagnetic actuator. It consist a stator with winding and rotor. The magnetic flux passes through stator, air gap and rotor. The rotor floats in magnetic flux. 0(b) is a cylindrical type magnetic actuator working same as C-type actuator. The cylindrical type actuator could be obtained by rotating C-type actuator through 360°. The winding is placed in the annulus space of stator. The magnetic flux passes through stator, air gap and rotor as shown in 0(b). The cylindrical type actuator could be used as bearing to support the Thrust loads 1. The lengths of flux path in the core (stator of bearing) are ℓ_1 and ℓ_2. The flux path length in the core is ℓ_3. The winding has N turns. The instantaneous current passing through winding is i, and the Magneto Motive Force is Ni. The air gap length at the nominal position is h (Fig. 1).

If B is flux density and H is magnetic field intensity, the magnetic energy W_m, and the magnetic co-energy W_{co} stored in magnetic fields could be expressed as [19–20]

$$W_m = \int_v \int_0^B H(B)dBdv \tag{1}$$

$$W_{co} = \int_v \int_0^H B(H)dHdv \tag{2}$$

The independent variables in a magnetic suspension system are normally the winding current and object displacement. If the system is moved by δx then it can be shown that the work done is equal to the change in co-energy of the system. Since the work done is the force $F \times \delta x$, the theoretically electromagnetic force F_{th} is given as the partial derivative of the magnetic co-energy [19]

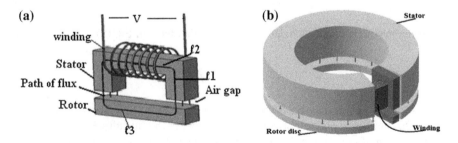

(a) V — winding, Stator, Path of flux, Rotor, ℓ_2, ℓ_1, Air gap, ℓ_3

(b) Stator, Rotor disc, Winding

Fig. 1 Magnetic actuators. **a** C-type. **b** Cylindrical type

$$F_{th} = \frac{\partial W_{co}}{\partial h} \tag{3}$$

Assuming a linear system, where the self-inductance L is constant and i input current, the magnetic co-energy W_{co}, derived as [19]

$$W_{co} = \int_0^i Lidi = \frac{1}{2}Li^2 \tag{4}$$

The electromagnetic force F_{th} could be derived from (3) and (4)

$$F_{th} = \frac{\mu_0 A_g N^2 i^2}{4h^2} \tag{5}$$

where μ_0 is relative permeability of air, A_g is area of air gap, h is height of air gap, N is number of coil turns and i is input current. A correction factor k is introduced to account for all the above said losses. The correction factor k could be calculated as

$$k = \sqrt{\frac{F_{th}}{F_a}} \tag{6}$$

where F_a is actual force developed between stator and rotor at particular air gap and current.

Then k is included in the Eq. (5) to calculate the force F accounting all losses as

$$F = \frac{\mu_0 A_g N^2 i^2}{4h^2 k^2} \tag{7}$$

The force generated by horizontal direction, F, could be considered up to the second order terms 1

$$F = F|_{(h0,i0)} + \frac{\partial F}{\partial h}|_{(h0,i0)}h + \frac{\partial F}{\partial i}|_{(h0,i0)}i \tag{8}$$

If the magnetic circuit is balanced, then the first term in Eq. (8) is equal to zero and

$$F = F_h h + F_i i \tag{9}$$

The quantity F_h is referred as the open loop stiffness and represents the change in the horizontal force due to horizontal displacement. The quantity F_i represents the actuator gain of the bearing. It represents changes in horizontal force due to current i. Expressions for the open loop stiffness and the actuator gain are determined by performing the appropriate differentiation of the force expression. In Eq. (7) μ_0, A_g

and N are constants then the product of these three terms could be taken as constant i.e. $C = \mu_0 A_g N$, then the force expression could be written as

$$F = C \frac{i^2}{4h^2 k^2} \qquad (10)$$

By differentiating the Eq. (10) with h considering k as constant the position stiffness and current stiffness could be obtained

$$F_h = \frac{\partial F}{\partial h} = -C \frac{i^2}{2h^3 k^2} \quad \text{and} \quad F_i = \frac{\partial F}{\partial i} = C \frac{i}{2h^2 k^2} \qquad (11)$$

3 Modeling of Actuator Parameters in RBFNN

Radial Basis Function Neural Networks (RBFNN)

A neural network is a parallel distributed processor with important virtue of the ability to learn from input data by using the learning algorithm. The ANN is made up of an inter connection simple processing unit, known as neurons. Neurons can be either linear or non linear. Usually, situated neurons in the hidden layer are selected non linear while the neuron in the output layer are chosen linear. In recent researches, neural networks have been considered as potential tools for modeling of complicated and unknown systems. Neural networks can adapt themselves with variations in the environment conditions and learn the characteristics of their input signals [14]. Radial Basis Functions (RBF) are embedded in a two layer neural network, where each hidden unit implements a radial activated function. The output units implement a weighted sum of hidden unit outputs. The input into an RBF network is nonlinear while the output is linear. Due to their nonlinear approximation properties, RBF networks are able to model complex mappings, which perception neural networks can only model by means of multiple intermediary layers. In this unit the correction factor (k), force (F), position stiffness (F_h) and current stiffness (F_i), and have been modeled [14].

RBFNN Architecture for Modeling of Correction Factor

The current (i), and air gap (h) are the two important factors influencing the correction factor (k). Then the correction factor is taken as a function of current and air gap. Hence the problem is modelled as a two input (current and air gap) and single output (correction factor) radial basis function neural network. A typical RBFNN architecture has been shown in 0 (Fig. 2).

The structure of an RBF networks involves three layers.

First layer (input layer) The input vector designed as $x_j = [h_j, i_j]$.

Second layer (hidden layer) The second layer is hidden layer which is composed of nonlinear units and are connected directly to all the nodes in the input layer. Each

Fig. 2 The structure of RBF
network for modelling
correction factor

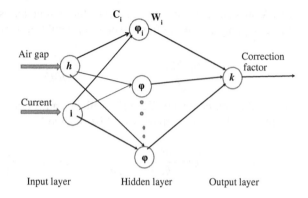

Input layer   Hidden layer   Output layer

hidden unit consists of a center (c_i), spread parameter (σ_i), and weight (w_i). The centers are selected from input data $c_i = [h_i, i_i]$, The distance measured from the centre to the data sample which is the Euclidean distance is expressed as

$$r_{ij} = ||x_j - c_i|| \tag{12}$$

Several base functions are used in RBF neural network, but out of all the Gaussian function is widely used function. Hence Gaussian function is considered as hidden layer base function ϕ_{ij} that could be expressed as

$$\phi_{ij} = \phi(r_{rj}) = \exp(-r_{ij}^2/\sigma_i^2) \tag{13}$$

σ_i is called the spread parameter which influence the bell-shape of the function, usually taken as constant value.

Third layer (output layer) The final layer performs a simple weighted sum with a linear output. Here output is correction factor k could be calculated as

$$k = \sum W_i \phi(|| x_j - c_i ||^2/\sigma_i^2) \tag{14}$$

The Eq. (14) could be expressed in matrix form as

$$\begin{pmatrix} \phi_{1j} & \cdots & \phi_{nj} \\ \vdots & \ddots & \vdots \\ \phi_{m1} & \cdots & \phi_{mn} \end{pmatrix} \begin{pmatrix} w_1 \\ \vdots \\ w_n \end{pmatrix} = \begin{pmatrix} k_1 \\ \vdots \\ k_n \end{pmatrix} \tag{15}$$

The weights (w_i) in Eq. (15) could be determined using pseudo inverse technique.

Modeling of Force

The expressions for force, linear parameters have been derived in equation from (10) and (11). However these parameters using RBFNN considering $k(h, i)$ include the derivatives of k. The expressions for parameters F, F_h and F_i have been derived.

The force developed between stator and rotor at a specified air gap and current input could be calculated by inserting correction factor which is estimated using RBFNN in Eq. (10)

$$F = C\frac{i^2}{4}\left(k^{-2}h^{-2}\right) \tag{16}$$

where $C = \mu_0 A_g N^2$.

Modeling of Actuator Parameters

The position stiffness (k_h) could be obtained by differentiating F w.r.t h considering k as function of h and i

$$F_h = \frac{\partial F}{\partial h} = -C\frac{i^2}{2h^3 k^2}\left[kh\left(\frac{\partial k}{\partial h}\right) + 1\right] \tag{17}$$

The current stiffness (k_i) could be obtained by differentiating F w.r.t i

$$F_i = \frac{\partial F}{\partial i} = C\frac{i}{2h^2 k^2}\left[1 - \frac{i}{k}\left(\frac{\partial k}{\partial i}\right)\right] \tag{18}$$

To determine $\frac{\partial k}{\partial h}$ and $\frac{\partial k}{\partial i}$ in Eqs. (17) and (18) the following expressions have been derived.

The value of correction factor depends on input current and air gap for specific actuator. Hence the correction factor could be considered as function of air gap h and current i, it could be written in RBFNN as

$$k = \sum W_i \phi_i(h, i) \tag{19}$$

By differentiating k with respect to h and i

$$\frac{\partial k}{\partial h} = \sum W_i \frac{\partial \phi_i}{\partial h} \quad \text{and} \quad \frac{\partial k}{\partial i} = \sum W_i \frac{\partial \phi_i}{\partial i} \tag{20}$$

The base function $\phi_i = e^{-r^2/\sigma_i^2}$ and $r^2 = (h - h_i)^2 + (i - i_i)^2$ then the partial derivatives of ϕ, with respect to h and i could be written as

$$\frac{\partial \phi_i}{\partial h} = -\frac{2(h - h_i)}{\sigma_i^2} e^{-r^2/\sigma_i^2} \quad \text{and} \quad \frac{\partial \phi_i}{\partial i} = -\frac{2(i - i_i)}{\sigma_i^2} e^{-r^2/\sigma_i^2} \tag{21}$$

Differentiating $\frac{\partial \phi_i}{\partial h}$ and $\frac{\partial \phi_i}{\partial i}$ with h and i respectively the second order terms could be obtained

Training of RBF Network

Training of an RBF neural network can be achieved with the selection of the optimal values for the parameters like center vector (c_i), spread parameter (σ_i), the hidden layer base function (ϕ_i), weights between the hidden layer and the output layer (w_i). In this section a procedure of finding the optimal parameters are explained in following flow chart and in training algorithm.

An Algorithm to Train the Network

An algorithm has been developed to obtain optimal parameters such as number of centers, spread, and mean square error. The procedure has been explained in following sub sections and in 0.

Training data and training target The experimentation has been done at all air gaps from 1.5 to 4.5 mm in steps of 0.5 mm. The force has been measured with increase of current at all air gaps. The air gap (h) and current (i) are considered as training data. As stated previously the correction factor (k) is treated as function of h and i. Hence k is considered as training target.

Spread parameter The air gap is varied from 1.5 to 4.5 mm and current is varied from 0 to 6 A. Hence the maximum range of current is considered as spread parameter, so that the dumbbell shape of the base function influences the total range. The effect of spread parameter on the network has been described in results and discussions.

Selection of centers Randomly one center was chosen and the mean square error with one center was determined and stored the mean square error as best error and center as best center. Find mean square error with another center, compare the error with the best error, if best error is greater than present error then the best error is present error and best center is present center. Else try another center as present center, continue the process until all the data is tried as center. Add the best center as center. Continue adding new centers to the network until the maximum number of centers has been reached or the convergence criteria is met (Fig. 3).

Training the weights Determination of weights is an aspect in training of network. Here the pseudo inverse method [14] has been used as explained in Eq. (15).

Convergence criteria The value of RMS error has been determined by increasing the value of maximum number of centers. The evaluation was stopped where the RMS value does not decrease though the number of centers increased.

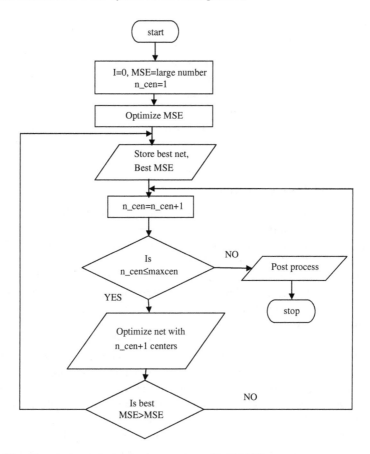

Fig. 3 Flow chart to determine optimal parameters with RBFNN

4   Preparation of Actuator and Testing Procedure

Preparation of Solid Cylindrical Magnetic Actuator

A magnetic thrust actuator has an electromagnetic stator and a rotor, which are separated by an air gap, as illustrated in 0. In its simplest form, the electromagnetic stator is formed by an inner and outer pole connected by a common base. 0(b), clearly shows the stator, shaft, winding, and thrust collar of a single acting magnetic thrust actuator. Mild steel is used to make stator and rotor of actuator. Initially the mild steel discs are collected and they turned into desired size on the lathe machine. The dimensions given in Table 1 are used to make stator and rotor disc. The inner and outer poles and annulus space are made on stator using lathe machine.

The winding, which occupies the annulus space between the inner and outer poles of the stator, produces the magnetic flux in the actuator.

Table 1 The dimensions of stator and rotor of actuator

Parameter	Symbol	Value (mm)
Inner radius of stator	r_i	20
Inner radius of coil gap	r_{ci}	30
Outer radius of coil gap	r_{co}	45
Outer radius of stator	r_o	52.5
Depth of coil gap	d	38
Axial length of thrust runner	h_t	10
Air gap	h	1:0.5:5

Test Setup and Testing Procedure

The stator and rotor disc (flotor) are held by frame as shown in 0. The stator part of bearing is hold by upper horizontal bar of the frame and it has nut and bolt arrangement to change the air gap between stator and rotor. The rotor and load are hold by the middle horizontal bar. The rotor shaft is free to move in the middle horizontal bar. The shaft of rotor passes through the middle horizontal bar and dead weights are attached to it. A dimmer stat is used to change the current input. An ammeter and a voltmeter are used to measure the input current and voltage respectively (Fig. 4).

As the difference between the relative permeability of air and aluminum is negligible, the air gap between stator and rotor is maintained by keeping aluminum plates of desired thickness. The thickness of aluminum plates ranging from 1.5 to 4.5 mm in the steps of 0.5 mm. The voltage is varied from 50 to 120 V. The input current is varied from 0 to 6 A. The rotor is allowed to contact with stator at value of current, the attractive force is measured by applying equivalent force in the opposite direction. The gap is maintained by keeping desired thickness of aluminum foil between stator and rotor disc. The experiment is repeated three times at each air gap and the average values have been taken as final values.

Fig. 4 Test setup

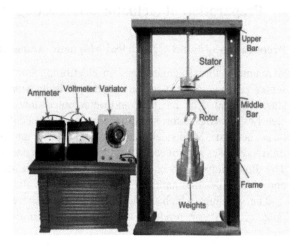

5 Results and Discussions

Determination of Parameters of Actuator Using Experimental Data

As explained in Sect. 4, the experimentation has been done at all air gaps from 1.5 to 4.5 mm in steps of 0.5 mm, the results of the magnetic actuator at the air gaps of 2 mm have been shown in Table 2 as model.

The voltage V, current A, the actual force F_a, the theoretical force F_{th}, the correction factor k, Force after considering correction factor F, the position stiffness F_h, and current stiffness F_i. The actual force between stator and rotor F_a has been measured using test setup at specified air gap and current. The parameters F_{th}, k, F, F_h, and F_i have been determined using the Eqs. (5)–(7), and (11), respectively. The current is varied from 1.88 to 3.85 A. The actual force measured between stator and rotor at each current value and it varies from 15.43 to 62.78 N. The correction factor has been determined using Eq. (6). Its value varies from 1.55 to 1.58. The position stiffness F_h(N/m) is varied with current from −16291.95 to −68689.78. The current stiffness F_i(N/A) varied from 17.38 to 35.68. Similar analysis has been carried out at all air gaps from 1.5 to 4.5 mm considered. The parameters (F, F_h, F_i) have been calculated using the experimental correction factor.

Estimation of Correction Factor Using RBFNN

As explained in flow chart and training algorithm the maximum number of centers and spread parameter has been decided. The decrement of MSE with centers has been shown in Table 3. In the present work, mean square error MSE is minimum at i.e. 0.0147 at 30 as maximum number of centers. Further increase of number of

Table 2 Bearing parameters with experimental correction factor at 2 mm air gap

Voltage (V)	Current (A)	F_a	F_{th}	k	F	F_h	F_i
50	1.88	15.43	37.27	1.55	15.56	−16291.95	17.38
55	2.00	18.36	42.40	1.52	17.70	−18536.62	18.54
60	2.13	22.12	47.87	1.47	19.99	−20926.11	19.70
65	2.33	25.38	57.30	1.50	23.93	−25050.51	21.55
70	2.50	28.30	66.25	1.53	27.66	−28963.48	23.17
75	2.63	32.20	73.04	1.51	30.50	−31932.23	24.33
80	2.78	35.77	81.63	1.51	34.08	−35685.90	25.72
85	2.95	40.36	92.25	1.51	38.52	−40328.74	27.34
90	3.15	44.34	105.18	1.54	43.92	−45982.41	29.20
95	3.30	45.98	115.43	1.58	48.20	−50465.96	30.59
100	3.37	49.21	120.15	1.56	50.17	−52525.58	31.20
105	3.52	51.06	131.09	1.60	54.74	−57310.35	32.59
110	3.67	53.66	142.51	1.63	59.50	−62303.65	33.98
115	3.82	59.82	154.41	1.61	64.47	−67505.49	35.37
120	3.85	62.78	157.12	1.58	65.60	−68689.78	35.68

Table 3 Variation of MSE with centers

Max no. of centers	1	5	10	15	20	25	30	35
MSE	0.1779	0.0692	0.0429	0.0291	0.0271	0.0196	0.0147	0.0147

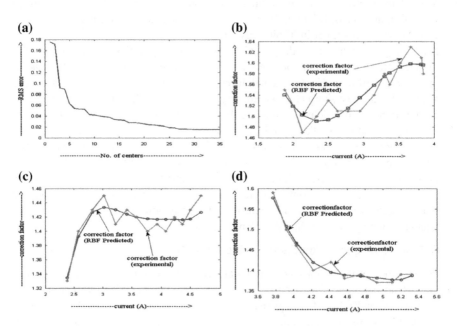

Fig. 5 **a** Variation of RMS with centers, **b** variation of correction factor at 2 mm air gap, **c** variation of correction factor at 3 mm air gap, **d** variation of correction factor at 4 mm air gap

centers is not decreasing the MSE value. Hence 30 has been considered as optimum number of centers for network.

The maximum range of current value has been taken as spread parameter. The 0(a) shows the variation of MSE with number of centers. Radial Basis Function Neural Networks (RBFNN) has been used to predict the correction factor at all air gaps considered but at 2, 3 and 4 mm air gaps have been shown in 0(b), (c) and (d) (Fig. 5).

A comparison is made between the experimental correction factor and RBF predicted correction factor. It is observed that the percentage of error between experimental correction factor and correction factor predicted by RBFNN is varies from 0.02 to 2.71 %. The 0 shows the variation of correction factor at all air gaps between 1.5 and 4.5 mm considered (Fig. 6).

Fig. 6 **a** Comparison of force prediction at all air gaps, **b** comparison of position stiffness at all air gaps, **c** comparison of current stiffness at all air gaps

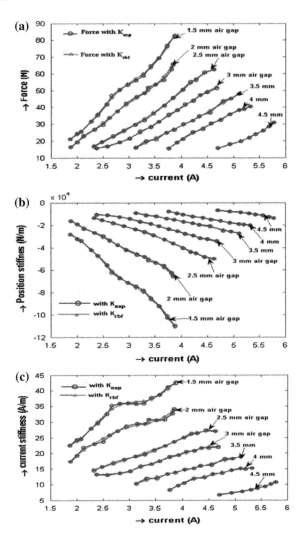

Estimation of Parameters Using RBFNN

The force F, position stiffness F_h, and current stiffness F_i are determined with experimental correction factor and RBF predicted correction factor using the equations given Sect. 2. The comparison of results at each parameter has been plotted in 0. The percentage of error between parameters determined with experimental correction factor to parameters predicted with RBF correction factor is from 1.09 to 4.5 %.

6 Conclusions

In the present work the parameters of solid cylindrical magnetic actuator system namely Force (F), Position stiffness (F_h), Current stiffness (F_i), have been estimated using Radial Basis Function Neural Networks and validated with the experimental values.

In this process an magnetic actuator has been made and tested on own test setup at all air gaps from 1.5 to 4.5 mm in steps of 0.5 mm. It is observed that a lot of difference between theoretical force F_{th} and actual force F_a due to leakage of flux and inherent losses in the system. A correction factor has been introduced to account for all these losses. An RBF network has been modeled to estimate the correction factor. An algorithm to train the network parameters namely the centers, spread and weights has been proposed and described. The RBF predicted correction factor has been validated with that of the experimental results. The maximum percentage of error between RBF predicted correction factor to the experimental one at different air gaps varies from 0.055 to 2.71 %.

The parameters of the magnetic actuator system namely F, F_h and F_i have been modeled using experimental correction factor and RBF predicted correction factor separately. All these parameters of magnetic actuator have been estimated using RBF network successfully and compared with the parameters obtained using experimental results. The maximum percentage of error between the experimental parameters to RBF estimated values are varied from 1.09 to 4.5 %.

The estimated parameters using RBF Neural Networks could be utilized in the active control of magnetic bearing systems. This method of modeling correction factor and determining the other parameters using neural networks could be applied in similar applications. Using the parameters determined by RBF Networks, a control system for a double acting active magnetic bearing system will be designed as a future work. This neural network method could also be utilized in the estimation of multiple loss factors and system parameters of hybrid magnetic bearings as a future work.

References

1. P.E. Allaire, A. Mikula, B.B.W. Banerjee, D. Lewis, J. Imlach, Design and test of magnetic thrust bearing. J. Franklin Inst. **326**(6), 831–847 (1989)
2. D.C. Meeker, E.H. Myounggyu, D. Noh, An augmented circuit model for magnetic bearings including eddy currents, fringing, and leakage. IEEE Trans. **32**(4), 3219–3227 (1996)
3. N.J. Groom, V.D. Bloodgood, A comparison of analytical and experimental data for a magnetic actuator, in *NASA-(2000)-tm210328*
4. S.I. Bekinal, T.R.R. Anil, S. Jana, Analysis of radial magnetized permanent magnet bearing characteristics. Progr. Electromagnet. Res. B **47**, 87–105 (2013)
5. S.I. Bekinal, T.R.R. Anil, S. Jana, S.S. Kulkarni, A. Sawant, N. Patil, S. Dhond, Permanent magnet thrust bearing: theoretical and experimental results. Progr. Electromagnet. Res. B **56**, 269–287 (2013)

6. M. Aenies, E. Knoof, R. Nordmann, Active magnetic bearings for the identification and fault diagnosis in turbo machinery. Mechatronics **12**, 1011–1021 (2002)
7. A. Guinzburg, F.W. Buse, Magnetic bearings as an impeller force measurement technique, in *Proceedings of the Twelfth International Pump Users Symposium* (1995), pp. 69–76
8. J.T. Marshal, M.E.F. Kasarda, J. Imlach, A multipoint measurement technique for the enhancement of force measurement with active magnetic bearing. Trans. ASME **125**, 90–94 (2003)
9. M.E. Kasarda, J.T. Marshal, R. Prince, Active magnetic bearing based force measurement using the multi-point technique. Mech. Res. Commun. **34**, 44–53 (2007)
10. T.P. Minihan, S. lei, G. Sun, A. Palazzolo, A.F. Kascak, T. Calvert, Large motion control for thrust magnetic bearings with fuzzy logic, sliding mode, and direct linearization. J. Sound Vib. **263**, 549–567 (2003)
11. J.S. Rao, R. Tiwari, Optimum design and analysis of thrust magnetic bearings using multi objective genetic algorithms. Int. J. Comput. Methods Eng. Sci. Mech. **9**, 223–245 (2008)
12. F. Lin, S. Chen, M. Haung, Adaptive complementary sliding-mode control for thrust active magnetic bearing system. Control Eng. Pract. **19**(7), 711–722 (2011)
13. F.J. Lin, S.Y. Chen, M.S. Haung, Nonlinear adaptive inverse control for the magnetic bearing system. J. Magn. Magn. Mater. **209**, 186–188 (2000)
14. S.Y. Chen, F.J. Lin, Decentralized PID neural network control for five degree-of-freedom active magnetic bearing. Eng. Appl. Artif. Intell. **26**(3), 962–973 (2013)
15. R. Achkar, M. Owayjan, Control of an active magnetic bearing with multi-layer perceptrons using the torque method, in *10th International Conference on Intelligent Systems Design and Applications (ISDA)* (2010), pp. 214–219
16. S. Haung, L. Lin, Stable fuzzy control with adaptive rotor imbalance compensation for nonlinear magnetic bearing. J. Chin. Inst. Eng. **28**(4), 589–603 (2005)
17. C.T. Lin, C.P. Jou, GA-based fuzzy reinforcement learning for control of a magnetic bearing system. IEEE Trans. Syst. Man Cybern. B Cybern. **30**(2), 276–289 (2000)
18. S.C. Chen, V.S. Nguyen, D.K. Le, M.M. Hsu, ANFIS controller for an active magnetic bearing system, in *IEEE International Conference on Fuzzy Systems (FUZZ)* (2013), pp. 1–8
19. A. Chiba, T. Fukao, O. Ichikawa, M. Oshima, M. Takemoto, D.G. Dorrell, Magnetic Bearings And Bearing Less Drives. Newnes, An imprint of Elsevier, Linacre House, Jordan Hill, Oxford OX2 8DP, 30 Corporate Drive, Burlington, MA 01803 (2005)
20. S.A. Nasar, *Electromagnetic Energy Conversion Devices and Systems* (Prentice-Hall, Englewood Cliffs, 1970)

An Efficient Audio Watermarking Based on SVD and Cartesian-Polar Transformation with Synchronization

N.V. Lalitha, Ch. Srinivasa Rao and P.V.Y. Jaya Sree

Abstract In the present years, abundant digital audio content is made available on public networks due to high speed internet and audio processing techniques. This is an advantage to our everyday life but it suffers from the ownership protection. Audio watermarking is the one of the solution for this problem. In this paper, total audio is divided into non-overlapping segments and Fast Fourier Transform (FFT) is applied to each segment. For embedding the watermark, low frequency coefficients are selected and Singular Value Decomposition (SVD) is applied to the low frequency coefficients of each segment. A binary watermark is embedded into the highest singular values using Cartesian Polar Transformation and embedding rule. The above approach suffers from de-synchronization attack and it is made resistant by including synchronization code. The results shows that this water-marking scheme is highly robust against various signal processing attacks such as re-sampling, re-quantization, cropping, MP3 compression, low-pass filtering, signal addition and subtraction. The algorithm is also compared with the state-of-art techniques available and is better in terms of SNR and payload.

Keywords Synchronization · Cartesian polar transformation · Singular value decomposition · Arnold transform

N.V. Lalitha (✉)
Department of ECE, GMR Institute of Technology, Rajam 532 127, A.P., India
e-mail: lalitha.nv@gmrit.org

Ch.Srinivasa Rao
Department of ECE, JNTU-K University, Vizianagaram, A.P., India
e-mail: ch_rao@rediffmail.com

P.V.Y. Jaya Sree
Department of ECE, GIT, GITAM University, Visakhapatnam, A.P., India
e-mail: pvyjayasree@gitam.edu

© Springer India 2016   365
S.C. Satapathy et al. (eds.), *Microelectronics, Electromagnetics
and Telecommunications*, Lecture Notes in Electrical Engineering 372,
DOI 10.1007/978-81-322-2728-1_32

1 Introduction

Due to rapid development in multimedia technology and communication networks, the protection of copyright is the major problem which is to be solved. Digital watermarking [1] has been widely used for protection of multimedia content (audio, image, and video). A watermark is an undetectable mark embedded into an audio content which should be imperceptible. According to International Federation of the Phonographic Industry (IFPI) [2], an effective audio watermarking [3] algorithm should meet the following requirements: (i) Imperceptibility: the embedded watermark should not affect the quality of the audio signal, (ii) Payload: the number of watermark bits that can be included into the host audio signal without losing the imperceptibility that is measured in bits per second (bps), (iii) Security: the person cannot detect the watermark without knowing the secret key, (iv) Robustness: algorithm is proposed in such a way that, any common signal processing attacks cannot disturb the embedded watermark. Achieving of inaudibility and robustness are much more difficult in audio watermarking compared to image and video watermarking because of Human Auditory System (HAS) is more sensitive than Human Visual System (HVS) [4]. Recently, audio watermarking algorithms achieved significant development. Generally, audio watermarking algorithms are categorized into two groups, i.e., time domain techniques and frequency domain techniques. Time domain audio watermarking techniques [3, 5] implementation is easy but robustness is somewhat less compared to frequency domain techniques. In frequency domain audio watermarking, Fast Fourier Transform (FFT) [6], Discrete Cosine Transform (DCT) [7] and Discrete Wavelet Transforms (DWT) [8] are frequently used transforms.

The rest of the paper is organized as follows. Section 2 focuses on FFT, SVD and CPT. Section 3 explains the embedding and detection processes. The simulation results and conclusions are discussed in Sects. 4 and 5 respectively.

2 Fundamental Theory

2.1 Fast Fourier Transform

Fast Fourier Transform (FFT) is most commonly used in audio signal processing to convert the data from the time domain into the complex frequency domain, means it contains both the amplitude and phase information. FFT has a property translation-invariant to resist small distortions [6].

2.2 Singular Value Decomposition (SVD)

The singular value decomposition (SVD) [9] matrix is very useful for decomposition. In SVD transform, a matrix M is decomposed into three matrices,

$$M = USV^T \tag{1}$$

where, U and V are the real or complex unitary matrices and S is the diagonal matrix. The diagonal elements $(\rho_1, \rho_2, ..., \rho_{n-1}, \rho_n)$ are called singular values (*SVs*). The SVD has an interesting property, that is any small changes on SVs, that doesn't influence the quality of the signal. Thus, SVD can be exploited in audio watermarking to meet the imperceptibility and robustness.

2.3 Cartesian-Polar Transformations

In polar coordinates system (r, θ), where r represents the distance from the origin and θ represents the angle [10] between a line of reference and the line through the origin and the point. To convert Cartesian coordinate system (x, y) to the polar coordinate system (r, θ), the following equations are used,

$$r = \sqrt{x^2 + y^2}, \quad \theta = \tan^{-1}\left(\frac{y}{x}\right) \tag{2}$$

To convert Polar coordinate system (r, θ) to the Cartesian coordinate system (x, y), the following equations are used,

$$x = r\cos\theta, \quad y = r\sin\theta. \tag{3}$$

3 Watermarking Scheme

3.1 Watermark Pre-processing

In order to increase the robustness and confidentiality, watermark should be pre-processed first. In this embedding scheme, the binary watermark W is scrambled using Arnold transform [11] as given below.

$$\begin{bmatrix} x' \\ y' \end{bmatrix} = \begin{bmatrix} 1 & 1 \\ 1 & 2 \end{bmatrix} \begin{bmatrix} x \\ y \end{bmatrix} (\text{mod}N) \tag{4}$$

where x, y are the original pixel positions, x' and y' are the scrambled positions and N is the size of the digital image.

3.2 Synchronization Code

De-synchronization attack is one of the problems in audio watermarking. De-synchronization attack means, even in the presence of watermark, watermark cannot be detected due to lack of synchronization [12]. Cropping and shifting attacks are examples for the de-synchronization attacks. In this embedding scheme, synchronization code is inserted into the starting of the audio signal to identify the correct position of the watermark. To generate the synchronization code, the logistic chaotic sequence with initial value in the range [0, 1] is used and denoted as,

$$y(n+1) = g(y(n)) = \delta \times y(n)(1 - y(n)) \tag{5}$$

where $3.57 < \delta < 4$, $y(n)$ is mapped into the synchronization sequence $\{syn(n)|n = 1, 2..., L_{syn}\}$ with the following rule,

$$syn(n) = \begin{cases} 1 & \text{if} y(n) > T \\ 0 & \text{otherwise} \end{cases} \tag{6}$$

where L_{syn} the synchronization code length and T is pre-defined threshold, sufficient value of T is 0.5.

3.3 Embedding Process

The original audio is divided into two parts. First part of the audio is used to embed the synchronization code which is generated from the Eq. (6). The pre-processed watermark is embedded in the second part of audio. In our method, the embedding process is chosen to be the same for synchronization code and watermark, the steps are detailed below.

Step 1 Audio signal x is segmented into non-overlapping frames based on size of watermark W_m, the length of each audio segment is N, in which embed a watermark, described as follows,

$$x_i(n), \ i = 1, 2, \ldots \ldots, Z, \quad n = 1, 2, \ldots N \tag{7}$$

$$N' = floor\left(\frac{L - L_1 + 1}{Z}\right), \ N = 2^{floor\left(log_2^N\right)} \tag{8}$$

where Z is the number of segments of size $M \times N$, L_1 is the initial position of embedding watermark.

Step 2 Select first l low frequency coefficients of segment $x_i(n)$ and perform SVD on square matrix R_i to decompose into three matrices,

$$R_i = U_i \times S_i \times V_i^T \qquad (9)$$

Step 3 Calculate the mean m_i, variance v_i and Euclidean norms e_i of singular values $(\rho_1, \rho_2, \ldots, \rho_N)$ of each matrix S_i,

$$m_i = \frac{1}{N}\sum_{q=1}^{N}\rho_q, \; v_i = \frac{1}{N}\sum_{q=1}^{N}(\rho_q - m_i)^2, \; e_i = \sqrt{\sum_{q=1}^{N}(\rho_i)^2} \qquad (10)$$

Step 4 Select the highest two singular values of each matrix S_i, which are assumed as the elements of polar coordinate system. Convert these elements into Cartesian transformation elements [10], formula as given below,

$$A_{ix} = S_i(1,1)\cos\theta_1, \quad A_{iy} = S_i(2,2)\sin\theta_1 \qquad (11)$$

$$B_{ix} = S_i(1,1)\cos\theta_1, B_{iy} = S_i(2,2)\sin\theta_1 \qquad (12)$$

where θ_1 and θ_2 are the predefined angle of decomposition, i.e. $45°$. The two singular values $S_i(1,1)$ and $S_i(2,2)$ of each matrix S_i are preserved and is treated as secret key K_2 and further can be used in the decoding process.

Step 5 The binary watermark image is pre-processed using Arnold chaotic map.

Step 6 Modify the elements A_{ix}, A_{iy}, and B_{ix}, B_{iy} by the following rule [10], If embedded bit is '1', the elements are modified as follows:

$$A'_{ix} = A_{ix} + \frac{m_i}{C_1} + \frac{v_i}{C_2} + \frac{e_i}{C_3}, \quad B'_{ix} = B_{ix} + \frac{m_i}{C_1} + \frac{v_i}{C_2} + \frac{e_i}{C_3} \qquad (13)$$

where A'_{ix} and B'_{ix} are modified x elements of $S_i(1,1)$ and $S_i(2,2)$, respectively C_1, C_2 and C_3 are user defined constants.

If embedded bit is '0', the elements are modified as follows,

$$A'_{iy} = A_{iy} - \left(\frac{m_i}{C_1} + \frac{v_i}{C_2} + \frac{e_i}{C_3}\right), \quad B'_{iy} = B_{iy} - \left(\frac{m_i}{C_1} + \frac{v_i}{C_2} + \frac{e_i}{C_3}\right) \qquad (14)$$

where A'_{iy} and B'_{iy} are modified x elements of $S_i(1,1)$ and $S_i(2,2)$, respectively.

Step 7 The modified singular values are obtained by using Cartesian-to-polar transformation:

$$S'_i(1,1) = \sqrt{A'_{ix} + A'_{iy}}, \quad S'_i(2,2) = \sqrt{B'_{ix} + B'_{iy}} \qquad (15)$$

Step 8 Each modified highest singular values $S'_i(1,1)$ and $S'_i(2,2)$ is reinserted into matrix S_i and apply inverse SVD.

Step 9 The modified FFT coefficients (l) are placed at the beginning of each segment and apply inverse FFT for obtaining the watermarked audio signal.

3.4 Watermark Detection Process

The detection process is reverse operation of embedding process, and steps are given below:

Step 1 Perform Steps 1–3 of embedding process.

Step 2 Polar-to-Cartesian transformation is applied to $S_i^*(1, 1)$ and $S_i^*(2, 2)$ of each matrix S_i^* of the attacked watermarked audio segment to calculate A_{ix}^*, A_{iy}^* and B_{ix}^*, B_{iy}^*.

Step 3 Watermark sequence is obtained as follows by using the secret key K_3:

$$W_m' = \begin{cases} 1 & if A_{ix}' > A_{ix} or A_{iy}' > A_{iy} \\ & B_{ix}' > B_{ix} or B_{iy}' > B_{iy} \\ 0 & otherwise \end{cases} \qquad (16)$$

where A_{ix}, A_{iy} and B_{ix}, B_{iy} are used as secret key K_3.

Step 4 Apply inverse Arnold transform to extract the binary watermark image.

4   Simulation Results

The performance of watermarking algorithm is evaluated based on imperceptibility, robustness and payload. For evaluation, five different classes of 16 bit mono audio signals (Pop, Rock, Folkcountry, Blue and Jazz) with sampling frequency 44.1 kHz of 10 s are used. In each frame (frame size is 64) of audio signal embed one binary watermark bit. A binary watermark image and the corresponding scrambled image of size M × M = 96 × 96 = 9216 are shown in Fig. 1. The low frequency FFT coefficients (l = 36) are selected from each frame of original audio signal. The selected FFT coefficients (l = 36) are arranged in a 6 × 6 matrix (N = 6) and SVD is applied.

Fig. 1 Original and encrypted watermark images

4.1 Imperceptibility Test

Imperceptibility test is an audio quality test. In order to measure the imperceptibility, following SNR equation is used:

$$SNR(x, x^*) = 10 \log_{10} \left(\frac{\sum_{i=0}^{length-1} x^2(i)}{\sum_{i=0}^{length-1} [x(i) - x^*(i)]^2} \right) \qquad (17)$$

where x and x^* are original and watermarked audio signals, respectively. International Federation of the Phonographic Industry (IFPI) state that, the watermarked audio should be imperceptible when SNR is over 20 dB [14]. The imperceptibility test on this algorithm shows that the average SNR value is 36.70 for all five classes of audio signals.

4.2 Robustness Test

To compare the similarities between the original watermark W and the extracted watermark W^*, the parameters Normalized Correlation (NC) and Bit Error Rate (BER) are used.

The Normalized Correlation (NC) is computed as:

$$NC(W^*, W) = \frac{\sum_{i=0}^{M-1} \sum_{j=0}^{M-1} w(i,j) * w^*(i,j)}{\sqrt{\sum_{i=0}^{M-1} \sum_{j=0}^{M-1} w^2(i,j)} \sqrt{\sum_{i=0}^{M-1} \sum_{j=0}^{M-1} w^{*2}(i,j)}} \qquad (18)$$

If $NC(W, W^*)$ is close to 1, then the correlation between W and W^* is very high. If $NC(W, W^*)$ is close to zero, then the correlation between W and W^* is very low.

The BER is computed as given below:

$$BER = \frac{\sum_{i=0}^{M-1} \sum_{j=0}^{N-1} w(i,j) \oplus w^*(i,j)}{M * M} \qquad (19)$$

To assess the robustness of algorithm, the following attacks are performed on watermarked audio signal.

1. Re-sampling: 44.1 kHz sampled watermarked audio signal is re-sampled at 22.050 kHz and then back to 44.1 kHz.
2. Re-quantization: 16 bit watermarked audio signal is quantized to 8 bits/sample and again re-quantized back to 16 bits/sample.
3. Random Noise: A random noise is added to the watermarked audio signal.
4. Low-pass filtering: A second order Butterworth filter with cut-off frequency 20 kHz is applied to the watermarked audio signal.

5. Echo addition: 0.1 % decay and 400 ms delayed audio is added to the water-marked audio signal.
6. Cropping: 1000 samples are replaced by zeros at the beginning, middle and end portions of the watermarked signal.
7. Additive Noise: A 60 dB additive white Gaussian noise is added to the watermarked audio signal.
8. MP3 compression: The watermarked audio signal is compressed using MP3 compression at the bit rate of 160 kbps and then back to the WAV format.
9. Signal addition: 2000 samples of original audio signal are added to the beginning of the corresponding samples of watermarked audio signal.
10. Signal subtraction: 2000 samples of original audio signal are subtracted from the beginning of the corresponding samples of watermarked audio signal.

The robustness results for the audio signal 'Pop','Rock', 'Folkcountry', 'Blue' and 'Jazz' respectively are summarized in Table 1. The table indicates that the BER

Table 1 BER and NC values for different audio signals

Audio signal	Type of attack	BER	NC
Pop	No attack	0	1
	Resampling	0.0006	0.9995
	Re-quantization	0.0058	0.9959
	Random noise	0	1
	LPF (20 k)	0.0002	0.9998
	Echo addition	0.0514	0.9641
	Cropping (front)	0	1
	Cropping (middle)	0.0029	0.9979
	Cropping (end)	0	1
	AWGN (60 dB)	0.0002	0.9998
	MP3 compression	0.0013	0.9991
	Signal addition	0	1
	Signal subtraction	0	1
Rock	No attack	0	1
	Resampling	0.0213	0.9847
	Re-quantization	0.00002	0.9998
	Random noise	0	1
	LPF (20 k)	0.0013	0.9991
	Echo addition	0.0077	0.9945
	Cropping (front)	0	1
	Cropping (middle)	0.0028	0.9980
	Cropping (end)	0	1
	AWGN (60 dB)	0	1
	MP3 compression	0.0405	0.9715
	Signal addition	0	1
	Signal subtraction	0	1

(continued)

Table 1 (continued)

Audio signal	Type of attack	BER	NC
Folkcountry	No attack	0	1
	Resampling	0.0053	0.9962
	Re-quantization	0.0021	0.9985
	Random noise	0	1
	LPF (20 k)	0.0012	0.9991
	Echo addition	0.0291	0.9794
	Cropping (front)	0	1
	Cropping (middle)	0.0027	0.9981
	Cropping (end)	0	1
	AWGN (60 dB)	0	1
	MP3 compression	0.0036	0.9974
	Signal addition	0	1
	Signal subtraction	0	1
Blue	No attack	0	1
	Resampling	0.0063	0.9955
	Re-quantization	0.0058	0.9959
	Random noise	0	1
	LPF (20 k)	0.0008	0.9994
	Echo addition	0.0497	0.9648
	Cropping (front)	0	1
	Cropping (middle)	0.0027	0.9981
	Cropping (end)	0	1
	AWGN (60 dB)	0.0001	0.9999
	MP3 compression	0.0120	0.9914
	Signal addition	0	1
	Signal subtraction	0	1
Jazz	No attack	0	1
	Resampling	0.0005	0.9996
	Re-quantization	0.0377	0.9736
	Random noise	0.0007	0.9995
	LPF (20 k)	0.0006	0.9995
	Echo addition	0.0753	0.9481
	Cropping (front)	0	1
	Cropping (middle)	0.0007	0.9995
	Cropping (end)	0	1
	AWGN (60 dB)	0.0049	0.9965
	MP3 compression	0.0012	0.9991
	Signal addition	0	1
	Signal subtraction	0	1

Table 2 Comparison with previous methods

References	Algorithm	SNR	Payload (bps)
Wu et al. [13]	Self-synchronization	–	172
Bhat et al. [14]	DWT-SVD	24.37	45.9
Chen et al. [15]	Optimisation based quantisation	29.50	172.41
Khaldi et al. [16]	EMD	24.12	50.03
Pranab et al. [10]	CPT-SVD	36.86	689.56
Proposed	Synchronization+Pranab et al. [10]	36.70	934.40

values ranges from 0 to 0.0753 and NC lies in between 0.9481 and 1, demonstrate that the scheme exhibit robustness against different attacks. For de-synchronization attack (cropping), the BER lies in between 0 to 0.0027 and NC values ranges from 0.9979 to 1 and is achieved due to the use of synchronization code.

4.3 Data Payload

Payload is defined as the number of data bits that can be embedded into the original audio signal per unit of time without losing the quality of audio. It can be measured in terms of bits per second (bps).

$$Payload = \frac{N_w}{t} \tag{20}$$

In this scheme, N_w = 9344 bits is embedded in 10 s audio signal, thus the pay load is 934.4 bps. This is relatively high payload, and typical value is 20–50 bps.

The Table 2 illustrates the payload and SNR of the present scheme is more when compared to [13–16]. The more value of payload indicates that more amount of information that can be embedded into the host audio signal. As well high SNR indicates better imperceptibility of the watermarked audio signal.

5 Conclusion

An efficient audio watermarking scheme is provided based on SVD and CPT including the synchronization code. The initial work [10] is based on SVD and CPT but suffers from de-synchronization attack. This problem is addressed in our proposed work by inserting a synchronization code into the beginning samples of the original audio and then binary watermark is embedded. Due to this, the algorithm is made robust to cropping and MP3 compression attack, also it is highly robust against various signal processing attacks such as re-sampling, re-quantization, low-pass filtering, signal addition and subtraction. The payload and SNR of the present scheme is more and compared with recent existing methods. The work can be extended by including error-correction codes and evaluating the algorithm with respect to stir-mark attacks.

References

1. I.J. Cox, J. Kilian, F.T. leighton, T. Shamoon, Secure spread spectrum watermarking for multimedia. IEEE Trans. Image Process. **6**(12), 1673–1687 (1997)
2. S. Katzenbesser, F.A.P. Penticolas, *Information Hiding Techniques for Steganography and Digital Watermarking* (Artech House Norwood Mass, USA, 2000)
3. P. Bassia, I. Pitas, N. Nikolaidis, Robust audio watermarking in the time domain. IEEE Trans. Multimedia **3**(2), 232–241 (2001)
4. E. Ergun, B. Leyla, Audio watermarking scheme based on embedded strategy in low frequency components with a binary image. Digit. Signal Process. **19**, 265–277 (2009)
5. W.N. Lie, L.C. Chang, Robust high-quality time-domain audio watermarking based on low-frequency amplitude modification. IEEE Trans. Signal Process. **54**(12), 46–59 (2006)
6. M. Fallahpour, D. Megias, Secure logarithmic audio watermarking scheme based on the human auditory system. Multimedia Syst. **20**, 155–164 (2014)
7. B.Y. Lei, L.Y. Soon, Z. Li, Blind and robust watermarking scheme based on SVD-DCT. Signal Process. **91**, 1973–1984 (2011)
8. X.Y. Wang, H. Zhao, A novel synchronization invariant audio watermarking scheme based on DWT and DCT. IEEE Trans. Signal Process. **54**(12), 4835–4840 (2006)
9. H.C. Andrews, C.L. Patterson, Singular value decomposition and digital image processing. IEEE Trans. Acoust. Speech Sig. Process. ASSP **24**(1), 26–53
10. D. Pramab Kumar, S. Tetsuya, Audio watermarking in transform domain based on singular value decomposition and cartesian-polar transformation. Int. J. Speech Technol. **17**, 133–144 (2014)
11. L. Wu, J. Zhang, W. Deng, D. He, Arnold transformation algorithm and anti-arnold transformation algorithm. IEEE Int. Conf. Inf. Sci. Eng. 1164–1167 (2009)
12. X.Y. Wang, H. Zhao, A novel synchronization invariant audio watermarking scheme based on DWT and DCT. IEEE Trans. Signal Process. **54**(12), 4835–4840 (2006)
13. S. Wu, J. Huang, D. Huang, Y.Q. Shi, Efficiently self-synchronization audio watermarking for assured audio data transmission. IEEE Trans. Broadcast. **51**(1), 69–76 (2005)
14. V.K. Bhat, I. Sengupta, A. Das, An adaptive audio watermarking based on the singular value decomposition in the wavelet domain. Digit. Signal Process. **20**(6), 1547–1558 (2010)
15. S.T. Chen, G.D. Wu, H.N. Huang, Wavelet-domain audio watermarking scheme using optimization. IET Signal Process. **4**(6), 720–727 (2010)
16. K. Khaldi, A.O. Boudraa, Audio watermarking via EMD. IEEE Trans. Audio Speech Language Process. **21**(3), 675–680 (2013)

A Proposal for Packet Drop Attacks in MANETS

Mahesh Swarna, Syed Umar and E. Suresh Babu

Abstract The packet drop attack is more familiar to protection risks in MANETS. These attackers implement loop-hole which bring with envious characteristics because the path finding procedure which is vital and unavoidable. Researchers and investigators have performed distinct recognition methods recommend various types of recognition schemes. AODV protocol which correctly appropriate redirecting method for the MANETS and it is more susceptible to dark gap strike by the envious nodes. A harmful node that wrongly delivers the RREP (route reply) that it has a newest path with lowest hop count to location and then it falls all the getting packets. In this paper we present four types of different protocols for detecting black whole attacks and discuss state of the art routing methods. We also perform different properties in collaborative packet drop attacks and analyze categories of propose protocols with specified features stored in wireless ad hoc networks. We analyze comparison of proposed protocol with existing protocols and their methods with respect to time and other features in wireless ad hoc networks.

1 Introduction

WMAHN (or basically MANETS in presented paper) are a self-configured system which consists of different portable customer devices. These wireless nodes connect with other nodes surrounding without any facilities; All of the transmitting hyperlinks are recognized as wireless Technology. According to the interaction method described earlier. MANETS are most commonly used in army objective, catastrophe place, individual place system and so on. Over previous times several

M. Swarna (✉)
C.S.E Department, K L University, Guntur, A.P., India
e-mail: mahesh.swarna1@gmail.com

S. Umar · E. Suresh Babu
Department of C.S.E, K L University, Guntur, A.P., India
e-mail: umar332@gmail.com

© Springer India 2016
S.C. Satapathy et al. (eds.), *Microelectronics, Electromagnetics and Telecommunications*, Lecture Notes in Electrical Engineering 372,
DOI 10.1007/978-81-322-2728-1_33

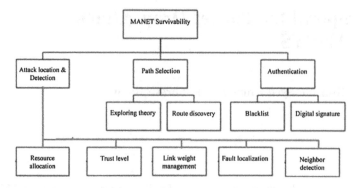

Fig. 1 Path selection in MANET specified by the AODV protocol hierarchy

years, there has been an increasing interest in wireless systems, as the price of portable devices such as PDA's, laptop computers, mobile phones, etc. have reduced drastically whereas the efficiency of these gadgets have improved significantly.

As shown in the Fig. 1 we process to develop route discovery and attack prevention in MANETS based on trusted authority with preferred resource allocation in fault localization. Mobile Ad hoc systems or MANETS are the type of wireless systems which do not need any set facilities or platform channels. They can be quickly implemented in locations where it is challenging to setup any wired facilities. In MANET each node acts as a wireless router which creates redirecting complicated when compared to Wireless LANs, where the main entry way functions as the router between the nodes. Protection is a primary significance in scenarios of implementation such as battleground in an ad hoc network. Since MANET has multiple hop hyperlinks, it is venerable against several strikes like dark gap strike, Byzantine attack, wormhole strike etc. This document reveals the comparison of AODV, OLSR and ZRP under packet drop attack.

2   Related Work

In this area, we existing the famous and well-known redirecting methods in MANET. Before a cellular node communicates to the focus on node, and will transmit its existing position to the neighbors due to the existing redirecting details are different. By the way how the information/message is obtained, by redirecting methods which are can be categorized into practical, sensitive and combination of both practical and sensitive redirecting.

2.1 Proactive Routing/Path Protocol (PPP)

The practical redirecting is also known as table-driven redirecting method. In this redirecting method, mobile nodes regularly transmit their redirecting details to the others who live nearby. Every node in the path redirects desk not only with details of the adjacent nodes and obtainable nodes but also with variety of trip. It also terms as, nodes in the path have to assess their communities provided that the system topology has modified [1]. Therefore, the drawback is that the expense improves as the system size increases, an important interaction expense within a bigger system topology. So, the benefits are as system position can be instantly shown if the envious attacker connects. Frequent acquainted kinds which are practical way of DSDV protocol [2] redirecting methodology with which OLSR protocol is laid.

2.2 (On-Demand) Reactive Routing Protocol (RRP)

On-Demand routing protocol is prepared with some more application known as On-demand routing protocol. Compared with practical redirecting, the sensitive redirecting is simply began when nodes wish to deliver information packages [3]. The durability is lost bandwidth induced from the cyclically transmitted can be decreased. Nevertheless, this might also be the critical injure when there are any harmful nodes in the system atmosphere. The weakness is that inactive redirecting technique results in some bundle loss. In this we gave a brief concept of two frequent on-demand redirecting methods which are AODV and DSR protocol [1]. In AODV, every node will saves the information of the next hop details in the redirecting desk and also preserves it for recovering a routing path from resource to location node.

2.3 Hybrid Routing Protocol (HRP)

The HRP is one of the distinct methods which will brings the advantages of Reactive and Proactive protocols. In this mostly there will be multiple redirecting methods which are designed as a requested packets or padded system. Some of the protocols which follows HRP are as follows ZRP, TORA etc.

3   A Packet Drop Attack in MANET

MANETS are accessible to distinct attacks, which are general attacks on the layers which functions as routing mechanism of the network or topology. Attacks will be caused by two purposes: it may change the parameters such as sequence number, count of hop or it may not forward the packets. In packet drop attack a envious node stops. And at that case that node will acts as one of the intermediate node through which the communication will be continues and laid a route from source to destination. As in the DSR that node will send a RREQ message and then waits until it get response from other neighbor nodes then from any other node it will get a false RREP packet from the source S which will modify its higher sequence number so by this source node will know that a new route has laid presently to the destination [4] So by this mainly the packets are attracted by the packet drop nodes, So it will not reach the destination or target node.

3.1   Distinct Packet Drop Attack

Single packet drop Attack   A dark gap issue indicates that one harmful node uses the redirecting method to self claimed as showing fastest direction to the location node, but falls the redirected messages/packets but does not ahead packages to others who live nearby. Only one dark gap strike is quickly happened in the cellular ad hoc systems [4] (Fig. 2).

Collaborative packet drop Attack   Many systems have been suggested for fixing the single dark gap attack in the past few years. So there are many recognition techniques are not supported for the cooperative black gap issues. Some harmful nodes work together in order to beguile the regular into their designed redirecting information, moreover, covers up from the existing recognition plan. Consequently,

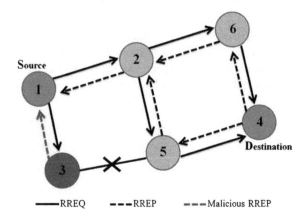

Fig. 2   A single packet drop system for processing events in real time networks

Fig. 3 Processing of collaborative packet drop attack

several supportive recognition techniques are proposed preventing the collaborative dark gap attacks [1] (Fig. 3).

The specific process is well recognized for the black gap problems. As a remedy the original backbone network is found which are designed as a set of powerful central source nodes (BBNS) over the ADHOC systems. These nodes are more reliable and permitted to spend the RIP when any new node becomes as member of the network. The resource node searches the closest BBN to assign a RIP before transfer information technologies/packages then delivering RREQ to the location node which deals with of RIP. If only the resource node gets the location node's RREP, it indicates that there is no black gap [5]. At this condition whenever the resource gets the RREP bundle from RIP, it indicates an adversary might be persisted in the system. Nodes around the RIPs will modify promiscuous mode due to resource node which will delivers the observed message or packets or information to aware of those. These nodes will observes the information of the assigned nodes which can also from the suspicious nodes, So the resource nodes transfers some phony information or packets to analyze the envious nodes.

4   Protocol Heirarchy

Destination Sequence Distance Vector   DSDV redirecting method is a pro-active, table-driven redirecting method for MANETs. Every node will sustain a desk record all the other nodes it has known either straight or through some others who live nearby. The access will have details about the node's IP deal with, last known series variety and the hop depend to achieve that node (Fig. 4).

Along with these details the desk also keeps a record of the Next-Hop next door neighbor to achieve the location node, the timestamp of the last upgrade obtained for that node. Simulation results of the DSDV protocol hierarchy in packet drop data processing and all the relevant data appearance in network data simulation in terms of transmission of data in network.

As shown in Fig. 5, the analysis of DSDV protocol analysis of through put based on number of hops present in similar processing and other communication details in data transmission and other proceedings in real time networks [6].

DSR Routing on Packet drop   DSR is one of the another redirecting methodology that has been selected to simulate the protocol. This technique will controls bandwidth by eliminating used periodic desk by upgrading. DSR will set up a path to target node form resource node, so it is not mandatory to send regular 'HELLO'

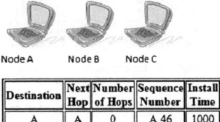

Node A   Node B   Node C

Destination	Next Hop	Number of Hops	Sequence Number	Install Time
A	A	0	A 46	1000
B	B	1	B 36	1200
C	B	2	C 28	1500

Fig. 4 Routing table for maintain of packet drop attacks in mobile ad hoc networks

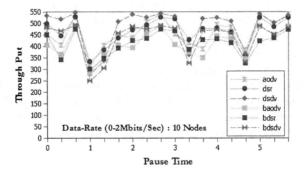

Fig. 5 Performance evaluation of DSDV protocol with respect to process the pausing time and through put levels in packet drop attack procession

messages to nodes to inform it's who live nearby about his existence [7]. The main factor of this technique is that neighbor nodes of MANETS do not get direction details which creates less load in the system and the direction is basically defined in details packages of resource node (Fig. 6).

Packet drop Attack on AODV Routing Protocol The packet drop attacks contains harmful nodes that forget the nodes to fall the details packages. Whenever resource node desires to connect with the nodes which are nearby or transfers the details packages to the location, it delivers a RREQ to its others who live nearby to know the real direction to the destination [4]. If there is one or more harmful node (black hole node), it gets the RREQ then delivers a bogus RRE to similar which reveals harmful node already has a real path to the location and this RREP concept contains false redirecting details and bogus greater series number that reveals it is a clean direction. When the similar of RREQ gets the RREP, it represents the harmful node as real node then it delivers the details packages within the route that specified by dark gap node [7]. Black gap nodes receive the details packages without delivering the packages to the location or the other nodes.

Fig. 6 Performance evaluation of DSR routing protocol with respect to through put

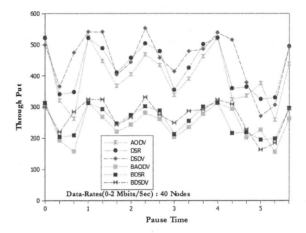

Fig. 7 Cooperative packet drop attack detection process

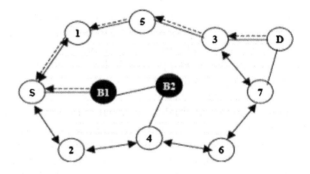

As shown in the Fig. 7, by developing various redirecting loops, system blockage and route argument, attackers degrades the system efficiency. The resource node transmits RREQ packages to its next door neighbor nodes "B" and "D" to finds a clean way or path to the location "F". The black gap node "M" instantly react to the resource node without verifying its redirecting desk to say it has a clean direction to the designed location which is done by sending a bogus RREP to the resource node "A". The resource node "A" views that the path finding has been done then denies other RREP concept from other nodes. Then, the enemy will fall the obtained packages without sending to the location "F".

Performance Analysis over AODV Various research and activities analytics can be used to assess the suggested redirecting methods with and without dark gap strikes. These matrices are important to display the efficiency research of system [8]. This section is targeted to describe the essential analytics that are used in this thesis.

Network Throughput A system throughput which is a common quantity at which message is efficiently taken between resource node and destination node. It is also generally known as the rate of the amount of information obtained from its mailer to

Fig. 8 Analysis of the
AODV protocol hierarchy
with comparison of through
put process

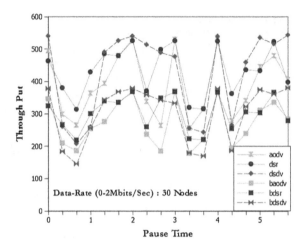

time the last bundle gets to its location. Pieces per second (bps), packages per second or bundle per time port can be considered to assess the throughput but OPNET deploys bits per second to assess the throughput [9]. A MANET system needs to perfect throughput which should be at advanced stage. The primary aspects that impact on the throughput are data transfer usage, restricted power, modify in topology and un-trusted interaction (Fig. 8).

End-to-End delay End-to-end delay is the common delay that begins at first node by producing packages until wait coming the messages in location node which will proves in a few moments. E2E delay contains the overall wait in the systems [10]. In MANETS, measurement of delay could be accumulated due to web link downs and/or indicates the weak factors between nodes [11]. With the measure of this delay parameter whenever decreases with an efficient redirecting method is set up in the system, caused by redirecting protocol determines a real path and every node had an information about path to its location so, the variety of packages is reduced (Fig. 9).

The flexibility of 10 m/s is regarded to nodes in the network. This simulator executed in 1000 × 1000 m. All the scenarios were run for 600 a few moments. The Packet Inter-Appearance Time (s) is regarded as rapid (1) and bundle size (bits) is rapid (1024). 11 Megabyte per second is taken for each cellular node as information quantity. A continuous rate of 10 m/s was assigned as unique way factor flexibility with stop time of continuous 100 a few moments. This stop time is taken after information gets to the location only. The primary objective was to discover out the better method against strikes in situation of black gap strike. AODV and DSR redirecting method which are sensitive methods respectively are chosen. In both methods, harmful nodes shield is reduced to a level which improve bundle fall.

Fig. 9 End to end time delay
performance may increase the
performance of the network

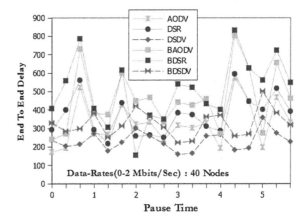

5 Conclusion

In the WSN there are distinct factors to assess efficiency of any redirecting method
which briefly explained as in past area. Both redirecting protocols AODV and DSR
reflects excellent amount to start information packets whenever node flexibility
availability. These two DSR protocols and AODV protocol are On demand
methods which primary behavior is confirmed in the form of its expense. To
recognize particular variations in the simulator outcomes which will evaluate outcomes, the simulator has some circumstances depending on various system
dimensions may have dark gap strike which will indicates first phase for frequent
function of MANETS and second phase for MANETS function underly with
supportive black gap strike. The tests show motivated results acquired some circumstances. The MANETS underly frequent function out performs the MANET
under cooperative dark gap strike with regards to throughput and network fill
in situations. The outcomes acquired which can be used to find the effect of the
supportive dark gap strike on MANETS because the system fill and throughput of a
good system should be high. However, the results in phrase of E2E wait show that
MANETS under supportive dark gap strike had a minor decrease because the dark
gap nodes declare to have at fast path to location by offering a fast RREP to
resource node which creates these nodes as harmless node and it is apparent that the
End-to-End wait will be reduced in the whole network.

References

1. P.-C. Tsou, J.-M. Chang, Y.-H. Lin, H.-C. Chao, J.-L. Chen, Developing a BDSRScheme to
avoid packet drop attack based on proactive and reactive architecture in MANETs, in *Paper
Presented at the 13th International Conference on Advanced Communication Technology*,
Phoenix Park, Korea, Feb 2011, pp. 13–16

2. E. Çayırcı, C. Rong, *Security in Wireless Ad Hoc and Sensor Networks*, 1st edn. (Wiley, New York, 2009), p. 10
3. S. Ci et al., Self-regulating network utilization in mobile ad-hoc wireless networks. IEEE Trans. Vehic. Tech. **55**(4), 1302–1310 (2006)
4. M.-Y. Su, Prevention of selective packet drop attacks on mobile ad hoc network through intrusion detection systems. Comput. Commun. 21–26 (2010)
5. H. Khattak, Nizamuddin, F. Khurshid, N. Amin, Preventing Black and Gray Hole Attacks in AODV Using Optimal Path Routing and Hash. 978-1-4673-5200-0/13/$31.00 ©2013 IEEE
6. Z. Ahmad, J. Ad Manan, K. AbdJalil, Performance evaluation on modified AODV protocols, in *IEEE Asia-Pacific Conference on Applied Electromagnetics*, 11–13 Dec 2012
7. K. Osathanunkul, N. Zhang, A Countermeasure to Packet Drop Attacks in Mobile Ad Hoc Networks. 978-1-4244-9573-3/11/$26.00 ©2011 IEEE
8. N. Arora, N.C. Barwar, Performance analysis of DSDV, AODV and ZRP under packet drop attack. Int. J. Eng. Res. Technol. (IJERT), **3**(04), (2014)
9. A. Mukija, Reactive routing protocols for mobile ad hoc network, in *IEEE Network Magazine, Special Issue on Networking Security*, vol. 14 (2001)
10. K. Netmesiter, Routing protocols in mobile ad hoc networks: challenges and solutions, in *IEEE Wireless Communications Magazine, Sponsored by IEEE Communications Society*, Vol. 11 (February 2010)
11. R.H. Jhaveri, S.J. Patel, DoS attacks in mobile ad-hoc networks: a survey. Second Int. Conf. Adv. Comput. Commun. Technol. **2**(2), 535–540 (2012)

Heptagonal Fractal Antenna Array for Wireless Communications

V.A. Sankar Ponnapalli and P.V.Y. Jayasree

Abstract Fractal antenna technology is geometry-based, not material based, that's why novel design methodologies and novel fractal shapes helpful for further improvement of antenna parameters. Side lobe level is one of the prominent challenges at higher expansion (S) levels of fractal antenna arrays. This report primarily focuses on the design and analysis of heptagonal fractal planar array antenna, based on concentric circular ring sub array mathematical generator. Due to this new structure a notable improvement has observed in Directivity, Side lobe level and Side lobe level angle. These fractal arrays are analyzed and are simulated using by MATLAB programming.

Keywords Fractal antenna array · Expansion factor · Array factor · Side lobe level · Directivity

1 Introduction

Antenna arrays have been used to fulfill high directivity requirements in communication systems [1]. Depending on the construction, antenna arrays can be divided into three types, they are linear, planar and conformal arrays [2]. Fractal antenna arrays have been used to fulfill the wide band and ultra wide band requirements in advanced communication systems [3, 4]. Multiband behavior can be achieved by fractal antennas and antenna arrays [5]. Broad band and multiband behavior of fractal antenna arrays depends on fractal shapes and how they are generated. That's why these are geometry based not material based. Some applications require

V.A. Sankar Ponnapalli (✉) · P.V.Y. Jayasree
Department of Electronics and Communication Engineering, GITAM University, Visakhapatnam, India
e-mail: vadityasankar3@gmail.com

P.V.Y. Jayasree
e-mail: pvyjayasree@gitam.edu

© Springer India 2016
S.C. Satapathy et al. (eds.), *Microelectronics, Electromagnetics and Telecommunications*, Lecture Notes in Electrical Engineering 372, DOI 10.1007/978-81-322-2728-1_34

antenna size to be very less for these needs fractal construction methods play a prominent role [6]. Concentric circular ring sub array geometric generator is one of the famous design methodology for the generation of fractal linear and planar arrays [7]. Any polygon shape can generate with this design methodology. Cantor and binomial fractal linear arrays and fractal square, triangular and hexagonal arrays are generated with this design methodology [7]. Pentagonal and octagonal fractal antenna arrays are investigated in [8] using the same design methodology.

Beyond this methodology another types of fractal arrays are also available, they are nature inspired fractal random arrays [9], and cantor linear array for even number of elements [10]. The cantor ring array is also the best example for fractal antenna arrays. These arrays are generated by polyadic cantor set and designed for less side lobe levels [11]. To avoid gaps and overlaps between the elements like in conventional fractal planar arrays a new class of fractal arrays having fractal boundaries named as fractile arrays are investigated in [12]. This paper examined heptagonal (i.e., seven elements) fractal planar array based on concentric circular ring sub array generator with uniform current excitations. Section 2 of this paper investigates design equation of the heptagonal fractal antenna array. Section 3 deals with analysis of output results for various conditions.

2 Design Equation of Heptagonal Fractal Antenna Array

In this paper, design and analysis of heptagonal fractal array using concentric circular ring sub-array generator is presented. This geometric process gives more flexibility in the expansion and analysis of fractal heptagonal antenna array, and executes the results in MATLAB programming.

The generalized array factor for fractal array of this type can be expressed as:

$$AF_P(\psi) = \prod_{P=1}^{P} GA\left(S^{P-1}(\psi)\right) \tag{1}$$

where G.A. (ψ) is the Array factor associated with generating sub-array, here array factor of heptagonal fractal antenna array. S is the Scale (or) Expansion factor that governs how large the array grows with each recursive application of the generating sub-array. The self-scalable heptagonal array is a fractal array is generated by a 7-element ring sub-array generator. Figure 1 shows the 7-element ring sub-array whose individual generating elements are located on each of the vertices of a heptagon. Similar to the case of the self-scalable hexagonal array [1], the self-scalable heptagonal array is generated in a way allowing stacking of some of the elements upon each other at higher stages of growth. Each stack of generated elements can be represented by a single element. This implementation can reduce the number of real elements, while the current distribution on the array becomes non uniform, leading to lower side lobe levels.

Fig. 1 7-element ring sub
array generator

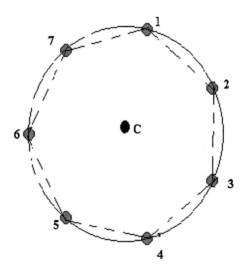

Figure 2 shows the self-scalable heptagonal antenna array for two stages of growth. The growth may be extended to infinite set of values. The array factor AF_P (θ, ϕ) of this array at stage p can be expressed using reference and plugging in $N = 7$ with uniform current amplitudes, as follows,

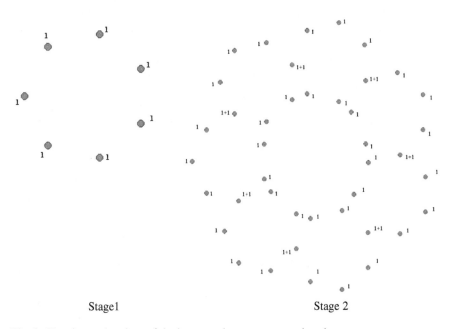

Stage1   Stage 2

Fig. 2 The element locations of the heptagonal array up to two iterations

$$AF(\theta, \phi) = 1/7^P \prod_{P=1}^{4} \left[\sum_{n=1}^{7} I_n e^{js^{p-1}} \psi_n(\theta, \phi) \right] \tag{2}$$

$$\psi_n(\theta, \phi) = \pi/2(\sin\theta\cos(\phi - \phi_n) + \alpha_n) \tag{3}$$

where P is the number of iterations, in this paper number of iterations considered up to four; α_n is the phase excitation on nth element; I_n is the excitation current amplitude of the nth element; $\phi_n = 2\pi(n - 1)/N_P$, here N_p is the total number of elements generated by respective iteration and in this case N_p starts with seven. In this report, four stages of recursively generated heptagonal array is investigated for radius, r = λ/2 and λ. the array factor associated with this seven-element generating sub array can be shown to have the following representation:

$$\text{For r} = \lambda/2, AF(\theta, \phi) = 1/7^P \prod_{P=1}^{4} \left[\sum_{n=1}^{7} e^{js^{p-1}} (\pi/2 \sin\theta\cos(\phi - \phi_n) + \alpha_n) \right] \tag{4}$$

$$\text{For r} = \lambda, AF(\theta, \phi) = 1/7^P \prod_{P=1}^{4} \left[\sum_{n=1}^{7} e^{js^{p-1}} (\pi \sin\theta\cos(\phi - \phi_n) + \alpha_n) \right] \tag{5}$$

Array factors (4) and (5) calculated for the unmodified (i.e.) 7-element heptagonal fractal array without center element for different radii, λ/2, λ, expansion factor (S) of 2, and different iterations (p = 1, 2, 3, 4) is observed in Figs. 3 and 4 respectively. Modified version i.e. heptagonal fractal antenna array with center element as shown in Fig. 3 and center element cab be excited with amplitude of two units. Like unmodified version of this array it can also be extended up to infinite set of iterations but the difference is the center element. We have observed a notable change in the parameters of this array while putting center element. The self-scalable heptagon antenna array can be modified by inserting an element at the center of the sub-array generator. The array factor for self scalable heptagonal antenna array for radii of λ/2, λ with an expansion factor of two and for four iterations (P) is given by,

$$\text{For r} = \lambda/2, \quad AF(\theta, \phi) = 1/7^P \prod_{P=1}^{4} \left\{ 2 + \left[\sum_{n=1}^{7} e^{js^{p-1}} (\pi/2 \sin\theta\cos(\phi - \phi_n) + \alpha_n) \right] \right\} \tag{6}$$

$$\text{For r} = \lambda, \quad AF(\theta, \phi) = 1/7^P \prod_{P=1}^{4} \left\{ 2 + \left[\sum_{n=1}^{7} e^{js^{p-1}} (\pi \sin\theta\cos(\phi - \phi_n) + \alpha_n) \right] \right\} \tag{7}$$

In this paper also observes the directivity of the heptagonal antenna array for both modified and unmodified systems. The directivity of N-element arrays for

Fig. 3 The element locations including center element of the heptagonal fractal antenna array up to two iterations

broadside operation can be determined by assuming that all elements are isotropic. The directivity for 2-D arrays may be expressed as [8],

$$D = \left(\sum_{n=1}^{N} I_n\right)^2 \Big/ \left(\sum_{n=1}^{N} I_n\right)^2 + \sum_{m=2}^{N} \sum_{n=1}^{m-1} I_n I_m \sin(k|r_n - r_m|)/(k|r_n - r_m|) \qquad (8)$$

where I_n is the current amplitude excitation of nth element; I_m is the current amplitude excitation of nth element on the mth ring; r_n is the position vector of magnitude of r_n; $K = 2\pi/\lambda$.

3 Results and Discussion

Design and analysis of a new type of fractal array, i.e. heptagonal fractal array proposed in this paper and directivity of this array compared with the octagonal fractal array. Figure 2 describes the first and second iterations of heptagonal fractal array and corresponding array factors up to four iterations of these arrays with different radii ($\lambda/2$ and λ) are shown in Fig. 4. Behavior heptagonal fractal array with center element is also observed in this paper for different radii ($\lambda/2$ and λ) as shown in Fig. 3 and corresponding array factors of these arrays are shown in Fig. 4.

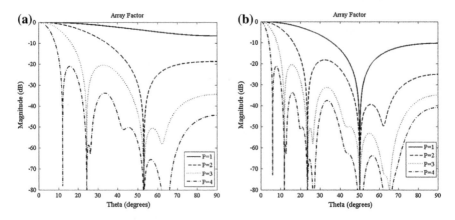

Fig. 4 Plots of the far-field radiation patterns produced by a series of four (P = 1, 2, 3, 4) heptagonal fractal arrays generated with an expansion factor of S = 2 and **a** r = λ/2 **b** r = λ

Maximum directivity of 26.1 dB achieved at fourth iteration of this array and this directivity is the highest value among the various iterations of modified and unmodified cases of heptagonal fractal antenna array. In this paper directivity of proposed array is compared with the octagonal fractal antenna array at stage three. The directivity of the heptagonal fractal array is nearly equal to the octagonal fractal array. This means the desired directivity achieved with the least number of elements. This will reduce the array complexity and antenna designer can be choose heptagonal fractal array than octagonal fractal array for this desired directivity. Zero

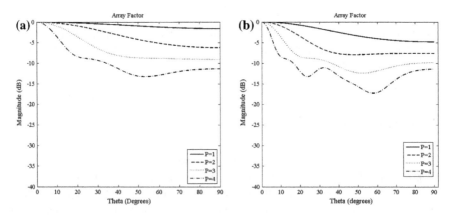

Fig. 5 Plots of the far-field radiation patterns produced by a series of four (P = 1, 2, 3, 4) heptagonal fractal arrays with center element generated with an expansion factor of S = 2 and **a** r = λ/2. **b** r = λ

Table 1 Directivity (dB), SLL (dB) and SLL angle of heptagonal fractal antenna array for unmodified and modified arrays

Element spacing	Iterations(p)	Un-modified array			Modified array		
		Maximum directivity (dB)	SLL (dB)	SLL angle (deg.)	Maximum directivity (dB)	SLL (dB)	SLL angle (deg.)
0.5	P = 1	8.3	−∞	–	7.5	−∞	–
0.5	P = 2	16.2	−∞	–	16.8	−∞	–
0.5	P = 3	22.3	−20.4932	2.3	23.1	−∞	1.72
0.5	P = 4	25.2	−21.0794	8.2	26.1	−∞	2.5
1	P = 1	8.2	−∞	–	7.7	−∞	–
1	P = 2	15.8	−18.2735	5.8	15.7	−∞	5
1	P = 3	22.0	−20.4932	11.9	22.3	−∞	5.2
1	P = 4	24.8	−21.0795	13.6	24.9	−11.1	7.23

Table 2 Comparison of maximum directivity for a stage 3 unmodified self-scalable octagonal array and a stage 3 modified self-scalable octagonal array [8]

Element spacing	Maximum directivity (dB)	
	Stage 3 unmodified self-scalable octagonal array	Stage 3 modified self-scalable octagonal array
0.5	21.51	26.11
1	23.41	23.90

side lobe levels achieved at maximum iterations of modified heptagonal fractal array. Less SLL and wide angle separation between the main lobe and side lobes can improves the signal-to-noise ratio. Heptagonal fractal array also exhibits good SLL angle (Fig. 5, Tables 1 and 2).

4 Conclusion

The research study is mainly focused on the design and analysis of heptagonal fractal antenna array based on concentric circular sub-array generator. The arrays are designed with and without center elements and for different radii. Directivity of the array improves by increasing order of iterations (P). Maximum directivity 26.1 dB and nearly zero side lobe levels achieved at modified section of the array having a radius of $\lambda/2$. Directivity of the heptagonal (7-Elements) fractal array at various iterations is nearly equal to the directivity of octagonal (8-Elements) fractal antenna array at same iterations, this means same directivity achieved with less number of total elements. This simple form of generation of the fractal array helps antenna designers to implement arrays with less side lobe level and good directivity. When comparison held between unmodified and modified fractal arrays, modified array shows better results.

Acknowledgments The authors are grateful to GITAM University, for the supporting and encouragement in the field of research.

References

1. M.T. Ma, *Theory and Application of Antenna Arrays* (Wiley & Sons, inc., Publication, 1974)
2. C.A. Balanis, *Antenna Theory, Analysis and Design*, (John Wiley & Sons, inc., Publication, 1997)
3. D.H. Werner, R. Mittra, *Frontiers in Electromagnetics* (IEEE Press, 2000)
4. A.J. Puente, C. Borja, *Fractal-shaped Antennas: A Review* (Wiley encycl. RF MICROW. Eng, 2005), pp. 1620–1635
5. R. Steven, Best, small and fractal antennas (chapter 10), in *Modern Antenna Handbook*, (John Wiley & Sons, inc., Publication)
6. D.H. Werner, S. Ganguly, An overview of fractal antenna engineering research. IEEE Trans. Antennas Propag. Mag. **45**(1), 38–57 (2003)
7. D.H. Werner, R.L. Haupt, P.L. Werner, Fractal antenna engineering: the theory and design of antenna arrays. IEEE Trans. Antennas Propag. Mag. **41**(5), 37–59 (1999)
8. W. Kuhirun, A new design methodology for modular broadband arrays based on fractal tilings, in *Thesis in Electrical Engineering* (The Pennsylvania State University, Aug 2003)
9. Y. Kim, D.L. Jaggard, The fractal random array. Proc. IEEE **74**(9), 1278–1280 (1986)
10. V. Srinivasa Rao, V.A. Sankar Ponnapalli, Study and analysis of fractal linear Antenna arrays, IOSR-JECE, **5**(2), 23–27 (2013). e-ISSN:2278-2834, p-ISSN:2278-8735
11. D.L. JAGGARD, A.D. JAGGARD, Cantor ring arrays, in *1998 IEEE International Symposium on Antennas and Propagation Digest*, vol. 2. (June 1998), pp. 866–869
12. D.H. Werner, W. Kuhirun, P.L. Werner, Fractile arrays: a new class of tiled arrays with fractal boundaries. IEEE Trans. Antennas Propag. Mag. **52**(8), (2004)

Multiplexer Based 2's Complement Circuit for Low Power and High Speed Operation

Kore Sagar Dattatraya, Belgudri Ritesh Appasaheb
and V.S. Kanchana Bhaaskaran

Abstract This paper presents a novel multiplexer based 2's complement circuit that can be used for the subtraction process using 2's complement method. The proposed multiplexer based 4, 8 and 16-bit 2's complement circuits are compared with the conventional subtractor circuits using 2's complement circuits for validating the proposal. Industry Standard EDA Tools and technology libraries have been employed.

Keywords 2's complement subtraction · Adders · Subtractors · Dividers

1 Introduction

The power dissipation of the microprocessors and microcontrollers has been a challenging problem in digital system design with the ever increasing performance requirements. The adder and subtractor units are two of the primary building blocks of the arithmetic and logic unit (ALU) of a general purpose microprocessor, microcontroller and digital signal processing architectures [1, 2]. Hence, the design of the adders and subtractors plays a greater role in low power operating capability and all the while concentrating on increased speed performance requirements.

2 The 2's Complement Subtraction

A subtractor can be designed using an analogous approach to that of an adder, in which the Boolean function of the 1-bit binary subtraction can be represented [2] as given below.

K.S. Dattatraya (✉) · B.R. Appasaheb · V.S. Kanchana Bhaaskaran
School of Electronics Engineering, VIT University Chennai Campus,
Chennai, Tamil Nadu, India
e-mail: sagarkore20@gmail.com

V.S. Kanchana Bhaaskaran
e-mail: vskanchana@gmail.com

© Springer India 2016   395
S.C. Satapathy et al. (eds.), *Microelectronics, Electromagnetics
and Telecommunications*, Lecture Notes in Electrical Engineering 372,
DOI 10.1007/978-81-322-2728-1_35

$$\text{Difference} = A \oplus B \oplus \text{Bin}$$
$$\text{Borrow} = A'.B + B.\text{Bin} + \text{Bin}.A'$$

where A, B, Bin are 1-bit inputs. For a multi-bit or an n-bit word subtraction process, n numbers of such 1-bit full subtractors are required.

This work aims at reducing the space and silicon area of an ALU (Arithmetic Logic Unit), while using the conventional adder module for the subtraction process. The 2's complement subtraction process is employed, in which the n-bit subtrahend is complemented and added with the n-bit minuend using the adder circuit present in ALU. During this addition process, the adder may produce a carry out bit, which may be either a one or a zero. If the carry out bit happens to be a one, then the carry is neglected and the remaining answer is taken as the result of the subtraction process. If the carry out bit is zero, then the 2's complement of the result will be considered the output of the subtraction process.

3   Conventional 2's Complement Circuit

This section explains the conventional 2's complement process [2–5] and the circuit normally employed. One of the circuit inputs is 1's complemented and the resultant binary word is incremented by 1 bit to obtain the 2's complement. The schematic representation of such a process is depicted in Fig. 1, where X0 X1 X2 X3 represent the binary input whose complemented value is added with a binary bit 1 as shown in Fig. 1.

For example, consider the input pattern X3 X2 X1 X0 is 1111, which when inverted produces 0000. The inverted (complemented or flipped) output binary word is added with bit 1 using the binary adder as depicted in the Fig. 1, which

Fig. 1 The conventional 4-bit 2's complement circuit

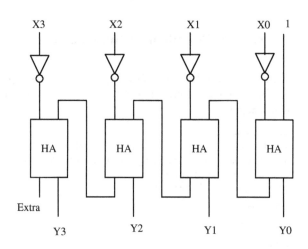

Table 1 The 2's complement values of the 4-bit binary numbers

Number (X3-X0)	Twos complement (Y3-Y0)
0000	0000
0001	1111
0010	1110
0011	1101
0100	1100
0101	1011
0110	1010
0111	1001
1000	1000
1001	0111
1010	0110
1011	0101
1100	0100
1101	0011
1110	0010
1111	0001

gives 0001. In this manner, the typical binary to 2's complement conversions for the 4-bit binary word are shown in the Table 1.

From the Table 1 shown below, the Boolean equations for the individual output bits of the 2's complemented outputs of the 4-bit input words have been derived as follows:

$$Y0 = X0$$
$$Y1 = X0.X1' + X0'X1$$
$$Y2 = X0.X2' + X0'.X1'.X2 + X0'.X1.X2'$$
$$Y3 = X2.X3' + X1.X2'.X3 + X0.X1'.X3' + X0'.X1.X2'.X3$$

4 The Proposed 2's Complement Using Multiplexer

This section presents a practical and simplified circuit arrangement for finding the 2's complement of a number. It employs the straightforward method as defined by the following algorithmic steps to convert the binary number into its 2's complement:

1. Start at the least significant bit (LSB)
2. If the LSB is zero followed by more 0 bits, then copy all the zero bits (working from LSB toward the most significant bit) until the first 1 is reached
3. When a bit 1 is reached in the path of travel from LSB to MSB, then copy that 1
4. Flip all the remaining bits that follow the 1 bit, till the MSB is arrived at

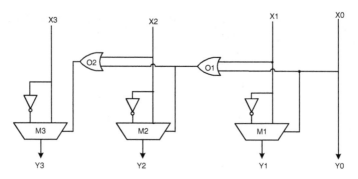

Fig. 2 The proposed multiplexer based 4-bit 2's complement circuit

Hence, the conversion of the number to its 2's complement can be accomplished without the two fold process of (1) finding the 1's complement and (2) the need of adding a bit 1 to the complemented value. Note that the second step may incur the maximum delay of the 1-bit addition process, while making the carry bit at LSB travel up to MSB based on the bits processed.

The proposed 4-bit multiplexer based 2's complement circuit operates as follows. Consider, the 2's complement of the 8-bit binary word 0011 1100 is to be found out, which is to be 1100 0100. It may be observed that the 3 LSB bits remain unchanged by the copying operation (while the rest of the digits were flipped). Figure 2 depicts the multiplexer-based 2's complement circuit for a 4-bit word. The bits X3-X0 are the input bits to be complemented and the bits Y3-Y0 are the output (complemented) bits. Note that the X0 is copied as it is to the output and it acts as the select bit of the multiplexer M1 and the output of OR gate O1 with inputs X0 and X1 acts as the select bit line of multiplexer M2 as so on, as shown in Fig. 2.

The internal structure of the multiplexer is elaborated in Fig. 3. Assuming A and B are the two input bits to the multiplexer, to be chosen by the select bit S. The select bit S and the S_bar bits control the transmission gates fed with the two input bits A and B, producing the Out bit based on S and S_bar bits. Considering Fig. 2, when the input is 0010, as per the steps shown above, the LSB bits 10 will be copied as it is, which will produce Y1 = 1 and Y0 = 0 at the output. In other words, the OR gate O1 checks the input (X1) which is a 1, to produce a bit 1at its output, which is connected to the selection line of the 2:1 multiplexer M2 and this will pass the inverted X2 to output. Here, note that when the first 1 is detected from LSB it will make all OR gate (O1, O2) output as 1, so automatically all the subsequent bits will get inverted or flipped.

Fig. 3 The internal circuit diagram and symbol of the multiplexer. **a** Circuit diagram of the multiplexer. **b** Symbolic representation

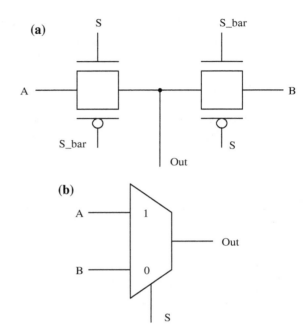

5 Simulation Results and Discussions

To validate the proposed Multiplexer based 2's complement circuit structure against the conventional method though comparisons, the 4, 8, and 16-bit circuit architectures are implemented using the industry standard cadence EDA tools and the UMC 90 nm technology library files. The simulation results are depicted in Table 2. The power dissipation of the circuits, the delay incurred across the two types and the numbers of transistors used in the designs are tabulated. The power-delay-product (PDP) and the product PDP X # of devices also are tabulated to demonstrate the impact of the proposed circuit.

From the simulation result it is observed that the average power dissipation of 4, 8 and 16-bit of multiplexer-based 2's complement circuit is 96, 37.60, and 32 % less compared to the its conventional 2's complement circuit counterpart. This is depicted in Fig. 4a. The delay values of the 4, 8 and 16-bit multiplexer-based 2's complement circuit is reduced by 46, 52, and 48 % compared to its conventional 2's complement counterpart circuit as depicted in Fig. 4b.

The power delay product (PDP) is an additional parameter that is considered for validating the proposed circuit. The product values of the power as measured in the conventional 2's complement circuit and the proposed multiplexer based circuit and the delay incurred by the respective circuits for the two types are justifiable estimates of PDP. Figure 4c shows the PDP values for the types. It can be seen that the PDP of the proposed circuit is less by 98, 70, 61.80 % for the 4, 8 and 16-bit

Table 2 Simulation results of the conventional and multiplexer based 2's complement circuits

Word size	2's complement type	Power (uw)	Delay (psec)	Number of transistor	PDP (10–15)	PDP X No. of Transistor (10–13)
4-bit	Conventional	18.03	257.1	88	4.626	4.06
	Multiplexer based	0.659	138	58	0.09	0.05
8-bit	Conventional	32.10	537	176	17.23	30.3
	Multiplexer based	20	257	126	5.14	6
16-bit	Conventional	59.82	1100	352	65.80	228.8
	Multiplexer based	44	571	298	25.124	74.8

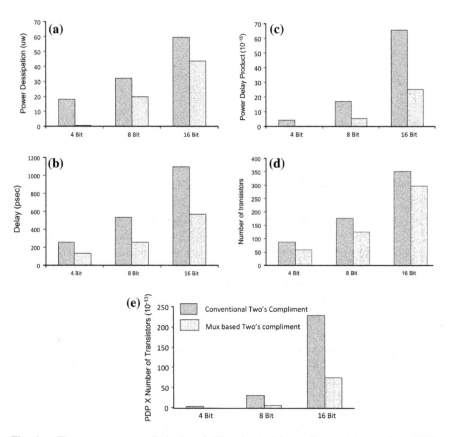

Fig. 4 **a** The average power dissipation. **b** The circuit delay. **c** Power delay product (PDP). **d** Number of transistors. **e** Product of power, delay and number of transistors

multiplexer based 2's complement circuits than their conventional counterparts respectively.

Additionally, the total number of transistors required for realizing the subtraction operation being the determining factor for the circuit layout area and hence the silicon area, the number of transistors employed for the 4, 8 and 16-bit circuit multiplexer based 2's complement circuits are compared against the conventional subtraction circuits. They are found to be reduced by 34, 28 and 15 % respectively as shown in Fig. 4d. An additional comparison has been made that takes into account the product of the PDP value and the transistor count of each circuit. As can be observed from Fig. 4e, the product of the PDP and the number of transistor is reduced by 98, 80 and 67 % respectively as against the conventional type of subtraction.

6 Conclusion

The novel circuit of a multiplexer based 2's complement circuit is presented in the paper. The design is validated through comparisons with the conventional 2's complementing circuit counterparts. The proposed multiplexer based 4, 8 and 16-bit 2's complement circuits result in reduced delay, number of transistors, power-delay-product values and hence resulting in reduced silicon area requirement. The product of power-delay-number of transistors of the proposed multiplexer based 4, 8 and 16-bit 2's complements are found to be 98, 80 and 67 less than the conventional complementing circuit counterparts. Industry standard EDA tools and foundry provided technology libraries have been employed to validate the designs.

References

1. J.F. Wakerly, *Digital Design Principles and Practices*, 3rd edn. (Prentice Hall, 2000), p. 47. ISBN 0-13-769191-2
2. I. Koren, *Computer Arithmetic Algorithms*. ed. by A.K. Peters (2002). ISBN 1-56881-160-8
3. http://www.cs.cornell.edu/ ~ tomf/notes/cps104/twoscomp.html
4. http://www.cs.uwm.edu/ ~ cs151/Bacon/Lecture/HTML/ch03s09.html
5. http://ecee.colorado.edu/ecen4553/fall12/intel_v1.pdf

Robust Hybrid Video Watermarking Using SVD and DTCWT

T. Geetamma and J. Beatrice Seventline

Abstract Watermarking is one of the widely used applications to provide security for the content shared over internet. Internet is a place where the data is not of assured security to the fullest. DTCWT is widely used in all image processing applications over a decade. A novel imperceptible video watermarking scheme is proposed. This method is implemented with the use of Dual Tree Complex Wavelet Transform (DTCWT) and Singular value decomposition (SVD) which helps in proof of ownership. In this method Singular Value Decomposition is applied to the Dual Tree Complex Wavelet Transform (DTCWT) coefficients of both watermark and original image and the singular Eigen values are interchanged. Because of advantages, Shift invariance and Directionality we prefer DTCWT. As SVD decomposes the matrix into 3 matrices U, S, V, we need U, V at the time of extraction. So we use U, V as watermark in audio. Watermark extracted from the audio gives the U, V matrices which makes this method blind watermarking. In order to prove the robustness of the method the results are compared with similar algorithms proposed in this paper but with DWT.

Keywords SVD · Arnold encryption · DTCWT · Blind and non–blind video watermarking

1 Introduction

Over the past decade internet spread widely, not just geographically but in terms of applications, has reached every corner of the world providing faster means of transferring information or data. However this spread has been exploited to stealing

T. Geetamma (✉)
Department of ECE, GMR Institute of Technology, Rajam, India
e-mail: tgeetamma@gmail.com

J.B. Seventline
Department of ECE, GITAM University, Vishakapatanam, India
e-mail: samsandra2003@yahoo.com

© Springer India 2016   403
S.C. Satapathy et al. (eds.), *Microelectronics, Electromagnetics
and Telecommunications*, Lecture Notes in Electrical Engineering 372,
DOI 10.1007/978-81-322-2728-1_36

of data to claim originality. In order to avoid theft of valuable data, security measures are employed, each method has its advantages. Watermarking is one of them, particularly suited for multimedia files. Watermarking can also be used for text files, but its wide use in Multimedia data made it a highly suited security measure. There are different types of watermarking methods; some are classified based on transform used and some on the type of watermark. Watermarking is a procedure in which a secrete file is embedded in a document (image, text, audio), as a precaution to save the document from attacks while transferring the data over a communication channel. File embedded is called *watermark*. Watermarking method employed in this paper uses DTCWT because of advantages like Shift invariance and Directionality which are not implemented in DWT. SVD is used for blind watermarking in this paper. Similar research in literature referred SVD watermarking implementation on images in [1] 2007 and in 2009 [2] were non blind. Simulation results are compared with DWT methods to analyse the performance DTCWT.

2   Proposed Method

This method uses Dual Tree Complex Wavelet Transform (DTCWT) [3], Singular Value Decomposition (SVD) [4] and Arnold encryption method [5]. In place of DTCWT we also have used DWT to compare the results. Whole process is divided into 4 steps they are:

1. Watermark pre-process
2. Video pre-process
3. Embedding
4. Extraction

2.1   Algorithm

Watermark Preprocess: In this process, Watermark is altered every second, considering 24 frames for a second, each of the 24 frames have similar watermark. A secret key K is used to modify each second, where $K \in \{1, -1\}$. K changes every second. w is an arary of 1's and −1's [6].

$$W = \begin{bmatrix} K(1) * w & K(2) * w \\ K(3) * w & K(4) * w \end{bmatrix}$$

Watermark Size is chosen based on the size of 2nd level DTCWT coefficient matrix of video frame [1], i.e. chrominance channel in YUV (4:2:0) representation of the video frame. Watermark is encrypted before embedding using Arnold encryption method.

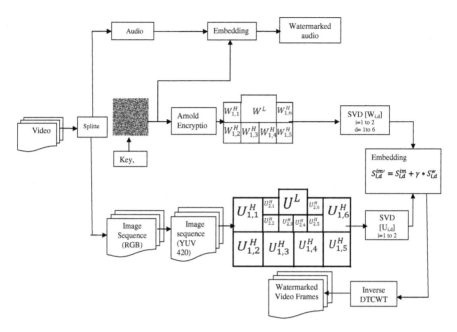

Fig. 1 Hybrid watermarking algorithm

$$[U^w, S^w, V^w] = SVD(W_{i,d})$$

Video preprocess: Video is mixture of both audio and images, we separate them using software (format factory). RGB video frame is converted to standard YUV (4:2:0) representation. 2-level DTCWT is applied on the YUV frame. On the higher band coefficients SVD is applied.

Embedding: DTCWT coefficients of watermark undergo SVD before embedding and S is embedded into coefficients of the Chrominance channel of video frame.

- Video frame is converted to YC_bC_r colour space (which was originally in RGB colour space) (Fig. 1).
- 2 level DTCWT is applied on blue Chrominance channel of video frame.
- SVD is applied on 2nd level coefficients which will give 3 matrices.

$$SVD(A_{i,d}) = [U^{im}, S^{im}, V^{im}]$$

- The singular values of $U_{i,d}$ are interchanged with the singular values of Watermark based on the following equation.

$$S_{i,d}^{im'} = S_{i,d}^{im} + \gamma * S_{i,d}^{w}$$

where $S_{i,d}^w$ is the singular vector of watermark and γ is the embedding strength
- In the last step, inverse SVD is applied on U^i, $S^{i'}$, V^i which finally generates watermarked frame.

$$ISVD\left[U^{im}, S^{im'}, V^{im}\right] = A_{i,d}^w$$

where $A_{i,d}^w$ is the watermarked coefficients.

Extraction: Extraction process follows reverse operation of embedding algorithm. First a watermarked frame is taken and 2-level DTCWT is applied. Later the matrix embedded is extracted by the following equation: $S_{i,d}^w = \left(S_{i,d}^{im''} - S_{i,d}^{im}\right)/\gamma$

Audio Watermarking: Watermark used for embedding is 'w'. Embedding algorithm uses FFT, The FFT coefficients are framed as matrices and SVD is applied on the coefficients. By analyzing the coefficients a particular threshold level is chosen and embedding is done. The embedding procedure is that the every coefficient is checked with watermark and if the bit is 1, threshold is added to coefficient, and if watermark is 0, threshold is subtracted. Extraction follows a similar process which is the inverse of the embedding process.

Video watermark Extraction: In the extraction process first the singular vector of watermark is extracted, giving us $S_{i,d}^w$, after that we extract the watermark from audio signal which is 'w' (Fig. 2).

Consider an attack similar to extraction attempt from audio and video: the attacker will never be able to get the linkage between the two styles of audio and video relationship. Hence extracted watermark will not permit the true watermark that can prove ownership. We replace $S_{i,d}^{wa}$ with $S_{i,d}^w$ and perform inverse SVD that gives same watermark. In case of non blind watermarking, SVD matrices are considered directly.

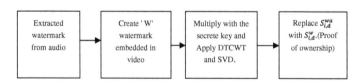

Fig. 2 Watermark extraction from video frame

Table 1 Non blind watermarking results

Frame number	Watermark (120 × 120)	Watermarked frame (480 × 480)	PSNR (dB)	Correlation factor
1			27.63	0.8739
30			27.6667	0.8749
60			27.65	0.8739
96			27.66	0.8741

3 Experimental Results

Both Blind and Non-Blind algorithms are implemented on a video with 480 × 480 resolution. "Watermark with the size of 120 × 120 is used. Watermarked results below are 4 frames (1, 30, 60, 96) each from a second re taken and the PSNR value of watermarked frame and original frame along with the correlation factor of extracted watermark". Same video is also considered for DWT approach (Table 1).

Embedding algorithm for blind watermarking is same as in Non blind. In case of audio watermarking we have (Table 2):

Table 2 Audio watermarking results

Number of samples present in the audio	Watermark (60 × 60)	Watermarked audio	PSNR (dB)	Correlation factor
Present: 217,728 Considered for watermarking: 215,681			103.6	1

Table 3 Comparing the results DTCWT of blind and non blind watermarking

Frame number	Correlation (non-blind watermarking)	Correlation (blind watermarking)
1	0.8739	0.9630
30	0.8749	0.9627
60	0.8739	0.9624
96	0.8741	0.9625

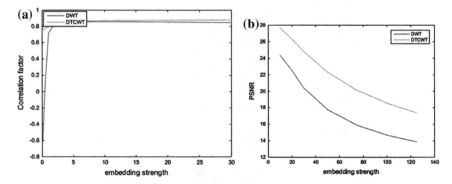

Fig. 3 Correlation and PSNR values for different embedding strength values for DWT and DTCWT

From the expected watermark from audio we generate watermark which is embedded in the video frames. 1-leve DTCWT is applied, followed by SVD that gives 6*3 matrices. Replace all the $S_{i,d}^w$ with the matrices extracted from the video frames. This is the process of embedding. Audio algorithm used is similar to both the methods one with DWT and another with DTCWT (Table 3).

In blind watermarking we have lower band coefficients of watermark, which increase the correlation factor. Now consider correlation factor for both DWT and DTCWT (Fig. 3).

Assume the video to be transmitted through a noisy channel. Comparison for DWT and DTCWT is as follows (Fig. 4).

In this method to face synchronisation attacks watermark is introduced in all of the video frames.

Fig. 4 Comparison between DWT and DTCWT for different noise factors

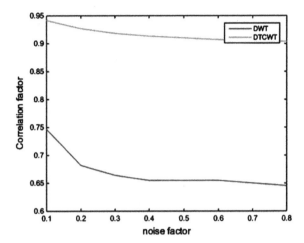

4 Conclusion

DTCWT is useful tool in signal processing applications. This method is secure compared to other pervious implemented methods of similar process. Results show improvement in performance to previous methods. This method can resist from any form of theft. The algorithm is better for noise induction, cropping, scaling but not so robust to geometrical distortions. Future work include increasing the robustness of audio watermarking algorithm and making the Video watermarking robust to all sorts of attacks present.

References

1. S. Mabtoul, E. Hassan, I. Elhaj, D. Aboutajdine, Robust color image watermarking based on singular value decomposition and Dual tree complex wavelet transform. In *ICECS 2007. 14th IEEE International Conference Electronics, Circuits and Systems* (2007)
2. H.A. Abdallah, M.M. Hadhoud, A.A. Shaalan, SVD-based watermarking scheme in complex wavelet domain for color video. in *ICCES 2009. International Conference on Computer Engineering & Systems*, (2009)
3. P. Loo, N. Kingsbury, Digital watermarking with complex wavelets. In *IEEE seminar on secure and image Authentication* (2000)
4. K. Baker, *Singular Value Decomposition Tutorial* (2005)
5. E. Chrysochos, V. Fotopoulos, M. Xenos, M. Stork, A.N. Skodras, J. Hrusak, Chaotic-correlation based watermarking scheme for still images. In *Proceedings of Applied Electronics 2008 International Conference*, (Pilsen, Czech Republic, 2008), pp. 10–11

6. M. Asikuzzaman, M.J. Alam, A.J. Lambert, M.R. Pickering, A blind digital video watermarking scheme with enhanced robustness to geometric distortion. In *International Conference on Digital Image Computing Techniques and Applications (DICTA)* (2012)

Efficiency Comparison of MWT and EWT all Backcontact Nanowire Silicon Solar Cells

Rakesh K. Patnaik, Devi Prasad Pattnaik and Ritwika Choudhuri

Abstract Nanowire Solar cells (NW SCs) have shown hopeful advance for light absorption and charge carrier transport because in a vertical structure absorption and carrier transport are orthogonal to each other. In this work high-efficiency back-contact back-junction (BC-BJ) structures like emitter wrap through and metallization wrap through Silicon (Si) SCs for one-sun applications were studied. The top portion of the device gets fully illuminated and there is no loss incorporated due to the absence of front metallization. Key parameters were extracted of proposed structure and a vertical NW SC structure with a front contact using 3D-TCAD simulation tool. First of all a comparison in the performance is observed between a planar solar cell (PSC) and a NW SC, then two different types of all back contact (ABC) structures are studied. The entire work is simulated using AM1.5G solar spectrum at room temperature. The key solar parameters like open circuit voltage (V_{OC}), short circuit current (I_{SC}), fill factor (FF), short circuit current density (J_{SC}) and conversion efficiency are calculated. The efficiency of the proposed structure is more comparable to a regular NW structure which is considered with similar device characteristics.

Keywords Solar cell · Back contact nano wire · EWT · MWT

R.K. Patnaik (✉) · R. Choudhuri
Department of EIE, NIST, Berhampur, Odisha, India
e-mail: rkshpatnaik@gmail.com

R. Choudhuri
e-mail: ritwika97@gmail.com

D.P. Pattnaik
Department of ECE, GITAM University, Vishkapatnam
Andhra Pradesh, India
e-mail: deviprasad.pattnaik@gmail.com

© Springer India 2016
S.C. Satapathy et al. (eds.), *Microelectronics, Electromagnetics and Telecommunications*, Lecture Notes in Electrical Engineering 372,
DOI 10.1007/978-81-322-2728-1_37

1 Introduction

Solar Cell is the best option to avoid fossil fuel consumption and for economic consideration [1]. More than 50 % of the total budget required to manufacture a Solar cell is utilized in primary processing of Si wafer. A metallurgical grade Si required very careful processing to convert it into solar cell grade Si wafer. Wafer with impurity reduce the conversion efficiency of the Solar cell. For this reason efficiency of bulk and thin film solar cell is limited maximum up to 25 % [2–5]. Conventional solar cell has planar p-n junction so loss occurs in carrier collection. In a planar solar cell more than 60 % of the solar radiation loss occurs due to reflection from the front the surface. So because of lack of light trapping mechanism, loss of carrier before collection and higher manufacturing cost first and second generation solar cell always create a tradeoff between cost and quality. Then third generation solar cells are developed by modifying the solar structure or by altering the material characteristics [6–10]. NW SC gives a promising approach for better light trapping and carrier collection. NWSC provides a vertical or radial junction trough out the length of the nanowire, so it easily separates the electron and hole. Also because of long length of the nanowire, all wavelength of solar spectrum are absorbed. When array of nanowires are grown on a substrate light also trapped, so less reflection loss. The efficient collection mechanism due to small diameter and vertical p-n junction it gives an opportunity to lower the amount effort required for preparation pure wafer. So with an impure wafer of low cost also higher conversion efficiency can be achieved with light trapping and efficient carrier collection. To further enhance the conversion efficiency of the NW SCs surface may be passivated or impurity can be added to the structure [11].

Figure 1 shows a Si NW SC with inside core doped with p-type impurity material and outer shell doped with n-type impurity material.

Among efforts to raise the performance of the NW Si SCs, the all-back-contact cell design is an appealing candidate. Placing the n-contact grid on the front surface is a common strategy. But one inherent demerit associated with this type of structure is Optical shading loss due to presence of front contact grid. It means that presence of front metalized grid deprive the portion of front surface in receiving sun light also it becomes quiet challenging to groove the front surface for better light absorption and trapping. In an ABC arrangement can be done to collect the carrier on the rear surface of the solar cell instead of placing the metalized grid on the front surface [12].

The all back contacted SCs, which exhibits both 'p' and 'n' of metal contacts on the back side, can be classified into three types: (i) Emitter Wrap Through (EWT) SCs [13], in this structure the front surface collecting junction is connected to the interdigitated contacts on the back surface via laser-drilled holes, (ii) Metallization Wrap Through (MWT) SCs, in this structure the front surface collecting junction and the front metallization grid are connected to the interconnection pads on the back surface via laser-drilled holes, (iii) Back Contact Back Junction (BC-BJ) SCs [14], also named Interdigitated Back Contact (IBC) SCs, this

Fig. 1 Cross sectional view
of the NW SC

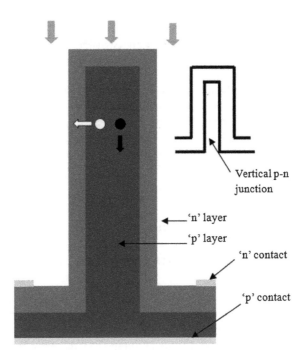

Vertical p-n
junction

'n' layer

'p' layer

'n' contact

'p' contact

structure both contacts and the collecting junction placed on the back side of the cell.

Here extension of n-type emitter vertically downward grown into the base to connect it to the n-contact present on the rear side. However, there are some challenges associated to the back-contact SC structure. The finger which connects n-type emitter to the n-contact forms a depletion region with p-type base. So the width of the n doped finger must be sufficient to facilitate the carrier transport means width must be more than the width of depletion region and minority carrier diffusion length. Another factor matters here is a lot of substrate area also contribute to conversion of solar energy into electrical energy. The placement of transparent contact on top of nanowire arrays suffers some issue.

In this work a PSC and a NW SC is designed, both the said structures were simulated and their efficiency was compared. It has been found that a NW SC proves to be more efficient than PSC. There are sufficient literature available claims that a vertical p-n junction SC like NW SC leads the PSC in terms of efficiency. Then on the same NW SC substrate or on the rear side three BC BJ structures are implemented and simulated. The efficiencies of all three BC BJ structures further improve the efficiency of a NW SC with both polarity contacts placed at front and rear surface.

Light which are fall on substrate also generate photo carriers and improves the solar current. So by placing front contact on rear side contribution of substrate can be further improved. As discussed above a connecting finger is used to connect the front surface and emitter part to the rear portion of the device where the required contact is present. Finger is an extension of emitter material or metal contact. So in the proposed structure the outer shell acts as an emitter and core as the base. So n-type connecting finger is made on the base of the device to connect the emitter portion to the rear surface where n-type contact is placed. This proposed structure a NW SC made of Si is simulated using TCAD software by keeping all contacts on back side of the structure or on rear surface. Conversion efficiency of proposed structure and the regular NW SC is compared and discussed. Out of three ABC structures only EWT and MWT are possible to implement on NW SC or co-axial p-n junction type SC.

2 Simulation Setup

2.1 Device Geometry of NW SC and PSC

This section describes the tool flow of the Synopsys TCAD simulation tool. For each step, the associated input parameters, extracted parameters, and important settings in the command file are discussed. First of all Si substrate is taken with a minimum dimension to accommodate the growth of a Si NW of the required diameter. The diameter of the NWSCs must be optimized for proper collection of photo generated carriers.

Figure 2a shows a Si NW of radius 100 nm and length 100 μm is grown using a circular mask, it is called core of the NW SC. Then again a layer of Si is deposited with a thickness of 100 nm called shell. Core is doped with p-type impurity material and shell is doped with n type material. Doping order in both core and shell is varied in the order of 1016–1019 cm^{-3} (Fig. 2b, c). Then the top surface is coated with anti-reflective coating (ARC). Aluminum metal contacts are made on p-core at the bottom and Silver contact on the n-shell developed on the substrate side (Fig. 2d). Likewise 20 such NWs are grown on a 3 × 3 μm^2 area substrate forming an array of Si NW SC (Fig. 3). A planar p-n SC of same dimensions with top n layer thickness of 100 nm and bottom p layer of thickness 100 μm is used for reference.

EWT SCs an extension of n-type layer grown towards base or rear surface of the structure then metal contacts are placed on p-type base and n-type emitter as shown Fig. 4a.

MWT SCs an extension of metal extended towards base or rear surface of the structure then metal contacts are placed on p-type base and n-type emitter as shown Fig. 4b. Though thick bus bar are not present on the front side more amount of front surface exposed to light and more photo carriers are generated.

Proposed structures are designed using process emulation commands of Sentaurus Structure Editor (SDE) of Synopsys TCAD suite [15]. For accurate simulation result tight meshing strategy is adapted (Fig. 2a). The general practice is

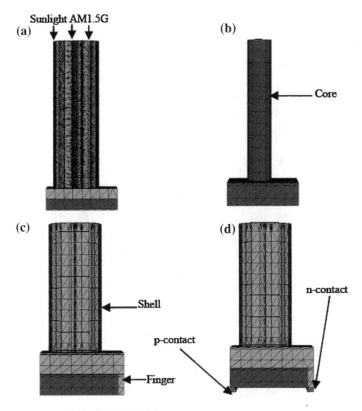

Fig. 2 Structure of a single EWT NW SC

to apply a coarse mesh to the whole region first, then focus into specific areas that require high resolution, and refine the mesh in those regions. A tight mesh is necessary in particular places such as material interfaces, p-n junctions, and contacts. And the programs do the analysis from top to bottom, therefore, it is recommended to refine the vertical mesh spacing towards the top surface to resolve the optical generation profile.

2.2 Physics Section

Physics section for the proposed structure developed and simulated using Sdevice tool of Synopsys TCAD. Here, AM1.5G solar spectrum is chosen as input file. Two electrodes are used as p-contact and n-contact connected to p-region and n-region respectively. In the physics section optical generation is calculated considering the optical absorption per volume of the devices. Transfer matrix method (TMM) is evaluated by layer wise extraction throughout the devices. Then by using Poisson

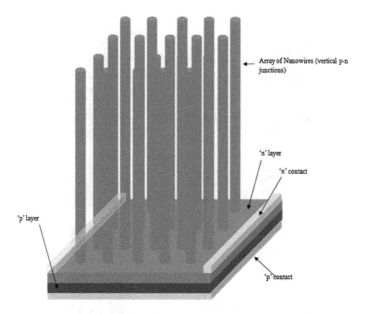

Fig. 3 Structure of array of nano wires

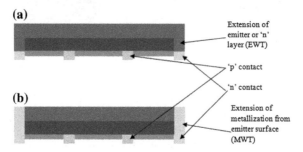

Fig. 4 **a** Structure of EWT NW SC (only lower base part is show). **b** Structure of MWT NW SC (only lower base part is show)

equation the electric field in the device calculated. Then the voltage is ramped across the contact to short circuit the device and current with respect to voltage is calculated. So by considering different voltage and current rating (V–I rating) V–I curve is plotted. The voltage between the terminal when no current is drawn known as open circuit voltage (V_{OC}) and the current through the SC when voltage across it is zero called short-circuit current (I_{SC}). Then maximum voltage (V_M), maximum current (I_M) and power rating are calculated. After extracting these parameters filling factor (FF) and efficiency is calculated.

It is difficult to incorporate the various recombination models appropriately in the proposed model, because the recombination rate is an essential factor in SC simulation. All said structures are subjected to tight meshing and on various small regions several recombination phenomenon like SRH, auger and surface recombination are considered.

2.3 Plot Section

When the device simulation is completed, the output includes one plot file and two geometry files which are the output files. I–V curve data can be visualized in the stored plot file and key parameters of SC are extracted. An input parameter Pin, representing the incidence intensity, is required for the cell efficiency calculation. Usually, an incidence intensity of 100 mW/cm^2 is assumed for the AM1.5G solar spectrum.

3 Result and Discussion

In Si solar cell it is requires photons of energy 1.12 eV or more. So photons with energy less than 1.12 eV unable to generate electron and hole pair. Photons with energy more than 1.12 eV will be lost because of thermalization process. In Si maximum open circuit voltage is less than band gap energy. Basically, separation of quasi-Fermi level defines the open circuit voltage [16].

The fill factor of a solar cell is up to 85 %, not a rectangular shape. It is due to various unavoidable recombination phenomenons. The current and voltage has an exponential relationship. The maximum power produced by a solar cell is less than the product of short circuit current and open circuit voltage.

When photon absorbed in a solar cell, it generates electron and hole pairs called photo generated carriers. So the concentration of minority carrier increases and these photo generated carriers flows through the depletion region towards the quasi-neutral region. The flow of photo generated carriers constitutes photo generated current it is added up with thermally generated current. In this condition if both 'n' and 'p' contacts are not connected to any external circuit then there is no net current flow inside the 'p' and 'n' junction. So by lowering the electrostatic potential barrier across the depletion region an opposite recombination current increases to balance the photo and thermally generated current by an amount of open circuit voltage.

The effect of n-layer doping variation on J_{SC}, V_{OC} and efficiency for PSC, NW SC, MWT NW SC and EWT NW SC structure shown in Fig. 5a–c respectively.

Figure 5 explains that by varying the doping order from 10^{16} to 10^{19} cm^{-3} raise the built in potential which increases the open circuit voltage. In NW solar cell structure more surface recombination occurs because of a large surface area than planar structure. So the planar solar cell offers larger V_{OC} than V_{OC} of nano wire

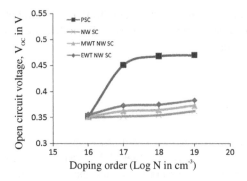

Fig. 5 Characteristics of planar, radial, radial MWT and radial EWT structures for different amount of doping in n-layer (V_{OC} vs. doping conc.)

solar cell. In case of MWT NW SC structure more amount of front surface is exposed to illumination compare to planar and regular NW SC. It creates a little more amount of photo carriers and a small variation of increase in V_{OC}. And in EWT NW SC structures the entire front surface illuminated. Out of four discussed SC structures, proposed EWT NW SC able to generate more amounts of photo carriers and little more increased variation in V_{OC} as doping order varies from 10^{16} to 10^{19} cm^{-3}.

As doping order varies from 10^{16} to 10^{19} cm^{-3} in case of radial structure gives rise to higher electric field and it increases the J_{SC} and it compensate the loss occurs due to recombination center. But recombination loss prominent in planar structure and reduces the J_{SC}. In case of EWT NW SC and MWT NW SC structure due to more amounts of generated photo carriers it provides higher current density (Fig. 6).

Further from Fig. 7 it has been observed that the best performance obtained from the EWT NW SC with n-layer with a doping order of 10^{19} cm^{-3}.

At above discussed dimension and doping density values of p-core and n-shell doping radial structure has 3 to 11 % times higher conversion efficiency over planar structure as shown by Fig. 7. Even though V_{OC} of NW structure is small as measured up to planar structure, but since NW structure has higher J_{SC} than planar

Fig. 6 Characteristics of planar, radial, radial MWT and radial EWT structures for different amount of doping in n-layer (J_{SC} vs. doping conc.)

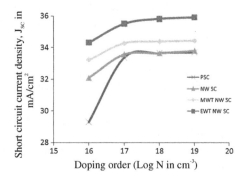

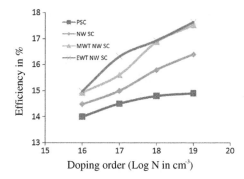

Fig. 7 Characteristics of planar, radial, radial MWT and radial EWT structures for different amount of doping in n-layer (efficiency vs. doping conc.)

structure, as a product of overall conversion efficiency of NW structure is higher than that of planar structure. One of the important characteristics of having radial junction is that even with poor quality material it gives satisfactorily high efficiency. Further improvement can be done on the substrate by implementing MWT or EWT technique and exposing the maximum or entire front surface and avoiding optical shading loss efficiency can be further improved. By using MWT NW SC efficiency can be increased up to 3 to 7 %, EWT NW SC efficiency can be increased up to 3 to 8 %. EWT NW SC provides better efficiency compare to other structures. Implementing ABC technique on NW SC, performance of radial geometry SC can be further improved and the shortcoming of top transparent contact can be avoided.

4 Conclusion

In conclusion, two different types of ABC NW SC were simulated. At first a NW which contains vertical or radial p-n junction is formed using Sentaraus device editor of 100 nm radius and 100 μm length. Both core and shell are doped with varying order from 10^{16} to 10^{19} cm^{-3}. Then 20 number of NW array formed. An equivalent PSC is designed. On the existing NW SC, two different ABC scheme is applied. One is called MWT where n layer surface of the base is connected or interdigitated towards the rear surface using metal finger. The other scheme is called EWT scheme where the metal finger is replaced with the emitter or shell material i.e., the n-type material doped with silicon. All four structures are subjected to simulation where AM 1.5G spectrum file is executed and output file is extracted. In NW SC, because of radial junction its conversion efficiency is higher compare to PSC. Then two said ABC schemes are applied on NW SC, and there performances are compared. The proposed two different ABC NW SCs shows higher conversion efficiency than regular NW SC. In AMC NW SC entire front surface of the base on which NWs are grown are exposed to illumination and more amount of photo carriers are generated and successfully collected. EWT NW SC shows higher efficiency then MWT because recombination loss at metal and semiconductor layers is reduced. Similarly V_{OC} and J_{SC} variation with respect to doping variation also found. All above results are found from above designed structure. This study serves as useful guideline for designing and developing more efficient NW based solar cells. These findings give new degrees of freedom for improvement in the field of third generation photovoltaic devices by introducing new approaches in designing NW SCs.

Acknowledgments The authors wish to thank Dr. Chayanika Bose, Associate Professor, Jadavpur University, for providing the Synopsys TCAD software for simulation of above mention devices. This software was funded by UGC.

References

1. M.I. Hoffert, K. Caldeira, A.K. Jain, E.F. Haites, L.D.D. Harvey, S.D. Potter, M.E. Schlesinger, S.H. Schneider, R.G. Watts, T.M.L. Wigley, D.J. Wuebbles, Nature **395**, 881–884 (1998)
2. M. Fawer-Wasswer, Solar energy-sunny days ahead? current status and outlook for photovoltaics and solar thermal power. Sarasin Sustainable InVestment Report, Nov 2004
3. M.A. Green, Sol. Energy **76**, 3–8 (2004)
4. Silicon, USGS Mineral Commodities SurVey, January 2008
5. M.G. Mauk, J.A. Rand, R. Jonceyk, R.B. Hall, A.M. Bamett, Solar-grade silicon: the next decade. In *Proceedings of the Third World Conference on Photovoltaic Energy Conversion*, vol. 1, Osaka, Japan, IEEE Publications, Los Alamitos, CA, 11–18 May 2003, pp. 939–942
6. R. He, P. Yang, Nat. Nanotechnol. **1**, 42–46 (2006)
7. A.I. Hochbaum, R.K. Chen, R.D. Delgado, W.J. Liang, E.C. Garnett, M. Najarian, A. Majumdar, P.D. Yang, Nature **451**, 163–167 (2008)
8. X.L. Feng, R.R. He, P.D. Yang, M.L. Roukes, Nano Lett. **7**, 1953–1959 (2007)
9. C.K. Chan, H.L. Peng, G. Liu, K. McIlwrath, X.F. Zhang, R.A. Huggins, Y. Cui, Nat. Nanotechnol. **3**, 31–35 (2008)
10. B.M. Kayes, H.A. Atwater, N.S. Lewis, J. Appl. Phys. **97**, 114302 (2005)
11. M.D. Kelzenberg, D.B. Turner-Evans, B.M. Kayes, M.A. Filler, M.C. Putnam, N.S. Lewis, H. A. Atwater, *Single-Nanowire Si Solar Cells*, (California Institute of Technology, Pasadena, CA 91125
12. M. Späth, P.C. de Jong, and J. Bakker, A novel module assembly line using back contact solar cells, in Technical Digest of the 17th International Photovoltaic Solar Energy Conference, Fukuoka, Japan, 436 (2007)
13. J.M. Gee, W. Kent Schubert, P.A. Basore, *Emitter Wrap-Through Solar Cell* (Sandia National Laboratories Albuquerque, NM 87185)
14. F. Granek, M. Hermle, C. Reichel, O. Schultz-Wittmann, High-efficiency back-contact back-junction silicon solar cell. In: 23rd European Photovoltaic Solar Energy Conference, September 2008 (Valencia, Spain), pp. 1–5
15. Synopsys, Sentauraus TCAD user manual (2010)
16. P. Würfel, *Physics of Solar Cells* (Wiley, Weinheim, 2005)

Simulation of Electrical Characteristics of Silicon and Germanium Nanowires Progressively Doped to Zener Diode Configuration Using First Principle Calculations

Mayank Chakraverty, P.S. Harisankar, Kinshuk Gupta, Vaibhav Ruparelia and Hisham Rahman

Abstract The effect of incorporating pairs of dopant atoms of opposite polarities into the nanowire lattice on the electrical behavior of nanowires has been presented in this paper. The dopants used are boron and phosphorus atoms. Intrinsic silicon nanowire is incapacitated with boron-phosphorus dopant atom pairs in a progressive manner, starting from one pair to nine dopant-atom pairs. The nanowire is simulated each time an additional dopant pair is introduced in the nanowire lattice to obtain current-voltage characteristics. These characteristics have been compared with that obtained by introducing similar dopants in an intrinsic germanium nanowire lattice. The power efficiencies of both intrinsic and doped silicon and germanium nanowires have been discussed towards the end of the paper.

Keywords MOSFET · First principle · NGEF · Incapacitation · CNT · ITRS · SCES

M. Chakraverty (✉) · P.S. Harisankar · V. Ruparelia
GlobalFoundries, N1 Block, Manyata Embassy Business Park,
Nagavara, Bangalore 560045, India
e-mail: nanomayank@yahoo.com

P.S. Harisankar
e-mail: harisps09@gmail.com

V. Ruparelia
e-mail: vaibhavaruparelia@gmail.com

K. Gupta
ISRO Satellite Center, PB No. 1795 Vimanapura Post, Bangalore 560017, India
e-mail: kinshuk.chandigarh@gmail.com

H. Rahman
Indian Institute of Science, CV Raman Ave, Bangalore 560012, India
e-mail: hishamrahman@gmail.com

© Springer India 2016
S.C. Satapathy et al. (eds.), *Microelectronics, Electromagnetics and Telecommunications*, Lecture Notes in Electrical Engineering 372,
DOI 10.1007/978-81-322-2728-1_38

1   Introduction

For the past several decades, a reduction in transistor size has remained the mainstay for performance improvements in electronic systems. The trend of scaling the transistor size to meet the performance requirements has now brought the transistor sizes to nanoscale regime. Scaling of transistors not only increases the packing density [1] but also brings reduction in operating voltages, increase in circuit speed and decreased power dissipation [2]. But reducing transistor sizes below 50 nm poses limitations of fundamental physics in the form of Short Channel Effects (SCEs) which prove to be a bottleneck for further downscaling. Some of the SCEs that contribute to being a barrier to downscaling are gate oxide leakage current, off state drain to source leakage current [1, 3–5], quantum confinement related threshold voltage increase [4] and random dopant induced fluctuations [1].

With decrease in gate length below 10 nm, degradation in subthreshold slope 'S' is observed. This is due to the fact that at such gate lengths, the control of gate over the channel decreases resulting in carrier tunneling from source to drain. This off-state leakage current is a limitation for device design from the point of view of operation and power budget. To deal with this limitation and ensure proper device turn-off, it is imperative that for sub 10 nm gate length devices, the channel layer should be thinner. Ultra-Thin Body (UTB) FETs fulfill this requirement as the MOSFET is fabricated on SOI substrate with a thin body region. But with such a structure, the electrons in the inversion layer are located away from the surface and occupy discrete energy levels in the channel. It leads to an increase in threshold voltage of the device as a larger surface potential is required to populate the inversion layer [6].

An approach to deal with the above limitations is to adapt to the Gate All Around (GAA) concept so that gate can exercise better control over the channel. GAA structure provides the dual advantage of improved device performance and reduction in SCEs. As Nanowires (NWs) are cylindrical single crystal structures with a diameter of a few nanometers, they provide a platform for unique wrap around structure where GAA concept can be implemented. Hence, when compared to planar and dual gate MOSFETs, NW FETs can offer superior performance. The silicon NW FETs have reported to provide excellent current drive and are compatible with conventional CMOS processing [1]. In view of these features, ultimate scaling of devices can be achieved by using NWs as part of the next generation device structures.

2   Overview of Nanowires

A NW is formally defined as an object with a 1D aspect such that the ratio of the length to the width is greater than 10 and the width does not exceed a few tens of nanometers [7]. But in the past few years, this definition has been extended to atomic and molecular wires, which do not possess the geometrical characteristics as mentioned above.

NWs belong to the category of one dimensional Carbon Nanotube (CNT) structures [2]. After the development of CNTs by Iijima [1]. NWs have been one of the most promising candidates for electronics industry. NWs have two quantum confined directions and an unconfined direction for electrical conduction. The ability of inorganic NWs to act as active components has led to a great interest in their synthesis and characterization [8]. The NWs also provide an attractive proposition for interconnects. Because of their unique properties, NWs are considered to have the capability to revolutionize nanoelectronics and nanotechnology applications [9, 10].

3   Simulation Methodology

QuantumWise Atomistix ToolKit (Ver.13.1) has been used for the purpose of silicon and germanium NW electrical behavior simulation. The calculator used is ATK-DFT (Device) and the exchange correlation used is Local Density Approximation.

In the present paper, work has been carried out to characterize nanowires of 3 nm diameter. First of all, using the above mentioned Toolkit, instances of intrinsic silicon and germanium nanowires are created. The bias arrangement for this work is such that at one end of nanowire, voltage is applied while the other end is kept at ground potential as shown in Fig. 1. Using the toolkit, the applied voltage is varied from −1 to 1 V in steps of 0.2 V and the resulting current-voltage characteristics of undoped instrinsic nanowires has been obtained.

Another set of simulations were performed with a slight modification in lattice structure of the nanowire. One atom of boron and one atom of phosphorus were incorporated in the lattice along the two ends to obtain the P-N diode configuration in the nanowire. The current-voltage characteristics so obtained have been recorded. The process of adding one boron and one phosphorus atom at a time has been carried out progressively until nine pairs of boron-phosphorus atoms were added to the lattice. After every addition of one pair of dopant atoms, current-voltage characteristics have been obtained. The same simulation procedure is followed by instantiating a silicon nanowire and a germanium nanowire, respectively, to obtain current-voltage characteristics with respect to silicon and germanium nanowires with Zener (highly P and highly N) configuration modeled on it.

GND ━━━━━━━━━━━━━━━━━━━━━━━━━━━━━━━━━━━━ BIAS VOLTAGE

Fig. 1 Schematic of simulated nanowire before doping

4   Simulation Results

The current-voltage characteristics of intrinsic silicon and germanium nanowires are shown in Fig. 2 (left and right respectively). The curves for the two intrinsic nanowires seem to be very close, the difference being that the curve for intrinsic germanium nanowire looks flatter than the intrinsic silicon nanowire curve. Also, germanium nanowire reports a comparatively lower current level as compared to its silicon counterpart with the same bias applied. This means that the power consumption in intrinsic silicon nanowire is more than that of intrinsic germanium nanowire, when these 1D structures are implemented in nanowire circuits.

A perturbation in the electrical neutrality of the one dimensional nanowires occurs as dopant atoms are incorporated in the nanowire lattice. The schematic of the nanowires, post addition of dopants is depicted in Fig. 3. The current-voltage curves pertaining to silicon nanowire for each dopant atom pair incorporated into the nanowire lattice is depicted in Fig. 4. And the Fig. 5 shows the current voltage curves for each of the dopant atom pairs introduced in the nanowire in a single plot. The configuration pertaining to 9 dopant atom pairs introduced in the nanowire lattice resulting in heavily doped P and N regions can be compared to a Zener diode configuration.

Unlike the characteristics of a normal silicon based Zener diode on a bulk silicon substrate, the feature observed in Fig. 4i with 9 dopant atoms pairs is not exactly identical but an increase in bias from 0.4 to 0.6 V and from 0.8 to 1 V results in negative resistance regions. Also, from the Fig. 5, it is evident that the addition of dopant atom pairs into the nanowire gradually shifts the intrinsic silicon nanowire characteristics towards right for every new pair of dopants incorporated into the lattice.

From the current and voltage values of Fig. 4i, the conductance has been calculated and plotted against the bias voltage. This is shown in Fig. 6. It is observed

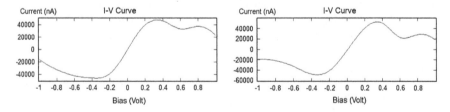

Fig. 2 Current-voltage characteristics of intrinsic (*left*) silicon nanowire; and (*right*) germanium nanowire

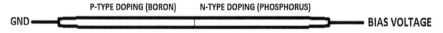

Fig. 3 Schematic of simulated nanowire after doping

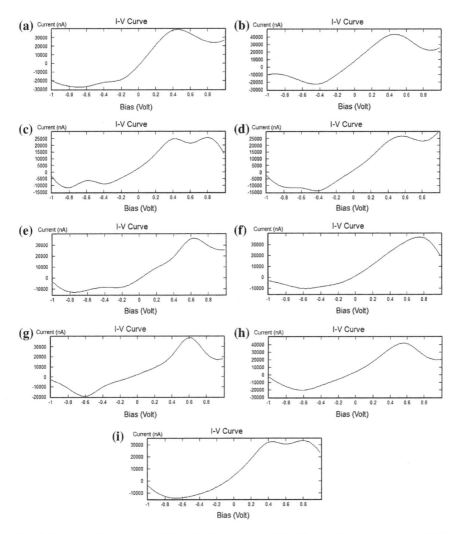

Fig. 4 Current-voltage characteristics of silicon nanowire for different dopant atom pairs. **a** NW with 1B and 1P Atom. **b** NW with 2B and 2P Atoms. **c** NW with 3B and 3P Atoms. **d** NW with 4B and 4P Atoms. **e** NW with 5B and 5P Atoms. **f** NW with 6B and 6P Atoms. **g** NW with 7B and 7P Atoms. **h** NW with 8B and 8P Atoms. **i** NW with 9B and 9P Atoms

that, there is a conductance peak at 0.2 V and the value drops gradually as the bias voltage increases.

Similar to the analysis done with silicon nanowire, the characteristics of germanium nanowire has also been simulated starting with an intrinsic Ge nanowire followed by incorporation of dopant atom pairs, one pair at a time, to obtain current-voltage curves. The current curves obtained by simulating the germanium nanowire with 9 pairs of dopant atoms is depicted in Fig. 7 while Fig. 8 shows the

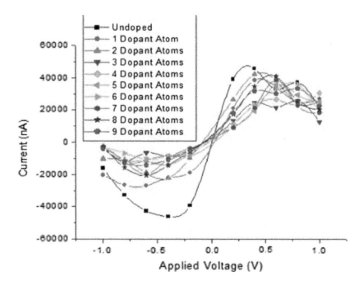

Fig. 5 Consolidated current-voltage characteristics of silicon nanowire with varying doping densities

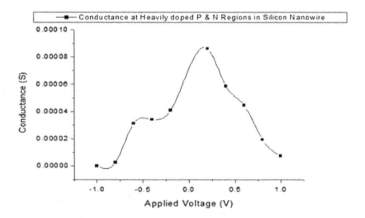

Fig. 6 Plot of conductance against bias voltage for silicon nanowire with 9 dopant atom pairs

consolidated current-voltage plot of germanium nanowire under different doping conditions. As evident from Figs. 5 and 8, there is a considerable variation in the current curves of silicon and germanium nanowires doped with nine pairs of dopants. Also, germanium nanowire with heavily doped (9 dopant atom pairs) P and N regions reports lower current levels when compared to its silicon counterpart. This infers that a Zener diode configured on a germanium nanowire reports lower power consumption as compared to the silicon nanowire based diode. The germanium nanowires are also more efficient in microwave applications, owing

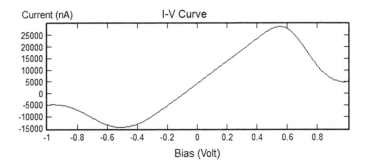

Fig. 7 Current-voltage characteristics of germanium nanowire with 9 dopant atom pairs

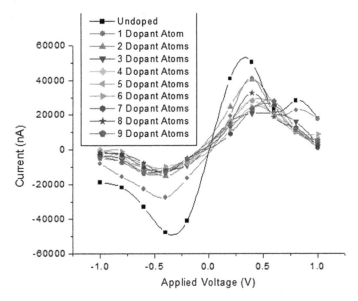

Fig. 8 Consolidated current-voltage characteristics of germanium nanowire with varying doping densities

to highly prominent negative resistance regions observed in doped Ge nanowire characteristics as compared to that reported in Figs. 4 and 5.

5 Conclusion

This paper analyzed the need and feasibility of the application of nanowires in nano electronic devices like nanowire field effect transistors to be used in post CMOS era electronic circuits that can circumvent the problems encountered due to short channel effects in bulk silicon and SOI technologies. An overview of different types of nanowires has been presented in this paper. Silicon and germanium nanowires have been simulated in their intrinsic forms to observe that the current levels with germanium nanowire are low as compared to silicon nanowire. Both the nanowires have been doped with boron and phosphorus atoms, one dopant atom pair at a time. Both silicon and germanium nanowires have been simulated each time a dopant atom pair is incorporated in the nanowire lattice. Introduction of dopant atom pairs has been continued till 9 pairs of dopant atoms have been introduced in the nanowire structure for it to attain highly P and highly N doped configuration that relates to Zener diode configuration. It is observed that germanium nanowire with heavy doping reports lower current levels than that reported with silicon nanowire with heavy P and N doping. To conclude, if these nanowires are modeled to be used as Zener diodes or normal PN junction diodes in CMOS circuits, germanium nanowires would result in higher power efficiency than their silicon counterparts.

References

1. J. Wang, Device physics and simulation of silicon NW transistors. PhD Thesis, Univ. Purdue, August 2005
2. E. Sangiorgi, A. Asenov, H.S. Bennett, R.W. Dutton, D. Esseni, M.D. Giles, M. Hane, C. Jungemann, K. Nishi, S. Selberherr, S. Takagi, Special issue on simulation and modeling of nanoelectronics devices. IEEE Trans. Electron Devices **54**(9), 2072–2078
3. Wang, Theoretical investigation of surface roughness scattering in silicon nanowire transistors. Appl. Phys. Lett. **87**, 043101 (2005)
4. C. Enz, E.A. Vittoz, Charge-based MOS transistor modeling. The EKV model for low power and RF IC design. Wiley (2006)
5. A.K. Sharma, S.H. Zaidi, S. Lucero, S.R.J. Brueck, N.E. Islam, Mobility and transverse electric field effects in channel conduction of wrap-around-gate NW MOSFETs. In: IEE Proceedings-Circuits Devices System, vol. 151, no. 5, pp. 422–430, October 2004
6. A. Chaudhry, J.N. Roy, MOSFET models, quantum mechanical effects and modeling approaches: a review. Semiconductor Technol. Sci. **10**(1), 20–27 (2010)
7. Y.S. Yu1, N. Cho, J.H. Oh, S.W. Hwang, D. Ahn, Explicit continuous current–voltage (I–V) models for fully-depleted surrounding-gate MOSFETs (SGMOSFETs) with a finite doping body. J. Nanosci. Nanotechnol. **10**(5), 3316–3320 (2010)
8. C.A. Richter, H.D. Xiong, X. Zhu, W. Wang, V.M. Stanford, W.-Ki Hong, T. Lee, D.E. Ioannou, Q. Li, Metrology for the electrical characterization of semiconductor NWs. IEEE Trans. Electron Devices **55**(11), 3086–3095 (2008)
9. B. Iñiguez, T.A. Fjeldly, A. Lázaro, F. Danneville, M.J. Deen, Compact-modeling solutions for nanoscale double-gate and gate-all-around MOSFETs. IEEE Trans. Electron Devices **53** (9), 2128–2142 (2006)
10. C. Mayank, G. Kinshuk, V.G. Babu et al., A technological review on quantum ballistic transport model based silicon nanowire field effect transistors for circuit simulation and design. J. Nanosci. Nanoeng. Appl. **5**(2), 20–31p (2015)

An Embedded Visually Impaired Reconfigurable Author Assistance System Using LabVIEW

R. Supritha, M. Kalyan Chakravarthi and Shaik Riyaz Ali

Abstract The advent of new technologies is dragging the attention of visually impaired people towards electronic gadgets. Though the conventional method is preferred by a few, visually impaired people are eager to explore the technological part of the braille system, which is the basic means of communication for them. A Visually Impaired Reconfigurable Author Assistance System (VIRAAS) using LabVIEW is proposed in this paper. The main objective is to convert a telugu language to speech using LabVIEW.

Keywords LabVIEW · Braille system · Visually impaired · Text to speech (TTS) NI USB 6211

1 Introduction

Braille system has been in demand by the visually impaired from the day it was invented until now. The Braille system has been the only means of communication among the visually impaired in the eighteenth century, which was invented by Louis Braille. These days there are many upcoming fields where the blind people are getting a chance to explore their talents. The shortage in the methods of teaching aids, resources systems, are few of the problems raised for poor learning of the braille language. Braille translation for educational purpose in the initial days of its

R. Supritha (✉) · M. Kalyan Chakravarthi
School of Electronics Engineering, VIT University, Chennai 600127, India
e-mail: supreeram@gmail.com

M. Kalyan Chakravarthi
e-mail: maddikerakalyan@gmail.com

S.R. Ali
Department of Electronics and Communication Engineering,
Chaitanya Engineering College, Kommadi, Visakhapatnam 530041, India
e-mail: riyazali06@gmail.com

© Springer India 2016
S.C. Satapathy et al. (eds.), *Microelectronics, Electromagnetics and Telecommunications*, Lecture Notes in Electrical Engineering 372,
DOI 10.1007/978-81-322-2728-1_39

invention, was very laborious since very few people were very well versed with the braille script. As the trend is changing swiftly, visually impaired people prefer electronic gadgets over the conventional way of communication these days.

Several improvements have been brought up in the basic braille system in the past. A method for an automatic document development of the braille text from word processed documents is discussed by Blenkhorn and Evans in [1]. As the technology is trending, lot of advancements are being implemented on the traditional braille system. Refreshable braille displays have been explored in the recent past by Yobas et al. [2, 3]. With the abundance in the availability of the resources, the refreshable braille system is getting improvised in a reasonable span of time. Systems for tracking the movement of fingers of a braille reader have also been developed as an aid to the research in this particular field. Aranyanak and Reilly studied about one such finger tracking system [4]. The conversion of any language to braille text and vice versa is a tedious task. The script should be standardised and should be accepted by all. Advance technologies in image processing have been used to convert the Braille code to Arabic language and its corresponding voice, which is elaborated in [5]. On the other hand, Japanese script is converted to braille using an Adaptive Knowledge Base (AKB) [6]. Blenkhorn has examined about a system for converting print to braille [7]. This system described by Blenkhorn can be extended for different languages. An efficient way of converting Chinese script to braille was introduced by Wang and few others [8]. Braille to text conversion for Hungarian is also developed [9]. With the availability of resources these days, systems which can act as a communication bridge between a visually impaired and a sighted person have also been developed [10]. Refreshable Braille cell involving piezoelectric motors, which were invented taking into consideration the portability factor of the gadget has been proposed [11]. The proposed system VIRAAS basically concentrates on the conversion of Telugu language to speech using LabVIEW. If a blind person is willing to write a poem or a prose, he can start typing his own document with the voice coming from LabVIEW as an aid to the text for error detection.

2 Methodology

The proposed system is a reconfigurable braille cell, which acts as an aid to the blind authors. The audio of a letter corresponding to a particular combination of inputs, is heard from LabVIEW. The sound for each combination of the cell, which is controlled by switch is obtained and stored in LabVIEW. The database is also managed in LabVIEW. The data is retrieved from the database dynamically. Each combination of cell calls the corresponding sound from the database and it is heard by the blind author. It serves as assistance to the blind person who wants to write his/her own prose or poem. If the blind authors want to make any changes in the document or if they have made any mistake in the document, they can make the changes themselves with the help of the audio from LabVIEW. There will not be any requirement of a normal person, to assist the blind author for pointing out the errors they are making while typing the paper. A single letter's audio can be extended to a letter with consonant like the sound "aa". This is done by introducing another control switch, the word Change button. The proposed system concentrates on the audio of a single letter. The extension of the work with the seventh switch is yet to be implemented.

2.1 Block Level Implementation of VIRAAS

The VIRAAS has 7 Single pole Single throw switches. Single pole single throw switch is the basic on-off switch. It is small in size and makes the Braille cell look compact. The switches can be accessed easily by the blind author. The data from the switches is transmitted to LabVIEW using NI DAQ 6211. The 7 analog inputs ports in the DAQ were used for the signal acquisition from the switches (Fig. 1).

The Braille cell with word change button, which is intended to produce the sound of letters with consonant, can also be extended to generate a sentence. Presence of an extra eighth button can act as a space between consecutive words. This can further increase the reliability of the system proposed.

2.2 Circuit Diagram of VIRAAS with Word Change Button

VIRAAS has been designed in such a way that it is handy in nature and easily accessible by the blind author. One end of the seven switches is given to positive slots of analog input in NI USB 6211 from a0–a7. The negative slots from (a0–a7) in the DAQ are grounded. The other end of the seven switches is also grounded. The voltage value coming from the DAQ is compared with a threshold value in LabVIEW. The comparison gives either a high or a low output which is connected indicator for a particular switch. Figure 3 gives the circuit diagram. Figure 2 explains the graphical program designed for implementing on VIRAAS.

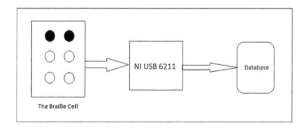

Fig. 1 Block diagram of VIRAAS

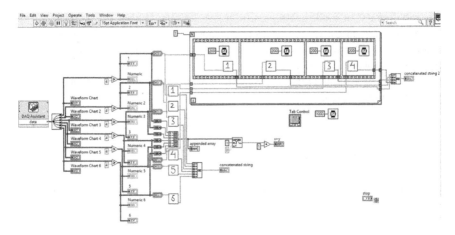

Fig. 2 Circuit diagram of VIRAAS with word change

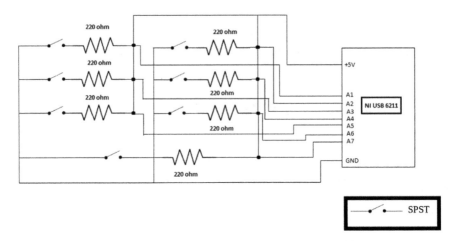

Fig. 3 Circuit diagram of VIRAAS with word change

2.3 Flowchart of the VIRAAS with the Word Change Button

The data flow in VIRAAS is dynamic in nature. The blind author can access the switches according to their necessity. The flowchart of VIRAAS with word change button is shown in Fig. 6. Initially a combination of keys is pressed by the blind author. A delay is given for the word change button to be pressed. If the word change button is not pressed within the specified time then the alphabet alone will be retrieved from the database. If the word change button is pressed, the next set of combination is given to braille cell.

This now corresponds to a consonant in the database like "k + aa = kaa", will be retrieved from the database. The seventh button in the circuit is just connected to the LabVIEW using NI USB 6211, its functionality and logic is yet to be defined in the system.

3 Results and Discussions

VIRAAS is designed with an intension to produce a compact electronic gadget for the visually impaired people. The sound corresponding to each combination of braille code is heard from the system's speaker via LabVIEW. The device can be handled by any visually impaired easily. The signal passing through the switches is compared with a threshold value. If the value is greater than the threshold, the LED in on, else it is off. By this comparison, a digital logic is created with produces a particular number for each combination of code.

The corresponding sound for each code is stored in the LabVIEW and data is retrieved dynamically each time the author presses a combination of switches. The LED's in the front panel is shown in Fig. 4. The waveform of the audio corresponding to the letter "na" is also seen in LabVIEW front panel.

The voltage passing through each switch is represented in Fig. 5. We can also see that the switches which are pressed have more voltage (near to 5 V) than the switches which aren't pressed. The first graph in the Fig. 6 signifies the presence of

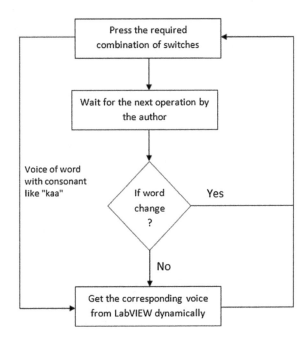

Fig. 4 Flow chart of VIRAAS with word change

Fig. 5 The VIRAAS system
set up with LabVIEW

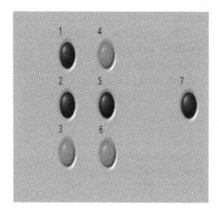
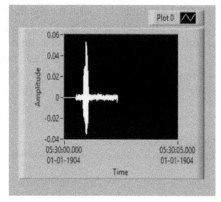

Fig. 6 The VIRAAS system set up and LabVIEW front panel

voltage greater than the threshold set by us, in that corresponding switch (switch 1). The LED connected to that switch via DAQ in LabVIEW glows. Similarly a sequence of switches will correspond to a particular letter which is already stored in LabVIEW.

4 Conclusions

The proposed reconfigurable braille cell for the virtually impaired (VIRAAS) has a facility for the virtually impaired to write their own prose or poem without any help of a normal person. The device is small and compact in nature. The device is designed in such a way that it can be transported from one place to another easily, i.e., it is portable. The audio from LabVIEW can be heard from the speakers of the laptop or the author can even put on a headphone and listen to whatever he/she is typing. The device makes the visually impaired person feel independent while they are writing their own peace of poem or prose.

References

1. P. Blenkhorn, G. Evans, Automated braille production from word-processed documents. IEEE Trans. Neural Syst. Rehabil. Eng. **9**(1) (2001)
2. L. Yobas, Dept. of Biomed. Eng., Case Western Reserve Univ., Cleveland, OH, USA , D.M. Durand, G.G. Skebe, F.J. Lisy, M.A. Huff, A novel integrable microvalve for refreshable Braille display system. J. Microelectromech. Syst. **12**(3) (2003)
3. J. Su Lee, Opt. & Semicond. Devices Group, Imperial Coll. London, UK, S. Lucyszyn, A micromachined refreshable Braille cell. J. Microelectromech. Syst. **14**(4) (2005)
4. I. Aranyanak, R.G. Reilly, A system for tracking braille readers using a Wii remote and a refreshable braille display. Behav. Res. **45**, 216–228. doi:10.3758/s13428-012-0235-8 (Springer)
5. S.D. Al-Shamma, S. Fathi, Arabic Braille recognition and transcription into text and voice. In: *2010 5th Cairo International Biomedical Engineering Conference Cairo, Egypt, December 16–18, 2010*, 978-1-4244-7170-6/10/$26.00 ©2010 IEEE
6. S. Ono, T. Yamasaki, S. Nakayama, Adaptive knowledge base with attribute weight and threshold adjustments for Japanese-to-Braille translations. Artif. Life Robot. **12**, 59–64 © ISAROB (2008). doi:10.1007/s10015-007-0442-z
7. P. Blenkhorn, Dept. of Comput., Univ. of Manchester Inst. Sci. & Technol., UK, A system for converting print into braille. IEEE Trans. Rehabil. Eng. **5**(2)
8. C. Wang, Lab. of Intell. Inf. Process., Chinese Acad. of Sci., Beijing, China; X. Wang; Y. Qian, S. Lin, C. Wang, Lab. of Intell. Inf. Process., Chinese Acad. of Sci., Beijing, China; X. Wang, Y. Qian, S. Lin, Accurate Braille-Chinese translation towards efficient Chinese input method for blind people. Pervasive Comput. Appl. (ICPCA)
9. A. Arato, T. Vaspori, G. Evans, P. Blenkhorn, Braille to text translation for Hungarian. Comput. Helping People with Special Needs Lecture Notes Comput. Sci. **2398**, 610–617 (2002)
10. T. Dasgupta, Indian Inst. of Technol., Kharagpur, India, A. Basu, A speech enabled Indian language text to Braille transliteration system. Inform. Commun. Technol. Dev. (ICTD) (2009)
11. P. Manohar, Anna Univ., Chennai, A. Parthasarathy, An innovative Braille system keyboard for the visually impaired. In: *11th International Conference Computer Modelling and Simulation UKSIM '09* (2009)

Estimation of RCS for a Perfectly Conducting and Plasma Spheres

Swathi Nambari, G. Sasibhushana Rao and K.S. Ranga Rao

Abstract This paper presents about plasma technology when the plasma is applied on a simple target, like Sphere and its Radar Cross Section (RCS) is computed with respect to the parameters like size, wave frequency and plasma frequency and compared its RCS with a perfectly conducting sphere (Gao Y et al, The calculation of back-scattering radar cross section of plasma spheres. Institute of Electronic Engineering, Hefei, IEEE (2000) [1]). The RCS of a perfectly conducting sphere has been computed using Mie scattering series with a relation given by Kerr DE, Propagation of short radio waves. McGraw-Hill, Newyork (1951) [2]. The analysis given in this is based on spherical polar scattering geometry (SPSG) in which the scattering parameters (a_n^s), (b_n^s) are defined. The physical interpretation of scattering coefficients aids in visualizing the mechanism of the scattering process. In this paper, not only the RCS of a perfectly conducting Sphere is computed at different frequencies with particular diameter but also the RCS of a perfectly conducting sphere is computed for various diameters at different bands of frequencies. Theoretically computed electron volume density and current density of plasma at a particular plasma frequency for an Argon gas and also RCS comparison is made for a plasma sphere and perfectly conducting sphere at standard dimensions (Skolnik MI, Introduction to Radar Systems. McGraw-Hill, Newyork (1962) [3]) in which RCS is very less for plasma sphere when compared to perfectly conducting sphere. The plasma cover on the targets helps in getting less RCS and also makes the target unseen by the enemy Radar called Active Stealth Technology. RCS treatment in this paper is based on Radar frequencies ranging from 0.1 to 40 GHz.

Keywords RCS · Plasma frequency · Spherical polar scattering geometry

S. Nambari (✉) · G. Sasibhushana Rao
Department of ECE, Andhra University College of Engineering,
Visakhapatnam, A.P., India
e-mail: swathiyadav23@gmail.com

K.S. Ranga Rao
Retd Sc-'F', RCS Division, NSTL, DRDO, Visakhapatnam, A.P., India

© Springer India 2016
S.C. Satapathy et al. (eds.), *Microelectronics, Electromagnetics
and Telecommunications*, Lecture Notes in Electrical Engineering 372,
DOI 10.1007/978-81-322-2728-1_40

Fig. 1 Radar cross section

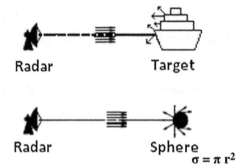

Radar   Target

Radar   Sphere
$\sigma = \pi\, r^2$

1 Introduction

Radar cross section is the measure of a target's ability to reflect radar signals in the direction of the radar receiver, i.e. it is a measure [1–3] of the ratio of backscatter power per steradian (unit solid angle) in the direction of the radar (from the target) [3–8]. The RCS of a target can be viewed as a comparison of the strength of the reflected signal from a target to the reflected signal from a perfectly smooth sphere of cross sectional area of 1 m^2 as shown in Fig. 1.

2 Scattering Regions of a Sphere

Rayleigh Region: When the wavelength is large compared to the object's dimension, scattering is called Rayleigh region where $2\pi a/\lambda \ll 1$. It is named after Lord Rayleigh who first observed this type of scattering in 1871, long before existence of radar, when investigating the scattering of light by microscopic particles. RCS in Rayleigh region is proportional to fourth power of the frequency and is determined more by the volume of the scatterer than by its shape. At radar frequencies echo from rain is usually described by Rayleigh scattering [3, 4] (Fig. 2).

$$\sigma = \left[\pi r^2\right]\left[7.11(kr)^4\right] \text{ where } k = 2\pi/\lambda \tag{1}$$

Mie (Resonance): In between the Rayleigh and Optical regions is the resonance region where the radar wavelength is comparable to the object's dimension i.e. $2\pi a/\lambda \approx 1$. The RCS of the Sphere in resonance region oscillates as a function of frequency or $2\pi a/\lambda$ and its maximum occurs at $2\pi a/\lambda \approx 1$ and is 5.6 dB greater than its value in the optical region. Changes in cross section occur with changing frequency because there are two waves that interfere constructively and destructively. One is direct reflection from face of the Sphere and the other is creeping wave that

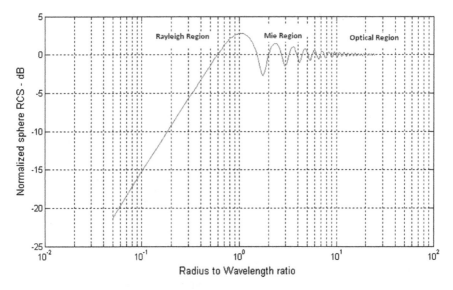

Fig. 2 RCS of a sphere representing the three regions

travels around the back of the Sphere and returns to the radar where it interferes with the reflection from front of the Sphere [3, 4].

$$\sigma = 4\pi r^2 \text{ at Crest (point A)} \tag{2}$$

$$\sigma = 0.26\,\pi r^2 \text{ at Trough (point B)} \tag{3}$$

Optical Region: In this region RCS of a Sphere is independent of frequency. When the wavelength is small compared to object's dimension is said to be optical region where $2\pi a/\lambda \gg 1$. Here radar scattering from a complex object such as an aircraft is characterized by significant changes in cross section when there is a change in frequency or aspect angle at which the object is viewed. In optical region RCS is affected more by the shape of the object than by its projected area. In Optical region, scattering does not take place over the entire hemisphere that faces the radar, but only from a bright spot at the tip of smooth sphere. It is more like what would be seen if a polished metallic sphere, such as a large ball bearing were photographed with a camera equipped with a flash. The only illumination is at the tip, rather than from entire hemispherical surface [3, 4].

$$\sigma = \pi r^2 \tag{4}$$

Creeping Waves: There is a region where specular reflected (mirrored) waves combine with back scattered creeping waves both constructively and destructively as shown in Fig. 3. Creeping waves are tangential to a smooth surface and follow

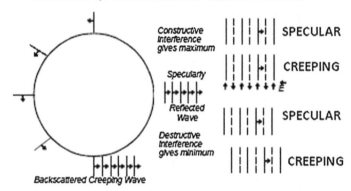

Fig. 3 Addition of specular and creeping waves

the "shadow" region of the body. They occur when the circumference of the sphere $-\lambda$ and typically add about 1 m^2 to the RCS at certain frequencies.

3   Mathematical Analysis on Backscattering RCS of a Perfect Sphere

In the case of the sphere, the scattered field must be represented in terms of θ and φ components, where θ is the bistatic angle subtended in the directions of incidence and scattering at the center of the sphere, and φ is the angle between the plane of scattering and the plane containing the incident electric field and direction of incidence. The components of the scattered field are [2, 9] (Fig. 4)

$$E_\theta^s = \frac{je^{-jka}\cos\emptyset}{kr} \sum_{n=1}^{\infty} (-1)^n \frac{2n+1}{n(n+1)} \left[b_n \frac{\partial P_n'(\cos\theta)}{\partial\theta} - a_n \frac{P_n'(\cos\theta)}{\sin\theta} \right] \qquad (5)$$

$$E_\emptyset^s = \frac{je^{-jka}\sin\emptyset}{kr} \sum_{n=1}^{\infty} (-1)^n \frac{2n+1}{n(n+1)} \left[b_n \frac{P_n'(\cos\theta)}{\sin\theta} - a_n \frac{\partial P_n'(\cos\theta)}{\partial\theta} \right] \qquad (6)$$

where, 'r' is the distance to the point of observation, 'a' is the radius of the sphere and $P_n'(\cos\theta)$ is the associated Legendre function of order 'n' and degree '1'. Where 'ka' represents the size of an object.

Kerr gives the following radar cross section σ of a sphere of radius 'a' from [2]

$$\frac{\sigma}{\pi a^2} = \frac{1}{\rho^2} \left| \sum_{n=1}^{\infty} (-1)^n (2n+1)(a_n^s - b_n^s) \right|^2 \qquad (7)$$

Fig. 4 Spherical polar
scattering geometry

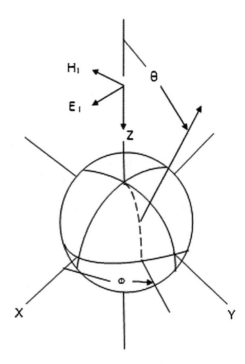

where $\rho = 2\pi a/\lambda$, λ is the wavelength, and the a_n^s and b_n^s are the terms of a
"multi-pole expansion". That is these terms are proportional to the amplitudes of
magnetic and electric multi-poles induced in the sphere by the incident wave. When
the sphere is perfectly conducting [5–8, 10]

$$a_n^s = -\frac{j_n(\rho)}{h_n^{(2)}(\rho)} \quad \text{and} \quad b_n^s = -\frac{[\rho j_n(\rho)]'}{[\rho h_n^{(2)}(\rho)]'} \tag{8}$$

where the primes denote differentiation with respect to the argument.

The functions j_n and $h_n^{(2)}$ are respectively the spherical Bessel function of the first
kind and spherical Hankel function of the second kind.

From [9, 11] it is given that
The coefficients a_n and b_n are

$$a_n^s = \frac{j_n(ka)}{h_n^2(ka)} \quad \text{and} \quad b_n^s = -\frac{[kaj_{n-1}(ka) - nj_n(ka)]}{\left[kah_{n-1}^{(2)}(ka) - nh_n^{(2)}(ka)\right]} \tag{9}$$

$$h_n^{(2)}(x) = j_n(x) - iy_n(x) \tag{10}$$

In which $j_n(x)$ and $y_n(x)$ are the spherical Bessel functions of the first and second
kinds, respectively (Fig. 5).

4 Mathematical Analysis on Backscattering RCS of a Plasma Sphere

The plasma is the fourth state of substance. It is a mixture of electrons, ions and neutral particles and is electrically neutral. Because the charged particles can interact with the electric and the magnetic field of the EM wave, the EM wave will be scattered, refracted and/or absorbed when it strikes the plasma. For a collisionless unmagnetized plasma, when the radian frequency of the incident wave is higher than the Langmuir frequency (also called the cut-off frequency or the plasma frequency) of the plasma, the EM wave can go into the plasma. On the other hand, when the frequency of the EM wave is lower than the Langmuir frequency of the plasma, the wave can't go into the plasma.

The complex index of refraction can be expressed by

$$n = \sqrt{\epsilon_r} = \sqrt{1 - \frac{\omega_{ne}^2}{\omega^2(1 - j\frac{v_e}{\omega})}} \tag{11}$$

Letting $n = n1 + jn2$, according to Mie's theory, the back-scattering RCS of the plasma sphere is given by [1, 12],

$$\sigma = \frac{\pi}{k^2} \left\{ \left[\sum_{l=i}^{\infty} (-1)^l (2l+1)(a_{lr} - b_{lr}) \right]^2 + \left[\sum_{l=1}^{\infty} (-1)^l (2l-1)(a_{li} - b_{li}) \right]^2 \right\} \tag{12}$$

where the subscripts r and i represent the real and imaginary part of a complex respectively, a_l and b_l are factors which can be obtained from Mie's theory.

Where k is a propagation constant given by,

$$k = \omega \sqrt{\mu_0 \epsilon_0 \left(1 - \frac{\omega_p^2}{\omega^2}\right)} \tag{13}$$

When $\omega < \omega_p$, Propagation constant k is imaginary, wave may be reflected in this case that is EM wave can't penetrate into the plasma as incident waves will be scattered and RCS is more in this case as it is somewhat like a metal sphere.

When $\omega > \omega_p$, Propagation constant k is real, wave propagates in the plasma that is EM signal penetrates the plasma shell, reflects off objects surface will drop in intensity while travelling through the plasma.

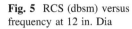

Fig. 5 RCS (dbsm) versus frequency at 12 in. Dia

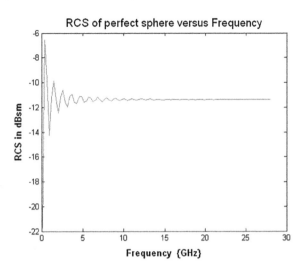

5 RCS Estimation of Perfectly Conducting Sphere

It is observed from Fig. 6 that RCS is changing with respect to frequency varying from 0.1 to 25 GHz and RCS of a sphere remains constant in the optical region and it is around −12 dbsm in optical region. This plot is obtained by programming the Kerrs relation [2] in Matlab. It is observed from Fig. 7 that RCS of a Sphere is computed for different diameters ranging from 1 to 10,000 in. at X band frequency.

From Fig. 7 it is observed that RCS of a Perfect Sphere in dBsm verses Diameter of a Sphere in inches is plotted for different frequency bands.

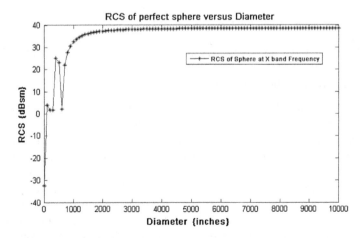

Fig. 6 RCS (dbsm) versus sphere diameter at X band

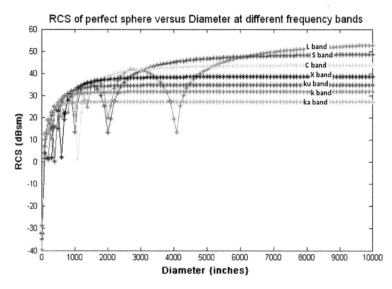

Fig. 7 Comparison plot for RCS of a perfectly conducting Sphere at different frequency bands

6 RCS Estimation and Comparison of Plasma Sphere and Perfect Sphere

Standard Spheres that are undergone for calibration tests with the dimensions 6"
Diameter with RCS value 0.018 mt² or −17.44 dBsm and 14" Diameter with RCS
value 0.1 mt² or −10 dBsm and 22" Diameter with RCS value 0.245 mt² or
−6.10 dBsm are considered for comparison. Condition implies that

(i) When EM wave frequency is less than plasma frequency that is f < fp EM
signal can't penetrate into plasma, waves may be reflected and it is like a metal
sphere
(ii) When EM wave frequency greater than plasma frequency f > fp EM signal
penetrates into plasma and reflects off objects surface will drop in intensity
while travelling through plasma.

It is observed from Figs. 8 and 9 that RCS in dBsm versus Frequency in GHz is
plotted for perfectly conducting sphere with standard dimensions 6 and 14 in.
compared with the same dimension of plasma sphere with respect to plasma frequency. It is observed from RCS of plasma sphere when wave frequency is less
than plasma frequency (f < fp) 'k' is imaginary that is em wave can't penetrate into
plasma, incident waves will be scattered, wave may be reflected and it is somewhat
like a Metal Sphere implies RCS is more and When EM wave frequency greater
than plasma frequency f > fp 'k' is real, EM signal penetrates into plasma and
reflects off objects surface will drop in intensity while travelling through plasma
implies RCS is less. In other words the real part of propagation constant is

Fig. 8 RCS of plasma sphere at 6 in. diameter

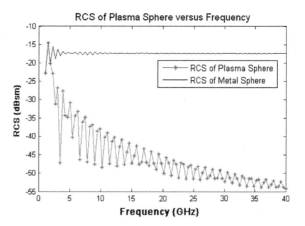

Fig. 9 RCS of plasma sphere at 14 in. diameter

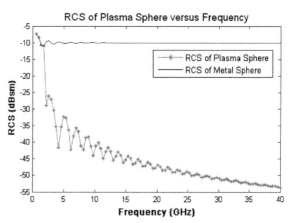

attenuation constant (α) where it is defined as the rate at which the amplitude of microwave reduces as it progresses in the medium.

$$\alpha = \omega \sqrt{\frac{\mu \varepsilon}{2} \left[1 + \frac{\sigma^2}{\omega^2 \epsilon^2} - 1 \right]} \quad \mathrm{dB/mt} \tag{14}$$

where α is directly proportional to frequency from Eq. (14). As frequency increases attenuation constant increases and hence RCS reduces.

7 Results and Discussions

The RCS of a Sphere with 12 in. diameter has been computed for frequencies ranging from 0.1 to 40 GHz. It is observed that some oscillations in the plot represent the Mie region. The oscillations are nothing but RCS variations due to amplitudes of magnetic and electric multipoles induced in the Sphere by the incident wave. It is observed from Fig. 7 that a comparison is made for RCS of Sphere at different frequency bands. At L band Frequency RCS is more where as at Ka band frequency RCS of sphere is less. It is due to the fact that RCS of a sphere is directly proportional to wavelength obtained from Kerrs relation as frequency increases, wavelength decreases and hence RCS decreases. It is observed from the comparison plots of plasma sphere and perfectly conducting sphere from Figs. 8 And 9 that when f < fp 'k' is imaginary and wave may be reflected and when f > fp 'k' is real, wave propagates in the Plasma. Practically it may not be possible to measure RCS of a Perfectly conducting Sphere at bigger dimensions say 10,000 in. but theoretically it is shown how RCS of a Perfectly conducting Sphere behaves at various frequencies with respect to size by programming the Kerr's relation in Matlab.

8   Conclusion

In this some investigations are made on RCS of a Sphere obtained from Polar Scattering Geometry by varying the diameters ranging from 1 to 10,000 in. at some specified frequencies as appropriate to applications in different frequency bands like L, S, C, X, Ku, K, Ka. Practically, oscillations occur in the Mie region due to creeping waves but theoretically the Kerr relation was formulated in such a way that RCS of a sphere was oscillating in Mie region due to increase/decrease of both a_n and b_n values. Theoretically Plasma Sphere was computed and compared with the Metal Sphere with standard dimensions. Results obtained through the formulations [1, 6] show a reasonable agreement with results as obtained individually, i.e., in isolation. The results are obtained through simulation carried out in Matlab.

References

1. Y. Gao, J. Shi, J. Wang, *The Calculation of Back-Scattering Radar Cross Section of Plasma Spheres* (Institute of Electronic Engineering, Hefei, IEEE, 2000)
2. D.E. Kerr, *Propagation of Short Radio Waves*. Radiation Laboratory Series, vol. 13, Chap. 6 (McGraw- Hill, Newyork, 1951)
3. M.I. Skolnik, *Introduction to Radar Systems*, 2nd edn. (Newyork, McGraw-Hill, 1962), p. 41
4. M.I. Skolnik, *Radar Handbook*, 2nd edn, Chap. 11, p. 11.1 (McGraw-Hill, New York, 1990)
5. S. Kingsley, *Understanding Radar Systems*, 2nd edn
6. B.R. Mahafza, *Radar System Analysis and Design Using Matlab*, 2nd edn, pp. 100–102 (2000)
7. N.C. Currie, *Techniques of Radar Reflectivity Measurement*, p. 27 (Archtech House, 1984)
8. D.C. Jenn, *Radar and Laser Cross Section Engineering* (Naval Postgraduate School, California, 2005)

9. E.F. Knott, *Radar Cross Section*, 2nd edn. (Archtech House, 1993)
10. H.G. Booker, *Cold Plasma Wave in Chinese* (Science Press, Beijing, 1985)
11. L.V. Blake, *Calculation of the Radar Cross Section of a Perfectly Conducting Sphere* (Naval Research Laboratory(NRL), Washington DC, July, 1972)
12. H.C. van de Hulst, Light Scattering by Small Particles (Dover Publication, Inc, 1981)

Fault-Tolerant Multi-core System Design Using PB Model and Genetic Algorithm Based Task Scheduling

G. Prasad Acharya and M. Asha Rani

Abstract This paper presents an innovative approach for testing the core components in the multi-core system-on chip framework by scheduling the tasks assigned for the Core-Under-Test to one of the remaining cores in the system. Real-time task scheduling in multi-core/processor systems always remains the NP-Hard problems. In multi-core Real time systems, a faulty core can be tolerated by way of executing two versions (primary and backup) of a task in two different cores. The Genetic Algorithms provides an innovative and heuristic approach of scheduling both primary and backup tasks. The work presented in this paper shows the optimal utilization of all the available cores for functional operation at a given time in a Multi-Core System environment by scheduling and executing all the tasks arrived for execution.

Keywords Multi-core system · Genetic algorithm · Task scheduling · Primary-Backup model

1 Introduction

Most of the VLSI designers use the modular and Core based design as the *de facto* design methodology as it facilitates the use of pre-designed and pre-verified IP cores from third-party vendors. Today's Silicon vendors have come up with Multi-Processor/Core systems in a single Si chip. The System-on Chip (SoC) technology

G. Prasad Acharya (✉)
ECE Department, Sreenidhi Institute of Science and Technology,
Hyderabad, Telangana, India
e-mail: gpacharya@sreenidhi.edu.in

M. Asha Rani
ECE Department, JNT University College of Engineering,
Hyderabad, Telangana, India
e-mail: ashajntu1@yahoo.com

© Springer India 2016   449
S.C. Satapathy et al. (eds.), *Microelectronics, Electromagnetics
and Telecommunications*, Lecture Notes in Electrical Engineering 372,
DOI 10.1007/978-81-322-2728-1_41

integrates one or more processors/cores like ARM, MIPS and DSP processors with associated peripherals, embedded memories, Memory and Network controllers, Analog and Mixed signal interfaces in a single Silicon chip. This architecture provides a platform wherein multiple tasks that arrive for execution will be scheduled for execution among one of the available cores so that Core-Under-Test can be made free from functional mode prior to switching into test mode.

The multi-core system under consideration is of homogenous type with the assumption that all the cores are identical and the execution time for a given task is same for all the cores. The test framework of System-on chip consists of Test Source, Test Access mechanism (TAM) and Test Sink. The cores may be switched into test mode using its dedicated built-in Test Wrappers. TAM delivers the test inputs from the test source to the Core-Under Test (CUT) and the test responses from CUT to the test sinks. In a multi-core environment, each of the cores may be scheduled for test by scheduling the tasks queuing up for execution as well as forthcoming tasks scheduled on the CUT to one or more of the available cores in the system. All the cores may be scheduled for test either periodically or on-demand by rescheduling its tasks before switching into test mode.

The organization of the paper as follows: The related previous work is presented in Sect. 2. Section 3 describes the concept of task scheduling in multi-core systems. A Test framework for multi-core systems is described in Sect. 4. The proposed periodic test scheduling algorithm for the multi-core systems is presented in Sect. 5. An experimental setup and results are discussed in Sect. 6.

2 Previous Work

[1] presents a framework for developing SoC test solution based on Test Access Mechanism and Test wrapper. [2] presents the task scheduling in a multi-core heterogeneous system wherein a task will initially be scheduled to the processor and with the help of dispatcher the task will be scheduled for execution in one of the available cores. [3] discusses the non-preemptive scheduling of 'n' independent tasks on a set of m parallel processors, and proposes a scheme that computes an approximate solution of any fixed accuracy in linear time. The researchers in articles [4, 5] present Pareto-based Genetic algorithm, which gives a family of solutions rather than a single solution for allocating and scheduling real-time tasks in a multi-processor the system with an aim to minimize the execution time of a task satisfying all the real-time constraints. [6, 7] proposes energy effective real-time task scheduling for low-power Dynamic Voltage Scaling (DVS) hand held devices considering the energy as minimizing the function. [8, 9] presents Genetic Algorithm based task scheduling algorithm wherein tasks will be assigned dynamically based on the utilization of the processor and fitness function. [10] presents primary-backup task scheduling approach for multiprocessor systems to ensure the timing deadlines are met, but at the cost of processing power. [11] presents real-time task scheduling using load-balancing approach among the

processors. [12, 13] presents the methodology for allocating and scheduling one of the co-processor for the execution of a task once the task gets a successful admission into the master processor.

3 Task Scheduling Mechanism

The Real time task scheduling can be of clock-driven, event-driven and hybrid. In clock-driven schedulers, clock edge determines the scheduling points whereas certain events and/or interrupts on certain signals define scheduling points in the event-driven schedulers.

The execution of next task will commence simultaneously on all the cores at a fixed time instants (t_0, t_1, t_2 etc. as shown in Table 1) once all the cores finish current tasks.

Genetic Algorithm Based Task Scheduling in Multi-Core Systems

Multi-core Systems contains either homogeneous or heterogeneous cores. In heterogeneous architecture, different cores take different execution time for the given task. In homogeneous architecture, all the cores are identical and hence take exactly the same time for execution of a given task. Assume 'n' independent tasks arrive to the system for execution at a given time. Task scheduling in Multi-Core system may be considered as a process of dynamically scheduling the tasks for execution among one of the available cores. For static task scheduling, the attributes of tasks i.e., task arrival time, worst computational time and deadline, are known priori. The tasks are scheduled for execution in a predefined order i.e., first come first serve or round robin basis. The static scheduling algorithms have the drawbacks of not optimally utilizing the processor resources. Moreover, tasks may be non-deterministic and may arrive dynamically for execution. In real-time systems, preemptive tasks may arrive at any instant during the execution of the scheduled tasks. These preemptive tasks must be scheduled for execution with minimum execution delay by optimizing the processor resources and rescheduling the tasks of a processor on which the preemptive tasks are to be scheduled.

Table 1 Illustration of task scheduling

Processors	Tasks					
P1	T1		T2		T3	
P2	T4		T5		T6	
P3	T7		T8		T9	
	t_0		t_1		t_2	

Notations The gray shades indicate a task is in running state on the scheduled processor and the blank shades indicate that the processor is in idle state after the completion of present task

The dynamic scheduling of tasks is a NP-hard problem. Hence, heuristic and hypothetical approaches are being adopted to get the optimal solution under the specified constraints. The main objective of these algorithms is to minimize the task execution time and maximize the utilization of processor resources.

The Genetic Algorithm (GA) has the potential of providing optimal solutions to the complex problems like scheduling the real time tasks and allocating the resources. It uses stochastic search techniques in finding an optimal solution and/or a family of solutions (called population) among the initial and iteratively updated solutions. The individual member of population is called chromosome. A new generation is formed by selecting and rejecting some of the members of the population based on the fitness values. After a number of iterations, the algorithm converges to the best chromosome, which represents the optimal solution for the given problem.

The GA based task scheduling algorithm [8] presents a primary-backup (PB) model for fault tolerant scheduling of dynamically arriving tasks. The PB model schedules a task T_i to a processor, P_j if $\sum \frac{Ci}{Di} < 1$ and its backup task TB_i to a different processor, P_k ($j \neq k$) where C_i, D_i and i are the worse computation time, deadline and index of a task scheduled on that processor respectively. The remaining utilization (R) is initially unity and is updated by $R - \frac{Ci}{Di}$ every time a task Ti is scheduled to the processor. The standard deviation of R defined as

$$fitness = \sqrt{\sum_{j=1}^{m} \left(\frac{(R_j - \overline{R})^2}{m - 1} \right)} \qquad (1)$$

is considered as the fitness function for the algorithm for the optimal utilization of the processors, where m is the number of processors, R_j is the remaining utilization of the processor j and $\overline{R}$ is the average remaining utilization.

4 Test Framework for Multi-core System

The proposed test framework for self-test and diagnosis of Core-based SoCs consists of Test wrapper, Test Access Mechanism and Test Schedular as shown in Fig. 1. A built-in self-checking logic module in a core is implemented using Berger code to detect the presence of any faults in it. All the cores are scheduled for test using Test scheduler. The Test Controller generates the timing control signals required for proper functioning of the test process. The on-chip ROM that stores the compressed test patterns generated by LFSR, acts as Test Source. The expected test responses after compaction are made available in on-chip ROM. The address for accessing test patterns as well as test responses from the respective ROMs is generated using sequence counter. The Test Access Mechanism delivers test patterns from Test Source to the Core-Under-Test (CUT). The Test Response Analyzer (ORA) compresses the responses delivered by TAM from CUT and compares it

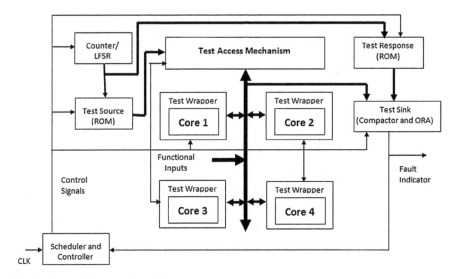

Fig. 1 Test architecture for multi-core system

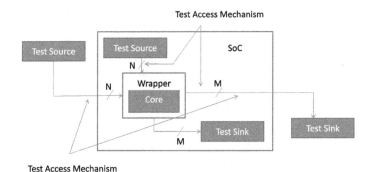

Fig. 2 Test framework using TAM architecture

with the expected reference available in ROM to determine the presence of faults in the CUT.

Test Access Mechanism (TAM)

The built-in TAM architecture, as shown in Fig. 2 delivers the test stimuli from test source to the CUT. The test source may be a built-in Test pattern Generator (TPG) or an external ATE. The TAM delivers the test responses to the test sink i.e., Output Response Analyzer (ORA). The TPG generates and delivers the test vectors through TAM to the CUT.

Test Wrapper

The IEEE Std 1500 test wrapper makes it possible to test the core-under-Test with minimum pinset by isolating it from rest of the system. The test wrapper, shown in

Fig. 2 is a DFT logic that provides an interface between CUT and TAM architecture. The wrapper also provides bandwidth adaptation between the core and TAMs if they have different bandwidths.

5 Proposed Test Scheduling Algorithm for Fault Tolerant Multi-core System

A Primary-Backup (PB) model is utilized for the realization of fault-tolerant systems. The PB model schedules a primary task (Ti) and its backup version (TBi) to two different cores satisfying the fitness function given by Eq. (1). The tasks Ti and TBi will not be scheduled for execution to the same core so that at least one of the cores will produce correct output at a given time.

The offline test procedure of one or more core(s) will be carried out periodically either during system boot-up or by suspending the tasks scheduled for execution on it. The following are the steps carried out for the offline test scheduling of cores:

Step 1: Select a core C_i having maximum value of R for test.

Step 2: Complete the execution of current task of C_i.

Step 3: Dynamically schedule all other tasks (primary as well as backup) to other cores using GA based algorithm as described in Sect. 3.

Step 4: Delete C_i from the list of processors participating and switch it into test mode.

Step 5: Run the BIST program on C_i.

   (a) If C_i passes the test operation, switch it back to normal operation so that it participates in normal function along with remaining processors. Set the check flag CF_i for the processor to indicate that the core C_i passes test.

   (b) If C_i fails in test operation, repeat step 5 for the second time to check whether the failure is permanent. If the failure still exists, run Built-in Self-repair procedure if the core is repairable otherwise permanently isolate it from the system.

Step 6: Select another core C_j ($j \neq i$) with maximum value of R whose CF_j bit is not set. Repeat steps 2–5 until all the cores are tested.

Step 7: Repeat step 1–6 after T seconds, where T is the predefined time duration between two successive tests.

6   Experimental Results and Discussion

The GA based task scheduling algorithm is implemented using Matlab. The experiment is conducted for 12 tasks (6 primary and 6 backup tasks) which are scheduled for execution in Four Cores (C1, C2, C3 and C4) of the system. Initially, the remaining utilization (R) of each core is assumed to be unity. The ratio C_i/D_i for each of the task is assumed to be as listed in Table 2.

Table 2 Attributes of tasks

Task	T1	T2	T3	T4	T5	T6
C/D	0.3	0.45	0.25	0.35	0.2	0.25
Task	TB1	TB2	TB3	TB4	TB5	TB6
C/D	0.3	0.45	0.25	0.35	0.2	0.25

Table 3 Sequence of task scheduling

Core	C1	C2	C3	C4	Fitness
First round of task assignment					
Tasks	T1	T2	T3	T4	0.85
R	0.7	0.55	0.75	0.65	
Second round of task assignment					
Tasks	TB2	TB4	TB1	TB3	0.12
R	0.25	0.2	0.45	0.4	
		R_{min}	R_{max}		

Table 4 Test Sequence for cores and affected task scheduling among the cores

Core	C1	C2	C3	C4	Fitness
First round of task assignment					
Tasks	T1	T2	T3	T4	0.85
R	0.7	0.55	0.75	0.65	
Second round of task assignment					
Tasks	Run BIST	TB4	TB1	TB3	0.12
R		0.2	0.45	0.4	
R at t = t1+		0.65	0.70	0.75	
Third round of task assignment					
Tasks	Run BIST	T6	T5	TB2	0.1
R		0.40	0.50	0.30	
R at t = t2+	1	Test mode	0.95	0.70	
Fourth round of task assignment					
Tasks	T6	Run BIST	TB5	TB6	0.1
R	0.80		0.75	0.50	
at t = t3+	0.80	1	0.80	0.80	

Note In the experimental setup, 16 vectors are applied as test vectors for a core and it is assumed that the test time is exactly equal to a single task execution time

All the primary and backup tasks are scheduled on one of four available cores based on the Genetic algorithm discussed in Sect. 3. Table 3 shows the task assignment, remaining utilization and fitness value of the respective cores.

Assume the core C1 is scheduled for periodic test after first round of task assignment. C1 executes T1 before switching into test mode. If C1 is found to be faulty, the output is taken from C3 after execution of TB1. This is illustrated in Table 4. The graphs in Fig. 3b, c also indicate that the remaining cores are optimally utilized when a CUT is under test.

The test framework for multi-core system architecture using GA based scheduling of cores is implemented using Verilog HDL in Xilinx 14.7 environment.

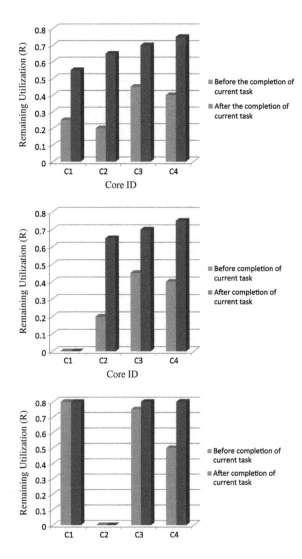

Fig. 3 **a** *Bar graph* indicating tasks scheduling among all cores. **b** *Bar graph* indicating C1 isolated from task execution for test. **c** *Bar graph* indicating C2 isolated from task execution for test

Table 5 Illustration of task execution time

No. of primary + backup tasks	Execution time (no core under test) (ms)	Execution time (single core under test) (ms)
4 + 4	3.2	4.8
6 + 6	4.8	6.4
8 + 8	6.4	9.6
10 + 10	8.0	11.2
12 + 12	9.6	12.8

The Add/Multiply Accumulate Macro (ADDMACC_MACRO) of DSP48 slice available in SPARTAN 6 FPGA has been considered as core of the architecture.

All the four cores in the architecture are scheduled for test periodically by scheduling the tasks assigned for execution on CUT to other core(s) available. Stuck-at faults are injected in the design for verifying the functionality of the design. Two cores are assigned a task, one for each of primary and backup version. In case of a core detecting a fault, the output is taken from the other core executing its backup task thereby providing the fault tolerance. The illustration of number of tasks and total task execution time is given in Table 5.

7 Conclusions

The Genetic Algorithm based task scheduling algorithm schedules the dynamically arriving tasks for execution. The experimental results obtained in this paper shows that the algorithm maximizes the utilization of a number of cores available in the system. The PB model presented in this paper enables the design of fault-tolerant system by executing a task in two different cores and ignoring the output from a faulty core. An experimental setup is carried out on a Multi-Core System with four homogenous cores and the obtained simulation results show that one of the core can be scheduled for test while the remaining cores functioning normally. However the latency in total execution time may be traded-off with online-BIST capability of Multi-core System architecture.

References

1. E. Larsson, Z. Peng, An integrated framework for the design and optimization of SOC test solutions. J. Electron. Test.: Theory Appl. **18**, 385–400 (2002)
2. P. Karande, S.S. Dhotre, S. Patil, Management for heterogeneous multi-core scheduling. Int. J. Comput. Sci. Inf. Technol. **5**(1), 636–639 (2014)
3. K. Jansen, L. Porkolab, A general multiprocessor task scheduling: approximate solutions in linear time. Soc. Ind. Appl. Math. SIAM J. Comput **35**(3), 519–530 (2005)

4. J. Oh, H. Bahn, C. Wu, K. Koh, *Pareto-Based Soft Real-Time Task Scheduling in Multiprocessor Systems* (IEEE, 2000)
5. G. Ascia, V. Catania, M. Palesi, An evolutionary approach for pareto-optimal configurations in SOC platforms, in *SOC Design Methodologies* (© Springer Science+Business Media, New York, 2002)
6. J.-J. Chen, C.-Y. Yang, T.-W. Kuo, C.-S. Shih, *Energy-Efficient Real-Time Task Scheduling in Multiprocessor DVS Systems* (IEEE, 2007), pp. 342–349
7. Y.-S. Chen, M.-Y. Chen, On-line energy-efficient real-time task scheduling for a heterogeneous dual-core system-on-a-chip. J. Syst. Archit.
8. G. Zarinzad, A.M. Rahmani, N. Dayhim, *A Novel Intelligent Algorithm for Fault-Tolerant Task Scheduling in Real-Time Multiprocessor Systems* (IEEE, 2008), pp. 816–821
9. M.H. Shenassaa, M. Mahmoodi, A novel intelligent method for task scheduling in multiprocessor systems using genetic algorithm. J. Franklin Inst. **343**, 361–371 (2006)
10. W. Jing, Y. Liu, W. Qu, *Fault-Tolerant Task Scheduling in Multiprocessor Systems Based on Primary-Backup Scheme* (IEEE, 2010)
11. K. Zhang, B. Qi, Q. Jiang, L. Tang, Real-time periodic task scheduling considering load-balance in multiprocessor environment, in *Proceedings of IC-NIDC*, ©2012 IEEE, pp. 247–250
12. Y.-S. Chen, H.C. Liao, T.-H. Tsai, Online real-time task scheduling in heterogeneous multicore system-on-a-chip. IEEE Trans. Parallel Distrib. Syst. **24**(1), (2013)
13. S.S. Poonam Karande, S.P. Dhotre, Task management for heterogeneous multi-core scheduling. Int. J. Comput. Sci. Inf. Technol. **5**(1), 636–639 (2014)

Investigation of Suitable Geometry Based Ionospheric Models to Estimate the Ionospheric Parameters Using the Data of a Ground Based GPS Receiver

K. Durga Rao and V.B.S. Srilatha Indira Dutt

Abstract The field of Navigation has been revolutionized with the advent of Global Positioning System (GPS). As GPS is a satellite based navigation system, transmitted GPS signals are affected by the refraction in the atmosphere which causes huge error in pseudorange information of a satellite. Of all the layers of the atmosphere, ionosphere is one of the major sources of error in ranging measurements, which reduces the accuracy of the navigation solution. Hence in this paper, in order to estimate the ionospheric parameters such as Total Electron Content (TEC), range delay and time delay, geometry based ionospheric models are investigated. The dual frequency receiver data of an IGS station, NGRI (Lat/Long:17°24′39″N/78°33′4″E), Hyderabad, Andhra Pradesh, India, provided by Scripps Orbit And Permanent Array Center (SOPAC) is considered for a typical day of 11th September 2014.

Keywords Mapping function · Total electron content · Refraction

1 Introduction

Electromagnetic waves received at the GPS receiver are affected by the refraction of the signals during their passage from atmosphere. Of all the atmospheric layers, refraction of the signal in the ionosphere causes a delay in the arrival of the signal at the receiver [1]. This delay in the arrival of the signal causes a huge error in ranging information of the satellite which in turn causes an error in navigation solution.

The ionosphere is dispersive in nature, caused by the interference of free electrons with the GPS satellite signals. These free electrons results in ionization pro-

K. Durga Rao (✉) · V.B.S. Srilatha Indira Dutt
Department of ECE, GIT, GITAM University, Visakhapatnam, A.P., India
e-mail: durgayedlapalli@gmail.com

V.B.S. Srilatha Indira Dutt
e-mail: srilatha06.vemuri@gmail.com

© Springer India 2016
S.C. Satapathy et al. (eds.), *Microelectronics, Electromagnetics and Telecommunications*, Lecture Notes in Electrical Engineering 372,
DOI 10.1007/978-81-322-2728-1_42

459

cesses of the ionosphere constituents, by the solar radiation. Using both GPS frequencies, the induced ionospheric delay can be eliminated. When measuring with a differential receiver set-up, the effect can be strongly reduced by forming error differences [2]. Fortunately, the GPS provides the possibility of continuously monitoring the state of the ionosphere by using GPS receivers.

The ionosphere induces time delay, which is related to the total number of electrons encountered by the radio wave on its path. By estimating the Total Electron Content (TEC) along the path of the signal, the delay in the signal arrival at the receiver can be estimated. With proper modeling of the satellite and receiver geometry, TEC as well as range delay and time delay of the satellite signals can be estimated.

The TEC is the total number of electrons integrated along the path from the GPS to receiver. TEC indicates ionospheric variability, from the GPS signal through free electrons. TEC is measured in TECU i.e. total electron content unit. $1 \text{TECU} = 10^{16}$ electrons per m^2

$$\text{TEC} = \int_p N(s) ds \tag{1}$$

where, N(s) is the electron content per unit volume and 'p' is the propagation path between the GPS satellite and the GPS receiver. With dual frequency pseudorange measurements on L1 (1575.42 MHz) and L2 (1227.60 MHz) frequencies, TEC can be estimated along the line of sight of the signal propagation and is called as slant TEC (STEC). To eliminate the influence of the Geometry, the STEC is converted into Vertical TEC (VTEC) using an appropriate mapping function.

$$F_{IPP} = \text{STEC}/\text{VTEC} \tag{2}$$

The estimated TEC is used to compute the Ionospheric delay

$$d_{ion} = \left(\frac{40.3}{f^2}\right) \text{TEC} \tag{3}$$

where 'f' is carrier frequency, which can be either L1 (1575.42 MHz) or L2 (1227.60 MHz).

2 Geometry Based Mapping Functions

In order separate the electron density distribution in horizontally and vertically, Shell approximation of the atmospheric layers is the general approach. The thin shell approximation and thick shell approximation are the two shell based approximation techniques which are significant in estimating the ionospheric parameters.

Fig. 1 Thin layer model with receiver (Rx) and transmitter (Tx)

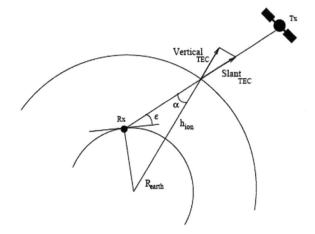

2.1 Thin Shell Approximation

The simplest model to define the effect of the ionosphere is to assume that the ionosphere electron content is confined in a spherical layer of infinitesimal width over the surface of the Earth. The maximum electron density is mainly located between 300 and 500 km within which this thin layer is centered (Fig. 1).

The thin mapping function F_{IPP} can be defined as:

$$F_{IPP} = 1/\left(1 - \left((R_e \cos(E)/(R_e + h_{IPP}))^2\right)^{1/2}\right) \qquad (4)$$

where 'E' is the elevation angle between the receiver and GPS satellite, 'hIPP' is the height of the thin layer, this height is about 350–450 km and 'Re' the Earth's radius.

2.2 Thick Shell Approximation

The thick shell model is an improvement over the thin layer model, where the ionosphere is confined to one spherical shell of finite thickness with a constant or variable electron density. In this paper, thick shell approximation with constant electron density is considered.

Consider a ray travelling along ionosphere with electron density 'Ne' with lower and upper threshold heights at 'h0' and 'h1' (Fig. 2).

Fig. 2 Thick layer model
where *Rx* denotes the receiver
while *Tx* denotes the
transmitter

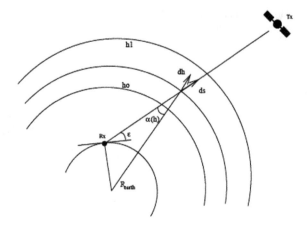

'Fthick' is the Mapping function of this thick layer model written as:

$$F\ thick = \frac{\sqrt{(Re + h1)^2 - \left((Re\ \cos(E))^2\right)} - \sqrt{(Re + h0)^2 - \left((Re\ \cos(E))^2\right)}}{h1 - ho}$$

(5)

where 'E' is the elevation angle, 'h0' is the lower height, 'h1' is the upper threshold
height and 'Re' is the Earth's radius.

3 Results and Discussions

The data of an IGS station NGRI Hyderabad is collected from SOPAC web site for
a typical day of 11th September 2014 and is processed using Converter software
tool to convert the data into RINEX format. The data is sampled at a rate of 30 s.

Figure 3 and 4 shows the variation of the thin shell approximation and thick shell
approximation mapping functions as a function of elevation angle and altitude for
Satellite SV2.

From Figs. 3 and 4, it can be observed that the both thick shell and thin shell
approximations are highly dependent on altitude for low elevation angles (<15°). It
can also be observed that thin shell approximation is more dependent on altitude
than thick shell approximation for low elevation angles.

From Fig. 5 it can be observed that for SV2 the maximum elevation angle is 70°
and minimum elevation angle is 10°.

The estimated ionospheric parameters of SV2 using thin shell approximation are
shown in Fig. 6. Using thick mapping approximation, the estimated range delay of
satellite SV2 is in between 6 and 24 m and the time delay estimated is ranging from
20 to 80 ns during its visibility period.

Fig. 3 Variation of thin shell approximation mapping function with altitude and elevation angle

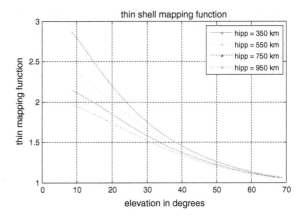

Fig. 4 Variation of thick shell approximation mapping function with altitude and elevation angle

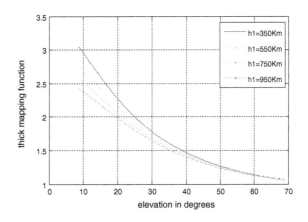

Fig. 5 Variation of the elevation angle for satellite SV2 during the period of its visibility

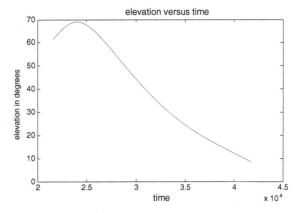

Fig. 6 Variation of time
delay and range delay for
satellite SV2 using thick shell
approx

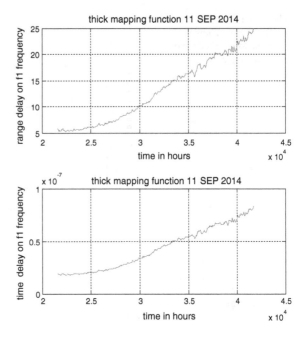

Fig. 7 Variation of time
delay and range delay for
satellite SV2 using thin shell
approx

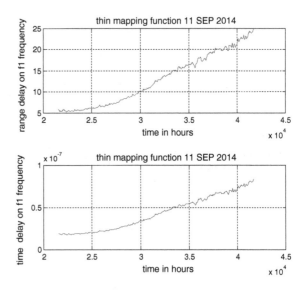

Figure 7 shows the variation of the time delay and range delay of satellite SV2 using thin mapping function. From Fig. 7, with thin shell approximation, the estimated range delay of satellite SV2 is in between 6.2 and 25 m and the time delay estimated is ranging from 22 to 81 ns during its visibility period (Table 1).

Table 1 Ionospheric parameters estimated with thin and thick shell approximations for satellite SV2 at lowest elevation angle 10°

Altitude (km)	Slant TEC (TECU)	Range delay (m)	Time delay (s)	Thin mapping function		Thick mapping function (h0 = 250 km)	
				Max. value of the mapping function	VTEC (TECU)	Max. value of the mapping function	VTEC (TECU)
350	153.075	24.8553	2.0247e-008	2.8695	53.6177	3.0463	52.6455
550	153.075	24.8553	2.0247e-008	2.4148	63.3901	2.7616	55.4293
750	153.075	24.8553	2.0247e-008	2.1449	71.3688	2.5650	59.6797
950	153.075	24.8553	2.0247e-008	1.9629	77.9859	2.4174	63.3220

4 Conclusion

The two geometry based ionospheric models, the thin shell approximation and thick shell approximation are investigated. These geometry based models simplifies the estimation of ionospheric parameters such as TEC, range and time delay by considering uniform electron density in the line of sight of the satellite and receiver, in which the vertical TEC depends only on the thickness of the shell. In thin shell approximation, the electron density depends on horizontal position of the receiver as the shell is of infinitesimal thickness and also heavily depends on the altitude for lower elevation angles (<15°). In case of Thick shell approximation, the electron density depends less on the altitude and more on the shell thickness. These two ionospheric models are investigated using the data of a ground based receiver and their performance can vary with Space based receivers and it is necessary to investigate the suitability of these two models using the data due to space based receiver.

References

1. Elliott D. Kaplan (ed.), *Understanding GPS: Principles and Applications* (Artech House Publishers, Bostan, 1996)
2. G. Sasi Bhushana Rao, *Global Navigation Satellite Systems* (Tata McGraw Hill publications, 2010)
3. A.J. Mannucci, B.D. Wilson, D.N. Yuan, C.H. Ho, A global mapping technique for GPS-derived ionospheric total electron content measurements, in *Radio Sci. The Scripps Orbit and Permanent Array Center (SOPAC)* (1998). http://sopac.ucsd.edu/dataArchive
4. R. Orus, M. Hernandez-Pajares, J.M. Juan, M. Garcia-Fernandez, Performance of different TEC models to provide GPS ionospheric corrections. J. Atmos. Solar Terr. Phys. **64**(2002), 2055–2062 (2002)

5. U. Tancredi, A. Renga, M. Grassi, Geometric total electron content models for topside ionospheric sounding, in *Environmental Energy and Structural Monitoring Systems (EESMS)*, 2014 IEEE Workshop (IEEE, Naples, Italy, 17–18 Sept 2014), pp. 1–6

A Novel Proposal of Artificial Magnetic Conductor Loaded Rectangular Patch Antenna for Wireless Applications

K.B.N. Girish, Pani Prithvi Raj, M. Vijaya Krishna Teja, S. Anand and D. Sriram Kumar

Abstract This paper presents a novel design of a rectangular patch antenna loaded with artificial magnetic conductor designed for WLAN applications. The proposed Artificial Magnetic Conductor (AMC) lends provision for operating at both the popular ISM bands. However, here the antenna is designed to operate at 6 GHz with a bandwidth of around 160 MHz which is suitable for the intended applications. The antenna has been analyzed for its performance in terms of return loss and VSWR. The performance of antenna loaded with the AMC is observed to possess better properties in comparison to the one without AMC. AMC loaded antenna shows 90.79 % improvement in VSWR at 6 GHz compared to its counterpart. For better comparison, their respective directivities, gains and 3D polar plots have been obtained and presented.

Keywords Rectangular patch antenna · Artificial magnetic conductor · 6 GHz · Voltage standing wave ratio · Return loss · Directivity · Gain

K.B.N. Girish · P. Prithvi Raj · M. Vijaya Krishna Teja · S. Anand (✉) · D. Sriram Kumar
Department of Electronics and Communication Engineering, National Institute
of Technology, Trichy, Tamil Nadu, India
e-mail: anand.s.krishna@gmail.com

K.B.N. Girish
e-mail: 108112052@nitt.edu

P. Prithvi Raj
e-mail: 108112040@nitt.edu

M. Vijaya Krishna Teja
e-mail: 108112099@nitt.edu

D. Sriram Kumar
e-mail: srk@nitt.edu

© Springer India 2016   467
S.C. Satapathy et al. (eds.), *Microelectronics, Electromagnetics
and Telecommunications*, Lecture Notes in Electrical Engineering 372,
DOI 10.1007/978-81-322-2728-1_43

1   Introduction

In the present day antenna world, several leaps of advancements are made to improve its directivity and gain. One of the most popular methods is to use the artificial magnetic conductor. It is actually made up of the naturally occurring materials but, the design of a unit of artificial magnetic conductor, its pattern of repetition, thickness and several other factors will enable it to act as a meta-material whose properties are not generally found in the naturally occurring materials. When this artificial magnetic conductor (AMC) is placed instead of a plain ground plane, the reflection of the electromagnetic waves from the antenna patch occurs without any phase difference, that is, in-phase reflection occurs. Owing to this, the backward radiation of the patch antenna is nullified and thus the directivity of the patch antenna is significantly improved.

The usual rectangular patch antenna is used to verify the influence of the artificial magnetic conductor on the radiation characteristics. After a several investigations of the possible structures for the artificial magnetic conductor an optimal design, as proposed has been obtained. In this paper, we present the detailed approach and study of the design and operation of the unit element of artificial magnetic conductor.

Also, the operation of the antenna without artificial magnetic conductor is presented in order to compare the effect of using an artificial magnetic conductor instead of a plain ground plane. Several previous works presents various designs of artificial magnetic conductor. However, the present paper attempts to optimize the size of a unit element. Besides, the choice of square shape enables the designer to make a regular structural repetition in order to easily fit it behind the antenna substrate.

This paper mainly focuses on the WLAN applications. There are several IEEE standards set up to offer the operating band of frequencies for the wireless applications. Of them 2.4 GHz is the most prominent ISM band. However, with the increasing number of devices operating at that band owing to its unlicensed operation, the wireless devices experience and cause interference with other devices operating at the same frequency. But the 5 GHz frequency band which stretches from 5.15 to 5.825 GHz, is relatively free as there are only few devices operating at that frequency. This paper presents a novel design for the artificial magnetic conductor backed antenna which resonates in the ISM frequency band thus supporting the WLAN applications. Several previous works have presented various designs for antenna [1–6] and artificial magnetic conductor [7] in order to work at or around 6 GHz for using it in WLAN or WiMAX [8] or short range communication [9]. Part of reference [10] presents the potential applications of 6 GHz frequency. Numerous works [11, 12] also present innovative designs AMC loaded antennas also.

Fig. 1 Dimensions of the patch

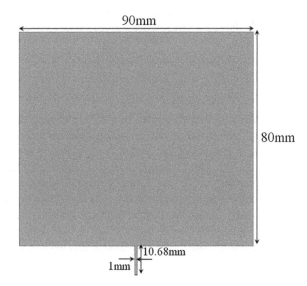

2 Design of Antenna

The patch used in the present work is a rectangular patch antenna with a direct feed. Figure 1 shows the design of the patch. A rectangle of 90 mm × 80 mm is used for radiation. The width of the feed line is obtained as 1 mm after several iterations of optimization. The equations for the design of the rectangular patch and the feed width is suggested in reference [13]. The substrate is chosen as Rogers RT/duroid 6010 as it has high dielectric constant which enables the substrate to absorb minimum energy while the back waves get reflected at the artificial magnetic conductor surface. This would add to the high gain and directivity obtained by the virtue of using the artificial magnetic conductor ground plane. The properties of the substrate are enlisted below:

- Relative permittivity = 10.2
- Relative permeability = 1
- Dielectric loss tangent = 0.0023

And the radiating patch and the AMC ground plane are chosen to be copper for its good bulk conductivity and radiation capabilities.

3 Design of Artificial Magnetic Conductor

The physical dimensions of the designed artificial magnetic conductor are as follows (Fig. 2).

a = 4.5 mm	b = 3.75 mm	c = 1.5 mm
d = 0.75 mm	e = 4.5 mm	f = 5.875 mm
g = 1.5 mm		

To study the characteristics of artificial magnetic conductor floquet port analysis is used. The master/slave boundary condition is assigned to each pair of the opposite side walls of the model. A Floquet port is placed above the structure and embedded into the artificial magnetic conductor surface. Figure 3 shows the refection phase curve of a normally incident plane wave. As it is evident, there are two kinds of special points in the refection phase curve, namely, AMC points and poles. AMC points are those frequencies where the refection phases are zeroes. Those frequencies where the refection phases curve exhibits abrupt discontinuities between −180° and 180° are denoted as poles.

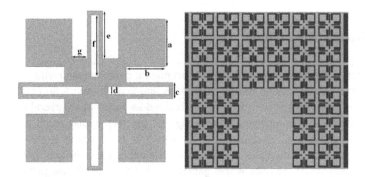

Fig. 2 Dimensions of the artificial magnetic conductor

Fig. 3 Reflection phase plot of the artificial magnetic conductor

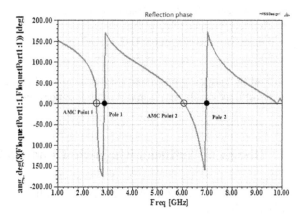

UC-EBG (Uni-planar Electromagnetic band gap) structures are essentially periodic frequency selective surfaces (FSS) [14, 15] printed on ground backed dielectric slabs. They can be analysed by the equivalent LC-network associated with the FSS. The analytical expression of the equivalent admittance Y_{FSS} (k, ω) of the network is given by [16]

$$Y_{FSS}(k,\omega) - \frac{j\omega C_o\left(\omega^2 - [\omega_{z_1}(k)]^2\right)\left(\omega^2 - [\omega_{z_2}(k)]^2\right)...\left(\omega^2 - [\omega_{z_n}(k)]^2\right)}{\left(\omega^2 - [\omega_{p_1}(k)]^2\right)\left(\omega^2 - [\omega_{p_2}(k)]^2\right)...\left(\omega^2 - [\omega_{p_n}(k)]^2\right)} - jY - |Y|e^{j\phi_y}$$

(1)

where

$$\omega_{p_1}(k) < \omega_{z_1}(k) < \omega_{p_2}(k) < \cdots < \omega_{p_n}(k) < \omega_{z_n}(k)$$

(2)

C_o is a constant independent on angular frequency ω and the wavenumber k. ω_{z_n} and ω_{p_n} are the nth zero and pole, respectively. The impendence Z_{FSS} (k, ω) of the network is the reciprocal of Y_{FSS} (k, ω), and is given by:

$$Z_{FSS}(k,\omega) - \frac{1}{Y_{FSS}(k,\omega)} - \frac{e^{-j\phi_y}}{|Y|} - \frac{e^{-j\phi_z}}{|Y|}$$

(3)

From Eqs. (1)–(3) two conclusions can be deduced,

(i) If $\omega_{z_{(n1)}} < \omega < \omega_{p_n}, Y > 0, \phi_y > 0$, hence $\phi_x + \phi_y < 0$
(ii) If $\omega_{p_n} < \omega < \omega_{z_n}, Y < 0, \phi_y < 0$, hence $\phi_x + \phi_y > 0$

From (i) and (ii), we can see that the phase is negative when the frequency is lower than the pole, while it is positive when the frequency is higher than the pole. So, there is sudden discontinuity in Z_{FSS} (k, ω) curve at the pole. Since there exists direct relation between phase of reflection coefficient and Z_{FSS} (k, ω) there will also be discontinuity in reflection phase curve.

4 Results and Discussions

The designed antenna with and without artificial magnetic conductor is separately simulated in High Frequency Structure Simulator (HFSS) tool. A radiation box is created which aids in the analysis of the antenna. The input port is defined as the lumped port with 50 Ω impedance. After simulating, the VSWR and return loss of

Fig. 4 Return loss plot of antennas

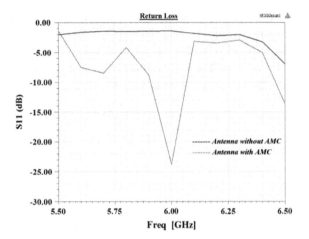

Fig. 5 VSWR plot of the antennas

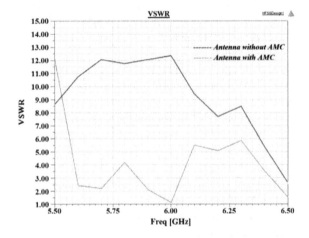

both antenna configurations are placed on a single plot respectively and compared for the relative merits (Figs. 4 and 5).

From the plots above, the advantage of using the AMC is obvious. Without the AMC the antenna designed would be unsuitable for the intended applications. There are a lot of reflection losses in the antenna at the ground. However, when the ground plane is replaced by the AMC, the back reflections at ground plane and the surface waves are nullified and consequently the return loss S_{11} to be less that −10 dB for a bandwidth of 160 MHz. Also, the directivity, gain and 3D polar plot of the two configurations of antennas are obtained from the simulation and compared as depicted by the Figs. 6, 7 and 8. The plots clearly indicate that the antenna loaded with the AMC possesses a greater directivity and gain in a particular direction. However, the antenna without AMC is omni-directional and of lesser gain owing to high losses occurring due to the reflections at the ground plane.

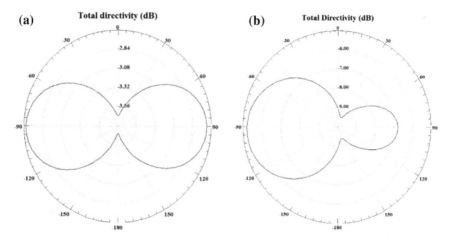

Fig. 6 Comparing total directivity of antenna without AMC (**a**) and antenna with AMC (**b**)

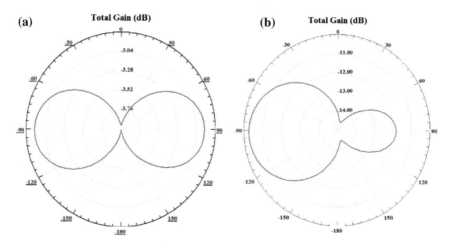

Fig. 7 Comparing total gain of antenna without AMC (**a**) and antenna with AMC (**b**)

In comparison to previous works, which are confined to use in frequency around 2.4 GHz this paper focuses on 6 GHz band which is widely employed in rapidly growing technologies. The proposed design provides feasible solutions for implementation of WiMAX at 5.8 GHz. This constitutes under IEEE 802.16 series officially called wireless MAN which offers attractive solutions in telecommunications (VoIP), IPTV, broadband communications, smart grid and metering.

In addition the proposed antenna is designed for application in vehicular communication which is a rapidly emerging technology. This technology employs 802.11p standard known as Wireless Access in Vehicular Environments (WAVE).

(a)   **(b)**

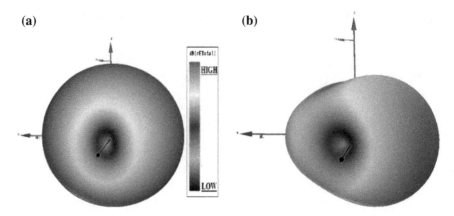

Fig. 8 Comparing 3D polar plot of antenna without AMC (**a**) and antenna with AMC (**b**)

Though much of research is dedicated to development of effective protocols for V2V systems, the proposed antenna is aimed to meet design aspects of the antenna system.

5   Conclusion

This paper presents a novel design for the AMC loaded antenna which operates at around 6 GHz for the WLAN applications. The design and simulation of the patch and AMC have been done in HFSS and the advantage of using AMC ground plane is established from the VSWR and the return loss plots. Also, for a better comparison, the directivity, gain and 3D radiation polar plots have been put forth. Thus, from this presentation, it is evident that the use of AMC as the ground plane enhances the gain and directivity of the antenna.

Acknowledgments The authors deeply owe their thanks to Dr. Anand S., for his extensive support in comprehending the innovative ideas and giving us the timely suggestions. Also, the authors are grateful to the Department of Electronics and Communication Engineering, National Institute of Technology, Tiruchirappalli for providing us with the facilities for working on this project.

References

1. M.N. Srifi, M. Meloui, M. Essaaidi, Rectangular slotted patch antenna for 5–6 GHz applications. Int. J. Microwave Opt. Technol. **5**(2), 52–57 (2010)
2. B.-K. Ang, B.-K. Chung, A wideband e-shaped microstrip patch antenna for 5–6 GHz wireless communications. Progr. Electromagnet. Res. PIER **75**, 397–407 (2007)

3. C. Bangera, M. Joshi, Design of V-slotted microstrip patch antenna for yielding improved gain bandwidth product. Int. J. Res. Eng. Technol. **03**(03), 344–348 (2014)
4. S.V.V. Sudhakar, K. Kranth Kiran, G. Babu Rao, K. Syamala Raju, M. Umamaheswara Rao, K.N.L. VamsiPriya, Design of triple-S shaped microstrip patch antenna for dual band applications. Int. J. Recent Innovation Trends Comput. Commun. **3**(3), 1190–1193 (2015)
5. M. Ali, R. Dougal, G. Yang, H.-S. Hwang, Wideband (5–6 GHz WLAN band) circularly polarized patch antenna for wireless power sensors, in *Wireless and Microwave Technology, WAMICON* 2005, IEEE Annual Conference. doi:10.1109/WAMIC.2005.1528389
6. H.F. AbuTarboush, H.S. Al-Raweshidy, R. Nilavalan, Compact wideband patch antenna for 5 and 6 GHz WLAN applications, in *International Symposium on Antennas and Propagation (ISAP)* (2008)
7. H.R. Raad, A.I. Abbosh, H.M. Al-Rizzo, D.G. Rucker, Flexible and compact AMC based antenna for telemedicine applications. IEEE Trans. Antennas Propag. **61**(2), 524–531 (2013)
8. H.F. AbuTarboush, D. Budimir, R. Nilavalan, H.S. Al-Raweshidy, Connected U-slots patch antenna for WiMAX applications. Int. J. RF Microwave CAE 1–18
9. N. Tiwari, S. Kumar, Microstrip patch antenna for 5.9 GHz dedicated short range communication system. Int. J. Adv. Electr. Electron. Eng. (IJAEEE) **3**(3), 1–4 (2014)
10. Frequency band review for fixed wireless service, in *Final Report Prepared for Ofcom*, AEgis Systems Limited, 29 Nov 2011, pp. 1–110
11. K. Agarwal, Nasimuddin, A. Alphones, Unidirectional wideband circularly polarised aperture antennas backed with artificial magnetic conductor reflectors. IET Microwave Antennas Propag. **7**(5), 338–346 (2013)
12. I.Y. Park, D. Kim, High-gain antenna using an intelligent artificial magnetic conductor ground plane. J. Electromagnet. Waves Appl. **27**(13), 1602–1610 (2013)
13. C.A. Balanis, *Antenna Theory*, 2nd edn. (Wiley Publications, 1997)
14. E.B. Tchikaya, F. Khalil, F.A. Tahir, H. Aubert, Multi scale approach for the electromagnetic simulation of finite size and thick frequency selective surfaces. Progr. Electromagnet. Res. M **17**, 43–57 (2011)
15. M.E. Cos, F.L. Heras, M. Franco, Design of planar artificial magnetic conductor ground plane using frequency selective surfaces for frequencies below 1GHz. IEEE Antennas Wirel. Propag. Lett. **8**, 951–954 (2009)
16. S. Maci, M. Caiazzo, A. Cucini, M. Casaletti, A pole zero matching method for EBG surfaces composed of dipole FSS printed on a grounded dielectric slab. IEEE Trans. Antenna Propag. **53**(1), 70–80 (2005)

Design of Dual Band Labyrinth Slotted Rectangular Patch Antenna for X Band Applications

K.B.N. Girish, Pani Prithvi Raj, M. Vijaya Krishna Teja, S. Anand and D. Sriram Kumar

Abstract This paper presents the design of a Labyrinth shaped antenna, with a dual operating frequency of 9 and 11 GHZ in the X band region. The structure being designed has an impressive Voltage Standing Wave Ratio (VSWR) of 1.38 at the operating frequency. As the X band demarcation of the microwave radio region is widely used in numerous communication based applications, the antenna being presented in this work gives us a viable alternative that can be used with an impressive performance in this region of microwave radio. Rogers RT/duroid 5880 (TM) is used as the substrate material. The Voltage Standing Wave Ratio (VSWR) plot obtained has been examined and elaborated in the subsequent sections of the paper.

Keywords Micro strip · Antenna · X band · Voltage standing wave ratio (VSWR) · Return losses · High frequency structure simulator (HFSS) · Rogers RT/duroid 5880

K.B.N. Girish · P. Prithvi Raj · M. Vijaya Krishna Teja · S. Anand (✉) · D. Sriram Kumar
Department of Electronics and Communication Engineering, National Institute
of Technology, Trichy, Tamil Nadu, India
e-mail: anand.s.krishna@gmail.com

K.B.N. Girish
e-mail: 108112052@nitt.edu

P. Prithvi Raj
e-mail: 108112040@nitt.edu

M. Vijaya Krishna Teja
e-mail: 108112099@nitt.edu

D. Sriram Kumar
e-mail: srk@nitt.edu

© Springer India 2016   477
S.C. Satapathy et al. (eds.), *Microelectronics, Electromagnetics
and Telecommunications*, Lecture Notes in Electrical Engineering 372,
DOI 10.1007/978-81-322-2728-1_44

1   Introduction

The X band segment of the microwave radio region of the electromagnetic spectrum is widely used in the present day scenario for a number of communication based applications such as satellite communication, terrestrial communication and networking and radar applications. In our work, we have come up with a micro-strip patch antenna designed at an operating frequency of 9 and 11 GHz that lie well within the X band frequency assortment. There are a number of research works going on in the X band segment, as this frequency range offers a wide array of commercial and military applications.

The Rogers RT/duroid 5880 (TM) [1] substrate being used here is a poly tetra fluoro ethylene (PTFE) composite reinforced with glass microfiber. These microfibers, being randomly oriented, enhance the reinforcement benefits in the direction most essential to the final micro-strip based circuitry application. Also, this substrate possesses several other advantageous properties such as low moisture absorption, strong chemical resistance, and uniform electrical properties over different frequencies, making it indispensable in the design of a micro-strip patch antenna.

The concise dimensions used in the design of the antenna, makes it advantageous. The elaborate descriptions of the various dimensions used in the patch antenna have been mentioned in the succeeding sections of the paper. Also, the subsequent sections of the paper have carried out elaborate analysis of the various reflection parameters of the intended antenna.

The present paper is organized in the following flow. The Sect. 2 which follows, primarily discusses the works of previous researchers pertaining to the development and applications of patch antenna using various innovative developments. Following this, Sect. 3 focusses on the design aspect of the proposed antenna. The section is organized to first present the general formulae used for designing an antenna and then about the design of antenna proposed in the present work. Section 4 analyses the results obtained while the Sect. 5 draws major conclusion of the proposal.

2   Previous Works

Almost all the works being carried out in this area of micro strip antenna, involves dual band characteristics, as it offers the advantage of tuning the antenna at two different frequencies in the limited frequency spectrum available.

One of the recent works [2] involved designing a dual band Micro strip antenna for the Global Positioning system (GPS) applications. In this work [2], different slot shapes were loaded with different patches of the same dimensions, making it appropriate for GPS applications. In [3, 4] different alphabetical shapes were used in the design of micro strip antenna. In [3], a dual band E shaped micro strip antenna that can be used for different applications was designed, with impressive gain and directivity. The simulations here were carried out using IE3D tool. The work [4] involves design of a U shaped slot on a patch antenna, fed by a broadband electromagnetic coupling probe called L-probe. Also, there are other works [3, 5–7] designed on similar lines, involving different applications such as ultra wideband applications, genetic optimization algorithm based applications and so on. Work [8] involves design of CPW fed slot antenna for use in wideband WLAN applications with variations in length and width of slot used for creating wideband of about 1.6 GHz. The influence of human body on UWB channel is investigated in [9] and results about the path loss and delay spread have been reported. The behavior of UWB antenna in a WBAN has also been presented in [10]. Moreover, a UWB antenna for a WBAN operating in close vicinity to a biological tissue is proposed in [11].

3 Design of the Micro Strip Patch Antenna with Slots

Based on simplified formulae, practical design steps of micro strip antenna are outlined from [12].

For the antenna to be efficient radiator, width formulated as

$$W = \frac{1}{2fr\sqrt{\mu_\circ\varepsilon_\circ}}\sqrt{\frac{2}{\varepsilon r+1}} \tag{1}$$

where f_r stands for the resonant frequency of operation and εr standing for relative permittivity. Effective dielectric constant of the antenna is calculated as

$$\varepsilon_{eff} = \frac{\in r+1}{2} + \frac{\in r-1}{2}\left(1+12\frac{h}{W}\right)^{-1/2} \tag{2}$$

where h is the height of the substrate. Once width and effective dielectric constant of antenna are designed, length of the patch is determined by

$$L = \frac{1}{2fr\sqrt{\in eff}\sqrt{\mu\in}} - 2\Delta L \tag{3}$$

where

$$\Delta L = 0.412h\left(\frac{(\in eff+0.3)(\frac{W}{h}+0.264)}{(\in eff-0.258)(\frac{W}{h}+0.8)}\right) \tag{4}$$

Fig. 1 Recessed micro strip line feed

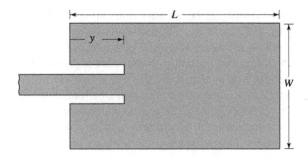

An inset feed that is recessed by a distance y can be used to effectively match the patch antenna using micro strip line feed. Using this technique the resonant input feed can be changed accordingly (Fig. 1).

$$R_{in}(y = y_0) = \frac{0.5}{G_1 \pm G_{12}} \cos^2\left(\frac{\pi}{L} y_0\right) \tag{5}$$

$$G_1 = \begin{cases} \frac{1}{90}\left(\frac{W}{\lambda_0}\right)^2 & W \ll \lambda_0 \\ \frac{1}{90}\left(\frac{W}{\lambda_0}\right)^2 & W \gg \lambda_0 \end{cases} \tag{6}$$

where G_1, G_{12} are conductance and mutual conductance of rectangular microstrip patch and R_{in} is real part of total resonant input impedance. 'y' stands for the length of recession of the feed into the patch.

The plus (+) sign is used for modes with odd (anti-symmetric) resonant voltage distribution beneath the patch and between the slots while the minus (−) sign is used for modes with even (symmetric) resonant voltage distribution.

The mutual conductance is defined, in terms of the far-zone fields, as

$$G_{12} = \frac{1}{120\pi^2} \int_0^\pi \left[\frac{\sin\left(\frac{kW}{2}\cos\theta\right)}{\cos\theta}\right]^2 J_0(kL\sin\theta)\sin^3\theta d\theta \tag{7}$$

J_0 is the Bessel function of the first kind of order zero.

The maximum value of R_{in} occurs at the edge of the slot ($y_0 = 0$) where the voltage is maximum and the current is minimum; typical values are in the 150–300 ohms. The minimum value (zero) occurs at the center of the patch ($y_0 = L/2$) where the voltage is zero and the current is maximum. As the inset feed point moves from the edge toward the center of the patch the resonant input impedance decreases monotonically and reaches zero at the center. When the value of the inset feed point approaches the center of the patch ($y_0 = L/2$), the $\cos^2(\pi y_0/L)$ function varies very rapidly; therefore the input resistance also changes rapidly with the position of the feed point (Fig. 2).

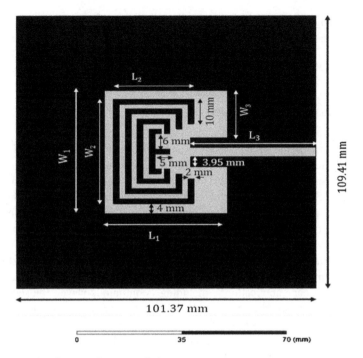

Fig. 2 Schematic of proposed antenna design

In the above figure depicting the design of the antenna, the various dimensions (in mm) are as follows:

L1 = 41.37, L2 = 27 mm, L3 = 42.49 mm, W1 = 49.41 mm, W2 = 42 mm, W3 = 18.8 mm.

The substrate used in the design of the above antenna is Rogers RT/duroid 5880 (TM). This substrate has a relative permittivity of 2.2 and relative permeability of 1

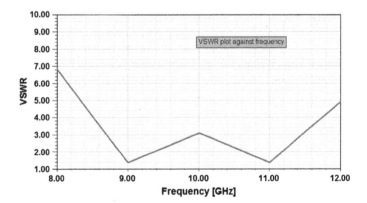

Fig. 3 VSWR plot against frequency

and a loss tangent of 0.0009 [1]. This substrate has been chosen, on account of its salient features as elaborated in the introduction. The radiating patch is made up of copper.

4   Results and Discussions

The simulation of the above design was carried out using the High Frequency Structure Simulator (HFSS) Tool. The results show that the antenna operates at two distinct frequencies of 9 and 11 GHz with a Voltage Standing Wave Ratio (VSWR) (Fig. 3) of 1.38 and return loss of −16 dB. This acceptable value of the VSWR ensures good performance of the antenna at the two operating frequencies.

Since, the operating frequencies fall within the X band region, this antenna can be used for various applications such as electron paramagnetic resonance (EPR) spectrometers, motion detectors, terrestrial communications and networking, satellite communications and radar applications. Also it is widely used by the military in radar applications including continuous-wave, pulsed, single-polarization, dual-polarization, synthetic aperture radar and phased arrays. X-band radar frequency sub-bands are used in civil, military and government institutions for weather monitoring, air traffic control, maritime vessel traffic control, defense tracking and vehicle speed detection for law enforcement.

The return loss plot shows that the bandwidth around the two operating frequencies is found out to be approximately 1 GHz. This larger bandwidth causes the signals to be spread thinly over a wide bandwidth resulting in extremely low power spectral densities leading to Low probability of intercept (LPI) and low probability of detection (LPD) that are very commonly used in military applications.

The previous works mentioned earlier focused mainly on development of different antenna structures for use in ISM band or in X-band for wearable medical

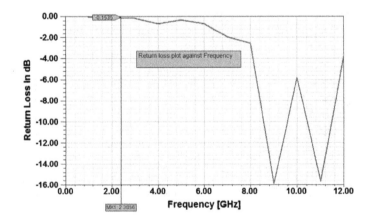

Fig. 4 Return loss plot against frequency

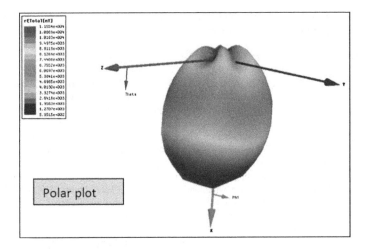

Fig. 5 Polar plot of radiation pattern

gadgets. But in this paper we present a simple antenna design in X-band for use in communication and military applications. The proposed antenna has approximate bandwidth of about 1 GHz which can be used in low power spectral density applications involving traffic control in air, water and defense tracking (Fig. 4).

The Fig. 5 depicts the radiation pattern obtained in terms of its high directivity and gain of the proposed antenna, thereby suggesting its promising utility.

5 Conclusions

Thus, a micro strip patch antenna with slots has been designed, operating at frequencies of 9 and 11 GHz with a bandwidth of around 1 GHz. The Voltage Standing Wave Ratio (VSWR) was found out to be 1.38. The structure has been simulated using the HFSS tool and the VSWR plot obtained has been analyzed and discussed. The applications of the antenna in the two frequencies have been discussed in the above sections.

References

1. Rogers RT/duroid 5880 (TM), https://www.rogerscorp.com/.../RT-duroid-5870-5880-Data-Sheet.pdf
2. H.A. Sabti, J.S. Aziz, Design of a dual band GPS micro-strip patch antenna. Int. J. Electr. Electron. Res. **2**(2), 92–95 (2014). ISSN:2348-6988

3. R.K. Prasad, A.K. Gupta, D.K. Srivastava, J.P. Saini, Design and analysis of dual frequency band E-shaped microstrip patch antenna, in *Conference on Advances in Communication and Control Systems* (2013)

4. J. Ghalibafan, F.H. Kashani, A.R. Attari, A new dual-band microstrip antenna with U-shaped slot. Progr. Electromagnet. Res. C **12**, 215–223 (2010)

5. W. Mazhar1, M.A. Tarar, F.A. Tahir, S. Ullah, F.A. Bhatti, Compact microstrip patch antenna for ultra-wideband applications, in *PIERS Proceedings*, Stockholm, Sweden, 12–15 Aug 2013

6. N. Hasan, S.C. Gupta, A dual band microstrip patch antenna with circular polarization, in *Conference on Advances in Communication and Control Systems* (2013)

7. O. Ozgun, S. Mutlu, M.I. Aksun, Senior Member, IEEE, L. Alatan, Member, IEEE, Design of dual-frequency probe-fed microstrip antennas with genetic optimization algorithm. IEEE Trans. Antennas Propag. **51**(8), (2003)

8. K. Nithisopa, J. Nakasuwan, N. Songthanapitak, N. Anantrasirichai, T. Wakabayashi, Design CPW fed slot antenna for wideband applications. PIERS Online **3**(7), (2007)

9. Y.P. Zhang, Q. Li, Performance of UWB impulse radio with planar monopoles over on-human-body propagation channel for wireless body area networks. IEEE Trans. Antennas Propag. **55**(10), 2907–2914 (2007)

10. K.Y. Yazdandoost, R. Kohno, UWB antenna for wireless body area network, in *Microwave Conference*, 2006. APMC 2006. Asia Pacific, 12–15 Dec 2006, pp. 1647–1652

11. M. Klemm, I.Z. Kovacs, et al., Comparison of directional and omnidirectional UWB antennas for wireless body area network applications, in *18th International Conference on Applied Electromagnetics and Communications* (ICECom, 2005), pp. 1–4

12. C.A. Balanis, *Antenna Theory Analysis and Design* (Wiley, 2005). ISBN:0-471-66782-X

Implementation and Analysis of Ultra Low Power 2.4 GHz RF CMOS Double Balanced Down Conversion Subthreshold Mixer

P.S. Harisankar, Vaibhav Ruparelia, Mayank Chakraverty and Hisham Rahman

Abstract This paper discusses the design of a 2.4 GHz operated, ultra-low power CMOS down-converting active mixer based on double balanced Gilbert-cell resistor-loaded topology fabricated in standard 180 nm RF CMOS low-power technology. All the MOS transistors of the mixer core have ideally been biased to sub-threshold region. Consuming only 500 μW of DC power using 1.0 V supply and minimal LO power of −16 dBm, this mixer demonstrates a simulated power conversion gain of 17.2 dB with Double Side Band (DSB) noise figure of 13.3 dB. With the same DC power dissipation and LO power, −11.7 dBm IIP3 and −20.1 dBm 1-dB point have been obtained as discussed in the paper. Pre-layout and post layout simulation results match very well. The ultra-low power consumption of the proposed mixer due to subthreshold region of operation and lower local oscillator power are the advantages of this subthreshold mixer.

Keywords Zigbee · Low power · RF CMOS · Mixer · DCR · Subthreshold · Impedance

P.S. Harisankar (✉) · V. Ruparelia · M. Chakraverty
GlobalFoundries, N1 Block, Manyata Embassy Business Park,
Nagavara 560045, Bangalore, India
e-mail: harisps09@gmail.com

V. Ruparelia
e-mail: vaibhavaruparelia@gmail.com

M. Chakraverty
e-mail: nanomayank@yahoo.com

H. Rahman
Indian Institute of Science, CV Raman Ave, Bangalore 560012, India
e-mail: hishamrahman@gmail.com

© Springer India 2016
S.C. Satapathy et al. (eds.), *Microelectronics, Electromagnetics
and Telecommunications*, Lecture Notes in Electrical Engineering 372,
DOI 10.1007/978-81-322-2728-1_45

1   Need for Low Power Mixers

To meet the stringent requirements of extending the battery life of latest wireless transceivers, there has been lot of research on novel low-power RF circuits. Frequency down-conversion mixer is one of the key RF front-end blocks, which consumes significant power while converting the RF signals to the baseband domain. In order to reduce power consumption, many novel RF CMOS mixer topologies, like circuit-stacking have been proposed [1–4]. However, these topologies require higher power supply voltages due to stacking of FETs and also require ideal AC ground and therefore require large-sized capacitors. Also, Cascode-mixer implementation have been proposed by [5, 6], whereby PMOS transistors are stacked on NMOS transistors. However, this topology can cause device breakdown and large power dissipation, due to high (1.8 V) dc supply. High-Q passive components in the (LC) tank [7] and transformers [8] can also be used to reduce the power dissipation. However, these approaches will increase the chip area and fabrication cost. Also, the band pass response of the transformer results in a smaller 3-dB bandwidth. As proposed in [9], the RF input can also be applied to the Body terminal of FET, which will avoid FET-stacking and allow lower supply voltage. The disadvantage of this approach is that, it requires a twin-well technology, wherein the back-gate coupling and other device-related parasitic diodes/capacitances are modelled accurately. Thus it is difficult to achieve low power CMOS RF mixer, without degrading other specifications or increasing cost. This paper proposes an active RF mixer design, wherein MOS transistors are biased in weak-inversion region. The advantage of this topology is significant reduction in DC power dissipation, without using large power supply voltages or expensive passive components.

2   The Gilbert Cell Multiplier Based Mixer Topology

Double balanced mixer topology is the most popular and simple active mixer topology, which is also called Gilbert cell multiplier. A simplified gilbert cell consists of three parts: gm-stage, switching stage and load stage. The principle of operation of mixer is multiplication of two time domain signals, which is effectively addition or subtraction in frequency domain. In gilbert cell, multiplication operation is based on switching of RF current by LO voltage.

Figure 1 shows the schematic of an active RF mixer called CMOS Gilbert Cell multiplier. The small signal RF and LO signal voltages are denoted by V_{RF} and V_{LO}. The dependency of these voltages on output current and voltage is given by the following expressions.

Fig. 1 Gilbert cell multiplier

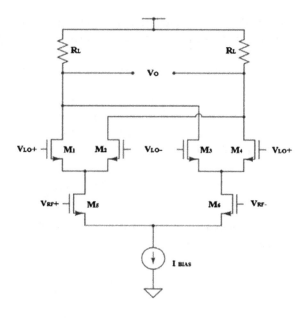

$$\Delta I = \sqrt{(K_x K_y)} V_x V_y; \quad V_o = \sqrt{(K_x K_y)} R_L V_x V_y \qquad (1)$$

Where

$$K_x = \mu_n C_{ox} \left(\frac{W}{L}\right)_{1-4} \qquad K_y = \mu_n C_{ox} \left(\frac{W}{L}\right)_{5-6}$$

From these equations, it is clear that the conversion gain of the mixer directly depends on load and transistors' aspect ratios. Differential output current is basically the modulated signal that varies proportional to the variation in V_{RF} and V_{LO}. This type of topology requires stacking of transistors and load stage, which leads to higher power supply.

3 MOS Transistors in Subthreshold Regime

Inversion region of MOS transistors is classified into weak inversion, moderate inversion and strong inversion. This section of the paper deals with the behavior of MOS transistor in subthreshold region, also called weak inversion region. This region of operation of MOS transistors requires very low gate voltages. Small variations in small signal voltage at gate produce huge variations in small signal current compared to subthreshold region of operation. Strong inversion model of MOSFET makes an assumption that the charge in the channel, when the gate voltage drops below the threshold voltage, goes to zero. But in reality, this is not true and charge density in the inversion region drops exponentially with decrease in gate voltage.

The current equation of MOS transistor operated in subthreshold is modeled as

$$I_D = I_o \frac{W}{L} \exp \frac{(kV_G - V_S)}{U_T} \left[1 - \exp \left(-\frac{V_{DS}}{U_T} \right) \right] \qquad (2)$$

where I_0 is the subthreshold current at $V_{GS} = V_{TH}$ and it is defined for NMOS as

$$I_{on} = \frac{2\mu_n C'_{ox} U_T^2}{k} \exp \left(-\frac{kV_{Ton}}{U_T} \right) \qquad (3)$$

where k (kappa) is called the gate coupling coefficient and it represents the coupling of the gate to the surface potential:

$$k = \frac{C_{ox}}{C_{ox} + C_{dep}} \qquad (4)$$

The significant difference between subthreshold and saturation mode of operation of MOSFET is that the drain current varies as V_{GS} increases. In subthreshold region of operation, current varies exponentially with respect to gate voltage. Whereas in active mode, drain current and gate voltage demonstrate a quadratic relationship. The relationship is evident from the plots in Fig. 2. The voltage headroom required for circuits is very low when transistors are biased in subthreshold regime, which results in lower power supply requirements and reduction in DC power dissipation. Smaller local oscillator signal power requirements for the subthreshold mixer due to sharp switching of transistor provides an additional reduction in DC power consumption of signal generation blocks like VCO, PLL, etc., which generates the local oscillator signal.

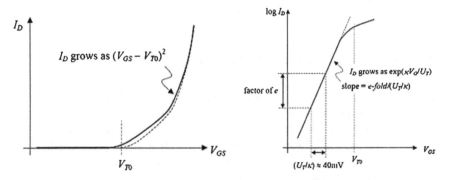

Fig. 2 Id-Vgs for transistor operating in saturation (*left*) and sub-threshold (*right*) regions

4 Mixer Implementation and Parameter Considerations

4.1 Mixer Implementation

Figure 3 (left) shows the proposed double balanced mixer, which is also well known as Gilbert cell mixer. In this topology, both local oscillator and RF signals are fed in differential modes. Differential mode of operation suppresses all even order harmonics of both LO and RF signals and leads to better isolation and linearity.

Figure 3 (right) shows the comparison of transfer characteristics of a normal sized transistor with a width W_1, operating in super-threshold region against a reasonably wider transistor of width W_2, which operates in subthreshold region. In super-threshold region of operation, the drain current is subjugated by drift current and results in a square dependence of gate voltage on drain current. In subthreshold region, due to dominant diffusion currents, the drain current exhibits an exponential dependence on gate voltage. Therefore, the g_m/I_D ratio required for the saturation mode mixer can be achieved in subthreshold mixer with larger transistor widths and lower drain currents. Therefore, it may be concluded that using oversized MOS transistors operated in subthreshold region is one good way of reducing the power dissipation of mixer. The transconductance stage of the mixer consisting of transistors M_1 and M_2 amplify the received RF signal and converts it into current. The transistors M_3 to M_6 are the gilbert cell switching quad. The purpose of switch quad is to modulate the current provided by the transconductance stages. These subthreshold switches have an upper hand over the switches used in conventional mixers in the context of steeper switching slope and correspondingly higher conversion gain. The conversion gain of the mixer is given in the equation below, where g_m is the transconductance of transistors M_1 and M_2 and R_{load} is the load resistance, which converts the modulated RF currents into voltages.

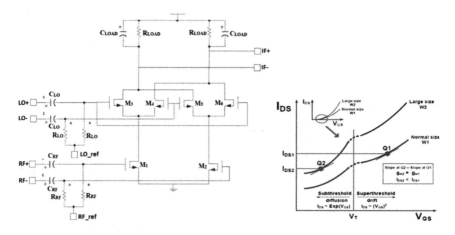

Fig. 3 Proposed mixer schematic (*left*) and I_D-V_{GS} for various widths (*right*)

$$G = \frac{2}{\pi} g_m R_{load} \tag{5}$$

Poly resistors are used as load resistances, because they exhibit higher resistance and lower flicker noise. In the proposed design, C_{load} is added along with load resistor, which acts as an LPF and its cutoff value is chosen in a way to avoid 2.4 GHz leakage component and 4.81 GHz ($f_{RF} + f_{LO}$) frequency components. The CS output stages are used as buffers which augments the conversion gain and offers a 50 Ω wideband matching over a bandwidth of 2.4 GHz by an appropriate choice of the load resistor (R_{IF}). Its effect of loading on switching stage is also very minimal due to high input impedance of CS stage and coupling capacitor (C_C) between them. The design parameters are tabulated as shown Table 1.

DC bias voltage of the transistors is fed through series RC network with output taken across resistor, which acts like HPF to filter out unwanted lower frequency components at the input ports. The HPF cutoff frequency has been chosen such that $f_{HPFI} \lll f_{RF}$ or f_{LO} at input ports and $f_{HPFO} \lll f_{IF}$ at output ports. There are four input terminals (two RF signals and two LO signals) and two output terminals for a double balanced mixer. Differential signals are generated using active BALUNs (Balanced Unbalanced Transformers). Matching is provided only to the single ended signal before baluns. Matching is done for 100 Ω, such that balun splits the impedances seen by ports to 50 Ω. Matching network also considers the impedance provided by the baluns.

4.2 Mixer Parameter Selection

Frequency Selection

This mixer is used to operate in ISM band. Hence, the RF frequency is selected as 2.41 GHz. We followed the ZigBee protocol and chose 2.4 GHz band of operation. 2.4 GHz band varies from 2400 to 2483.5 MHz with 16 channels and has a separation of 5 MHz IF frequency is selected based on the available channel band width and value is chosen as 10 MHz. Local oscillator frequency is selected as $f_{LO} = f_{RF} + f_{IF} = 2.4$ GHz.

Signal Power Selection

RF signal power: Sensitivity of ZigBee transceiver is −85 dBm, it's the minimum signal level that the transceiver can recognize as input. The maximum signal level that the receiver can withstand has been reported to be −20 dBm. In the receiver section, the mixer is preceded by LNA. Minimum and maximum signal levels that can be available to the mixer are:

Minimum signal level = Sensitivity + Gain of LNA
Maximum signal level = Maximum affordable input level + Gain of LNA

Table 1 Design parameters

Parameters	$W_{1,2}$	$W_{3,4,5,6}$	W_{7-8}	R_{LOAD}	R_{IF}	R_{LO}	R_{RF}	R_{BUFF}	C_{LOAD}	C_{LO}	C_{RF}	C_{BUFF}
Values	105 μm	75 μm	200 μm	1.77 KΩ	113 Ω	1.1 KΩ	1.1 KΩ	15 KΩ	1.2 pF	4.92 pF	4.92 pF	4.92 pF

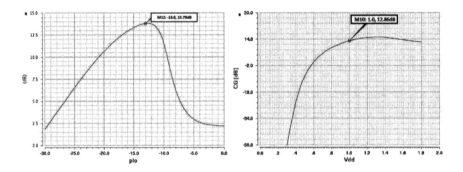

Fig. 4 Conversion gain versus local oscillator power (*left*) and V_{DD} (*right*)

Gain of LNA is chosen as 15 dB = 45 dBm. Hence, the possible input range is −40 to 25 dBm. RF signal power has been chosen as worst case value of −40 dBm.

LO signal power: It is chosen by sweeping conversion gain over a feasible range of LO power as shown in Fig. 4 (left) below and the value where conversion gain gets maximum has been selected. The proposed mixer shows maximum gain at lower local oscillator power of −13 dBm. This is the main advantage of sub-threshold mixer when compared to mixers operating in saturation mode.

Power Supply Selection

Power supply is selected by simulation of conversion gain versus power supply. The plot for conversion gain is almost constant from 1.0 to 1.8 V. Hence, we chose V_{DD} as 1.0 V. Figure 4 (right) above shows the variation of gain with supply voltage reduction. From this, we can observe that the conversion gain of the mixer is almost constant through the range of 1 to 1.8 V. Whereas, when the supply voltage goes below 1 V, the structure attenuates the signal.

5 Simulation Results

5.1 Plots for Subthreshold Mixer with Output Buffer

Transient Analysis

Figure 5 demonstrates the transient simulation results of the proposed mixer. The first signal indicates the output IF signal ($V_{IF} = V_{IF+} - V_{IF-}$) of voltage level 20 mV and its frequency is 10 MHz; Next two waveforms are input RF and LO signals respectively. The input RF signal is chosen as 2.41 GHz and its signal level is −40 dBm ($\sim$2.5 mV); the local oscillator signal is 2.4 GHz and its signal level is −13 dBm ($\sim$400 mV). From the figure, it is clear that the power conversion gain

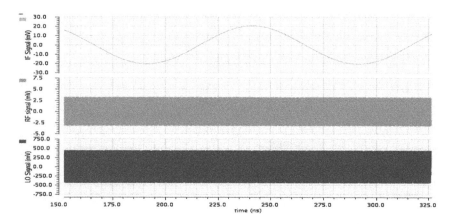

Fig. 5 Transient analysis

and voltage conversion gain are matching, which indicates that the input and output ports of the mixer are properly matched to 50 Ω.

Voltage Conversion Gain Versus RF Signal Frequency and Versus IF Signal Frequency (Swept PSS with PXF)

To demonstrate high frequency performance, conversion gain has been measured while RF signals were simultaneously swept to obtain down-conversion gain as a function of RF frequency. The Fig. 6 (left) shows a down conversion gain of 13.7 dB at an RF input frequency of 2.41 GHz. The Fig. 6 (right) shows the plot of voltage conversion gain plotted against IF signal frequency.

IIP3 Calculation (QPSS)

The IIP3 has been measured by applying a two tone test while sweeping the LO frequency and maintaining a constant RF frequency of 2.41 GHz. The Fig. 7 shows input referred third order intercept point, IIP3, of −8 dBm. This value can be considered as moderately good compared to mixers that will be operated in

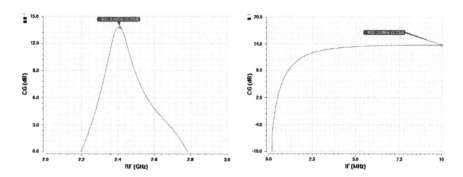

Fig. 6 Voltage conversion gain versus RF frequency (*left*) and IF frequency (*right*)

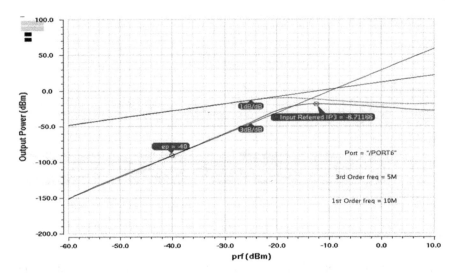

Fig. 7 IIP3 calculation

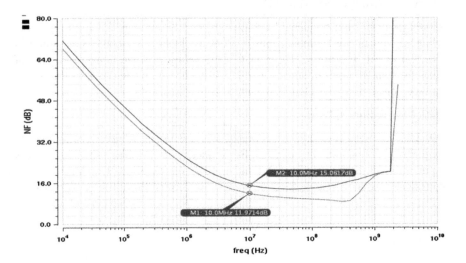

Fig. 8 SSB and DSB noise figure

saturation region. This problem can be alleviated by applying non-linearity improvement techniques proposed with a tradeoff between gain and noise figure. The output 1 dB compression point measured at an IF output frequency of 10 MHz has been found to be −17 dBm which is 9 dBm less than IIP3.

SSB and DSB Noise Figure (PSS and PNOISE)

Figure 8 shows Single side band (SSB) and Double side band (DSB) noise figures as a function of IF output frequency. The proposed mixer has a SSB noise figure of 15 dB and a DSB noise figure of 11.97 dB.

6 Conclusion

This paper presents an ultra-low power down conversion subthreshold mixer operating at 2.4 GHz for ZigBee transceiver applications and has been implemented in standard 0.18 μm RF CMOS technology. The subthreshold region of operation of the proposed mixer helps in achieving reduced DC power consumption of 500 uW under 1.0 V power supply as well as lower LO signal power of −13 dBm. The oversized transistors used in transconductance and switch quad stages help to achieve lower current and higher g_m values. This proposed mixer achieves a higher power conversion gain of 17.2 dB with a reasonable DSB noise figure of 13.3 dB. This subthreshold mixer has better overall performance than conventional mixers operating in saturation mode. Linearity of the proposed mixer can further be improved by either increasing bias current or by introducing degeneration inductors and cross coupled capacitors. Noise figure can further be increased by adapting current bleeding techniques. Further power reduction in the proposed mixer can be obtained by choosing advanced RF CMOS technology nodes.

References

1. H. Lee, S. Mohammadi, A 500 μW 2.4 GHz CMOS subthreshold mixer for ultra low power applications, in *IEEE Radio Frequency Integrated Circuits Symposium*, Jun 2007, pp. 325–328
2. C. Hermann, M. Tiebout, H. Klar, A 0.6-V 1.6-mW transformer-based 2.5-GHz downconversion mixer with +5.4-dB gain and -2.8-dBm IIP3 in 0.13-um CMOS. IEEE Trans. Microwave Theory Tech. **53**(2), 488–495 (2005)
3. C.-H. Li, C.-N. Kuo, Design optimization of a 1.4 GHz low power bulk-driven mixer, in *Proceedings of Asia-Pacific Microwave Conference*, Dec 2010, pp. 1035–1038
4. H. Lee, S. Mohammadi, A 3 GHz subthreshold CMOS low noise amplifier, in *Proceedings of RFIC Symposium*, June 2006, pp. 494–497
5. A. Zolfaghari, B. Razavi, A low-power 2.4-GHz transmitter/receiver CMOS IC. IEEE J. Solid-State Circuits **38**, 176–183 (2003)
6. T. Wang et al., A low-power oscillator mixer in 0.18-um CMOS technology. IEEE Trans. Microwave Theory Tech. **54**(1), 88–95 (2006)
7. V. Vidojkovic et al., A low-voltage folded-switching mixer in 0.18-um CMOS. IEEE J. Solid-State Circuits **40**, 1259–1264 (2006)
8. L. Liu, Z. Wang, Analysis and design of a low-voltage RF CMOS mixer. IEEE Trans. Circuit Syst. II **53**(3), 212–216 (2006)
9. H. Qi, J. Chen, M. Jian, P. Hao, A 1.2 V high performance mixer for 5.8 GHz WLAN application, in *International Conference on Wireless Communications, Networking and Mobile Computing*, Sept 2007, pp. 700–702

TDMA Based Collision Avoidance in Dense and Mobile RFID Reader Environment: DDFSA with RRE

K.R. Kashwan and T. Thirumalai

Abstract Densely deployed mobile RFID readers are bound to have collisions and interference due to inherence limitations. In this paper, authors have addressed the reader to reader collision using modified similar frame slotted ALOHA technique, called Dynamic Demand-oriented Frame Slotted ALOHA (DDFSA) based on a deterministic approach to avoid the collision. The readers form discrete adhoc network autonomously and resources are distributed using the super frame. The frame size can dynamically change according to number of adjoining readers in the network. Readers can play a role of master, backup, participating or hanging node in the network. At the start of super frame resources are allocated. The redundant reader is eliminated using reader capability. The master disseminates the frame plan and resources. This improves the network throughput by reducing number of readers and utilizing all the times slots. The simulation tests results, using *omnetpp+* software, show that resources are utilized up to 98 % with system efficiency at 99 % and nil unused time slots.

Keywords RFID reader · Tag · Anti collision · TWA · Multi channel · RRE · DDFSA · LBT · BAN

1 Introduction

Radio Frequency Identification (RFID) comprises of a reader and a tag which are called interrogator and transponder respectively. The tags can be of two types. First is active tag which is self powered and the second is passive tag, which receives

K.R. Kashwan (✉)
Department of Electronics and Communication Engineering, Sona College
of Technology (Autonomous), Salem 636005, Tamil Nadu, India
e-mail: kashwan.kr@gmail.com

T. Thirumalai
Electronics Corporation of India Limited (Government of India Enterprise),
IGCAR, Kalpakkam, Chennai 603102, Tamil Nadu, India
e-mail: ttm_jayala@yahoo.com

© Springer India 2016 497
S.C. Satapathy et al. (eds.), *Microelectronics, Electromagnetics
and Telecommunications*, Lecture Notes in Electrical Engineering 372,
DOI 10.1007/978-81-322-2728-1_46

energy from reader while backscattering the data for communicating with reader. The readers can read tag data even without in-line-of-sight communication. It then, forwards the data to central server. The RFIDs have wide applications in the areas of supply chain, security system, hospital management, body area network, internet of things (IOT) and indoor locating systems. RFIDs can be integrated with sensor networks easily. Assembly area and supply chain may have static and pre planned installations of RFID systems.

The classification of RFIDs is normally based on the frequency of operation and distance range. The regulatory authorities of respective countries govern the operations of RFID. The frequency of operations in range of 125–134 kHz is called low, 13.56 MHz is high and 860–960 MHz is called Ultra High Frequency (UHF). The frequency of operations in range of 2.48, 5.72 and 5.85 GHz are all called microwave frequency range. The UHF band passive tags find wide applications due many advantages [1]. EPC global Class-1 Gen-2 specifies the standards to be adopted for compatibility amongst customized RFID products [2]. The readers share the common wireless channels allowed for ISM applications. The readers may interfere with tags or with each other leading to degraded data communication.

The RFID system may suffer by three types of interferences. Tag to tag interference, reader to tag interference and reader to reader interference. Tag to tag interference happens when multiple tags tries to respond to a query from a reader. It causes collision. A technique called *singulation* is performed to avoid collision amongst multiple tag's response using Framed slotted ALOHA and binary tree walking algorithms [3]. Reader to tag interference happens if more than one reader tries to read the same tag located within reading range of both readers. The EPC global Class-1 Gen-2 standard provides guidelines for different frequency channels for reader to tag (forward) and tag to reader (reverse) communication. It addresses the issue of reader interference with tag's response. However the hidden terminal problem persists wherein tag may not respond to a specific reader for a query. Frequency hopping and TDMA schemes are promising solution for avoiding collision.

Figure 1a and b show the reader to tag interference with read range overlapping and non overlapping cases respectively. Figure 1c shows reader to reader

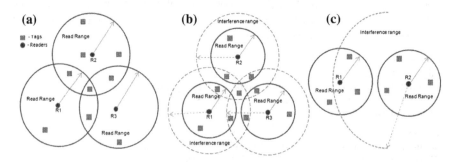

Fig. 1 Interference **a** reader to tag for overlapping of read range, **b** reader to tag for non-overlapping of read range, **c** reader to reader interference

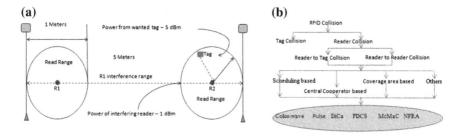

Fig. 2 **a** Transmit power of 1 W and 6 dBi antenna (assumed) show that the interference range can affect read range and **b** the hierarchical tree of RFID anti collision protocols

interference. The overlapping leads to hidden terminal problem. Figure 1a and b show that readers R1, R2, R3 read the same tag which is located in common read range or interference range. The reader to reader collision occurs if a reader which is not in its read range interfere the function of wanted reader as shown in Fig. 1c. Reader to reader collision also occurs if a reader reads a tag in its read range, while the other reader interference is stronger than backscattered energy of tag as shown in Fig. 2a.

2 Literature Review on Existing Work

A number of protocols have been developed for collision problem in RFID using multiple access techniques like TDMA. Each technique has a few limitations and advantages. The reader to reader anti-collision protocols are classified as shown in Fig. 2b. A few of protocols improve the frame utilization by adopting Redundant Reader Elimination (RRE) [4] and dynamic frame size [5]. Better utilization of time slot leads to reduced collisions. Another technique concentrates on a layered elimination optimization technique to reduce collision [6, 7]. NFRA neighbor friendly reader anti-collision is a server based anti-collision algorithm using TDMA. There are many other techniques employed to minimize reader collisions based on probabilistic models. These, however, can address the collision only after occurrence of collision

The TDMA based technique is used to avoid reader collision [8] by allotting randomly different colors and time slots for each reader trying to communicate simultaneously. Distributed collision avoidance protocol uses only one data channel [9]. It does not address reader to reader interference. The multi channel MAC is another form of distributed collision avoidance protocol, which uses control channel for exchange of data packets [10]. For multi-channel MAC the range is set such that reader can communicate through control channel. The EPC global class-1 Gen-2 uses odd channels for tag communication and even channels for reader communication [11]. Hierarchical queue learning algorithm is used for networks. It is based on reinforcement learning [12]. It adoptively learns the readers' collision patterns; however, it is not efficient for reader to tag collision. Anti-collision mobile RFID [13] is an improvement over PDCS [14]. The total number of time slots is controlled within a frame. The frame size is based on the number of readers in its interference range [15, 16].

3 Proposed Work

In this paper, DDFSA with RRE is proposed to reduce the collisions while trying to improve the network performance and utilization ratio. It is assumed that there is no centralized control on readers and the topological distribution of readers and tags is not considered and that the reader-tag communication is not homogeneous. The transmission range of a reader is shorter than interfering range. The proposed work involves notification and setting up the node status. It also involves resource allocation, synchronization, exchange of information and redundant reader elimination. It is based on TDMA frame consisting of multiple time slots. A time slot is assigned for Reader to Reader (R-R) and Reader to Tag (R-T) communication, as shown in Fig. 3.

The super frame as shown in Fig. 3 has multiple frames and at the start of super frame. The master starts sending frame plans and information to neighboring readers. In a dense reader environment the readers form a discrete adhoc network with ID. An adhoc network consists of master reader, back up reader, participating reader and others are hanging readers. The reader does not have MAC or IP address but has UPC/EPC or product serial number, of size 24 bits or more. Based on EPC each reader is assigned a temporary identification (TID).

The first reader enters the zone of consideration and act as a master with TID1. The next reader entering the network assumes backup master role with TID2. The subsequent readers may act as participating. If no EPC is heard, the reader sets its ID and start acting as master. The master maintains a table consisting the list of TIDs and their allotted resources, control flags, status flags etc. It updates and copies to back up regularly. Based on the number of readers in the zone the master

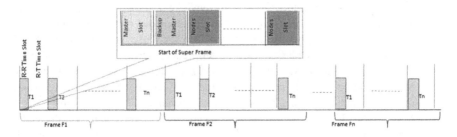

Fig. 3 Block diagram of super frame architecture

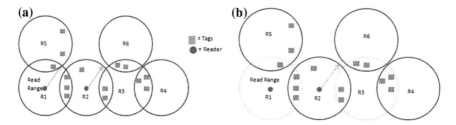

Fig. 4 **a** Readers R1 and R3 are found redundant. **b** Readers R1 and R3 are eliminated

distributes the frame plans to other readers for communication round. The reader can demand for additional resource in case of multiple tags. It can also relieve the resource if there are no tags to read. The reader is a part of network and if it is redundant it notifies the same to master for surrendering the resource.

The reader redundant elimination (RRE) works as follows. Figure 4a shows the normal readers participating in network and their tags' range coverage. After RRE is implemented, as shown in Fig. 4b, the readers R1 and R3 becomes redundant and are eliminated from the network. This facilitates less resource consumption, less contention and spatial separation aiding less interference. A reader maintains a table of tag IDs in its range for query round and is ready to be read. In a server control, this information is communicated to central server for further command and process. In new proposal, the matching tag IDs with neighboring readers in the network are checked for during the reader's home communication slot. If the IDs match, it declares itself as redundant reader and notifies the master for surrendering the resource allotted to it. The reader then waits for a new tag to arrive and request master during super frame for resource allotment. The proposed autonomous network formation is shown in Fig. 5a. The anti-collision algorithm consists of initial contention and three cases. The first case is no echo, second one echo and third case is multiple echoes.

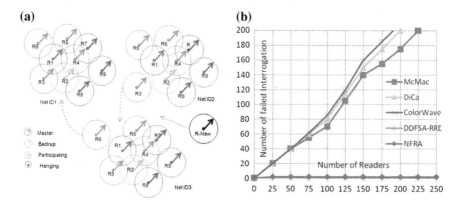

Fig. 5 **a** Discrete adhoc network formation by readers and **b** failed interrogation results

3.1 Case 1

It is assumed that initially, all reader start sending *keep alive* packets periodically during reader to reader communication. If there is no echo received by a reader then the reader sets its ID as TID1 and acts as master. If a collision is heard then it waits for time T_{wait}. It is given by Eq. (1). If two readers happen to have the same T_{wait} then the reader having higher EPC sets its own ID as TID1 and acts as a master.

$$T_{wait} = e \bmod n \tag{1}$$

where e is EPC/UPC number and n is the number of adjoining nodes or collisions detected. This ensures that the number of contention slots is less than the number of readers participating in contention during initial setup.

3.2 Case 2

Reader hears only one echo, TID1. Then ID is set as TID2 and the reader acts as backup master. Only if TID2 is heard, then the network is formed. It sets its TID as $TID_{MAX} + TID_{GAP} + 1$ and acts as a hanging node.

Only TID > TID2 is heard, then it sets its ID as $TID_{MAX} + TID_{GAP} + 1$ and acts as hanging node. These nodes copy the resource table, parameters from the heard TID. These can communicate in the slots other than the entry allowed for heard TIDs, till TID1 is heard.

3.3 Case 3

If multiple echoes having only EPCs are heard then collision contention and set up of master state is done similar to Case-1. If only TID1 and remaining EPCs are heard then set ID as TID2 and act as back up master. If both TID1 and TID2 are heard apart from other TIDs or EPCs then set as participating node (TID = $TID_{MAX} + 1$), wait for master to distribute frame plans and run query process. If there is no TID1 but only TID2 and remaining other TIDs/EPCs are heard, then set TID as $TID_{MAX} + TID_{GAP} + 1$ and act as hanging node till hearing TID1. The echoes consists only TID > TID2 with or without EPCs, then the hanging node activity is performed.

The participating readers keep on setting TIDs as $TID_{MAX} + 1$, whereas the hanging nodes set IDs as $TID_{MAX} + TID_{GAP} + 1$ and notifies through the adjoining readers. Any reader leaving the network in middle other than master and backup carries NID and it does not sense *keep alive* packets. The master empties the corresponding entry in the table. This creates a temporary gap in TID table and

TID$_{\text{TEMP}}$ flag is set to indicate that there is a vacant TID in table, which should be utilized by the next participating node. Similarly if a hanging node leaves in the middle, then TH$_{\text{GAP}}$ flag is set to notify the participants. TH$_{\text{Limit}}$ is set to indicate that the hanging nodes joining the network in nested fashion has reached the ID limit specified and free to set their IDs. The adhoc network is formed as illustrated in Fig. 5a, wherein various ID assignment and node movements are shown.

4 Simulation and Result Analysis

The simulation tests are performed on OMnetPP++ on windows 7 platform on a PC system. A field of 500 m^2 area is chosen with densely populated 250 readers and 5000 tags distributed over it. The readers' mobility is set to follow a random path. A simple path loss model is employed. For experiments, 15 frequency sub-bands are selected with a simple hopping sequence model. The performance comparison is carried out with existing protocol, TDMA representative Colorwave, DiCa, McMac, AC-MRFID, and NRFA. The Transmission range is set at 4 m and the interference range is kept at 20 m. Various simulation comparison parameters of frame utilization, throughput, system efficiency, network overheads and interrogation time with competitive protocols are observed and studied for result analysis.

The simulation result of interrogation failed is shown in Fig. 5b. This indicates the number of failed interrogation if the readers joining the network increase. Colorwave, Dica, McMac shows a steady increase in failed interrogation, while NFRA and newly proposed algorithm for this research work shows steady state irrespective of number of readers in network. Figure 6a shows the overheads and the number of message exchanges between readers during contention and interrogation period. The McMac shows higher overheads and other protocols show comparatively less overheads. The proposed algorithm consumes high overheads, but less compared to McMac, if the number of readers increases in the network. This indicates tight time synchronization and healthy network function. Other algorithms include message exchanges between readers and central server and message exchanges are also adopted amongst readers.

Result in Fig. 6c shows the total interrogation time for all readers in network, assuming that each reader has to identify equal number of tags within range. The McMac and DiCa have high collisions, keeping their interrogation time less even though the number of readers increases. NFRA avoids total collision but if the number of readers increases in the network the interrogation time also increases exponentially. DDFSA algorithm shows again steady state, keeping collision free and interrogation time does not depend on number of readers in the network. The frame utilization is shown in Fig. 6b. The proposed algorithm has the utilization at the maximum at par with DiCa protocol and compared to other protocols. The simulation result shown in Fig. 6d illustrates the network throughput for the proposed work with CC-RFID, ACHA which are suitable for single channel and multi channel network. The throughput and efficiency depends on the number of channels

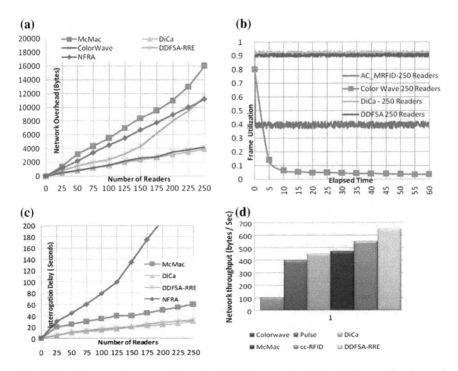

Fig. 6 Simulation results **a** network overheads, **b** interrogation delay, **d** frame utilization and **d** network throughput

used, and increases proportionally with increase in number of channels. DiCa and CC-RFID shows better throughput for single channel, ACHA shows higher throughput for multi channel. The proposed new algorithm shows better result and throughput in both the cases, single and multi channel networks.

5   Conclusion

The proposal DDFSA addresses the reader to reader collision problem in a deterministic model. It is implemented for elimination of redundant reader using readers' capabilities. DDFSA does not need additional hardware resources and coordination by a central server which is the present application requirements. Autonomous formation of discrete adhoc network is employed for new algorithm implementation. The simulation results show that the system efficiency is improved by 99 % and the throughput by 90 %. The frame utilization is 98 % compared to probabilistic approach adopted in Colorwave, PDCS, NFRA, McMac.

The future work shall concentrate on improving overheads used for establishment of network and methodology to control the nesting of hanging nodes. The future work will also involve design and fabrication of multi-resonant, composite, compact and light weight antenna based on strip line and patch principles. The patch or strip antenna can be integrated with RFID system to build a robust and complete stand alone system and protocol.

References

1. ETSI EN 302 208-1 VI. 1.1, www.etsi.org, EPC™ Radio-Frequency Identity Protocols Class-1 Generation-2 UHF RFID Protocol for Communications at 860 MHz–960 MHz Version 1.0.9 (2009)
2. Electronic product code. http://www.epcglobalinc.org
3. G. Shu-qin, W. Wu-chen, H. Li-gang, Z. Wang, Anti-collision algorithms for multi-tag RFID, INTECH, Croatia, p. 278 (2010). ISBN 978-953-7619-73-2
4. N. Irfan, M.C.E. Yagoub, Efficient algorithm for redundant reader elimination in wireless RFID networks. IJCSI Int. J. Comput. Sci. Issues **7**(3), 11 (2010)
5. K.C. Shin, S.B. Park, G.S. Jo, Enhanced TDMA based anti-collision algorithm with a dynamic frame size adjustment strategy for mobile RFID readers. Sensors **9**, 845–858 (2009). doi:10.3390/s90200845, ISSN 1424-8220
6. J.-B. Eom, S.-Y. Yim, T.-J. Lee, An efficient reader anticollision algorithm in dense RFID networks with mobile RFID readers. IEEE Trans. Ind. Electron. **56**(7), 2326–2336 (2009)
7. K.-I. Hwang, S.-S. Yeo, J.H. Park, Distributed tag access with collision-avoidance among mobile RFID readers, CSE '09. in *International Conference on Computational Science and Engineering*, vol. 2 (2009), pp. 621–626. doi:10.1109/CSE.2009.342
8. J. Waldrop, D.W. Engles, S.E. Sarma, Colorwave: a MAC for RFID reader networks. in *Proceedings of the IEEE International Conference on Communications, Anchorage, AK, USA* (2003), pp. 1701–1704
9. M. Victoria Bueno-Delgado, R. Ferrero, F. Gandino, P. Pavon-Marino, Member, and Maurizio Rebaudengo, a geometric distribution reader anti-collision protocol for RFID dense reader environments. IEEE Trans. Autom. Sci. Eng. **10**(2) (2013)
10. H. Dai, S. Lai, H. Zhu, A multi-channel MAC protocol for RFID reader networks. WiCom **2007**(21–25), 2093–2096 (2007)
11. K. Finkenzeller, *RFID handbook: fundamentals and applications in contactless smart cards and identification*, 2nd edn. (Wiley, New York, 2003)
12. J. Ho, D.W. Engels, S.E. Sarma, HiQ: a Hierarchical q-learning algorithm to solve the reader collision problem. in *International Symposium on, Applications and the Internet Workshops* (2006), pp. 23–27
13. F. Gandino, R. Ferrero, B. Montrucchio, M. Rebaudengo, DCNS: an adaptable high throughput RFID reader-to-reader anti-collision protocol. IEEE Trans. Parallel Distrib. Syst. **24**(5) (2013)
14. F. Gandino, R. Ferrero, B. Montrucchio, M. Rebaudengo, Probabilistic DCS: an RFID reader-to-reader anti-collision protocol. J. Netw. Comput. Appl. (2010). doi:10.1016/j.jnca.2010.04.007
15. D.-H. Shih, P.-L. Sun, D.C. Yen, S.-M. Huang, Taxonomy and survey of RFID anti-collision protocols. Comput. Commun. **29**, pp. 2150–2166 (2006). www.sciencedirect.com
16. F. Nawaz, V. Jeoti, A. Awang, M. Drieberg, Reader to reader anti-collision protocols in dense and passive RFID environment. in *IEEE 11th Malaysia International Conference on Communications, Kuala Lumpur, Malaysia* (2013)

Perturbed Elliptical Patch Antenna Design for 50 GHz Application

Ribhu Abhusan Panda, Suvendu Narayan Mishra and Debasis Mishra

Abstract A beam forming patch antenna is proposed for 50 GHz millimeter wave application for wireless personal area network (WPAN). The design involves some modifications of the geometry of prototype Rotman lens structure. The simulation result using HFSS and CST infers some outcomes like reflection co-efficient, VSWR directivity and surface waves, etc. which can be further improved by optimum choice of size and material.

Keywords Rotman lens · VSWR · HFSS · CST · 50 GHz WPAN · Directivity

1 Introduction

Competitive advantages of millimeter waveband (frequency > 40 GHz) includes large spectral capacity, compact antenna structure, light equipment etc. Designing a compact low profile antenna using printed circuit technology such as microstrip antenna in this high frequency band for 58 GHz mobile backhaul links, 50 GHz WPAN, 50 GHz RLAN, 77 GHz collision avoidance radar, is a highly challenging task. In this paper a different type of planar antenna particularly meant for 50 GHz applications is proposed. One of the attractive motivation for this type of implementation is Rotman lens which was developed as a mircostrip type lens in 1967 by Archer and May bell at Raytheon [1]. Success of Rotman lens as a beam forming lens antenna fabricated on high-resistivity silicon (HRS) wafer for IEEE802.153c

R.A. Panda (✉) · S.N. Mishra · D. Mishra
Department of Electronics and Telecommunication Engineering,
V.S.S. University of Technology, Burla 768018, Odisha, India
e-mail: ribhupanda@gmail.com

S.N. Mishra
e-mail: susoveny@gmail.com

D. Mishra
e-mail: debasisuce@gmail.com

© Springer India 2016

507

S.C. Satapathy et al. (eds.), *Microelectronics, Electromagnetics and Telecommunications*, Lecture Notes in Electrical Engineering 372, DOI 10.1007/978-81-322-2728-1_47

50 GHz WPAN indicates possibility of different variant of lens structure suitable for same type of applications [2]. From optics it is known that conic section structure possesses property of excellent beam converging at focus. Depending on the eccentricity the elliptical surface is much more suitable to get multiple focusing points. In the present work, the original geometrical equation invented by Rotman and Turner in 1963 is used as a 2D patch antenna that can be easily fabricated [3]. The proposed antenna is designed and simulated using HFSS and CST microwave studio suite.

2 Mathematical Formulation of Patch Geometry

2.1 Overview of the Prototype Roman Lens Equations and Suggested Modifications

The original lens geometry parameters are based upon the positions of the bema ports, receiving ports and the transmission line lengths. These parameters depend on many other interdependent design parameters that affect the phase error performance. The design of the lens is governed by the Rotman-Turner design equations that are based on the geometry of the lens, shown in Fig. 1 [3]. The left contour of the lens is a circular are. Based on Gent's equations for optical path length equality the co-ordinates of the points of positions of the antenna ports on right inner 'array curve' are derived with respect to three perfect focal points (G, F1 and F2) [3]. The lower case letters represent their upper case variable normalized to the focal length F and w is the phase delay in wavelengths between the antenna port on inner lens contour $\Sigma 1$ and antenna on outer contour $\Sigma 2$. Some of the predefining parameters of the Rotman lens are the internal scan angle a, focal length F, distance of on axis focal length from origin G.

From Fig. 1 the two symmetrical off-axis focal points F1 and F2 have coordinates $(-F\cos \alpha, F\sin \alpha)$, $(-F\cos \alpha, -F\sin \alpha)$ and on-axis focal point G has $(-G,0)$ relative to point O_1 All parameters of Rotman lens are defined as $n = N/F$, $w = (W-W_0)/f$,

Fig. 1 Geometry and design parameters of a Rotman lens [3]

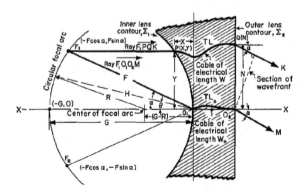

$x = X/F$, $y = Y/F$, $g = G/F$, $a_0 = \cos \alpha$, $b_0 = \sin \alpha$. Final design equations of the array curve of the lens can be summarized as

$$y = \eta(1 - w) \tag{1}$$

$$x^2 + y^2 + 2a_0 x = w^2 + b_0^2 \eta^2 - 2w \tag{2}$$

$$x^2 + y^2 + 2gx = w^2 - 2gw \tag{3}$$

To determine the value of 'w' we can solve the quadratic equation given as

$$Aw^2 + Bw + C = 0 \tag{4}$$

For fixed values of design parameters α and g, w can be computed as a function of η. The value of w from Eq. (1) when used in Eq. (3) the conic curves looks like

$$\frac{(x+p)^2}{a^2} + \frac{(y+q)^2}{b^2} = 1 \tag{5}$$

Which is the equation of an ellipse where

$$p = -g \tag{6}$$

$$q = \left(\frac{1}{\eta}\right) \frac{(g-1)}{1 - \frac{1}{\eta^2}} \tag{7}$$

$$a = (g-1)\left(\frac{\eta^2}{\eta^2 - 1}\right)^{1/2} \tag{8}$$

$$b = (g-1)\left(\frac{\eta 2}{\eta^2 - 1}\right) \tag{9}$$

In original lens design equations to avoid overall phase aberrations, the optimum value of "g" (ratio of on-axis to off-axis focal length G/F) is found from a simple relation given as a function of "α" [3].

$$g = 1 + (\alpha^2/2) \tag{10}$$

2.2   Feasibility of Eliptical Patch

To check the shape of the proposed patch, distance of on-axis focal-length from origin is first calculated from operating wavelength (G = 5 mm) [4]. The value of scan angle α is selected as 30° ($\alpha = \pi/3$) the value of g comes out to be 1.1 from (8)

Fig. 2 Layout of shape of elliptical patch using MATLAB

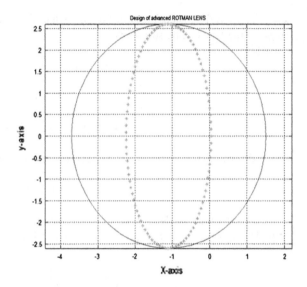

which gives a fine approximation to the optimum design [2, 3]. From value of ratio of focal length "g" and on axis focal length "G" the value of focal length "F" is calculated to be 4.545 mm. The two values of w is evaluated = 1.5016 and 0.0381 from previous relations. Using the equations values of all parameters and the co-ordinates of the outer contour are calculated and plotted using MATLAB, represented in Fig. 2.

3   Design of Perturbed Parabolic Patch Antenna

3.1   Selection of Substrate Material

The prime research in this design is selection of substrate material in the required frequency band (57–64 GHz) having minimum attenuation as low resistivity may cause attenuation. The relation between loss tangent and conductivity of a material is given as [5]

$$\tan \delta = (\sigma + \varepsilon')/\omega\varepsilon' \tag{11}$$

Symbols are self-explanatory. The attenuation "α" (dB/mm) of microstrip line on a wafer can be estimated from the following equation [2, 6].

$$\alpha = 8.686\pi \left[\frac{\varepsilon_{eff}-1}{\varepsilon_{r-1}}\right] \left[\frac{\varepsilon_r}{\sqrt{\varepsilon_{eff}}}\right] \tan \delta/\lambda g \qquad (12)$$

where λ_g is the guided wavelength (mm), ε_r is the actual dielectric constant of the substrate at the design frequency and ε_{eff} is the effective dielectric constant due to the microstrip and air dielectric above the substrate [7]. From (9) and (10) it is clear that attenuation is also inversely proportional to resistivity so high resistivity silicon (HRS) ($\rho \geq 10$ kΩ cm, $\varepsilon_r = 11.7$) has loss tangent (tan δ) is about 0.003 dB/mm in millimeter wave band, results very small attenuation [2, 8]. The second step of deign of patch antenna starts with consideration of thickness of the wafer as its increment may cause surface waves. As a solution HRS has dielectric constant approximately 12 this leads the substrate size to be small and thin [2]. The cutoff frequency (f_t) in GHz of a dielectric substrate as function of thickness is given by [2]

$$f_t = 150/h\pi \left[\frac{\sqrt{2}}{\sqrt{\varepsilon_{r-1}}}\right] \tan^{-1}\varepsilon_r \qquad (13)$$

Where "h" is thickness of substrate in millimeters. According to (11) the thickness of the HRS is selected to be h = 0.3 mm, which can support up to 100 GHz. As high dielectric constant may cause increase in surface waves an additional silicon dioxide (SiO$_2$) layer with dielectric constant ($\varepsilon_r = 3.9$) was developed upto 1.5 μm height [2]. The overall dimension of the substrate was kept 10 mm × 15 mm copper was deposited up to 8.52 μm on both sides of the substrate for both groundplane and patch formation Fig. 3

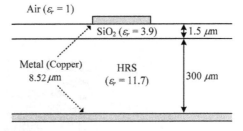

Fig. 3 Cross-section of substrate with ground plane and patch [2]

3.2 Design and Simulation Results of Elliptical Patch Antenna

The proposed planar perturbed antenna is designed and simulated using HFSS (High Frequency Structure Simulator) and CST (Computer Simulation Technology). The Feed type to patch antenna is selected as symmetrical microstrip line of (width = 0.4 mm) at the intersection point of axis of ellipse and circular arc Figs. 4 and 5.

From the simulation the resonant frequency of the designed antenna is found to be 50 GHz and the reflection co-efficient S_{11} at the corresponding frequency is −20.566 dB using HFSS and −25.1274 dB using CST. The S_{11} using HFSS is shown in Fig. 6 and S_{11} using CST is shown in Fig. 7.

Fig. 4 Planar Rotman lens antenna design using HFSS

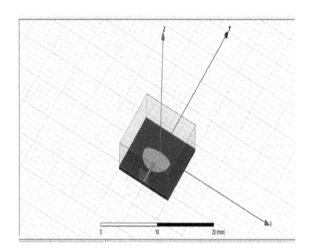

Fig. 5 Planar Rotman lens antenna design using CST

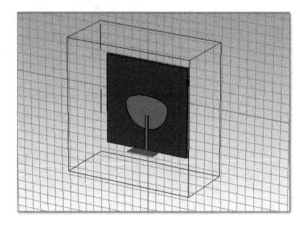

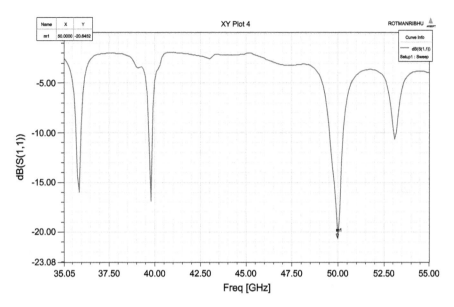

Fig. 6 S_{11} in dB using HFSS

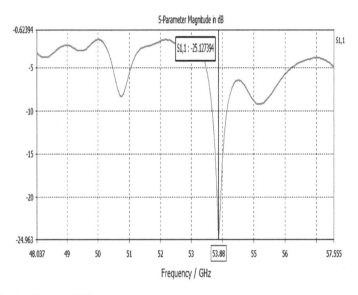

Fig. 7 S_{11} in dB using CST

In general the voltage standing wave ratio (VSWR) of an antenna should be ideally 1. For the designed antenna at resonant frequency the VSWR comes 1.61 using HFSS and 1.1173 using CST which is a very good agreement with the desired value. The VSWR is shown in Figs. 8 and 9.

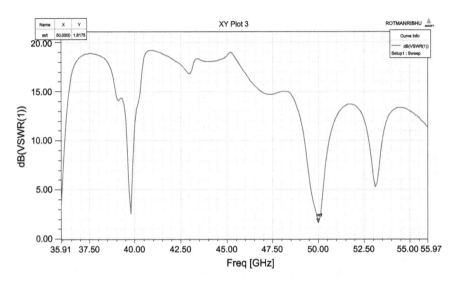

Fig. 8 VSWR using HFSS

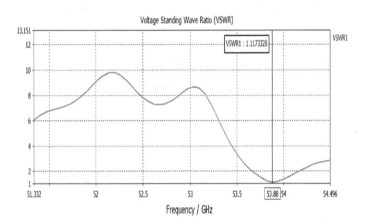

Fig. 9 VSWR using CST

The peak gain is found out to be 5.8 dB using HFSS and 4.039 dB using CST. The Peak Directivity is found out to be 8.51 dB using HFSS and 5.72 dB using CST. The radiation efficiency is found out to be 0.5381 which is 53.81 %. These are shown in following Figs. 10, 11, 12, 13, 14, 15 and Table 1.

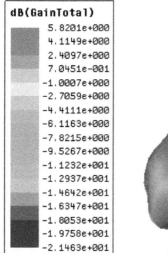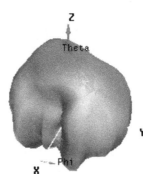

Fig. 10 3D gain using HFSS

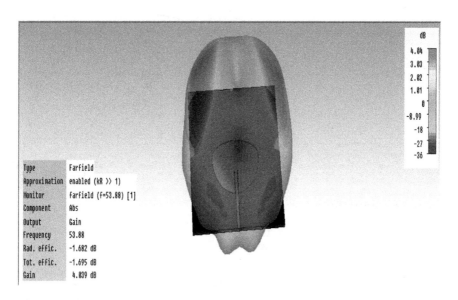

Fig. 11 3D gain using CST

Fig. 12 Peak gain using
HFSS

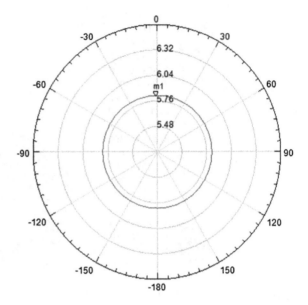

Fig. 13 Peak directivity
using HFSS

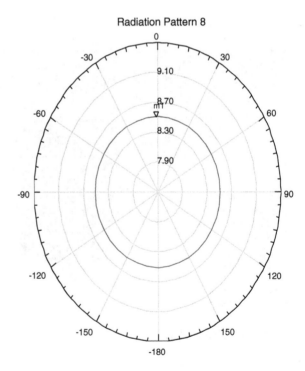

Fig. 14 Surface current distribution of proposed antenna

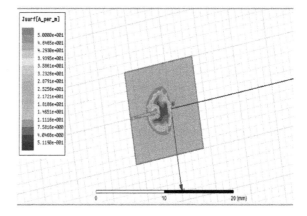

Fig. 15 Radiation efficiency using HFSS

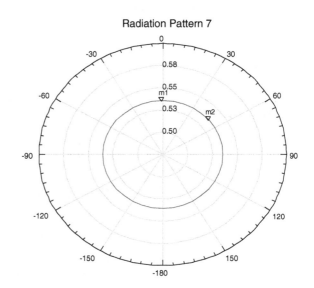

Table 1 Comparision of parameters measured in HFSS and CST

Parameters measured	HFSS	CST
Resonant frequency	50 GHz	53.88 GHz
VSWR	1.61	1.11
Directivity	8.51 dB	5.72 dB
Gain	5.82 dB	4.039 dB
S_{11} (return loss)	−20.6452 dB	−25.1274

4 Conclusion

A perturbed elliptical patch antenna is designed for 50 GHz WPAN application and simulated using HFSS and CST. In HFSS the resonant frequency was found out to be around 50 GHz and in CST it was found out to be 53.88 GHz which is in our desired range. In HFSS simulation the directivity was found out to be 8.51 dB with gain 5.82 dB which are very good values. With this value of directivity this proposed antenna can be said as a highly directive antenna.

5 Future Scope

To operate the proposed antenna in wide range of frequencies the log periodic implementation of the proposed patch can be done.

References

1. D.H. Archer, M.J. Maybell, Rotman lens development history at Raytheon electronic warfare systems 1967–1995. in *Proceedings of IEEE AP-S International Symposium*, vol. 2B (2005), pp. 31–34
2. W. Lee, J. Kim, C.S. Cho, Y.J. Yoon, Beam forming lens antenna on a high resistivity silicon wafer for 60 GHz WPAN. IEEE Trans. Antennas Propag. **58**, 706–713 (2010)
3. W. Rotman, R. Turner, Wide-angle microwave lens for line source applications. IEEE Trans. Antennas Propag. **11**, 623–632 (1963)
4. B. Diwedi, 60 GHz Perturbed Parabolic Patch Antenna, M.Tech Dissertation, VSSUT Odisha, Burla, 2013
5. D.M. Pozar, *Microwave Engineering*, 2nd edn. (Wiley, New York, 1996) ch. 1
6. E.J. Denlinger, Losses of microstrip lines. IEEE Trans. Microw. Theory Tech. **MTT-28**(6), 513– 522 (1980)
7. C.A. Balanis, *Antenna Theory: Analysis and Design*, 3rd edn. (Wiley, New York, 2005) ch. 14
8. R.C. Hansen, Design trades for rotman lenses. IEEE Trans. Antennas Propag. **39**, 464–472 (1991)

Design of an Analog Fault Model for CMOS Switched Capacitor Low Pass Filter

K. Babulu, R. Gurunadha and M. Saikumar

Abstract In this paper an analog fault model for a CMOS switched capacitor low pass filter is tested. The switched capacitor (SC) low pass filter circuit is modularized into functional macros. These functional macros include OPAMP, switches and capacitors. The circuit is identified as faulty if the frequency response of the transfer function does not meet the design specification. The signal flow graph (SFG) models of all the macros are analyzed to get the faulty transfer function of the circuit under test (CUT). A CMOS switched capacitor low pass filter for signal receiver applications is chosen as an example to demonstrate the testing of the analog fault model. The responses of the CUT are simulated using MENTOR GRAPHICS tool with 0.13 μm technology.

Keywords Switched capacitor circuit · Analog fault model · Dynamic range scaling and capacitor scaling

1 Introduction

Digital circuit fault model and analog circuit fault model are the two types of practical fault models and those are used to simplify the testing problems like quality and cost of the test. In this paper concentrate only on analog circuit fault model because it's a challenging research problem. The major practical issues in

K. Babulu (✉)
Department of ECE, JNTUK UCEK, Kakinada, A.P., India
e-mail: kapbbl@gmail.com

R. Gurunadha · M. Saikumar
Department of ECE, JNTUK UCEV, Vizianagaram, A.P., India
e-mail: gururavva@gmail.com

M. Saikumar
e-mail: saivenky25@gmail.com

© Springer India 2016
S.C. Satapathy et al. (eds.), *Microelectronics, Electromagnetics and Telecommunications*, Lecture Notes in Electrical Engineering 372,
DOI 10.1007/978-81-322-2728-1_48

519

developing this fault models are long fault simulation time and unverified fault assumptions [1].

Analog fault models are of two types parametric faults and catastrophic faults. Parametric fault occurs due to change in the parameter of the component and catastrophic fault occurs due to short or opens wire. Most of the CMOS analog circuits are implemented by using switched capacitor (SC) circuits [2]. Reference [3] proposed a static linear behavior (SLB) analog fault model for linear and time invariant sampled SC circuits. In this SLB fault model considered as within every clock cycle the transient responses of the circuit under test (CUT) are fully settled. Hence a fixed z-domain transfer function of CUT is taken. The SLB fault model decides circuit has a faulty one by comparing the design parameters with the frequency responses of the tested z-domain transfer function whose frequency responses are obtained from the test results of the circuit [4]. From the circuit of reference [3] inject some faults of catastrophic and parametric and to verify the assumption of the SLB analog fault model. Finally compare the test results with the estimated transfer function. If they fit each other, then it justifies no fault in CUT of the SLB fault model.

Fifth order low pass filter using SC circuit is chosen as an example to demonstrate the testing of analog fault model for signal receiver applications [5]. To pass the required signal in receiver circuits a SC filters are generally used and depending on the capacitor ratios in SC circuits only the base band channel selection with accurate bandwidth is measured.

2   Static Linear Behavior Fault Model

SLB fault model is an efficient fault model to test the SC circuits. The transient response of the CUT can be obtained by using SLB fault model. The transfer function of the CUT is useful for examining the accuracy of the model. The important parameters that effect the faultiness of the CUT are OP-AMPS, Capacitors and Switches [6]. The schematic circuit diagram for switched capacitor biquad is shown in Fig. 1

The specifications of the circuit are pass band from 0 to 20 kHz, clock frequency of 2.5 MHz and capacitance's C_D is 11.47096pF and C_B is 16.36952. By using the signal flow graph analysis calculate the overall input and output relationships of the circuit under test [3, 6–8].

$$V_{02}(z) = STF_{H_2}(z)V_i(z) + OTG_{12}V_{OS_1} + OTG_{22}V_{OS_2} \tag{1}$$

Equation (1) contains the signal transfer function (STF) and basic blocks of the SC biquad filter while the OTG's refer to the offset transfer gains of the corresponding stages of the biquad. By considering Eq. (1) and SFG models, the design parameters of the STF of the biquad filter is,

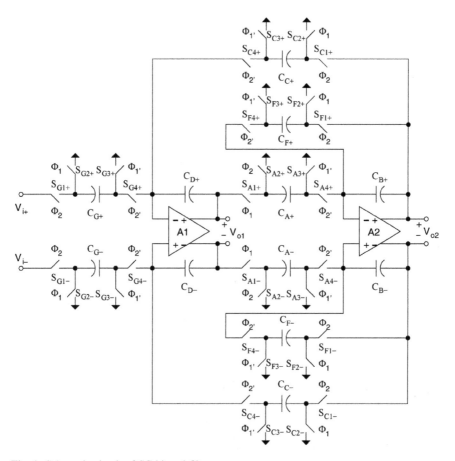

Fig. 1 Schematic circuit of SC biquad filter

$$V_{02}(Z) = (-(C_A/C_B)(C_G/C_D)Z^{-1}V_i(z))/DEN(z) \qquad (2)$$

where the denominator term is expressed as:

$$DEN(z) = (1+(C_F/C_B)) - (2-(C_AC_C/C_BC_D)+(C_F/C_B))\,z^{-1}+z^{-2} \qquad (3)$$

Assuming that the fault free Op-Amps have zeroed offsets. There are different kinds of faults can be injected in the biquad. From the Eqs. (2) and (3) the faults in the CUT changes the values of the coefficients of the signal transfer function which alter nothing but the capacitor ratios. Hence it is important to maintain the capacitor ratio instead of the absolute capacitance values. Capacitors may introduce a parametric fault. Parametric faults in Op-Amps include its open loop gain (OPG) and input referred offset while the catastrophic faults in Op-Amps are fatal due to its sensitivity in the design. These faults are covered by the parametric faults itself. The

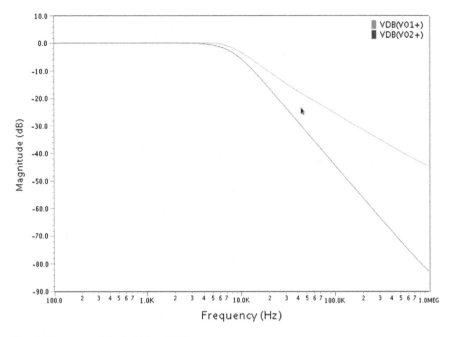

Fig. 2 Response of the fault free CUT

open or short faults of capacitors are equivalent to zero infinite capacitance values. Faults due to switches are mostly same as in digital circuits i.e., stuck-at faults.

The fault free response of the CUT is simulated using MENTOR GRAPHICS tool with 0.13 μm technology. The result is as shown in the Fig. 2. Here the simulated waveform of 80 dB Op-Amp OPG. In this simulation obtain the output at both the output nodes in the two stage SC biquad filter.

As seen from Fig. 2 outputs V_{02} and V_{01} illustrating outputs at first and second stages. Using EZ Wave tool in MENTOR GRAPHICS, This response can be used to verify the faultiness of the circuit just by comparing the results of the faulty and fault free CUT's.

3   Fault Injection to the Cut

The Fleischer-Laker SC biquad is used to realize a Butterworth low pass filter whose pass band is 20 kHz with a sampling frequency of 2.5 MHz and stop band attenuation is 30 dB with stop band frequency of 400 kHz. These are the design specification of this filter [9].

Table 1 Capacitor ratios and values of the CUT

Parameter	Fixed value (pF)	Parameter	Fixed value (pF)
C_A	1.288	C_H	0.184
C_B	6.348	C_I	0.023
C_C	0.460	C_J	0.046
C_D	6.348	C_{Apf}	2.576
$C_E = C_{E1} + C_{E2}$	5.152	C_{E1}, C_{E2}	2.576
C_G	0.644	C_{Gpf}	1.288

The SLB analog fault model covers both parametric and catastrophic faults. Parametric faults are the results of parameter deviations of components due to process, voltage, and temperature variations. Capacitance values deviations are typical parametric faults in SC circuits. Hence, add two differential capacitor pairs C_{Apf} and C_{Gpf} to the biquad and decompose C_E into two capacitors C_{E1} and C_{E2}. By issuing the control signals A_P and G_P, the total capacitance of the corresponding capacitors are enlarged respectively to model practical parametric faults in a controlled manner.

Similarly, activating the control signal E_P will reduce the effective capacitance of C_E by C_{E1} and C_{E2} to model another kind of parametric fault. The capacitance values of these capacitors are listed in Table 1. The SLB fault model assumes only the delay-free and delayed SC branches may suffer from catastrophic faults.

To verify the deductions of the SLB fault model about the catastrophic faults, add three switches including SGs, SHs, and SGo to the design. The aspect ratios of the switches SGs and SHs are designed to be much larger than those of the switches SGn and SHn which are in parallel with them. By keeping SGs or SHs turn-on, a short fault is injected to the corresponding switch. The design allows injecting a stuck-open fault to the biquad, too. Fifth Order Switched Capacitor Low Pass Filter (Fig. 3).

The schematic circuit diagram of fifth order SC low pass filter is shown in Fig. 4. To design the higher order low pass filter, it consists of first and two second order circuits and the transfer function can be obtained from the second order continuous time low pass filter by using SFG analysis.

Use extra switches to understand the parasitic insensitive integrator. Although in biquad the number of switches is increased due to the immunity of parasitic capacitance. To reduce the number of switches in circuit we use switch sharing technique. At the same time the capacitors C_1 and C_2 connected to virtual ground and true ground are shown in Fig. 4. One pair of these switches can be eliminated here due to virtual ground and true ground. Observe the voltages at output terminal of the OPAMP that should be same magnitude in the frequency range. So dynamic range scaling is used to equalize the dynamic range of each OPAMP output. The capacitance value of the capacitors of the circuit is listed in Table 2.

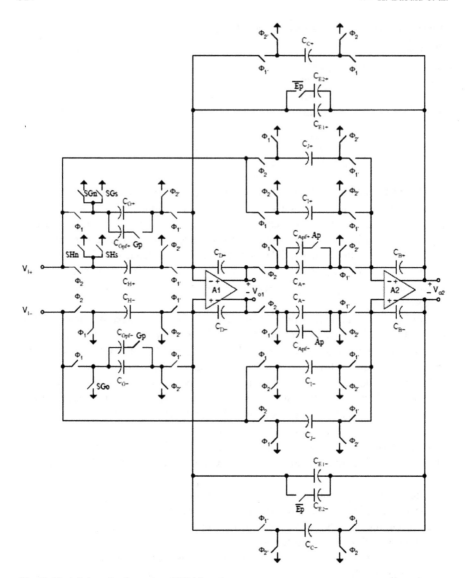

Fig. 3 Fault injected schematic of SC biquad

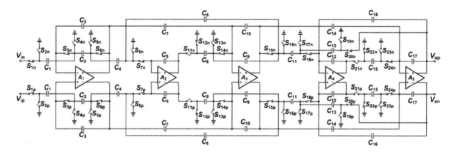

Fig. 4 Schematic circuit of fifth order low pass filter

Table 2 Capacitor values of the fifth order SC low pass filter

Capacitor (pF)	Minimum capacitor scaling	Capacitor (pF)	Minimum capacitor scaling
C_1	0.27	C_{10}	2.03
C_2	0.65	C_{11}	1.69
C_3	2.03	C_{12}	4.03
C_4	1	C_{13}	1.84
C_5	0.23	C_{14}	0.54
C_6	1	C_{15}	1.84
C_7	0.65	C_{16}	0.65
C_8	1	C_{17}	4.03
C_9	0.78	$C_{max/min}$	6.25

4 Simulated Results

If the test responses fit in the frequency responses of the faultless CUT, it justifies
the basic assumption of the SLB fault model. All the circuit simulations are done
using Mentor Graphics with 0.13 μm technology EDA tools.

Figure 5 shows the simulation results of the CUT when the parametric fault of
C_{Gpf} is injected. This parametric fault makes the pass band gain and the stop band
attenuation of the CUT out of the design specification. Even though the CUT fails
in the test, the estimated TF still accurately depicts the faulty behavior of the CUT.
Figure 6 illustrate the simulation result of the CUT after enabling the parametric
fault control signal E_p, respectively. The enlarged leads to too small minimum stop
band attenuation, while the CUT passes the test with the reduced CE. Similar to the
previous case, the estimated TF well predicts the frequency responses of the CUT in
both cases no matter the CUT passes or fails. Figure 7 depicts the experimental
results of the CUT when the catastrophic fault activated. Both catastrophic faults
fail the CUT in the test. Yet the estimated TFs successfully depict the faulty

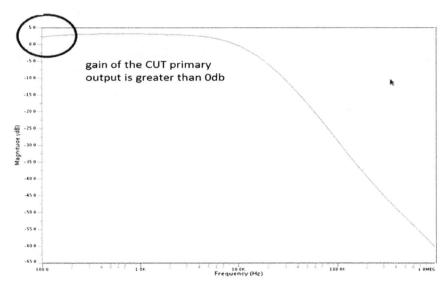

Fig. 5 Experimental result after injecting parametric fault C_{Gpf}

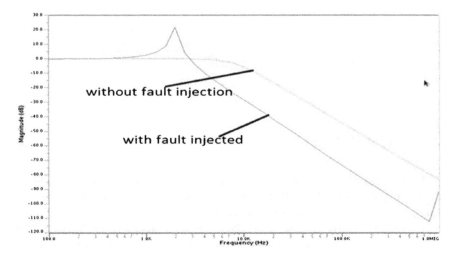

Fig. 6 Experimental result after injecting parametric fault E_p

frequency responses of the CUT. Figure 8 shows the frequency response of fifth order low pass filter, from that the pass band frequency of 8 MHz with 80 MHz clock frequency at the supply voltage of 1.8 V and stop band attenuation is higher than 40 dB.

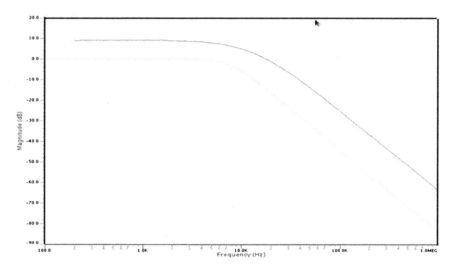

Fig. 7 Experimental result when switch SHs is short

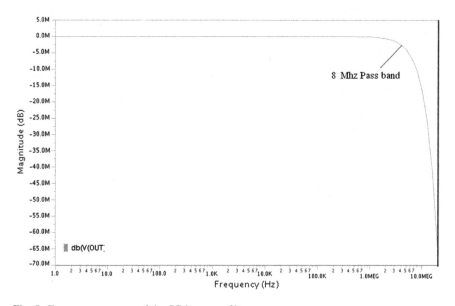

Fig. 8 Frequency response of the SC low pass filter

5 Conclusion

In this paper the SLB analog fault model for linear SC circuits have been verified
and SC low pass biquad filter is taken as an example to demonstrate the effectiveness of the SLB fault model. Conduct simulations with other stimulus tones and

check if the test responses fit in those predicted by the retrieved TF. Experimental results verify that the fixed TF template assumption holds for all the faults that are injected to the biquad filter circuit. Taking the fifth order SC low pass filter circuit for verifying signal receiver applications. When the low-pass filter is biased at 1.8 V, the proposed SC low-pass filter achieves a pass-band frequency of 8 MHz the presented filter has characteristics suitable for signal processing in a signal receiver circuit. Extending the fault model to include the timing related faults would be an interesting topic for the future research work.

References

1. L.-T. Wang, C.-W. Wu, X.-G. Wen, VLSI test Principles and Architectures. 500 Sansome St., Suite 400, San Francisco, CA 94111: Morgan Kaufmann Publishers (2006)
2. M. Soma, V. Kolarik, A design-for-test technique for switched capacitor filters. in *Proceedings of IEEE VLSI Test Symposium* (1994), pp. 42–47
3. H.C. Hong, A static linear behavior analog fault model for switched capacitor circuits. IEEE Trans. Comput. Aided Des. Integr. Circ. Syst. **31**(4), 597–609 (2012)
4. N. Nagi, A. Chatterjee, A. Balivada, J.A. Abraham, Fault-based automatic test generator for linear analog circuits. in *Proceedings of IEEE/ACM International Conference Computer-Aided Design (ICCAD)* (1993), pp. 88–91
5. U.-K. Moon, CMOS high-frequency switched-capacitor filters for telecommunication applications, IEEE J. Solid-State Circuits **35**(2), 212–220 (2000)
6. A. Petraglia, J. Canive, M. Petraglia, Efficient parametric fault detection in switched-capacitor filters. IEEE Des. Test Comput. **23**(1), 58–66 (2006)
7. L.-Y. Lin, H.-C. Hong, Design of a fault-injectable fleischer-laker switched capacitor biquad for verifying the static linear behavior fault model. in *22nd Asian Test Symposium* (2013)
8. M. Fino, J. Franca, A. Steiger-Garcao, Automatic symbolic analysis of switched-capacitor filtering networks using signal flow graphs. IEEE Trans. Comput.-Aided Des. Integr. Circuits Syst. **14**(7), 858–867 (1995)
9. S.F. Hung, L.Y. Lin, H.-C. Hong, A study on the design of a testable fleisher-laker switched-capacitor biquad. in *Proceedings of International Mixed-Signals, Sensors, and System Testing Workshop (IMS3TW)* (2012), pp. 119–122

Image Authentication Using Local Binary Pattern on the Low Frequency Components

Ch. Srinivasa Rao and S.B.G. Tilak Babu

Abstract Detection of copy move forgery in images is helpful in legal evidence, in forensic investigation and many other fields. Many Copy Move Forgery Detection (CMFD) schemes are existing in the literature. However, most of them fail to withstand post-processing operations viz., JPEG Compression, noise contamination, rotation. Even if able to identify, they consumes much time to detect and locate. In this paper, a technique is proposed which uses Discrete Wavelet Transform (DWT) and Local Binary Pattern (LBP) to identify copy-move forgery. Features are extracted by using LBP on the LL band obtained by applying DWT on the input image. Proper selection of similarity and distance thresholds can localize the forged region correctly.

Keywords Copy move forgery detection · Discrete wavelet transform · Local binary pattern

1 Introduction

Sophisticated digitized cameras and user friendly photo editing tools made it comparatively easy for creating digital forgeries. The integrity of digital images has important role in so many fields; criminal investigation, forensic investigation, journalism, surveillance systems, medical imaging and intelligence services. The image forgeries can be detected in two approaches: Active and Passive. The active approach of forgery identification requires preprocessing of original media data before its usage or at the time of creation. The second approach passive does not

Ch.Srinivasa Rao (✉) · S.B.G. Tilak Babu
Department of ECE, JNTUK-UCEV, Vizianagaram 535003
Andhra Pradesh, India
e-mail: chsrao.ece@jntukucev.ac.in

S.B.G. Tilak Babu
e-mail: thilaksayila@gmail.com

© Springer India 2016
S.C. Satapathy et al. (eds.), *Microelectronics, Electromagnetics and Telecommunications*, Lecture Notes in Electrical Engineering 372,
DOI 10.1007/978-81-322-2728-1_49

Fig. 1 Example for copy move forgery: **a** the original image, **b** the forged image and **c** forged region marked image (from *left* to *right* **a–c**)

require any preprocessing of the digital data or information and it relies on the image statistical features.

However, a Passive approach does not require any pre-processing of the digital data or information and it relies on the image statistical features. Copy-Move forgery in digital image tampering is that, some part of the image which is copied and pasted onto some other part of the same image. The intention is to hide some region or an object in that image. In cloning (also called copy-move forgery), as the copied part is from the same image, the color palette, the texture, noise components, intensity variations and most of the other properties will be compatible with the rest of the image. Hence, it becomes difficult for human visual system to identify the forgery. Exhaustive search method of copy-move detection is time consuming and computationally very complex. An example of Copy-move forgery [1] can be seen in Fig. 1.

In DWT and Singular Value Decomposition (SVD) are used to identify forgery [2]. The image is decomposed using DWT, SVD is applied on decomposed image and the speed is increased because of reduction of number of block. Khan et al. [3], Ghorbani et al. [4] used discrete wavelet transform to decrease complexity of CMFD and the duplicated blocks are identified with the similarity condition of phase correlation. Yang et al. [5] systematically combined discrete wavelet transform and discrete cosine transform to identify the forged region. Even though the method has the ability to identify various copy move forgeries, it consumes much time. Li et al. proposed an algorithm [6]. In this, the image overlapping blocks are represented by Local Binary Pattern, and a lexicographic sort is done to identify the tampered regions based on the textural similarity between the regions. As LBP is to be applied for the entire testing image which results in more number of overlapping blocks, thus affecting the computation complexity. More or less all these techniques suffers from computational complexity and common signal processing attacks like JPEG compression, AWGN, contrast, rotation and flipping, etc. The key idea of our method is to apply LBP on the LL band of the testing image, so that the number of overlapping blocks can be reduced and in turn the computational complexity.

In the proposed algorithm, the image is first decomposed through DWT into different sub bands and then the low frequency component of the image is converted into overlapping blocks. Each block is applied to LBP and LBP histograms are set to different bins according to their values. The paper is organized as follows: The proposed algorithm and its flow diagram is presented in Sect. 2. The experimental outcomes are presented in Sect. 3 and the conclusions are discussed in Sect. 4.

2 Proposed Algorithm

The important part of CMFD algorithm is detecting duplicated regions for unknown size and shape of the forged regions. This problem is addressed in this algorithm by applying DWT on the test image and LBP operator is applied on the overlapping blocks of fixed size for LL band to obtain texture feature. Figure 2 represents the flow diagram for proposed algorithm.

Initially, the test image C of M * N size is converted to gray scale image I of size M * N by the following Eq. (1). In Eq. (1) R represents red, G represents Green, B represents Blue.

$$I = 0.2989 * R + 0.5870 * G + 0.1140 * B \tag{1}$$

2.1 Discrete Wavelet Transform

Discrete wavelet transform is a multilevel decomposition technique of an image, which results in both spatial and frequency details [7]. In present algorithm, the image is decomposed with the help of discrete wavelet transform into four sub bands, which are HL, HH, LH and LL, here LL is low frequency part of the image

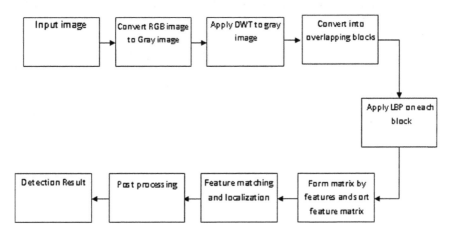

Fig. 2 flow diagram for proposed algorithm

and finest scale wavelet coefficient are represented by HL, HH, LH. The sub band LL contains most of the energy whose size is reduced to 3/4 of the given image. Then the LL band is converted into overlapping blocks and then applied to LBP to extract the features of all overlapping blocks.

In this method, only one level (j = 1) is considered, so that LL band size can be given as K = M/2, L = N/2 and is represented as LL_{K*L}. In further process of this algorithm, only LL band is used which itself indicates the reduced computation effort up to 3/4 of the actual process.

2.2 Dividing Image to Fixed Size Overlapping Blocks

Proposed algorithm is basically a block-based method. In this the image is divided into overlapping blocks of fixed size x * x pixels. Here, the block size is chosen in such a way that its size is smaller than the forged area which to be identified. The fixed size blocks are slide along the row and column direction by one pixel difference. As LL_{K*L} band is divided into overlapping blocks the sliding process will generate (K − x + 1) * (L − x + 1) number of overlapping block.

2.3 Local Binary Pattern

Local binary pattern is a kind of texture operator for gray scale image which is used for describing image texture in the spatial structure [8].

To handle textures at various scales, different sizes of local neighborhoods are available The notation (P, R) is used to indicate pixel neighborhood which means P number of sampling points on a circle of radius of R, shows that there are several uniform patterns of a rotated version though in the case of P = 8. So, a new pattern can be defined as

$$LBP_{P,R}^{riu2} = \begin{cases} \sum_{p=0}^{P-1} s(g_p - g_c) & \text{if } U(LBP_{P,R}^{riu2}) \leq 2 \\ P+1 & \text{otherwise} \end{cases} \tag{2}$$

In Eq. (2) $U(LBP_{P,R}^{riu2})$ represents number of spatial transitions and is given by Eq. (3)

$$U(LBP_{P,R}) = |s(g_{P-1} - g_c) - s(g_0 - g_c)| + \sum_{p=1}^{P-1} |s(g_p - g_c) - s(g_{p-1} - g_c)|$$

(3)

In Eq. (2) $s(g_p - g_c)$ represents sign of difference between the neighbor pixel gray value and centre pixel gray value.

Where $LBP_{P,R}^{u2}$ have P + 2 independent output values. The histogram for image K * L, after operated with $LBP_{P,R}^{u2}$ is given below

$$H^{riu2}(m) = \sum_{i=1}^{K} \sum_{j=1}^{L} f(LBP_{P,R}^{u2}(i,j), m)$$

(4)

Here $m \in [1, M]$

$$f(x, y) = \begin{array}{ll} 1 & if \quad x = y \\ 0 & otherwise \end{array}$$

(5)

where M is the number of $LBP_{P,R}^{u2}$ independent values P + 2. In the proposed method, we employed the operator in the Eq. (3) as our feature vector of the image blocks.

The rotation invariant features are extracted from the overlapping blocks of fixed size by applying rotation invariant uniform LBP. In this work, LBP is applied on the LL and (K − x + 1) * (L − x + 1) number of features can be extracted from it arrange the features of each block into a row array and set all the rows (K − x + 1) * (L − x + 1) as a matrix S and sort the features using lexicographical sorting.

2.4 Feature Matching

All the features are lexicographically sorted and are arranged in matrix. It is obvious that similar blocks should have similar features which indicate that rows of the matrix are to be compared to know the similarity. As the features are lexicographically sorted, there is no need to compare all the rows of the matrix, only r range of rows are considered for similarity measure. As well, only the blocks with distance larger than are compared. In this work, the search corresponding to the blocks is made by calculating the Euclidean distances between the feature vectors. To detect the forged region correctly, the distance threshold T_d, similarity threshold

T_s and are predetermined. The matching of the features starts from the first row of the matrix S. For a feature located in the ith row S_i, the distance with the following r rows will be computed and the smallest distance denoted by $D(i, \beta)$, between the ith row and the following βth can be obtained [6].

2.5 *Post-processing and Localization*

All the matched block pairs are available in set σ, the forged and the copied regions can be identified by marking the regions. In order to achieve this, the matched locations are stamped on a binary image to generate detection map. In our proposed method, we mark the binary image with gray intensity of 255 of size P. There are some falsely detected blocks stamped on the initial detection map. These false detected blocks can be removed by performing morphological operations like erosion and dilation on the initial detection map.

3 Experiment Results

For experimentation, the authenticated and forged images are taken from the standard database TIDED v.2.0 generated by Institute of Automation at Chinese Academy of Sciences CASIA [9]. The copy-move forgery and various post-processing attacks like rotation, flipping, JPEG compression are performed using Adobe photo shop tool. All the simulations are carried with MATLAB 2013 on a personal computer with 2.4 GHz Core i3 processor and 4 GB RAM. The over-lapping block size is varied from 12 to 18 and the neighborhood of size is selected 16 and 24 with radius 2 and 3 respectively. The threshold of similarity is more than 3 is selected and threshold of distance is selected more than the overlapping block size. The experiment results aredisplayed in (Table 1).

3.1 *Robustness of the Method*

The existing algorithms are useful only for general copy move forgery and some rotation angle of forged regions but not for attacks of JPEG compression, noise contamination and blurring. The proposed work works even the forged region JPEG compressed, noise contaminated or blurred. The robust results showed in Table 2.

Table 1 Results for different forged images

Attack type	Forged image	Initial detected forged region	Final forged region
Copy-move			
Rotate right 90°			
Rotate left 90°			
Horizontal flip			
Vertical flip			
Rotation 180°			

3.2 Comparative Analyses

Comparison of the three approaches in terms of number of overlapping blocks and feature dimension for a gray level image block of 512 * 512 dimension with overlapping block size 18 * 18. The results and comparative analyses show the computational complexity decreases 3/4 times of remaining two approaches. The comparative analyses displayed in Table 3.

Table 2 Results for JPEG compressed, noise added and blur forged images

Attack type	Forged image	Initial detected forged region	Final detected forged region
JPEG compression quality 90°			
JPEG Compression quality 80°			
JPEG Compression quality 70°			
JPEG Compression quality 60°			
Salt and pepper noise			
Motion blur			

Table 3 Comparative results

Algorithm	Feature representation	Number of overlapping blocks	Feature dimension for a block
Yang et al. [5]	DWT and FWHT	245,025	26
Leida [6]	LBP (24,3)	245,025	26
Proposed	DWT and LBP (24,3)	57,121	26

4 Conclusion

Many algorithms are existing to identify copy move forgery, some of them are unable to address the attack rotated or flipped before being pasted. Even others could address, the computational effort is more. An approach depends on DWT and

LBP is proposed to identify copy-move forgery. The features are extracted by using LBP on the LL band obtained by applying DWT on the input image. It has been observed that, proper selection of similarity and distance thresholds can localize the forged region even if the copied portion is JPEG compressed, noise contaminated, blurred, region rotated and flipped. Standard image database CASIA TIDED v.2.0 [9] is used to evaluate the performance of proposed approach.

References

1. Y. Cao, T. Gao, L. Fan, Q. Yang, A robust detection algorithm for copy-move forgery in digital images. Forensic Sci. Int. **214**(2012), 33–43 (2012)
2. G. Li, Q. Wu, D. Tu, S. Sun, A sorted neighborhood approach for detecting duplicated regions in image forgeries based on DWT and SVD, in *Proceedings of IEEE International Conference on Multimedia and Expo.* (2007), pp. 1750–1753
3. S. Khan, A. Kulkarni, An efficient method for detection of copy-move forgery using discrete wavelet transform. Int. J. Comput. Sci. Eng. **2**(5), 1801–1806 (2010)
4. M. Ghorbani, M. Firouzmand, A. Faraahi, DWT-DCT (QCD) based copy move image forgery detection, In *18th International Conference on Systems, Signals and Image Processing (IWSSIP 2011)* Sarajevo (2011), pp. 1–4
5. B. Yang, X. Sun, X. Chen, J. Zhang, X. Li, An efficient forensic method for copy-move forgery detection based on DWT-FWHT. Radio Eng. **22**(4),(2013)
6. L. Li, S. Li, H. Zhu, An efficient scheme for detecting copy-move forged images by local binary patterns. J. Inf. Hiding Multimedia Signal Process. Ubiquitous Int. **4**(1), (2013)
7. G. Amara, An introduction to wavelets. IEEE Comput. Sci. Eng. **2**(2), 50–61 (1992)
8. T. Ojala, M. Pietikainen, T. Maenpaa, Multiresolution gray-scale and rotation invariant texture classification with local binary patterns. IEEE Trans. Pattern Anal. Mach. Intell. **24**(7), 971–987 (2002)
9. CASIA, Image tampering detection evaluation database (2010), http://forensics.idealtest.org

An Adaptive Filter Approach
for GPS Multipath Error Estimation
and Mitigation

N. Swathi, V.B.S.S. Indira Dutt and G. Sasibhushana Rao

Abstract The positional accuracy of Global Positioning System (GPS) is affected by several errors such as delay due to atmosphere, satellite-receiver geometry, receiver clock error and multipath error. Along with the atmospheric errors ($\approx$10–40 m), multipath can cause error ($\approx$2–4 m) in the ranging measurements of the GPS receiver which degrades the positional accuracy. Since, the reception of multipath can create a significant distortion to the shape of the correlation function leading to an error in the receiver position estimate, Multipath is undesirable. The multipath disturbance is largely dependent on the receiver environment since satellite signals can arrive at the receiver via multiple paths, due to reflections from nearby objects such as trees, buildings, vehicles, etc. Although the multipath effect can be reduced by choosing sites without multipath reflectors or by using choke-ring antennas to mitigate the reflected signal, but it is difficult to eliminate all multipath effects from GPS observations. By using data processing schemes such as different adaptive filters, the effect of multipath error can be minimized to centimeter level. Estimation of the effect of multipath interference at the receiver antenna is the objective of this work for which both code range and carrier phase measurements are considered. Along with multipath error, ionospheric delay is also estimated. By using a dual frequency GPS receiver code ranges and carrier phases are extracted and corresponding differences called Code Carrier Difference (CCD) is performed, which results in cancellation of all effects except multipath and measurement noise. Least Mean Square (LMS), Normalized Least Mean Square (NLMS) and Recursive Least Squares (RLS) adaptive filters are considered to mitigate the multipath error. In order to carry out this work, dual frequency data of an IGS station (NGRI, Hyderabad, Lat:/Long:78°33′4″E/17°24′39″N) is collected

N. Swathi (✉) · V.B.S.S.I. Dutt
Department of ECE, GIT, GU, Visakhapatnam, India
e-mail: nadipineniswathi@gmail.com

V.B.S.S.I. Dutt
e-mail: srilatha06.vemuri@gmail.com

G. Sasibhushana Rao
Department of ECE, AUCE (A), Visakhapatnam, India

© Springer India 2016
S.C. Satapathy et al. (eds.), *Microelectronics, Electromagnetics
and Telecommunications*, Lecture Notes in Electrical Engineering 372,
DOI 10.1007/978-81-322-2728-1_50

from the website of Scripps Orbit and Permanent Array Centre (SOPAC) for the entire day of 11th September 2014.

Keywords Carrier phase · Code range · GPS observables · Receiver position estimate

1 Introduction

Global Positioning System is the satellite based navigation system which works on the principle of trilateration. Trilateration is method of finding the receiver position using the ranging information from three satellites. In GPS, distance between the satellite and the receiver is measured by Time of Arrival (TOA) of the GPS signal at the receiver [1]. So, if the signal travel time from the satellite to the receiver is known, the distance from the satellite to the receiver can be determined precisely. As the positional accuracy depends on the accuracy of the ranging measurements, it is necessary to receive the signal without delay due to the atmosphere and multipath [2].

There are several prominent methods available in literature to estimate and mitigate the multipath effect. Data processing schemes are used in most of the mitigation techniques. Carrier smoothing is one among those techniques [3]. The advantage of this technique is carrier phase measurement error is typically negligible compared to code multipath. Perfect combination of code and carrier phase measurements can efficiently reduce the code multipath. In this paper, the estimation and mitigation of multipath error is implemented based on data processing with software methods using predominant Least Mean Square (LMS), Normalized Least Mean Square (NLMS), and Recursive Least Squares (RLS) adaption algorithms.

2 Estimation of GPS Multipath Error

The GPS system provides two ranging measurements namely code range measurements and carrier phase measurements to provide the positioning information [4].

The code range observable on L1 frequency (1575.42 MHz) is expressed as Eq. 1

$$\rho_1 = P + c(dt - dT) + d_{ion} + d_{tro} + MP_{\rho L1} + \varepsilon_{\rho L1} \tag{1}$$

The carrier phase observable on L1 frequency is expressed as Eq. 2

$$\phi_1 = P + c(dt - dT) - \lambda_1 N_1 - d_{ion} + d_{tro} + MP_{\phi L1} + \varepsilon_{\phi L1} \tag{2}$$

where ρ_1 is measured pseudo range, ϕ_1 is carrier phase, P is geometric range, c is velocity of light, dt is receiver clock error, dT is satellite clock error, 'd_{ion}' is ionospheric error, 'd_{tro}' is tropospheric error, $MP_{\rho L1}$, $\varepsilon_{\rho L1}$ and $\lambda_1 N_1$ are multipath error(meters), measurement noise(meters) and integer ambiguity respectively on L1 frequency. Carrier phase measurements are represented with subscript ϕ, of which $MP_{\phi L1}$ and $\varepsilon_{\phi L1}$ are assumed to be small and negligible.

Except ionospheric refraction and multipath error all the other errors are independent of frequency and influences code and carrier phases by the same amount [5]. By using a dual frequency GPS receiver code ranges and carrier phases are extracted and corresponding differences called Code Carrier Difference (CCD) is performed, which results in cancellation of all effects except multipath and measurement noise [6]. Hence the CCD on L1 frequency is represented by Eq. 3

$$CCD = \rho_1 - \phi_1 + K1 \cong 2d_{ion} + MP_{\rho L1} + \varepsilon_{\rho L1} \tag{3}$$

The multipath and measurement noise on L1 carrier frequency is given as,

$$MP_{\rho L1} + \varepsilon_{\rho L1} \cong \rho_1 - \phi_1 - 2d_{ion} + K1 \tag{4}$$

The constant K1 is due to the integer ambiguity.

The ionospheric delay on L1 frequency by using dual frequency receiver data can be estimated as

$$d_{ion} = \frac{f_{L2}^2}{f_{L1}^2 - f_{L2}^2} (\phi_{L1} - \phi_{L2}) \tag{5}$$

By substituting d_{ion} in Eq. 4, code multipath error (including measurement noise) on L1, i.e., MP_{L1} is given as [7]:

$$MP_{L1} = \rho_{L1} - \frac{f_{L1}^2 + f_{L2}^2}{f_{L1}^2 - f_{L2}^2} (\phi_{L1}) + \frac{2f_{L2}^2}{f_{L1}^2 - f_{L2}^2} (\phi_{L2}) + K1 \tag{6}$$

Similarly, the code multipath error (including measurement noise) on L2 can be given as:

$$MP_{L2} = \rho_{L2} + \frac{f_{L1}^2 + f_{L2}^2}{f_{L1}^2 - f_{L2}^2} (\phi_{L2}) - \frac{2f_{L1}^2}{f_{L1}^2 - f_{L2}^2} (\phi_{L1}) + K2 \tag{7}$$

By using above equations, for all the samples of dual frequency GPS data the multipath error on L1 (1575.42 MHz) and L2 (1227.60 MHz) frequencies can be estimated. K1 and K2 are functions of unknown integer ambiguities and measurement noise, which can be assumed constant if there is no cycle slip in the carrier phase data [8].

The Multipath error on L1 and L2 is estimated for all the samples of dual frequency GPS data. GPS data was obtained from the archives of Scripps Orbit and Permanent Array Center (SOPAC) for the entire day of 11th September 2014 at 30 s interval. By using MATLAB, the multipath error on L1 and L2 carriers is

Fig. 1 Multipath error for
SVPRN 7 on carrier
frequencies L1 and L2

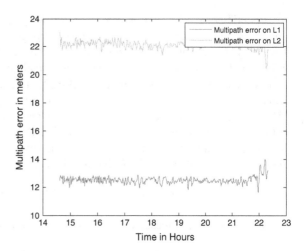

calculated for individual satellite vehicle numbers and is plotted for Satellite
Vehicle SV7.

Figure 1 shows the multipath error estimated using the ranging measurements of
Satellite Vehicle Pseudo Random Number (SVPRN) 7. The multipath error
observed on L1 carrier frequency ranges from 11.60557 to 13.98202 m and on L2
carrier frequency it ranges from 20.44175 to 23.047192 m. The results obtained are
promising and it is required to minimize the obtained multipath error using LMS
and NLMS adaptive filters.

3 Mitigation of Multipath Error with Adaptive Filtering Approach

An Adaptive Filter is a computational device. It models the relationship between
two signals in real time in an iterative manner. Adaptive filter consists of (1) A
transversal filter that produces output in response to the input sequence, (2) filter
output is compared with a desired signal to generate an error, (3) an adaptive update
algorithm such as LMS and RLS for controlling the adjustable parameters. The
pseudo range multipath errors are computed by using Eqs. 6 and 7 and are applied
as input to Finite Impulse Response (FIR) filter, which is also known as transversal
filter. The combination of these three steps together constitutes a feedback loop, as
shown in Fig. 2.

Where, $x(n)$ = input signal, $y(n)$ = filter output, $e(n)$ = error signal, $d(n)$ = desired
signal and $c(n)$ = tap-weights.

Here a 32 stage FIR filter consisting of 31 delay elements is used. Output of this
filter is obtained by multiplying the input sample with weight coefficient of each
delay element. Filter output is compared with a desired signal to generate an error

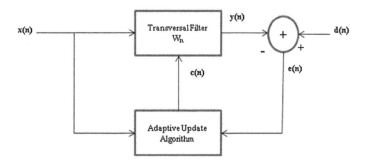

Fig. 2 Block diagram of an adaptive transversal filter

signal. This error signal is applied to an adaptive update algorithm so as to compute new coefficients. These new coefficients are multiplied with the FIR filter weights so as to generate new weights. Based on the new weights output of the filter changes with respect to desired signal. This is continues process in a feedback loop. For instance, the response of LMS adaptive filter is

$$\text{Filter output, } y(n) = \sum_{k=0}^{M-1} x(n-k)w_k(n)$$

$$\text{Estimation error, } e(n) = d(n) - y(n)$$

$$\text{Tap-weight adaption, } w_k(n+1) = w_k(n) + \mu x(n-k)e(n)$$

where μ is learning rate. Likewise each of the different adaptive filters have their own response and in this paper the analysis of Least Mean Square (LMS), Normalized Least Mean Square (NLMS) and Recursive Least Squares (RLS) algorithms has been done and results are compared.

4 Results

The estimated multipath error using CCD technique is considered as the input to the FIR filter. For the analysis, data due to a dual frequency GPS receiver of an IGS station, NGRI, Hyderabad is collected from SOPAC web data and is processed at a sampling rate of 30 s. The ranging information of all visible GPS satellites over a typical day of 11th September 2014 is processed to implement the proposed adaptive filtering method of multipath mitigation. As an illustration, multipath error in ranging information of satellite vehicle SV7 and mitigation of multipath using adaptive filters is presented in this paper. Here, the performance of 3 types of adaptive filters in mitigating the multipath error is compared. The amount by which input error signal is reduced at the output is known as the multipath mitigation

Table 1 Comparison of LMS, NLMS and RLS algorithms outputs for different satellites

S. no.	Filter type	SVPRN	Mean of MP1 (B_1)	Mean of MP2 (B_2)	Filter efficiency (%) $[(A_n - B_n)/A_n] * 100$, $n = 1, 2$	
					MP1	MP2
1	LMS	7	−0.0724	−0.0524	99.41	99.77
		8	−0.1011	−0.0770	98.87	99.45
		29	−0.2625	−0.1777	92.65	97.23
2	NLMS	7	−0.0453	−0.0318	99.63	99.86
		8	−0.0639	−0.0478	99.28	99.66
		29	−0.1523	−0.1069	95.73	98.33
3	RLS	7	−0.1794	−0.1762	98.51	99.23
		8	−0.2131	−0.2082	97.62	98.52
		29	−0.2248	−0.2377	93.71	96.30

efficiency of the adaptive filter. Table 1 shows the comparison of minimized multipath error mean values of satellite vehicles SV7, SV8 and SV29 using LMS, NLMS and RLS adaptive filters. The percentage of multipath error reduced at the output of the filter can be estimated as $[(A_n - B_n)/A_n] * 100$, $n = 1, 2$. Here, A_n and B_n are the mean multipath errors before and after filtering respectively. Here the suffix 1 and 2 refers to errors on L1 and L2 frequencies respectively. From Fig. 1, it is observed that A_1 and A_2 are 12.4697 m and 23.0472 m respectively for SVPRN 7. B_1 and B_2 are given in Table 1. It is observed that multipath error is more on L2 than on L1 (Fig. 1).

From Figs. 3, 4 and 5 it is observed that the mean multipath error values before filtering on L1 and L2 frequencies are 12.4697 and 23.0472 m respectively for

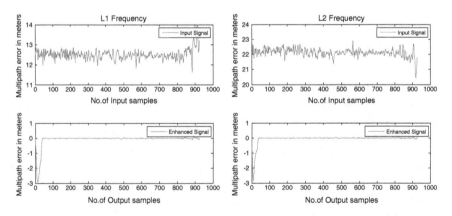

Fig. 3 Mitigation of multipath error on L1 and L2 frequencies of SVPRN 7 using LMS adaptive filter

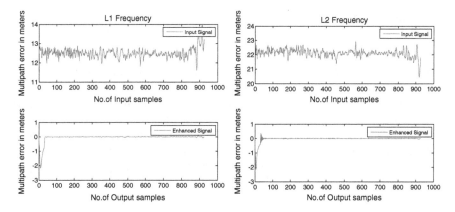

Fig. 4 Mitigation of multipath error on L1 and L2 frequencies of SVPRN 7 using NLMS adaptive filter

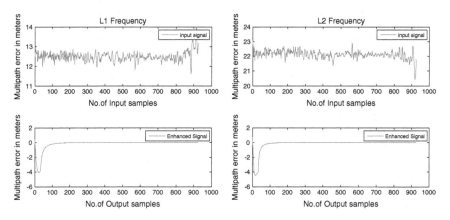

Fig. 5 Mitigation of multipath error on L1 and L2 frequencies of SVPRN 7 using RLS adaptive filter

SVPRN 7 and these errors are reduced to almost zero after filtering with either one of the three adaptive filters LMS, NLMS and RLS.

Table 1 presents the performance analysis of LMS, NLMS and RLS adaptive filter algorithms in mitigating the multipath error present in ranging information of three satellite vehicles SV7, SV8 and SV29. The filter efficiency of NLMS algorithm on L1 and L2 frequencies is 99 % (approximately) for SV7 and SV8 and is better than the RLS and LMS algorithms.

5 Conclusion

Multipath is the most predominant error that affects the accuracy of GPS applications. In this paper, multipath error effect is presented with simulated results. Multipath error is estimated Using Code Carrier Difference (CCD) method with experimental data and the results are presented for few satellites. Compared to LMS and RLS algorithms, by using Normalized Least Mean Square (NLMS) adaptive filter, the multipath error is mitigated by 99.63 % on L1 frequency and 99.86 % on L2 frequency for SVPRN 7. Rate of convergence is also faster compared to LMS algorithm. Therefore NLMS adaptive filter best suits for mitigation of multipath error on GPS signals and also in wireless communication for noise cancellation.

References

1. E.D. Kaplan (ed.), *Understanding GPS: Principles and Applications* (Artech House Publishers, Bostan, 1996)
2. M.S. Braasch, Multipath effects, in *Global Positioning System: Theory and Applications*, vol. 1, ed. by B.W. Parkinson, J.J. Spilker (AIAA, 1996)
3. B. Hofmann-Wellenhof, H. Lichtenegger, J. Collins, *Global Positioning System: Theory and Practice*, 5th revised edn. (Springer, Berlin Heifelberg, New York, 2007)
4. G. SasiBhushana Rao, *Global Navigation Satellite Systems* (Tata McGraw Hill publications, 2010)
5. V.B.S.S.I. Dutt, G.S.B. Rao, S.S. Rani, S.R. Babu, R. Goswami, C.U. Kumari, Investigation of GDOP for precise user position computation with all satellites in view and optimum four satellite configurations. J. Ind. Geophys. Union **13**(3), 139–148 (2009)
6. K. Yendukondalu, A.D. Sarma, V. Satya Srinivas, Multipath mitigation using LMS adaptive filtering for GPS applications, in *IEEE-Applied Electromagnetics Conference (AEMC-2009)*, Kolkata, India, Dec 2009, pp. 14–16
7. K. Yendukondalu, A.D. Sarma, A. Kumar, K. Satyanarayana, Spectral analysis and mitigation of GPS multipath error using digital filtering for static applications. IETE J. Res. **59**(2), (2013)
8. K. Yendukondalu, A.D. Sarma, V. Satya Srinivas, Estimation and mitigation of GPS multipath interference using adaptive filtering. J. Progr. Electromagnet. Res. M (PIER-M), U.S.A. **21**, 133–148 (2011)

Generation of Optimized Beams from Concentric Circular Antenna Array with Dipole Elements Using BAT Algorithm

U. Ratna Kumari, P. Mallikarjuna Rao and G.S.N. Raju

Abstract The Dipole is a simple antenna element used for communication purpose. The Planar Concentric Circular Antenna Array with Dipole radiators is practical antenna array to scan the entire azimuthal plane. In the present paper an attempt is made to reduce the Side Lobe Level and Beam width of such an Antenna Array. A very efficient optimization technique BAT algorithm is applied for the optimization of the antenna array. As it is well known that thinning process has lot of advantages the thinning is applied on the antenna array. These results are compared with uniform amplitude distribution of fully populated array.

Keywords Planar concentric circular antenna array · Dipole antenna · Narrow beams · BAT algorithm · Thinning

1 Introduction

Similar type of antenna elements arranged on a circle is known as circular antenna array (CAA). This Arrangement is used to obtain perfect beam pattern in every ϕ cut and to scan the entire azimuthal plane. If different circles with different radii share a common centre then such type of arrangement is known as Planar Concentric Circular Antenna Array (PCCAA) [1–7]. The antenna elements considered to be of any type. Non isotropic antenna like dipole is very simple antenna that is considered as antenna element in the present paper. Geometrically all the elements will lay on the perimeter of different circles. These concentric circular arrays are best suited for

U. Ratna Kumari (✉) · P. Mallikarjuna Rao · G.S.N. Raju
Department of ECE, AU College of Engineering, Andhra University,
Visakhapatnam, Andhra Pradesh, India
e-mail: ratnakumari.u@gmail.com

P. Mallikarjuna Rao
e-mail: pmraoauece@yahoo.com

G.S.N. Raju
e-mail: drrajugsn@yahoo.co.in

© Springer India 2016   547
S.C. Satapathy et al. (eds.), *Microelectronics, Electromagnetics
and Telecommunications*, Lecture Notes in Electrical Engineering 372,
DOI 10.1007/978-81-322-2728-1_51

wide bandwidth microwave direction finders, space communications, direction finding applications and wrap—around shipborne communications.

The radiation pattern of Dipole Planar Concentric Circular Antenna Array (DPCCAA) depends upon different parameters like dipole length, excitation currents. The side lobe level (SLL) and beam width can be reduced optimizing anyone or both the parameters. In the present paper the excitation currents and dipole length are optimized using very efficient optimization technique like BAT algorithm. These optimized excitation currents are considered further on which thinning is applied. The results are compared with antenna array with isotropic elements.

2 The DPCCAA Geometry

There is a small difference between the radiation patterns of PCCAA geometry considered with normal isotropic radiator and practical dipole radiator. The radiation pattern of PCCAA with practical dipole radiator is product of PCCAA array factor and individual field pattern of practical dipole radiator. The normalised power pattern of the DPCCAA is calculated as follows

$$P(\theta, \phi) = 20 \log_{10} \left[\frac{|E|}{|E|_{\max}} \right] \tag{1}$$

$$E = E_D(\theta) * E_C(\theta, \phi) \tag{2}$$

Here $E_D(\theta)$ is considered to be radiation pattern of single dipole and is given by

$$E_d(\theta) = \frac{\cos(\beta L \cos(\theta)) - \cos \beta L}{\sin \theta} \tag{3}$$

$E_C(\theta, \phi)$ is radiation pattern of the PCCAA given by the following equations

$$E_C(\theta, \phi) = \sum_{m=1}^{M} \sum_{n=1}^{N} I_{mn} e^{j \beta r_m [\sin \theta \cos(\phi - \phi_{mn}) - \sin \theta_o \cos(\phi_o - \phi_{mn})]} \tag{4}$$

$$rm = \frac{N d_m}{2\pi} \tag{5}$$

ϕ_{mn} is given by

$$\phi_{mn} = \frac{2n\pi}{N} \tag{6}$$

where
L   is the length of the dipole
I_{mn} is excitation currents

r_m is radius of mth ring
d_m is inter element spacing
β is the wave number, $\beta = 2\pi/\lambda$
θ is elevation angle
M   is no of rings
N   is no of elements in each ring
θ_0,   direction at which main beam achieves its maximum. For the designing
ϕ_0 problems presented here $\theta_0 = 0°$ and $\phi_0 = 0°$ are considered
ϕ is the azimuth angle between the positive x-axis and the projection of the far
        field point in the x–y plane as shown in Fig. 1.

Figure 1 shows DPCCAA with fully populated Dipole elements where as Fig. 2 shows Thinned DPCCAA.

Fig. 1 Fully excited DPCCAA

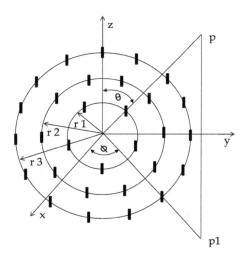

Fig. 2 Thinned DPCCAA

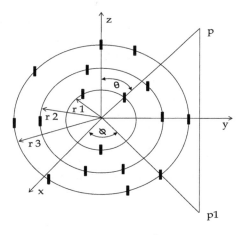

3 BAT Algorithm

A very efficient metaheuristic evolutionary algorithm developed by Xin-She Yang in 2010 is BAT algorithm [7–9]. This algorithm mainly depends upon the echolocation behaviour of micro bats with varying pulse rates of emission and loudness.

In the process of BAT algorithm BAT positions indicate the optimum solution of the problem. Depending on the best position obtained the quality of the solution is represented. The process echolation is used by the BATs to sense the food or prey and the background barriers. With the initial conditions such as minimum frequency, initial velocity, wave length and initial loudness they fly randomly searching the food or prey. Depending on the closeness of the food or prey they adjust the pulse emission rate and wave length and reach the target.

The process of the BAT algorithm is indicated below. The new best solution, velocities and frequencies used in the algorithm are given by

$$f_i = f_{min} + (f_{max} - f_{min})\beta \tag{7}$$

$$v_i^t = v_i^{t-1} + \left(x_i^t - x_*\right)f_i \tag{8}$$

$$x_i^t = x_i^{t-1} + v_i^t \tag{9}$$

where f_i is frequency, v_i is velocity of bats and x_i is new best required solution. The BAT parameters that are considered in the present paper are

Frequency minimum = 0.1, Frequency maximum = 0.9, Loudness = 0.3, Pulse rate = 0.8.

From the above equations and concept of BAT algorithm the fitness function is formulated and is presented in the Eq. 10. Using this equation the DPCCAA is optimized.

$$\text{Fitness} = \text{MIN}(\text{MAX}(20 * \log P(\theta, \phi)/P(\theta, \phi))) \tag{10}$$

where $P(\theta, \phi)$ is considered to be power of the DPCCAA.

All the multidimensional problems are easily solved using BAT algorithm. The following steps describe how the BAT algorithm works.

1. In every optimization problem if the fitness function is formed then half of the goal achieved. Here the fitness function is formed first to achieve the goal.
2. Initialisation of BAT Population and velocities is done as the second step.
3. Initialisation of pulse frequency, location of the BATs and loudness of the BATs are done. As all these parameters are to be initialised for the smooth running of the algorithm, they are initialised at the starting of the algorithm.
4. The entire algorithm has to be repeated for certain number of iterations. Hence a fixed number of iterations are considered and initialised.

5. All the initialised values are used to evaluate the fitness function. On evaluating the fitness function it is observed whether the initialised values will give the best results or not.
6. The BAT velocities and BAT locations are updated and the best solutions are selected.
7. New solutions are generated and accepted varying the Loudness and Search space.
8. With the effective stopping criteria the steps 4–7 are repeated and the final goal is achieved.

4   Array Design Using Thinning

With the thinning [10–15] process the antenna array with high dimensions is fabricated and built easily. Without degrading the performance of the DPCCAA systematically eliminating or turning on and off the elements which are responsible for more SLL and broad beam width is known as thinning. The excitation currents I_{mn} is made 1 to turn on the antenna element and made 0 to turn off the antenna element. In the present paper the DPCCAA is optimized using BAT algorithm to find out the optimized current excitation coefficients and dipole length for reduced sidelobe levels. Later thinning is applied using the optimized dipole length and excitation currents.

5   Results and Discussions

The total work is carried out in three different steps. In the first step the radiation patterns of Planar Concentric Circular Antenna Arrays were carried out considering non isotropic element such as Dipole using the Eqs. 1–6. In the second step BAT algorithm is applied to find the excitation currents for uniformly increased dipole length and BAT algorithm generated dipole lengths using the Eqs. 7–10. In the third step thinning is applied on the DPCCAA that resulted in very good performance.

Three rings DPCCAA in which the number of elements considered are multiples of 5. The dipole length and inter element spacing considered here is 0.5λ. All the results are compared with radiation patterns of PCCAA having isotropic radiators. The power pattern of 3 rings DPCCAA with uniformly increased dipole length and BAT excitation currents is presented in Fig. 3. From the figure it is observed that The SLL is −21.7 dB and the beam width is 13.7°. As compared with the uniform excitation and uniform spacing of DPCCAA the SLL and beam width are reduced considerably to appreciable values.

Fig. 3 Power pattern for
three rings DPCCAA for
uniform incremental dipole
length and BAT excitation
coefficients

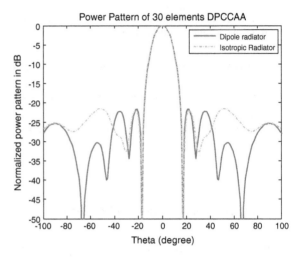

Fig. 4 Thinned power
pattern for three rings
DPCCAA for uniform
incremental dipole length and
BAT excitation coefficients

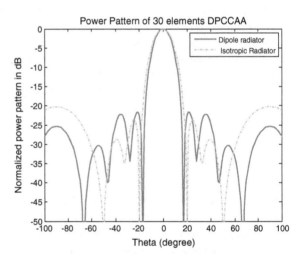

The thinned power pattern of 3 rings DPCCAA with uniformly increased dipole
length and BAT excitation currents is presented in Fig. 4. From the figure it is
observed that The SLL is −21.3 dB and the beam width is 13.7°. With 30 %
thinning of the antenna array the results were effectively achieved. Table 1 indicates
the SLL and Beam Width and Thinning percentage related to Figs. 3 and 4.

Figures 5 and 6 indicate the power patterns for 3 rings DPCCAA with BAT
dipole length and BAT excitation currents. Table 2 indicate the corresponding SLL
and Beam Width and Thinning percentage. It is observed that retaining the array
parameters the array is effectively thinned and the goal is achieved. Here The SLL
and beam width achieved are −22.7 dB and 13.8° respectively. With 40 % thinning

Table 1 Excitation co-efficients, SLL, beam width and thinning percentage comparison values for three rings DPCCAA for incremental dipole length

Radiation parameters	Uniform excitation	Uniform incremental dipole length and BAT excitation currents	Uniform incremental dipole length and thinned BAT excitation currents
MAX SLL	−15.82 dB	−21.7 dB	−21.3 dB
3 dB beam width	21.4°	13.6°	13.7°
Thinning percentage	0 %	0 %	30 %

Fig. 5 Power pattern for three rings DPCCAA for BAT dipole length and BAT excitation coefficients

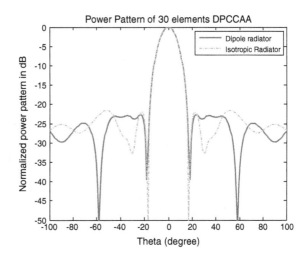

Fig. 6 Thinned power pattern for three rings DPCCAA for BAT dipole length and BAT excitation coefficients

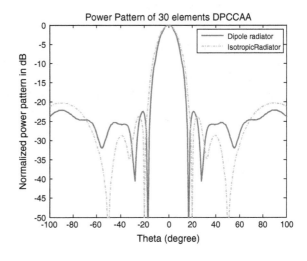

Table 2 Excitation co-efficients, SLL, beam width and thinning percentage comparison values for three rings DPCCAA for BAT dipole length

Radiation parameters	Uniform excitation	Uniform incremental BAT dipole length and BAT excitation currents	Uniform incremental BAT dipole length and Thinned BAT excitation currents
MAX SLL	−15.82 dB	−22.7 dB	−22.1 dB
3 dB beam width	21.4°	13.8°	13.4°
Thinning percentage	0 %	0 %	30 %

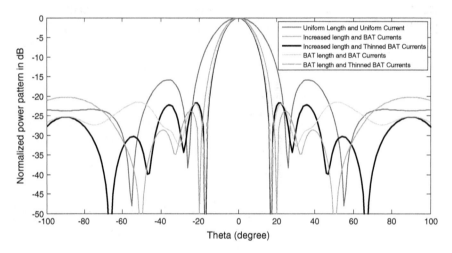

Fig. 7 Power pattern for three rings DPCCAA for uniform dipole length and uniform excitation currents with optimized dipole length and optimized excitation currents

the SLL is −22.1 dB and beam width is 13.3°. Figure 7 indicates Power pattern for three rings DPCCAA for Uniform Dipole Length and Uniform excitation Currents with Optimized Dipole Length and Optimized Excitation Currents. Tables 3 and 4 indicate Variation BAT excitation currents and thinned BAT excitation currents for incremental dipole length and BAT dipole length respectively.

Table 3 Incremental dipole length, BAT excitation currents and thinned excitation currents for three rings DPCCAA

Dipole length				BAT excitation currents				Thinned BAT excitation currents			
0.2000	0.2241	0.2483	0.2724	0.0406	0.6266	0.4477	0.1865	0.0000	0.0000	0.4477	0.1865
0.2966	0.3207	0.3448	0.369	0.7763	0.6626	0.1342	0.1502	0.7763	0.6626	0.1342	0.1502
0.3931	0.4172	0.4414	0.4655	0.1672	0.4085	0.2898	0.2046	0.0000	0.4085	0.2898	0.0000
0.4897	0.5138	0.5379	0.5621	0.8958	0.7649	0.8182	0.2846	0.8958	0.7649	0.8182	0.2846
0.5862	0.6103	0.6345	0.6586	0.4799	0.0384	0.3417	0.588	0.4799	0.0000	0.3417	0.588
0.6828	0.7069	0.731	0.7552	0.448	0.5367	0.1883	0.344	0.448	0.0000	0.1883	0.344
0.7793	0.8034	0.8276	0.8517	0.3788	0.5776	0.5856	0.568	0.3788	0.5776	0.0000	0.568
0.8759		0.9000			0.5309	0.6543			0.0000	0.0000	

Table 4 BAT dipole length, BAT excitation currents and thinned excitation currents for three rings DPCCAA

Dipole length				BAT excitation currents				Thinned BAT excitation currents			
0.9747	0.8481	0.2458	0.3053	0.0193	0.2791	0.7174	0.2338	0.0000	0.279	0.0000	0.2338
0.4172	0.9555	0.2103	0.5864	0.6086	0.778	0.7676	0.3685	0.0000	0.778	0.0000	0.3685
0.6667	0.8618	0.0577	0.4887	0.187	0.3475	0.4335	0.2605	0.187	0	0.0000	0.2605
0.4969	0.7902	0.6474	0.5668	0.0534	0.6025	0.2991	0.6122	0.0000	0.6025	0.2991	0.6122
0.7908	0.9606	0.4036	0.4505	0.2878	0.1069	0.8975	0.8147	0.0000	0	0.8975	0.8147
0.7546	0.4971	0.096	0.8803	0.0455	0.1379	0.8681	0.7616	0.0000	0.1379	0.8681	0.7616
0.8743	0.6782	0.8049	0.5936	0.8695	0.8923	0.6908	0.0102	0.8695	0.8923	0.6908	0.0000
0.5659		0.1685			0.8286	0.8682			0.8286	0.0000	

6 Conclusions

Present paper describes the study of Planar Concentric Circular Antenna Arrays with practical radiating elements like Dipole radiators. Application of BAT algorithm enhances the results of the array. It gives a very nice performance with less number of iterations. The plots are converged with in a very less time and in very less number of iterations. The produced plots can be used for point to point communication and for direction finding purpose also.

References

1. C.A. Balanis, *Antenna Theory Analysis and Design*, 2nd edn (Wiley, 1997)
2. G.S.N. Raju, *Antennas and Wave Propagation* (Pearson Education, 2005)
3. M.T. Ma, *Theory and Applications of Antenna Arrays* (Wiley, New York, 1974)
4. R.S. Elliott, *Antenna Theory and Design* (Prentice-Hall, USA, 1981)
5. A. Ishimaru, Member IRE, Theory of unequally-spaced arrays. IRE Trans. Antennas Wave Propag. 691–702 (1962)
7. C. Stearns, A. Stewart, An investigation of concentric ring antennas with low sidelobes. IEEE AP-S Trans. **13**(6), 856–863 (1965)
8. X.-S. Yang, *A New Metaheuristic Bat-Inspired Algorithm*, 23 Apr 2010. arXiv:1004.4170v1 [math.OC]
9. I.H. Osman, An introduction to metaheuristics, in *Operational Research Tutorial Papers*, ed. by M. Lawrence, C. Wilson (Operational Research Society Press, Birmingham, 1995)
10. R.L. Haupt, *Thinned Concentric Ring Arrays* (IEEE AP-S Symposium, San Diego, 2008)
11. B. Basu, G.K. Mahanthi, Thinning of concentric two-ring circular array antenna using firefly algorithm. Sci. Iranica Trans. D **19**(6), 1802–1809 (2012)
12. S.P. Ghoshal. et al., Optimized radii and excitations with concentric circular antenna array for maximum side lobe level reduction using wavelet mutation based particle swarm optimization techniques. Springer Telecommun. Syst. **52**(4), 2015–2025 (2013)
13. N. Pathak, P. Nanda, G.K. Mahanti, Synthesis of thinned multiple concentric circular ring array antennas using particle swarm optimization. J. Infrared Millimeter Terahertz Waves **30**(7), 709–716 (2009)
14. U. Ratna Kumari, P. Mallikarjuna Rao, G.S.N. Raju, Generation of narrow beams from concentric circular antenna array for wireless communications using BAT algorithm. Res. J. Appl. Sci. Eng. Technol. **10**(1), (2015)
15. U. Ratna Kumari, P. Mallikarjuna Rao, G.S.N. Raju, Synthesis of thinned planar concentric circular antenna array using evolutionary algorithms. IOSR–JECE **10**(2), 57–62 Ver. II (2015)

Design and Analysis of Single Feed Dual Band Stacked Patch Antenna for GPS Applications

Palleti Ramesh Raja Babu and M. Nirmala

Abstract An efficient double band with single feed GPS frequencies 1.22760 GHz (L2) and 1.57542 GHz (L1) such as Surveying, Mapping and Geo informatics. By trimming opposite sides of two different pair of square and isosceles corners of square patch in dielectric substrates, stacked patch antennae for double band operations are obtained. Both the upper and lower patches have 2 pairs of square and isosceles truncated corners each. The inculcation of corner truncation has achieved better return loss and gain at L1, L2 bands. A concise antenna is designed and simulated using IE3D simulation tool.

Keywords Global positioning system · Stacked patch · Truncated corners · Dual band

1 Introduction

Global Positioning System (GPS) receivers are used to find the exact location, accurate position and also determines the appropriate time which can be used in many applications. GPS is a dual system, which provides many applications for civil and military services. With the rapid growth in wireless communication GPS functions have become very important in day today life. As the demand for more accuracy and high reliability of the GPS. The GPS antenna should cover multi frequencies and should operate at same time [1, 2].

P. Ramesh Raja Babu (✉) · M. Nirmala
Department of Electronics Communication and Engineering (ECE), Anil Neerukonda Institute of Technology and Sciences (ANITS), Sangivalasa-Bheemunipatnam, Visakhapatnam, India
e-mail: rameshraja112@gmail.com

M. Nirmala
e-mail: nirmala.mattaparthi@gmail.com

© Springer India 2016
S.C. Satapathy et al. (eds.), *Microelectronics, Electromagnetics and Telecommunications*, Lecture Notes in Electrical Engineering 372,
DOI 10.1007/978-81-322-2728-1_52

With the increasing demand for accurate time and perfect position information by using Global navigation system i.e. GPS which has exalted its requirement in its development and research [3]. For designing of single feed stacked-patch antennas ratio should be less than 1.5 then only it covers both the L1 and L2 Bands. Stacked patch antennas has a very less permittivity with an air gap, which increases antenna size, cost, volume. To avoid these problems two patches are trimmed on opposite sides of square patches and a-dielectric constant used as a substrate [3–9].

Here, a single probe stacked patch antennas with trimmed corners on the all sides of both square patches for GPS bands L1 and L2 is explained. The proposed design has the benefit such that it will not have any air gap while stacking patches and also for both the patches only one material substrate is used. The proximity, modesty and specific frequency bands enable this new designed antenna for Surveying, Mapping, Geo informatics and other GPS applications [10–12].

2 Antenna Structure and Design

The top and side view of single feed double band stacked antennae for GPS bands L1 (1.57542 GHz), L2 (1.22760 GHz) is shown in Fig. 1.

The antenna has a two stacked square patches with radiating elements, both the patches are placed on the same side with the jeans cotton substrate with relative permittivity 1.67 of h1 = 2.84 mm and h2 = 5.68 mm. The lower and upper patches of the designed antenna produces L1 (1.57542 GHz) and L2 (1.227 GHz). The upper patch is denoted with Up = 71.33 mm which has truncated corners on all

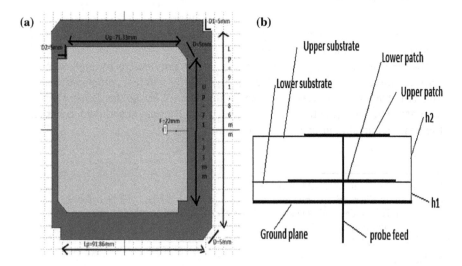

Fig. 1 Configuration of the single feed double-band stacked antennae. **a** Top view of designed antennae. **b** Side view of designed antennae

sides of the patch with D = 5 mm. The lower patch with length Lp = 91.86 mm has a trimmed corners square patches with D = 5 mm. By choosing proper feed point in the X axis i.e. (F = 22 mm) trimmed corner cuts on both the patches the designed antenna will have better impedance and good bandwidth. Specifications of the proposed single feed double band antenna.

Parameters	Lp	Up	D	h1	h2	F
Values (mm)	91.86	71.33	5	2.84	5.68	22

where

h1, h2   represents thickness of upper and lower patches

Up, Lp   represents length and width of upper and lower patches

D   represents truncated isosceles and square corners

F   represents feed point location

ϵ_r represents the relative permittivity of the material

2.1 Design Procedure

Step 1: Calculation of the Width (W):

$$w = \frac{v_o}{2f_r} \sqrt{\frac{2}{\varepsilon_r + 1}} \tag{1}$$

v_0 Velocity of light

f_r Frequency of resonance

ϵ_r Permittivity of the material

Step 2: Calculation of Effective permittivity (ε_{reff}):

$$\varepsilon_{reff} = \left(\frac{\varepsilon_r + 1}{2}\right) + \left(\frac{\varepsilon_r - 1}{2}\right)\left[1 + 12\frac{h}{w}\right]^{-1/2} \tag{2}$$

Step 3: Calculation of effective length extension (ΔL)

$$\frac{\Delta L}{h} = 0.412 \frac{(\varepsilon_{reff} + 0.3)\left(\frac{w}{h} + 0.264\right)}{(\varepsilon_{reff} - 0.258)\left(\frac{w}{h} + 0.8\right)} \tag{3}$$

Step 4: Actual length of the patch is calculated by (L):

$$L = \frac{1}{2f_r\sqrt{\varepsilon_{reff}}\sqrt{\mu_o\varepsilon_o}} - 2\Delta L \tag{4}$$

And Other parameters are calculated by

$$\text{VSWR is given by VSWR} = \frac{1+\rho}{1-\rho} \tag{5}$$

where ρ = Reflection coefficient
Radiation Pattern is related to Maximum gain given by

$$\text{Gain} = 4\pi\frac{U(\theta,\varphi)}{P} \tag{6}$$

where
$U(\theta,\varphi)$ Radiation intensity
P Total input power

Directivity

$$D = \frac{U}{Uav} = 4\pi\frac{U}{Prad} \tag{7}$$

where
U radiation intensity
Uav Average radiation intensity
$Prad$ Radiated power

3   Results and Discussion

The S parameters of the simulated antenna shown in above Fig. 2. The S parameter of GPS antenna bands (L2) and (L1) are at -12.07 dB and -13.15 dB. For a better performance of an antenna, return loss must be nearly -10 dB. But from the result it is clear that we get the better performance which is beneficial.

Figure 3 is the VSWR of antenna. A good real impedance matched antenna will have VSWR equal to one. The value of VSWR for the frequency L1 at 1.575 GHz is $1.78 < 2$ and L2 at 1.227 GHz is 1.62 also less than 2 which shows antenna has better impedance and less radiation losses at these bands.

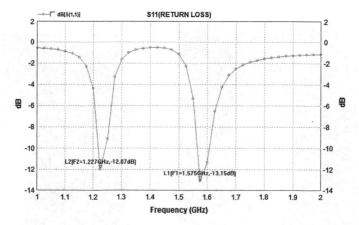

Fig. 2 S parameters of the dual band

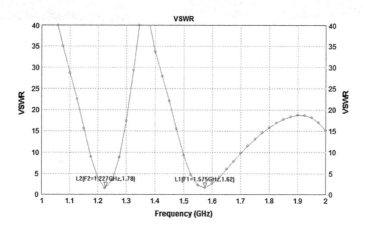

Fig. 3 VSWR at L1 and L2 bands

3D Radiation pattern of designed antennae at L1 (1.57542 GHz) and L2 (1.225 GHz) are shown in Fig. 4. The Gain is 7.2 dBi for L1 (1.575 GHz) frequency band and 7.7 dBi for L2 (1.227 GHz) frequency band.

The simulated peak gain antennae for the L1 and L2 band of GPS antenna in polar plot are 7. 7 and 7.2 dBi, respectively in both elevation and azimuth patterns are shown in Fig. 5.

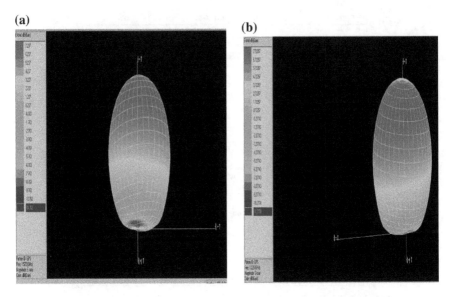

Fig. 4 a 3D radiation pattern at 1.5754 GHz. **b** 3D radiation pattern at 1.225 GHz

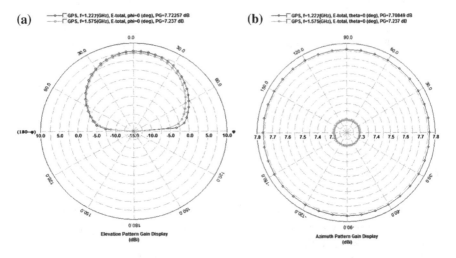

Fig. 5 Polar plots of 2D radiation patterns at 1.575 and 1.227 GHz. **a** Elevation pattern. **b** Azimuth pattern

4 Conclusion

An efficient dual band stacked patch antennae operating at GPS frequencies L1 (1.57542 GHz) and L2 (1.22760 GHz) bands are designed. Antennas use the same substrate material for both the patches. The designed stacked antennae has best emission characteristics of L1 (1.57542 GHz) and L2 (1.22760 GHz) GPS bands. An enhancement in return loss and gain are obtained due to corner truncation. The designed antenna have produced satisfactory performance in terms of return loss, VSWR, and the far field patterns which indicate that antenna have a significant potential in GPS application.

References

1. M. Ramkumar Prabhu, U.T. Sasikala, Designing a switchable stacked patch antenna for GPS application. IOSR J. Electron. Commun. Eng. (IOSR-JECE), **9**(2), 164–168 (2014)
2. M. Fadhil Hasan, A new compact dual band GPS patch antenna design based on minkowski-like pre-fractal geometry. Eng. Tech. J. **27**(7), 1370–1375 (2009)
3. W.-T. Hsieh, T.-H. Chang, J.-F. Kiang, Dual-band circularly polarized cavity-backed annular slot antenna for GPS receiver. IEEE Trans. Antennas Propag. **60**(4), 2076–2080 (2012)
4. S.C. Chen, G.C. Liu, X.Y. Chen, T.F. Lin, X.G. Liu, Z.Q. Duan, Compact dual-band GPS microstrip antenna using multilayer LTCC substrate. IEEE Antenna Wirel. Propag. Lett. **9**, 421–423 (2010)
5. Takafumi, Daisuke, Kouhei, Mitsuo, Dual–band circularly polarized microstrip antenna for GPS application. in *IEEE Conference Antennas and Propagation Society International Symposium*. E-ISBN 978-1-4244-2042-1, pp. 1–4 (July 2008)
6. A.A. Heidari, A. Dadgarnia, Design and optimization of a circularly polarized microstrip antenna for GPS applications using ANFIS and GA. in *IEEE Conference General Assembly and Scientific Symposium XXXth URSI*, Aug 2011. ISBN: 978-1-4244-6051, pp. 1–4
7. L. Boccia, G. Amendola, G. Di Massa, A dual frequency microstrip patch antenna for high-precision GPS applications. IEEE Trans. Antennas Propag. **3**(1), 157–160 (2004)
8. L. Boccia, G. Amendola, G. Di Massa, A shorted elliptical patch antenna for GPS applications. IEEE Trans Antennas Propag. **2**(1), 6–8 (2003)
9. Y.F. Wang, J.J. Feng, J.B. Cui, X.L. Yang, A dual-band circularly polarized stacked microstrip antenna with single-fed for GPS applications. IEEE Int. Symp. Antennas Propag. 108–110 (2008)
10. D.M. Pozar, S.M. Duffy, A dual-band circularly polarized aperture- coupled stacked microstrip antenna for global positioning satellite. IEEE Trans. Antennas Propag. **45**(11), 1618–1625 (1997)
11. H.M. Chen, Y.K. Wang, Y.F. Ling, C.Y. Lin, S.C. Pan, Microstrip-fed circularly polarized square-ring patch antenna for GPS applications. IEEE Trans. Antennas Propag. **57**(4), 1264–1267 (2009)
12. K.L. Wong, *Compact and Broadband Microstrip Antennas*. (A Wiley-Inter science Publication 2002, Chapter 9)

Emotion Recognition Model Based on Facial Gestures and Speech Signals Using Skew Gaussian Mixture Model

M. Chinna Rao, A.V.S.N. Murty and Ch. Satyanarayana

Abstract This paper addresses an approach for identification of the emotion and conforming the emotions by fusing to the facial gestures. Various Techniques have been floated in the area of emotion recognition based speech signals. However these speech signals that are generated may not being in coherent with the actual inner feelings. Therefore in this paper a model is proposed by fusing the facial expression and the uttered speech voices. In ordered to tested developed model, synthesized data set is considered and performances evaluated using the metrics like precision and recalled.

Keywords Emotion recognition · Skew gaussian mixture models · Speech samples · Facial expressions · Fusion

1 Introduction

Emotion Recognition has void range of applications in which forensic applications take the leading role. Many models have been developed using speech signals and several authors have proposed systems based on SVM [1], Neural Networks [2], Principal component analysis [3], Mel Frequency Cepstral Coefficients [4–6] Skew Gaussian Mixture models [7–9]. In most of these models only the speech signal is considered. In ordered to identify the emotion the recorded voice sample is generated and most of these signals that are recorded in the WAV format or mostly

M. Chinna Rao (✉) · Ch.Satyanarayana
Computer Science Department, JNTUK, Kakinada, India
e-mail: chinnarao.mortha@gmail.com

Ch.Satyanarayana
e-mail: chsatyanarayana@yahoo.com

A.V.S.N.Murty
Mathematics Department, VIT University, Vellore, India
e-mail: avsn.murty@vit.ac.in

© Springer India 2016   567
S.C. Satapathy et al. (eds.), *Microelectronics, Electromagnetics*
and Telecommunications, Lecture Notes in Electrical Engineering 372,
DOI 10.1007/978-81-322-2728-1_53

through the active sequences. The inherent expressions may not be reflected through the facial expressions or the voice occurred may not reveal the exact feeling. Therefore it becomes a mandate to identify the exact emotion of a speaker. This recognition has a significant contribution in many areas like teleconferencing, Criminal identification, and BPO etc. In particular, during the crime analysis, in order to assess the witness or to consolidate the witness statement, it is necessarily to identifying the genuinely of the speaker. The speaker mainly try to cancel is inherent expressions and may project in a descent manner or the responses may not truly signify the actual feelings. Therefore to identify the exactness it is customary to reduce meaningful information.

Feature extraction takes place a vital role in this context. Also among the speech signals some signals may be of vital information. Therefore these speech samples are to be extracted for proper identification. Hence, in this paper an attempt is made to develop a novel concept of emotion recognition varying the feature are extracted from the speech signals and fused with the facial expressions. A Skew Gaussian mixture model is considered for this purpose. The main advantage of considering the Skew Gaussian mixture model is that it can identify the feature from the speech sample even in the presence of noise or low frequency domain the rest of the paper is organized as follows.

In Sect. 2 of the paper, the work presented by the earlier research is highlighted. In Sect. 3 of the paper presence on Skew Gaussian mixture models. In Sect. 4 of the paper the feature extraction and the dimensionality deduction using Principle component analysis proposed. The Sect. 5 of the paper presents the experimentation and together with the results derived. The evaluations based on precision and recalled are presented. Considering Sect. 6 summarize the paper.

2 Related Work

Many authors have attempted to identify contains state of mind by proposing different model HIROYASU. M has classified the emotion into active, cool and certain. And there by author has tried to classify the emotion accordingly. Robat. P has considered the Emotion recognition has a part of schcology and classified the emotion into five different classes via Happy, Sad, Neutral, Angry and boredom. Weig has proposed a model by using Hidden Markov model and Support vector machines for identification of the emotion. The Emotion Recognition system based on dialects using Gaussian mixture model is proposed by N. Muralikrishna et al. In this model the authors have utilized Gaussian mixture mode for effective recognition of the dialects. In this paper the authors presented the need for considering Gaussian mixture model.

K. Suribabu et al. has proposed a model based on Centralized Gamma distribution for recognizing the speaker's emotion and the authors have experimented their work by considering a data set. However in most of these models the recognition is nearly based on speech samples and no attempt is made to correlate the facial feelings together with the speech sample for effective recognition. Hence in this paper a model is proposed in this direction.

3　Skew Gaussian Mixture Models

The main advantage of considering Skew Gaussian mixture model is that it can identify the speech signals effectively which by nature are asymmetric. The speech samples uttered may be having infinite range but to identify the emotions, every signal has its important. Therefore to interpret the signals more accurately a Skew Gaussian mixture model is proposed.

The probability density function of the Skew Gaussian mixture model is given by

$$f(z) = 2 \cdot \varphi(z) \cdot \Phi(\alpha z); \quad -\infty < z < \infty \tag{1}$$

where,

$$\Phi(\alpha z) = \int_{-\infty}^{\alpha z} \phi(t) dt \tag{2}$$

And,

$$\phi(z) = \frac{e^{-\frac{1}{2}z^2}}{\sqrt{2\pi}} \tag{3}$$

Let,

$$y = \mu + \sigma z \, z = \frac{y - \mu}{\sigma} \tag{4}$$

Substituting Eqs. (2)–(4) in Eq. (1) (Fig. 1)

$$f(z) = \sqrt{\frac{2}{\pi}} \cdot e^{-\frac{1}{2}\left(\frac{y-\mu}{\sigma}\right)^2} \left[\int_{-\infty}^{\alpha\left(\frac{y-\mu}{\sigma}\right)} \frac{e^{-\frac{1}{2}\left(\frac{t-\mu}{\sigma}\right)^2}}{\sqrt{2\pi}} dt \right] \tag{5}$$

In order to update the initial parameter of the Skew Gaussian mixture model, EM algorithm is used. The updated equations are presented [8–10].

Fig. 1 Frequency curves of
Skew normal distributions

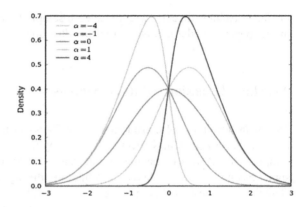

4 Feature Extraction

In order to identify the emotion more preciously features play a vital role. In this
paper we have considered two features, one feature extracted from the wave signal
of the speech signal and the other feature extracted from facial expression for this
Eigen values are considered. The Eigen values are obtained by using Principle
component analysis.

4.1 Algorithm for Recognition of Emotion from Speech Sample

Step 1: Record these speech signals in wav format.
Step 2: preprocess the voices generated by the speaker to eliminate noise.
Step 3: Extract amplitude sequences from these speech samples using MFCC
coefficient.
Step 4: Identify the range of each emotion based on the amplitude data.
Step 5: Categorize the score each emotions and store each emotion against the
speaker.
Step 6: Repeat the process for both gender.

4.2 Principal Component Analysis

This is one of the dimensional reductions of methodology and feature extraction
method consider mostly in the literature in particular for the facial verifications. The
algorithms for PCA given as follows [11–13].

Algorithm identifying the Eigen faces:

Step 1: Normalize the faces in the data bases.
Step 2: Eliminate the background information.
Step 3: Eliminate the noise by subtracting the mean value of all the images.
Step 4: Calculate the co variant matrix.
Step 5: Formulate characteristic equation which gives Eigen vectors.

This Eigen vectors that are generated are stored against each focus in the data bases.

5 Experimentation

In order to experiment the model data base of the facial expressions and the emotion speech samples extracted from the students of Kakinada Institute of Engineering Technology College is considered. This database consists of arranged 300 voice samples obtain from acting sequences from both gender with 5 different emotions namely angry, sad, neutral, happy and boredom each of the faces are normalized into a square. Only the frontal portion of the faces are considered and the background is, the Eigen vectors corresponding to each face are extracted are stored in the database.

MFCC features are extracted against each speech template pertaining to a particular emotion, the procedure is repeated for the speech template these features are given as input to the skew Gaussian distribution [13, 14].

The probability density function obtain against each of speech template are stored. the concept of fusion is acquired to normalize the emotional state of mind by fusing the PDF values obtained against the MFCC features, the main advantage of this model is that the emotional state of mind of an individual can either the understand by the inward physical appearance or outward inherent feelings, To validate the results derived using the metrics are True positive (+ve), False (−ve) negative, Precision and Recall.

(i) The true positive considers all the emotions and are classified as particular emotions among all the emotions which truly have these particular emotions.

$$\text{Tp Rate} = \text{Tp}/(\text{Tp} + \text{FN}) \tag{6}$$

where FN denotes False Negative.

(ii) False Negative, are those emotions which are wrongly classified and it is denoted as
FP = (Sum of the total emotions/the particular emotion) * Total No of emotions.

(iii) Precision: it is denoted as (The rate of a particular emotion)/(Sum of remaining emotions + the emotion).

Table 1 Comparison of confusion matrix to identify different emotions of male-using databases-1 (Kiet Engg College Students)

Stimulation	Recognition emotion (%)/proposed model				
	Angry	Boredom	Happy	Sadness	Neutral
Angry	92	0	0	8	0
Boredom	5	85	0	10	0
Happy	0	0	90	0	10
Sadness	0	10	0	86	4
Neutral	0	10	0	8	82

Table 2 Precision and recall values of male emotion-using databases-1 (Kiet Engg College Students)

	Precision	Recall
Angry	0.92	0.94
Boredom	0.85	0.8
Happy	0.9	1.00
Sad	0.86	0.76
Neutral	0.82	0.85

(iv) The Recall is probability that the particular speech signal belongs to a particular emotions as being in a class X, if it is actually belongs to that class.

The developed model is tested for recognition accuracy on both the databases-1 (KIET ENGG COLLEGE Students) (Table 1).

The recognition rate of about 90 % is recorded for Happy and Sad and about 80 % for other emotions. The performance of the derived results are tested for accuracy using metric precision and Recall, the results derived are presented below (Table 2).

6 Conclusion

This paper presents a new methodology based on multimodal emotion recognition system using Skew Gaussian Mixture Model. In this paper a model is proposed by fusing the facial expression and the uttered speech voices. The speech signals recorded in WAV format are used for investigation, together with the facial expressions. The emotions of both genders are extracted using MFCC. The testing is carried out using the database generated from the students of KIET Engineering College, Kakinada. The derived results are evaluated using metrics like Precision and Recall. The derived results show that the developed model performs efficiently in recognizing the emotions.

References

1. M. Chinnarao, A.V.S.N. Murthy, Ch. Satyanarayana, Emotion recognition system based on skew gaussian mixture model and MFCC coefficients. I. J. Inf. Eng. Electron. Bus. **4**, 51–57 (2015). Published Online July 2015 in MECS (http://www.mecs-press.org/). doi:10.5815/ijieeb.2015.04.07
2. B. Nayak, M. Madhusmitha, D.K. Sahu, Speech emotion recognition using different centred GMM. Int. J. Adv. Res. Comput. Sci. Softw. Eng. **3**(9), 646–649 (2014)
3. R.B. Lanjewar, D.S. Chaudhari, Speech emotion recognition: a review. Int. J. Innovat. Technol. Exploring Eng. **2**(4), (2013)
4. N. Thapliyal, G. Amelia, Speech based emotion recognition with gaussian mixture model. Int. J. Adv. Res. Comput. Eng. Technol. **1**(5), 65–69 (2012)
5. H. Miwa, H. Takanobu, A. Takanishi, Human-like robot head that has personality based on equations of emotion, in *Preprints of the Sixth Symposium on Theory of Machines and Mechanisms* (2000), pp. 1–8
6. A.C. Johnson, Characteristics and tables of the left-truncated normal distribution. Int. J. Adv. Comput. Sci. Appl. (IJACSA) 133–139, (2001)
7. M. Forsyth, M. Jack, Discriminating semi-continuous HMM for speaker verification. IEEE Int. Conf. Acoust. Speech Signal Process. **1**, 313–316 (1994)
8. M. Forsyth, Discrimination observation probability hmm for speaker verification. Speech Commun. **17**, 117–129 (1995)
9. A. George, K. Constantine, Phonemic segmentation using the generalized gamma distribution and small sample bayesian information criterion. Speech Commun (2007). doi:0.1016/j.specom.2007.06.005
10. D. Gregor, et al., Emotion recognition in borderline personality disorder-a review of the literature. J. Pers. Disord. **23**(1), 6–9 (2009)
11. Y.L. Lin, G. Wei, Speech emotion recognition based on HMM and SVM, in *4th International Conference on Machine Learning and Cybernetics*, vol. 8, Guangzhou, 18 Aug 2005, pp. 4898–4901
12. K. Meena, U. Subramaniam, G. Muthusamy, Gender classification in speech recognition using fuzzy logic and neural network. Int. Arab J. Inf. Technol. **10**(5), 477–485 (2013)
13. A. Prasad, P.V.G.D. Prasad Reddy, Y. Srinivas, G. Suvarna Kumar, An emotion recognition system based on LIBSVM from telugu rural dialects of Andhra Pradesh. J. Adv. Res. Comput. Eng. Int. J. **3**(2), (2009)
14. A. Tawari, M. Trivedi, Speech based emotion classification framework for driver assistance system,in *2010 IEEE Intelligent Vehicles Symposium University of California*, San Diego, CA, USA, June 21–24 2010, pp. 174–178

Analysis and Detection of Surface Defects in Ceramic Tile Using Image Processing Techniques

N. Sameer Ahamad and J. Bhaskara Rao

Abstract A large amount of ceramic tiles are constructed in the ceramic tile industry and it is very problematic to monitor the quality of each and every tile manually. It is very difficult to monitor the quality of each and every tile manually. This paper addresses a new technique to avoid such in detecting tile defects. Quality jurisdiction is an important task in the ceramic tile executive. The cost of ceramic tiles also depends on freshness of arrangement, truthfulness of colour, format etc. In ceramic tile factory, the manufacturing process has now performed automatically by industrial computerization system, apart from the observation procedure for ceramic quality classification which is still organize hand-operated. Tile's surface commonly suffers from cracks, holes, spots and corner defects. Classification process is achieved using the human visual appraisal to find and to analyze defect, where human perception is dependent thoroughly on experience and expertise. This operation wants a mechanical arrangement which can furnish an estimate of the ceramic condition precisely and frequently.

Keywords Computer vision · Defect detection · Image processing · Lab VIEW · Feature extraction · Fresh tiles · Morphological operation

1 Introduction

Now a day's Computer technology has been growing and expanding its use in helping to resolve the problems on various aspects of human activity, Due to technology development, the image processing is used in research for scientist and

N. Sameer Ahamad (✉) · J. Bhaskara Rao
Department of ECE, ANITS, Sangivalasa, Bheemunipatnam [M],
Visakhapatnam, India
e-mail: sameerahamad405@gamil.com

J. Bhaskara Rao
e-mail: janabhaskar@gmail.com

© Springer India 2016   575
S.C. Satapathy et al. (eds.), *Microelectronics, Electromagnetics
and Telecommunications*, Lecture Notes in Electrical Engineering 372,
DOI 10.1007/978-81-322-2728-1_54

engineering in the computer technology. Image processing has three steps for dispose of any image. First, it importing the real time sensor or scanner or digital camcorder, second in the image we are maneuver, improvement, condensation, feature eradication from the ceramic tiles, at minimum achieve result after implement various investigation on the image [1].

The construction operation at ceramic tiles manufactory have been done undoubtedly by instrument through industrial computerization system apart from the cross-examination process for ceramic quality arrangement which is still implement mutually by the interference of individual operator. This approach has excessive imperfection, similar as time performance, veracity and courage, which are much overpriced, since it depends upon laborer to task in conversion. The alternative hitch is individuality since it is totally motivate by the involvement and judgment of the employee. This deficiency can start to misstep in the stage of ceramic trait recognition [2, 3].

Analysis involves variety evaluation such as color search, element authentication, and surface defect detection. Hence in this paper we are introducing a capable defect detection approach which can not only shorten the employs task but also certification production quality. Ceramic tile manufacturing is a very complicated process. The production process is associations of chemical, mechanical, thermo-dynamical and other processes which must be deftly conform for good and prosperous manufacturing [4, 5].

2   Existing Defect Detection Methods

Some prospective various fault detection methods that have been prospective to find out the disparate type of image defects. Most of them are to deduce texture feature for detecting defect on ceramic, defect detection on ceramic tile using to deduce shape feature is proposed [1].

Elbehiery et al. [3] confer a few methods to recognize the glitch in the ceramic tiles. They separate their approach into two things. In the early part, Existing design accommodate with the captured images of tiles as input. As the output, they present the excitement regulate or histogram modify image. Subsequently they used the output of early factor as input for the second factor. In the second factor of their design, various specific interdependent image processing procedure have been used in form to analyze various variety of defects. Win out effort focal point the human visual check-up of the defects in the production. But their system is not computerized which is very enough mandatory in the construction process. Again their suggested method is procedure irrelevant because they practice their second unit on whole evaluation image to determine various groups of defects. Moreover, their recommended method is very time utilize.

Rahaman and Hossain [6] gives study for the preparation of the analyzed image noise removal, which was done using median filter and sobel edge detection. In this study, the proposed algorithm is intended to separate tiles into defect or no defect category. Separation had done by comparing the number of defect pixels on the analyzed image with the reference image. Morphological operations were then applied to the defective tile image during the defect classification process.

Hozenki and Keser [7] introduce a ceramic edge defect detection analysis based on contour description. To prepare the image for analysis, Thresholding method was used by the histogram of the foreground image with the background image. For edge detection, canny kernel was used. Afterwards, contour search was performed by tracing five angles (0, 45, −45, 90, −90 degree). The results were clear visible contour based on the shape and geometric structure.

3 Morphological Techniques

Morphological image processing is a collection of non-linear operations related to the shape or morphology of features in an image, such as boundaries, skeletons, etc. In any given technique, we probe an image with a small shape or template called a structuring element, which defines the region of interest or neighborhood around a pixel. Originally it was developed for binary images, and was later extended to grayscale functions and images. The main purpose is to remove imperfections added during segmentation. It is a set of image processing operations that process images based on shapes. Dilation and erosion are two basic operations [8].

4 Methodology Used

The proposed algorithm can be described in the Following step wise operations:

Step 1: Image Pre-processing Operations: The image processing operations includes Acquiescing an image, Re sampling of input image, Colour to Gray Image conversion, noise reduction using median filter.

Step 2: Image Segmentation (Morphology): The image Segmentation includes edge detection, Thresholding, basic and advanced morphology.

Step 3: Feature Extraction: This includes the Feature extraction step to be followed.

Step 4: Defect Measurement: This includes the particle filter measurement and histogram to find the number of pixels.

Step 5: Ceramic Quality Classification: This includes the Fuzzy logic to classify the ceramic tiles.

The following is the flowchart of the system designed in Lab VIEW (Fig. 1).

Fig. 1 Flow chart of the
proposed method

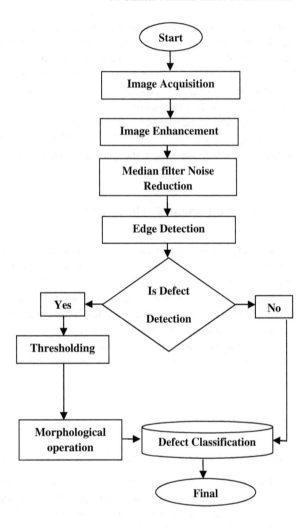

4.1 Fuzzy Logic Grading

In fuzzy logic grading, we can define the number of grades we want we want to break the output into. More the number the number of grades, more will be the accuracy in output grading. Let us take two examples where we define parameters for each case separately and obtain the membership functions graph. The function used to grade the samples here is a triangular waveform. The other waveforms that can be used are trapezoid, singleton, sigmoid, Gaussian waveform shapes. The user can also define the waveform that is to be used for fuzzy logic (Fig. 2 and Table 1).

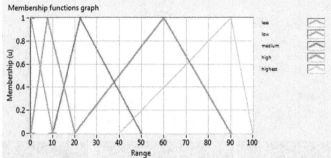

Fig. 2 Membership function graph

Table 1 Fuzzy cases

Least	Low	Medium	High	Highest
0–10	5–20	15–30	20–50	40–100

4.2 Fuzzy Rules

The fuzzy rules are used to classify the ceramic tiles. By using the membership functions in the fuzzy logic we can group the defect percentage into grades. The grades are nothing but the quality of ceramic tile based on defect analysis. The following are the membership functions using fuzzy logic.

1. IF 'percentage' IS 'percentage a' THEN 'grade' IS 'grade a'
2. IF 'percentage' IS 'percentage b' THEN 'grade' IS 'grade b'
3. IF 'percentage' IS 'percentage c' THEN 'grade' IS 'grade c'
4. IF 'percentage' IS 'percentage d' THEN 'grade' IS 'grade d'
5. IF 'percentage' IS 'percentage e' THEN 'grade' IS 'grade e'

5 Results and Discussion

See Figs. 3 and 4.

5.1 Area Calculation

We have re-sampled by considering x-resolution as 600 and y-resolution as 400 and this resolution was made fixed throughout our project. As the resolution was fixed,

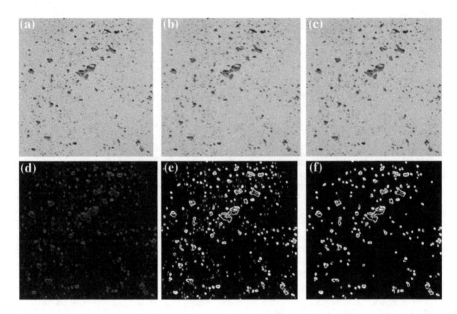

Fig. 3 a Original image. **b** Resampling. **c** Color plane extraction. **d** Edge detection filter.
e Threshold. **f** Advanced morphology to remove small objects

Fig. 4 LabVIEW output of
defect area calculation and
classification

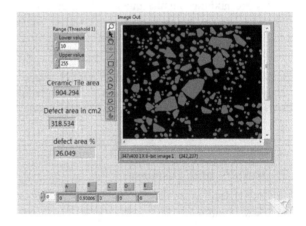

the no. of pixels is confined to 2, 40,000 though the sample was taken at different
heights.

From histogram analysis, we obtained the no. of pixels existing in both defective
and non-defective portion of ceramic tile. As area calculation is required for grading
of our sample, they can be calculated by using the following formulae. [9]

Total sample area	= Total No. of Pixels × Calibration value
Area of sample (Non-Defective Portion)	= No. of Pixels in Non - Defective Portion
	(obtained from Histogram Analysis) × Calibration value
Area of sample (Defective Portion)	= No. of Pixels in Defective Portion
	(obtained from Histogram Analysis) × Calibration value

5.2   Percentage Defect Area

$$\text{Defect Area percentage} = \frac{\text{Area of the defected portion} \times 100}{\text{Total Area of the input sample}}$$

6   Conclusion

This paper is mainly focus on one of the problem generally occurs in this result proves that, Morphological method can be used to detect rectangularity defects in the process of ceramic classification. By using this technique we can to increase the efficiency rate and reduced the total computational time for defect detection from the images of the ceramic tiles. This method plays an important role in ceramic tiles industries to detect the defects (industrial automation) and to analyse the quality of ceramic tiles.

References

1. R. Gonydjaja, Bertalya, T.M. Kusuma, Rectangularity defect detection for ceramic tile using morphological techniques. ARPN J. Eng. Appl. Sci. 9(11) (2014)
2. R. Singh, G.C.Yadav, Corner defect detection based on inverse trigonometric function using image of square ceramic tiles. Int. J. Eng. Comput. Sci. 03(09), 8047–8055 (2014). ISSN:2319-7242
3. H. Elbehiery, A. Hefnawy, M. Elewa, Surface defects detection for ceramic tiles using image processing and morphological techniques. World Academy of Science, Engineering and Technology. Int. J. Comput. Control Quantum Inf. Eng. 1(5) (2007)
4. Y. Zhao, D. Wang, D. Qian, Machine vision based image analysis for the estimation of pear external quality. In Second International Conference on Intelligent Computation Technology and Automation, IEEE, Oct 2009, pp. 629–632. ISBN: 978-0-7695-3804-4
5. R. Mishra, D. Shukla, An automated ceramic tiles defect detection and classification system based on artificial neural network. Int. J. Emerg. Technol. Adv. Eng. 4(3) (2014). www.ijetae. com; ISSN 2250-2459, ISO 9001:2008 Certified Journal
6. G.M.A. Rahaman and M. Hossain, Automatic defect detection and classification technique from image a special case using ceramic tiles. Int. J. Comput. Sci. Inf. Secur. (IJCSIS). 1(1–22) (2009)

7. Ž. Hocenski, T. Keser, Failure detection and isolation in ceramic tile edges based on contour descriptor analysis. In *Proceedings of the 15th Mediterranean Conference on Control and Automation*, 27 July 2007, pp. 1–6. ISBN: 978-1-4244-1282-2
8. R.C. Gonzales, R.E. Woods, *Digital Image Processing*. 2nd edn. (Prentice Hall, 2002). ISBN 0-201-18075-8
9. National Instruments, NI, ni.com, and Lab VIEW are trademarks of National Instruments Corporation. Refer to the Overview of the PID and Fuzzy Logic Toolkit, June 2009 Part Number 351286B-01

Lifeline System for Fisherman

Addanki Sai Charan, Vegesna S.M. Srinivasaverma
and Sk. Noor Mahammad

Abstract This paper provides a novel method to prevent the fishermen from knowingly or unknowingly trawl (fishing) across the International Maritime Boundary Line (IMBL). This method makes use of Global Positioning System (GPS) which provides reliable positioning, navigation and timing services under on a continuous basis. The existing system does not satisfy the safety requirement of fishermen during the fishing because the maritime boundaries between the countries cannot be visible. The main motive behind this paper is to implement an efficient scheme to preserve the safety of the fishermen using GPS.

Keywords RF communication · Global positioning system · Fisher man and IMBL

1 Introduction

In the current world, each county has defined its clear maritime boundaries. Crossing these boundaries will cause imprisonment or death penalty. This is the major problem faced by the fishermen in daily life because the maritime boundaries of two countries can't be identified during the fishing. So it is necessary to identify the maritime boundary during the fishing. The proposed system is not only prevents the fisherman from crossing the International Maritime Boundary Line but also

A. Sai Charan (✉) · V.S.M. Srinivasaverma
Department of Electronics, Indian Institute of Information Technology,
Design and Manufacturing (IIITD&M), Kancheepuram, Chennai 600127, India
e-mail: eds13m003@iiitdm.ac.in

V.S.M. Srinivasaverma
e-mail: eds13m017@iiitdm.ac.in

Sk. Noor Mahammad
Department of CSE, IIITDM, Kancheepuram, Chennai 600127, India
e-mail: noor@iiitdm.ac.in

© Springer India 2016   583
S.C. Satapathy et al. (eds.), *Microelectronics, Electromagnetics*
and Telecommunications, Lecture Notes in Electrical Engineering 372,
DOI 10.1007/978-81-322-2728-1_55

enables the fishermen to report to the Coast Guard on spotting an intruder. This increases the overall security of coast line and also reduces the necessity for periodic patrolling of sea by the coast guard.

2 Literature Review

Main drawback in GSM (Global System for Mobile communications) is communication breaks between the sea vessels, if the link provided by the network provider fails then message cannot be transmitted. We cannot expect proper signal strength in the mid of the sea. The coverage at sea is achieved through setting up an Ad hoc network between the fishing boats. The transceiver unit present in each fishing boat act as signal repeater/regenerator until it reaches the base station in water or in land. Here the communication fails if the intermediate boats are not present at a particular distance from the transmitting boat. Where the popular satellite phones are expensive to install in fishing boats, moreover they are meant for talking purpose hence activities can't be monitored by coast guards. Individual monitoring of all boats through RADAR (Radio Detection and Ranging) is practically not possible. Thus in this proposed method is to achieve reliable communication at sea through RF (Radio Frequency) communication. The proposed system sends a note to fisherman and coastal guard, which saves the fisherman from the imprisonment and death penalty. A RADAR based fisherman communication system is proposed in [1].

The Ultra High Frequency (UHF) transmission provides short wavelength with high frequency. The size of transmission/reception antennas is dependent on the size of the radio wave. The conspicuous antennas with small in size can't be used in the applications of high frequency ranges [2]. The broadcast range of the UHF is limited and which is the major drawback of UHF. For example, the distance between the transmission and reception antennas (line-of-sight) of the UHF based applications like two-way radio systems and cordless telephones is small. UHF can be widely used in Public safety, GSM, business communications and personal radio services such as GMRS, PMR446, UHF CB, and UMTS cellular networks.

3 Proposed System

Radio frequency transceiver architecture is designed with all the practically possible parameters of each and every component in the system is thoroughly considered to proceed for practical implementation. This system acts as self-guiding system for the Fisherman not to rely on other sources at hard times. This system may include:

3.1 Modulation Scheme

In QPSK (Quadrature Phase Shift Keying), the phase of the carrier wave is modulated according to the phase change of the digital information. In simple, QPSK refers to the phase shift keying (PSK) with four states. The binary modulations like OOK and FSK have more peak error rate (PER) than QPSK because QPSK uses temporally shorter packets. Therefore the energy of the transmitter/receiver is reduced due to the less active time for transmission/reception of data/acknowledgements.

3.2 System Requirements

1. The RF System is designed to work for a frequency range, f = 433.05–434.790 MHz, such that it results in providing a Bandwidth (BW) of 1.74 MHz. One of prior requirement is that no other communication devices or systems operating for the same frequency bands should be present. Presence of such systems/devices will result as interference for our RF system [3].
2. One of the goals of our system is to cover a maximum distance of 25 km; therefore the channel considered for wireless transmission will be air over sea as well as land. The calculated free space path loss and sea attenuation loss are 114 dB (approx) and 6 dB respectively. This sums out the channel loss to be 120 dB. Thereby the system is designed keeping this loss figure in mind, and if by any reason or circumstances this loss figure rises and cross a peak of 120 dB attenuation level, then it will result in high rise of BER (Bit error rate) for the resultant transmission.
3. The performance of a system is inversely related to the BER of that system. Therefore we need to strive hard to make sure that our BER of our system always remains below a certain BER value. In our system, we are using QPSK modulation scheme and for it the BER is related to SNR (signal to noise ratio) i.e. BER is a function of SNR, thereby we can calculate a SNR value below which our system should never fall to avoid going below a certain BER. After calculating this SNR value, we require to get the noise floor and thereby calculate noise figure. For this noise floor, we calculate the total amount of signal power which we need to transmit from the transmitting end and from the source end such that after every stage the SNR doesnot go below a certain bed. On the other hand the noise figure is required to calculate the sensitivity for the LNA used at the receiver end [4, 5].

4. For the stated transceiver system considering the power constraints as per regulations of Telecom Regulation Authority of India (TRAI) the minimum power values required at Transmitting end i.e. Base Station (Tx) and Receiving end i.e. Boat side (Rx) needs to be:

Transmitting Input power (Tx) = −10 dBm
Antenna Gain (Tx) = 25 dBi
Antenna Gain (Rx) = 0 dBi

3.3 IMBL Locations

According to our system every fishing boat must consist of a Boat Unit. The Boat Unit consists of a GPS receiver, a buzzer and a user interface to alert during intrusion. The satellite sends the signal to GPS receiver. Then, the GPS converts the received signal into desired data. The data is retrieved by the microcontroller. The IMBL (Latitude and Longitude values) are noted as a boundary line values and are stored as reference values in microcontroller. Real-Time GPS values are compared with that stored values consistently, such that if the value goes beyond that reference stored values, an alarm is set as indication of warning.

In distress condition, fisherman can use the emergency button provided with in the hand-held module such that the Latitude and Longitude coordinates of the current location of the fisherman obtained from GPS receiver are transmitted to the shore station using RF Transmitter embedded along with the GPS module. Accordingly, the shore station people can locate the fisherman and can do rescue operations needed.

Figure 1 shows the maritime boundary between India and Sri Lanka, which is known as Gulf of Mannar. These boundaries are defined by latitude and longitude.

3.4 Infrastructure Requirements

The proposed system supports 25 km RF communication, hence, it proposed that for every 50 km of coastal line distance one base station for RF transmitter setup. All the base stations are interconnected using Internet and finally it is connected to centralized help-line center. Help-line center will give the weather updates and other important information about sea.

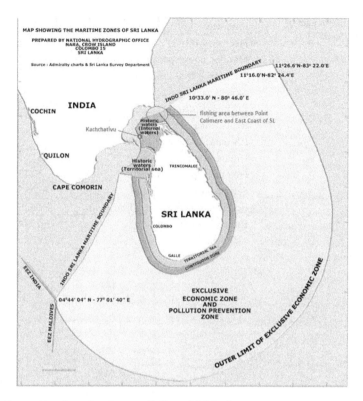

Fig. 1 The maritime boundary between India and Sri Lanka

4 Implementation Details and Results

4.1 Link Budget Analysis

Link budget is defined as the calculating and shaping the gain/loss of the power in the transmission system. It defines the total power required by the transmitter to transmit the signal with a corresponding Signal to Noise Ratio (SNR) and satisfactory Bit Error Rate (BER) [6]. During the link budget estimation, the path loss, distortion, failure by rain, connector losses, cable losses, and antenna gain are considered. Link budget analysis is made considering the required parameters as shown in Table 1.

The proposed system is implemented and verified using ADS software. Transmitter and receiver implementation details are shown in Figs. 2 and 3 respectively. Proposed system block diagram is shown in Fig. 5 and its operation is explained in flowchart as shown in Fig. 4. Enclosure for the proposed system is designed using IP65 standard, Solid-works 3D model is shown in Fig. 6. We can not use any metal material for outer casing of the product as RF module is

Table 1 Link budget of proposed system

Parameter	Value
Frequency (MHz)	433–434.8
Max Tx. power per carrier (dBm)	25
Tx antenna feed loss (dB)	1.5
Tx ant. gain (max) (dBi)	25
EIRP max (dBm)	29.60
Rx. antenna gain (dBi)	0
RX antenna cable loss (dB)	1.5
Distance (km)	25
Free space path loss (dB)	113.13
Sea attenuation loss (dB)	6
Received signal power (dBm)	−96.03
Rx sensitivity (dBm)	−104.8
Link margin	122.4

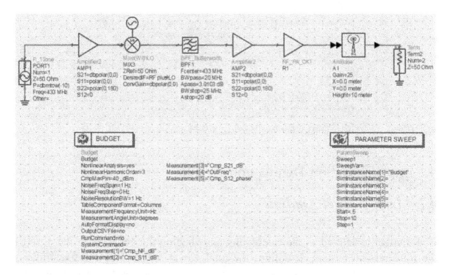

Fig. 2 Proposed transmitter block diagram

incorporated which leads to shielding. Hence, it is recommended to used a low cost engineering plastic (ABS plastic) for outer casing. The proposed system component wise simulation is shown in Fig. 7.

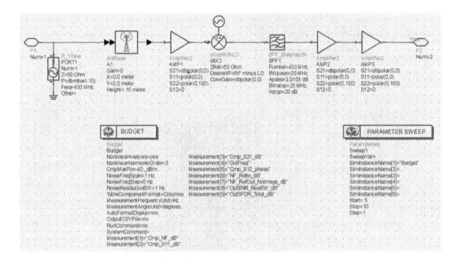

Fig. 3 Proposed receiver block diagram

Fig. 4 Flowchart

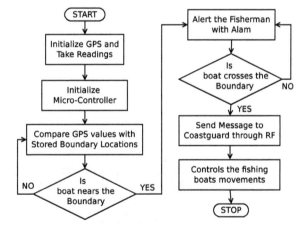

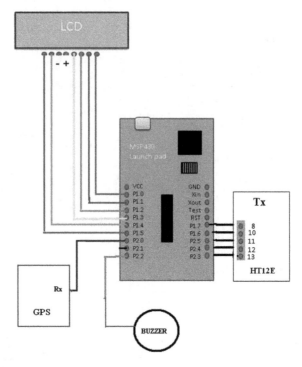

Fig. 5 Block diagram

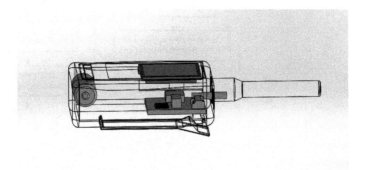

Fig. 6 Prototype

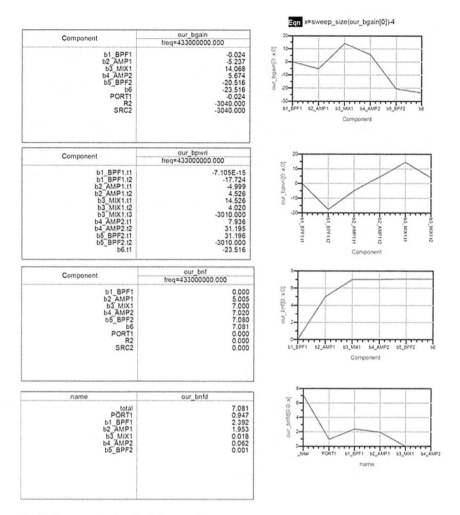

Fig. 7 Component wise simulation results

5 Conclusion

This paper proposed an efficient and cost effective solution for fisherman. The proposed systems warns the fisherman against the crossing the international maritime boundary line. Fisherman will get the sea weather updates to cautioned against the bad weather. In distress situation fisherman can press the emergency button of the proposed system. Where the GPS coordinates will be communicated to help-line center using RF communication, to rescue the fisherman with help of coast guards. The proposed system solution is cheaper than the satellite phones and this system will not give any source for misuse.

References

1. A. Senthilkumar, Portable life protection system for fishermen using global positioning system. Int. J. Emerg. Technol. Adv. Eng. **3**(9), 60–64 (2013)
2. M.S. Vasanthi, T. Rama Rao, RF transceiver design and simulation for indoor wireless applications. In *IEEE Third International Conference on Computing Communication & Networking Technologies (ICCCNT)*, pp. 1–6 (2012)
3. Y. Reza, R. Mason, An energy-efficient transceiver architecture for short range wireless sensor applications. In *IEEE Radio and Wireless Symposium*, pp. 458–461 (2009)
4. A.Y. Wang, C.G. Sodini, On the energy efficiency of wireless transceivers. In *Proceddings of IEEE International Conference on Communication (ICC)*, pp. 3783–3788 (2006)
5. D. Cornell, Modern communications receiver design and technology, 1st edn. (Artech House, 2010)
6. S.R. Bullock, *Transceiver and System Design for Digital Communications*, 3rd edn. (SciTech Publishing, 2000)

Characterisation of Mobile Radio Channel

Lavanya Vadda and G. Sasibhushana Rao

Abstract In densely built-up areas, the transmitted signal from the base station mostly arrives at the mobile station as a multitude of partial waves from different directions. This is known as *multipath propagation*. This effect gives rise to *multipath fading*. Due to this, received signal strength decreases and sometimes unable to recognise. So characterisation and modelling of wireless channel is highly important. The received signal strength in terms of power is measured using RF recorder for analysis at the mobile station at certain time intervals and the signal (in dBm) assumed to be received in multipath environment and is composed of fast fading caused by local multipath propagation and slow fading due to shadowing. In this paper, the real time data is analyzed by separating slow fading and fast fading components using moving average filter and then individual approximation of their cumulative distribution functions (CDF) are compared with the theoretical Rayleigh distribution and lognormal distribution.

Keywords Multipath · Fading · Rayleigh fading · Lognormal distribution · CDF

1 Introduction

Wireless communication refers to the transfer of information using electromagnetic (EM) waves over the atmosphere rather than using any propagation medium. Radio signals exist as a form of electromagnetic wave. During the propagation of the radio signal through the medium, it accounts for all possible propagation paths as well as

L. Vadda (✉) · G. Sasibhushana Rao
Department of Electronics and Communication Engineering,
Andhra University College of Engineering, Visakhapatnam
Andhra Pradesh, India
e-mail: lavanyavadda@gmail.com

G. Sasibhushana Rao
e-mail: sasigps@gmail.com

© Springer India 2016   593
S.C. Satapathy et al. (eds.), *Microelectronics, Electromagnetics
and Telecommunications*, Lecture Notes in Electrical Engineering 372,
DOI 10.1007/978-81-322-2728-1_56

the effects of absorption, attenuation, reflection losses, scintillation, delay spread, angular spread, Doppler spread, dispersion, interference, motion and fading.

The atmosphere reflects, absorbs or scatters radio waves; these waves travel from transmitter to receiver over more number of paths and is referred to as *multipath* propagation, which causes fluctuations in the received signal's amplitude, phase and angle of arrival, giving rise to *multipath fading* [1]. Fading is the deviation of the signal attenuation (the gradual loss in intensity). It is induced by multipath propagation or by shadowing [2, 3], which increases the error rate of received data. The straight path from transmitter to receiver is called line-of-sight (LOS) path and remaining all are non line-of-sight (NLOS) paths because of obstacles. These signals from multiple paths may interfere constructively or destructively at the mobile station. The effect of multipath causes fading.

There are many types of fading that may occur in our real life mobile environment today. There are mainly two types of fading effects characterize mobile communication [4]. They are large-scale and small-scale fading.

Large Scale Fading: Large-scale fading represents the average signal power attenuation or path loss due to motion over large areas. This occurs as the mobile moves through a distance of the order of the cell size, and is typically frequency independent. Signal attenuation due to penetration through buildings and walls is called shadowing. Shadow fading is called slow fading because the duration of the fade may last for multiple seconds or minutes.

Small scale fading: If there is a small change in the spatial separation between the receiving station and transmitting station causes dramatic change in the signal's phase and amplitude is referred as small-scale fading. Small scale fading can be statistically described by Rayleigh and Rician fading models [5].

The mobile signal is combination of fast fading component caused by multipath propagation and slow fading component caused by shadowing. Then envelope of the received signal is expressed as

$$S_u(t) = f(t) + s(t) \tag{1}$$

where f(t) is the fast fading envelope, which closely follows Rayleigh distribution and s(t) is the slow fading component which is lognormally distributed. The envelope $S_u(t)$ has a Suzuki distribution [3]. The Suzuki distribution is rather complicated and not easy to handle mathematically. It would be very useful to have an approximation by a lognormal distribution [3].

The statistical properties of received signal are important for the planning of mobile radio networks and development of digital communication systems. The main intention of this paper is to separate slow variations and fast variations from the received mobile signal using a moving average filter in order to perform an independent study of shadowing and multipath effects.

2 Analysis of the Received Mobile Radio Propagating Signal

In order to analyze the mobile radio propagating signal, the signal is recorded using RF recorder. Total measured data consists of one lakh samples in 5000 s. The measured data is under the multipath environment and is shown in Fig. 1 which illustrates the signal variation at certain time intervals in multipath environment.

In order to normalize the measured data in terms of voltage, first power levels are converted into dBs. To model the voltage, load resistance, R, of 50 Ω is assumed, the power and the voltage are related by the following equation

$$v = \sqrt{2RP} \tag{2}$$

Figure 2 shows the received voltage verses elapsed time using Eq. (2). The received signal is composed of fast fading caused by multipath propagation and slow fading caused by shadowing. In order to study the variations caused by slow and fast fading independently, they are separated by means of a moving average filter.

2.1 Moving Average Filter

The moving average is the most common digital filter in signal processing. The moving average filter is *optimal* for a common task. As the name implies, the moving average filter averages number of points (values) from the input signal to produce each point (value) in the output signal. In equation form,

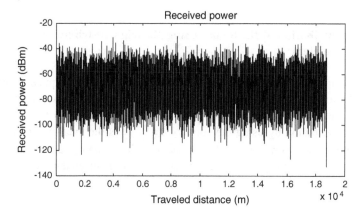

Fig. 1 Received power versus elapsed time

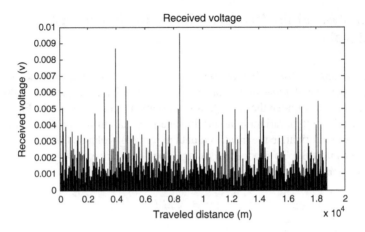

Fig. 2 Received voltage versus elapsed time

$$p[i] = \frac{1}{N}\sum_{j=0}^{N-1} q[i+j] \tag{3}$$

where q[ ] is the input signal, p[ ] is the output signal, N is the number of points in the average, i is the point at which the output is considered and j is an index which run from 0 to N − 1 [6].

The separation of fast and slow variations is performed by moving average filter which implements a rectangular window that is slid through the received signal series. After this process the output of the filtered series contain unreliable samples at the starting and at the last, which can be discarded. For this a window size of 10λ and sample spacing of $\lambda/4$ (samples per wavelength) has been used for separating the fast variations and slow variations.

2.1.1 Normalisation of the Signal Using Rayleigh Distribution Parameters

Measured data is normalized with respect to one of its parameters. Here the expression for the Probability Density Function (PDF), is taken as a function of the mode for the measured data. Mode is estimated from the mean in the case of a Rayleigh distribution by using the equality $\sigma\sqrt{\frac{\pi}{2}}$. Thus the measured data of v has been normalized with respect to its estimated modal value σ, i.e., $v' = v/\sigma$ for fast variations is shown in Fig. 3. The new measured or normalized data, v', should have a modal value equal to one.

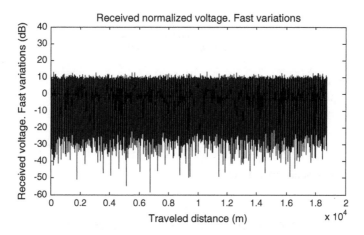

Fig. 3 Voltage time measured data normalized with respect to its estimated modal value

3 Generation of Cumulative Distribution Function (CDF) for Rayleigh Distribution and Lognormal Distribution

In a multipath environment, where the LOS is blocked, then the signal variation are usually represented by a Rayleigh distribution expressed in units of voltage. The fast variations are analyzed by Rayleigh distribution [7]. The probability distribution function (PDF), of the Rayleigh distribution is given by the following equation.

$$f(x) = \frac{x}{\sigma^2}\exp(-\frac{x^2}{2\sigma^2}) \quad for \quad x \geq 0$$
$$= 0 \qquad\qquad otherwise \tag{4}$$

where x is a voltage which actually represents, $|x|$, the magnitude of the complex envelope. This distribution has a single parameter, its mode or modal value, σ. The cumulative distribution function (CDF), is the integral of the PDF. In computing outage probabilities in link budgets, CDF plays a important role.

$$CDF(X) = \int_0^R f(x)dx = 1 - \exp\left(-\frac{X^2}{2\sigma^2}\right) \tag{5}$$

For the analysis of slow variations a lognormal or normal for the variation in dB is used. The probability distribution function, of the log normal distribution is given by [8]

$$f(x) = \frac{1}{x\sigma\sqrt{2\pi}}\exp\left(\frac{-(\ln(x)-\mu)^2}{2\sigma^2}\right) \tag{6}$$

where x is a random variable, μ is mean and σ is standard deviation and the cumulative distribution function is

$$F(x) = \frac{1}{2}\left[1 + erf\left(\frac{x-\mu}{\sigma\sqrt{2}}\right)\right] \tag{7}$$

where, erf represents error function.

4   Comparison of Theoretical CDF and Obtained Sample CDF

To know whether the normalized measured data follows a Rayleigh distribution (for fast fading) and lognormal distribution (for slow fading), the theoretical and the sample CDF's were plotted and observed that the match is reasonably good as

Fig. 4 Time measured data and $\sigma = 1$ Rayleigh CDFs

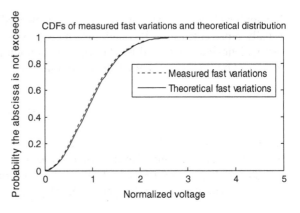

Fig. 5 Time measured data and $\sigma = 1$ lognormal CDFs

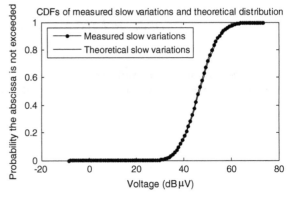

shown in Figs. 4 and 5. The theoretical CDF's have been computed using Eqs. (5) and (7) and the sample CDF using function fCDF.

5   Conclusion

The cumulative distribution function of a signal is used to compute the outage probability in link budget. In this paper fast fading and slow fading are separated and theoretical CDF's of fast fading effect and slow fading effect are compared with the CDF's obtained from real-time data. The standard deviation of the shadowing usually varies from 3 to 12 dB in macro cellular, microcellular and indoor environments. From the plots it is observed that the fading pattern for real-time data is in line with theoretical values. The separation of both types of fading is useful in the development of a proper mitigation technique individually.

References

1. J.D. Parson's, *The Mobile Radio Propagating Channel*, 2nd edn. (Willey, 2000)
2. B. Sklar, Rayleigh fading channels in mobile digital communications, Part I: characterization. IEEE Commun. Mag. **35**(7), 90–100 (1997)
3. H. Suzuki, A statistical model of mobile radio reception. IEEE Trans. Commun. COM **25**, 673–680 (1977)
4. G. Vijaygar, *Wireless Communications and Networking* (Elsevier Inc, 2007)
5. W.C. Jakes, *Microwave Mobile Communications* (WileyInterscience, 1974)
6. S.W. Smith, *The Scientist and Engineer's Guide to Digital Signal Processing* (California Technical Publishing, San Diego, CA, USA ©1997)
7. G. Sasibhushana Rao, *Mobile Cellular Communication* (Pearson Education, New Delhi, 2013)
8. Z. Peyton, J.R. Peebles, *Probability Random Variables and Random Signal Principles*, 4th edn. (McGraw-Hill, 2002)

Implementation of Reversible Arithmetic and Logical Unit and Its BILBO Testing

Sk. Bajidbi, M.S.S. Rukmini and Y. Ratna Babu

Abstract Reversible logic is gaining more importance day by day, because of its feature of low power dissipation which is the basic need in designing nano electronic devices, bioinformatics, low power CMOS designs and quantum computing. Reversible logic is one which realizes n-input n-output functions that map each possible input vector to a unique output vector. It is a promising computer design paradigm for constructing arithmetic and logic units which are the basic building blocks of computer that do not dissipate heat. After designing a system, it is also equal important to test it. In this paper reversible ALU (Arithmetic and Logical Unit) performing four operations (Addition, Multiplication, Subtraction and Bit wise- AND) is implemented and the simulated results like power consumed, delay and area obtained are compared with that of conventional ALU. Testing is also done on proposed reversible ALU by using BILBO (Built—in Logic Block Observer) blocks, which was the first BIST (Built-in Self Test) architecture to be proposed and undergo wide spread use. The proposed reversible ALU is implemented and simulated using Verilog HDL in Xilinx 13.4 version.

Keywords Reversible logic · Low power dissipation · Reversible ALU · BIST (Built-in self test) · BILBO (Built-in logic block observer)

Sk. Bajidbi (✉) · M.S.S. Rukmini · Y. Ratna Babu
Department of ECE, Vignan's University (VFSTR University),
Vadlamudi, Guntur 522 213, A.P., India
e-mail: skbajidbi12@gmail.com

M.S.S. Rukmini
e-mail: mssrukmini@yahoo.co.in

Y. Ratna Babu
e-mail: ratna2k5@yahoo.com

© Springer India 2016
S.C. Satapathy et al. (eds.), *Microelectronics, Electromagnetics and Telecommunications*, Lecture Notes in Electrical Engineering 372,
DOI 10.1007/978-81-322-2728-1_57

601

1 Introduction

In general computers which are irreversible, there is loss of one bit of information for each logical operation carried out in it. But Landauer's principle says that for every one bit information loss, KTln2 (Where k is Boltzmann's constant and T is absolute temperature) joules of energy is dissipated in the form of heat [1]. Although this amount of energy is negligible in case of simple circuits, it becomes very significant in huge circuits [2]. This power dissipation decreases the reliability and lifetime of circuits.

Hence there is a need for typical kind of logic instead of conventional logic which reduces this power dissipation. i.e. Reversible logic [3, 4]. Reversible logic is that in which the charge stored on the cells which consists of transistors is not allowed to lost but it can be reused though reversible computing thus reducing the power dissipation [5]. Due to this unique feature of lowering power dissipation, reversible logic has gain several applications in different fields like quantum computing, nano electronics etc. [6].

While designing a reversible circuit, it is also important to test the circuit for detecting faults and obtain the fault coverage of the circuit [7]. Significant contributions have been made in the literature towards the testing of reversible logic gates and arithmetic units [8–10]. But still there are very little efforts done towards the testing of reversible ALUs [11, 12].

In this paper 16 bit reversible ALU with three arithmetic operations i.e. addition, subtraction, multiplication and a logical operation i.e. bitwise-and is designed and testing is done on reversible ALU using BILBO (Built-in Logic Block Observer) blocks [7].

The reversible ALU with four operations using reversible gates are designed first and it is placed between two BILBO blocks [13, 14]. The test pattern generated by the BILBO block is applied as input to circuit under test i.e. reversible ALU and output obtained is compared with fault free circuit. If the output of circuit under test (reversible ALU) is same as that of fault free circuit then the circuit is said to be fault less circuit [15, 16].

2 Reversible Logic

Reversible logic is one in which there are equal number of inputs and outputs. A reversible circuit can be achieved by adding some inputs called constant inputs and some outputs called garbage outputs to the conventional circuit to make the number of inputs and outputs equal. But these constant inputs and garbage outputs should be maintained low for the better circuit performance.

The basic reversible logic gates encountered during the reversible design are Feynman gate, Toffoli gate, Fredkin gate, Peres gate and Double Peres gate which are discussed in reference papers.

Fig. 1 ALU design

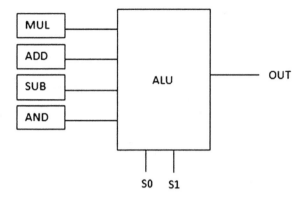

3   16 Bit Reversible ALU Design

3.1   Proposed Work

The below Fig. 1 shows the design of ALU with four operations i.e. Multiplication, Addition, Subtraction and Bitwise-AND. Depending on the two select lines of ALU, the operation to be performed is selected. When s0s1 = 00 multiplication operation will be performed, when s0s1 = 01 addition operation is performed, when s0s1 = 10 subtraction operation is performed, and when s0s1 = 11 logical Bitwise-AND operation is performed. To implement 16-bit reversible ALU each of these sub modules should be implemented using reversible logic.

3.2   Reversible Gates Used in Proposed Work

Instead of fundamental reversible gates like Feynman, fredkin, peres etc., some other reversible gates are used in the proposed work. The reversible gates are:

(1) *HNG gate*
   HNG gate is a 4 * 4 gate with inputs (A, B, C, D) and outputs P = A, Q = B, R = A^B^C, S = ((A^B)&C)^(A&B)^D (Fig. 2).
(2) *HNG2 gate*
   HNG2 gate is a 4 * 4 gate with inputs (A, B, C, D) and outputs P = A, Q = B, R = A^B^C, S = (((~A)^B)&C)^((~A)&B)^D (Fig. 3).
(3) *F2g gate*
   F2g gate is a 3 * 3 gate with inputs (A, B, C) and outputs P = A, Q = A^B, R = A^C (Fig. 4).

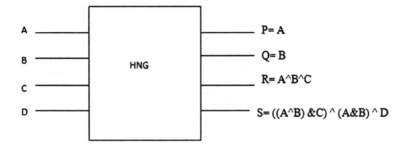

Fig. 2 HNG gate

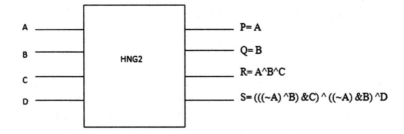

Fig. 3 HNG2 gate

Fig. 4 F2g gate

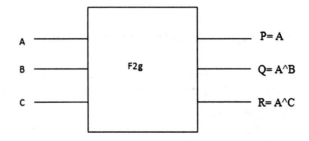

(4) *NFT gate*

NFT gate is a 3 * 3 gate with inputs (A, B, C) and outputs P = A&B, Q = ((~B)&C) ^ (A&(~C)), R = (B&C) ^ (A&(~C)) (Fig. 5).

(5) *IG gate*

IG gate is a 3 * 3 gate with inputs (A, B, C) and outputs P = A, Q = A^B, R = (A&B) ^C, S = (B&D) ^ ((~B) & (A^D)) (Fig. 6).

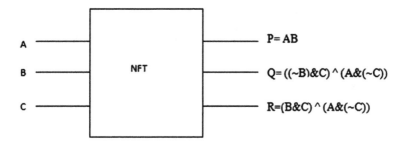

Fig. 5 NFT gate

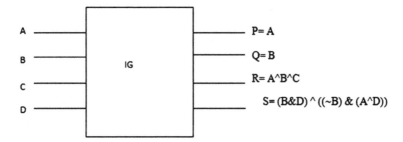

Fig. 6 IG gate

3.3 Design of 16 Bit Reversible Adder

In below Fig. 7 reversible 16 bit adder is shown with 16 HNG gates. The two inputs a, b along with a carry in (c_{in}) and fourth input is maintained as zero (constant input) in each HNG gate. The sum is obtained at third output and carry is obtained at fourth output which is added to next input bits sum whereas the remaining outputs are garbage outputs. Thus the final sum and carry output (c_{out}) are obtained at third and fourth outputs of 16th HNG gate.

This 16 bit reversible adder is designed using verilog HDL, simulated and synthesized using Xilinx 13.4 version.

3.4 Design of 16 Bit Reversible Subtractor

In below Fig. 8 reversible 16 bit subtractor is shown with 16 HNG2 gates. The two inputs a, b along with a carry in (c_{in}) and fourth input is maintained as zero (constant input) in each HNG2 gate. The difference (diff) is obtained at third output and borrow (b0) is obtained at fourth output which is given to next input bits difference whereas the remaining outputs are garbage outputs. Thus the final

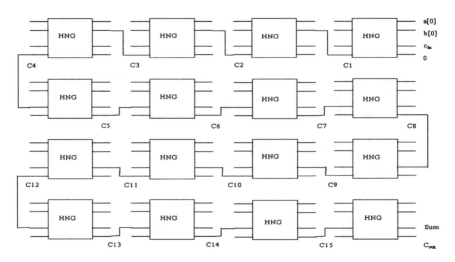

Fig. 7 16 bit reversible adder

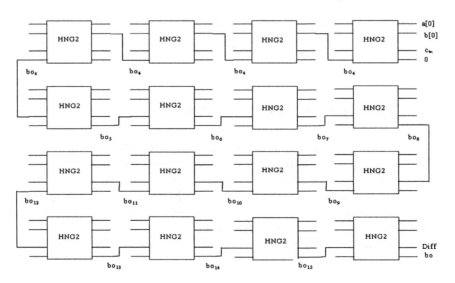

Fig. 8 16 bit reversible subtractor

difference and borrow output (b_o) are obtained at third and fourth outputs of 16th HNG2 gate.

This 16 bit reversible subtractor is designed using verilog HDL, simulated and synthesized using Xilinx 13.4 version.

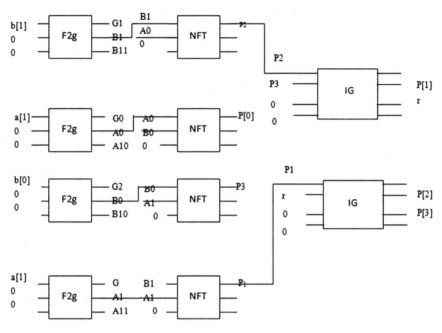

Fig. 9 2 bit reversible multiplier

3.5 Design of 16 Bit Reversible Multiplier

In below Fig. 9 reversible 2 bit multiplier is shown with 4 F2g gates, 4 NFT gates and 2 IG gates. Using this 2 bit reversible multiplier 16 times (like a[0]a[1] × b[0]b [1], a[2]a[3] × b[0]b[1], a[4]a[5] × b[0]b[1] and soon), 16-bit reversible multiplier is obtained. Because of space complexity only 2-bit reversible multiplier is shown. In 2 bit reversible multiplier, each bit of two inputs a, b along with two zeros (constant inputs) are given to four F2g gates and outputs P[0], P[1], P[2] and P[3] are obtained from second NFT gate and two IG gates. The remaining outputs are garbage outputs.

This 2 bit reversible multiplier is designed using verilog HDL, simulated and synthesized using Xilinx 13.4 version.

3.6 Design of 16 Bit Reversible Bitwise-AND

In below Fig. 10 reversible 16 bit Bitwise-AND operation is shown with 16 NFT gates. The two inputs a, b along with third input is maintained as zero (constant input) in each NFT gate. The output 'c' of 16 bits is obtained at first outputs of each NFT gate whereas the remaining outputs are garbage outputs.

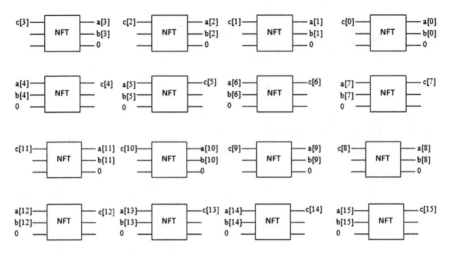

Fig. 10 16 bit reversible bitwise-AND

This 16 bit reversible Bitwise-AND is designed using verilog HDL, simulated and synthesized using Xilinx 13.4 version.

3.7 BILBO Testing of 16 Bit Reversible ALU

It is quite important to test the circuit after designing it. Of the testing methods employed for fault simulation, BILBO (Built-in Logic Block Observer) technique is the best one. It was the first BIST (Built-in Self Test) architecture to be proposed and got huge significance in testing VLSI chips with less performance degradation.

In the below Fig. 11, the block diagram of BILBO testing of reversible ALU is shown. In this, the reversible ALU is placed in between two BILBO blocks which

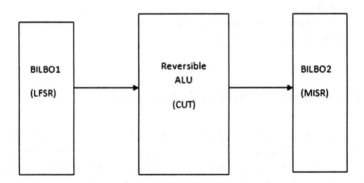

Fig. 11 Block diagram of BILBO testing of reversible ALU

acts as LFSR (Linear Feedback Shift Register) and MISR (Multiple Input Shift Register). During LFSR mode BILBO1 acts as TPG (Test Pattern Generator) and generates test patterns which are applied as inputs to CUT (Circuit Under Test) i.e., Reversible ALU. The response of CUT is given as input to BILBO2 which acts as ORA (Output Response Analyzer) and performs signature analysis and compares the output of circuit with that of fault free circuit. If both the responses are same then the CUT is said to be fault less circuit.

In this proposed work reversible ALU using BILBO blocks is designed using verilog HDL, simulated and synthesized using Xilinx 13.4 version and results are shown in simulation results.

4 Simulation Results

All four operations of reversible ALU are implemented using Verilog HDL in Xilinx 13.4 version and simulations results obtained are as shown below.

4.1 16 Bit Reversible Adder Output

In the figure shown below the addition operation is performed and output is obtained as sum = "0000000000000001" and $c_{out} = 1$ when the inputs are a = "1000000000000000" and b = "1000000000000000" and $c_{in} = 1$ (Fig. 12).

4.2 Reversible Subtractor Output

In the figure shown below the Subtraction operation is performed and output is obtained as diff = "0000000000000010" and borrow b0 = 1 when the inputs are a = "0000000000000001" and b = "1111111111111111" and $c_{in} = 0$ (Fig. 13).

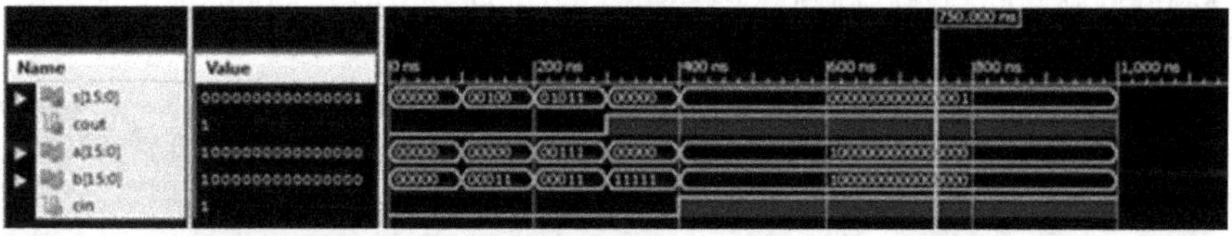

Fig. 12 Output response of 16-bit reversible adder

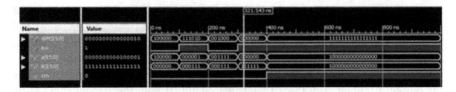

Fig. 13 Output response of 16-bit reversible subtractor

Fig. 14 Output response of 16-bit reversible multiplier

4.3 Reversible Multiplier Output

In the figure shown below the Multiplication operation is performed and output is obtained as c = "00000000000000000000000000001000" when the inputs are a = "0000000000000100" and b = "0000000000000010" (Fig. 14).

4.4 Reversible Bitwise-AND Output

In the figure shown below the Bitwise-AND operation is performed and output is obtained as c = "0000000000000000" when the inputs are a = "0000000000000000" and b = "0000000000000000" (Fig. 15).

4.5 Reversible ALU Output

In the figure shown below the reversible ALU output is obtained as out = "0000 000000000000000000000000000001" when the inputs are a = "0000000000000001",

Fig. 15 Output response of 16-bit reversible bitwise-AND

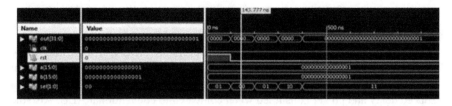

Fig. 16 Output response of 16-bit reversible ALU

b = "0000000000000001", clk = 0, rst = 0, sel = 00 (i.e. multiplication operation) (Fig. 16).

4.6 BILBO Testing Output

In the figure shown below the BILBO testing output is obtained as out = 0 when clk = 1, rst = 0 and sel = 01 (i.e. addition operation). The output obtained here is same as the output of reversible ALU when select line is '01'. Hence 16-bit reversible ALU designed is a fault free circuit (Fig. 17).

5 Synthesis Results

Table 1 shows the analysis of parameters like slices utilization, LUT's utilization, delay and power of conventional 16-bit ALU and that of reversible 16-bit ALU. From this table it is proved that reversible circuits are more beneficial than conventional ones in terms of delay, power and area.

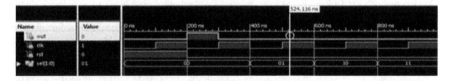

Fig. 17 Output response of BILBO testing of 16-bit reversible ALU

Table 1 Comparision table of 16-bit conventional ALU and reversible ALU

Type of architecture	Slices (utilization)	LUTs (utilization)	Delay (ns)	Power (mW)
Conventional ALU	479 (11 %)	892 (10 %)	39.135	2.31
Reversible ALU	346 (7 %)	650 (6 %)	28.346	0.81

6 Conclusion and Future Scope

In this paper design and synthesis of 16-bit reversible ALU and its testing using BILBO blocks is proposed. The reversible logic significantly decreases the loss of information bits and hence reduces power dissipation. The proposed ALU showed a reduced delay and power consumed when compared to conventional ALU. Also the BILBO testing done on reversible ALU is the best technique for fault simulation with faster diagnosis and less effort of designing the testing process. Further the BILBO technique on reversible ALU proposed in this work can be applied to various digital circuits for fault simulation.

References

1. H. Thapliyal, N. Ranganathan, S. Kotiyal, Design of testable reversible sequential circuits. IEEE Trans. Very Large Scale Integration (VLSI) Syst. **21**(7) (2013)
2. B. Raghu Kanth, B. Murali Krishna, M. Sridhar, V.G. Santhi Swaroop, A distinguish between reversible and conventional logic gates. Int. J. Eng. Res. Appl. (IJERA) **2**(2), 148–151 (2012). ISSN: 2248–9622
3. H.R. Bhagyalakshmi, M.K. Venkatesha, An improved design of a multiplier using reversible logic gates. Int. J. Eng. Sci. Technol. **2**(8), 3838–3845 (2010)
4. H. Thapliyal, N. Ranganathan, Cirucit for reversible quantum multiplier on binary tree optimizing ancilla and garbage bits. In: *Proceedings of the 27th International Conference on VLSI Design*, Mumbai, India, Jan 2014
5. A. Jamal, J.P. Prasad, Design of low power counters using reversible logic. Int. J. Innov. Res. Sci. Eng. Technol. **3**(5) (2014). An ISO:3297:2007 Certified Organization
6. J.P. Hayes, I. Polian, B. Becker, Testing For missing gate faults in reversible circuits. In *Proceedings of 13th Asian Test Symposium*, 2004, pp. 100–105
7. M. Morrison, N. Ranganathan, Design of a reversible ALU based on novel programmable reversible logic gate structures. To Appear In *The IEEE International Symposium on VLSI* (2011)
8. M. Suresh, A.K. Panda, M.K. Sukla, Design of arithmetic circuits using reversible logic gates and power. In *International Symposium on Electronic System Design* (2010)
9. A. Banerjee, A. Pathak, Reversible multiplier circuit. In *Proceedings of 3rd International Conference on Emerging Trends in Engineering and Technology*, 2010, pp. 781–786
10. M. Perkowski, J. Biamonte, M. Lukac, Test generation and fault localization for quantum circuits. In *Proceedings of 35th IEEE International Symposium on Multiple Valued Logic*, 2005, pp. 62–65
11. B. Premananda, Y.M. Rravindhranath, Design and synthesis of 16 bit ALU using reversible logic gates. Int. J. Adv. Res. Commun. Eng. **2**(10), 2278–1021 (2013). ISSN(Print):2319–5940. ISSN(Online):Business Media New York 2014
12. Y. Ratna Babu, Y. Syamala, Implementation and testing of multipliers using reversible logic. Int. Conf. Adv. Recent Technol. Commun. Comput. (2011)
13. Y. Syamala, A.V.N. Tilak, Synthesis of multiplexer and demultiplexer circuits using reversible logic. Int. J. Recent Trends Eng. Technol. **4**(3), 34–35 (2010)

14. B. Konemann, J. Mucha, G. Zwiehoff, Built-in logic block observation techniques. In *Proceedings of IEEE Test Conference*, 1979, pp. 37–41
15. I. Polian, T. Fiehn, J.P. Hayes, A family of logical fault models for reversible circuits. In *Proceedings of 14th Asian Test Symposium*, 2005, pp. 422–427
16. J. Chen, D.P. Vasudevan, E. Popovici, M. Schellekensm, Reversible online bist using bidirectional Bilbo. In *Proceedings of ACM International Conference on Computing Frontiers*, 2010, pp. 257–266

Evaluation of Radiation Characteristics of Dipoles in the Presence of Earth

U. Jaya Lakshmi, M. Syamala and B. Kanthamma

Abstract The radiation characteristics of dipoles are evaluated by assuming that the dipoles are isolated in free space or isolated from meatalic bodies. The assumptions are carried out that the dipoles are far away from conducting bodies and reflecting surfaces. In this paper, the radiation characteristics of vertical dipole is carried out in the presence of earth and are presented in sin-θ domain and 3 dimensional patterns are also presented for various heights from the ground.

Keywords Antenna · Far field pattern · Losy earth · Radiation intensity

1 Introduction

Marconi explained the propagation of elctromagnetic waves beyond the line of sight distance to the horizon are also affected by the properties of earth [1, 2]. Sommerfeld considered Marconi's view that the electromagnetic wave was guided along the surface of earth and given a detailed analysis of the radiation problem for vertical diploe over lossy earth [3, 4].

An antenna is an electrical device which converts electrical power into radio waves and vice versa. It is usually used with radio transmitter or radio receiver. The radiation of dipoles are evaluated by assuming that the dipoles are isolated in free space. In other words, the assumptions are carried out that dipoles are far away from the conducting bodies and reflecting surfaces [5, 6]. The earth is regarded as an

U. Jaya Lakshmi (✉)
Department of Technical Education, Visakhapatnam 531001, India
e-mail: u_jayalakshmi@yahoo.co.in

M. Syamala
ANITS College of Engineering, Visakhapatnam 530040, India
e-mail: syamala_misala@yahoo.in

B. Kanthamma
College of Engineering, Andhra University, Visakhapatnam, India

© Springer India 2016
S.C. Satapathy et al. (eds.), *Microelectronics, Electromagnetics and Telecommunications*, Lecture Notes in Electrical Engineering 372, DOI 10.1007/978-81-322-2728-1_58

infinite ideally conducting plane surface due to its high conductivity. The radiation field intensity is calculated by assuming the radiation of dipole is located over the plane earth. The antenna elements or dipoles are always installed with in a few wave lengths away from the surface of the earth or some other reflecting surface. In such cases current slope in the reflecting surface and the radiation patterns are modified. The induced currents are found to depend on frequency, conductivity, permittivity of the reflecting surface. It is interesting to note that the earth behaves like a perfect reflector at low and medium frequencies and it behaves differently at high frequencies [7, 8].

In view of the above facts, an attempt is made to obtain the radiation charac-teristics for vertical dipoles above the ground. The geometry of wire antenna is considered in this paper and the far field strength is obtained. The radiation patterns of vertical dipole antennas in the presence of reflecting earth is presented in polar and 3D form.

2   Far Field Pattern

Any antenna can be successfully measured on either a near-field or far-field range, with appropriate implementation. There are significant cost, size, and complexity details which will lead to a recommendation of one type over the other. In general, farfield ranges are a better choice for lower frequency antennas.

The omni-directional antenna radiates or receives equally well in all directions. It is also called the "non-directional" antenna because it does not favor any particular direction. Figure 1 shows the pattern for an omni-directional antenna, with the four cardinal signals. This type of pattern is commonly associated with verticals, ground planes and other antenna types in which the radiator element is vertical with respect to the Earth's surface.

Fig. 1 Radiation Pattern of Isotropic Radiator

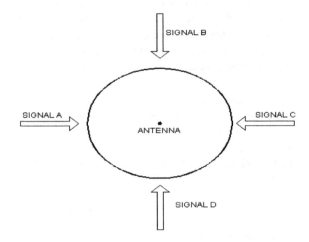

3 The Current Distribution of Randomly Oriented Wire Antenna

When a randomly oriented wire antenna in free space is not isolated and is close to the earth, there is considerable effect on the patterns due to the earth. The earth behaves like an image antenna.

Consider the geometry of a typical wire antenna element as shown in Fig. 2. Assuming xy-plane is flat, homogeneous and lossy earth. The current distribution in the element is given by [9].

$$I(\ell) = \frac{jV}{60X \cos \beta L} \left[\sin \beta(L - |\ell|) + T_U(\cos \beta \ell - \cos \beta L) + T_D \left(\cos \frac{\beta \ell}{2} - \cos \frac{\beta L}{2} \right) \right], \ \beta L \neq \frac{\pi}{2}$$

(1)

or

$$I(\ell) = -\frac{jV}{60X} \left[\sin \beta|\ell| - 1 + T'_U \cos \beta \ell - T'_D \left(\cos \frac{\beta \ell}{2} - \cos \frac{\pi}{4} \right) \right], \quad \beta L = \frac{\pi}{2} \quad (2)$$

Here,

V excitation voltage
ℓ distance along the antenna axis measured from the feeding point
$2L$ length of the antenna

$$\beta = \frac{2\pi}{\lambda}$$

Fig. 2 Isolated wire antenna

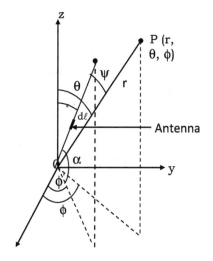

The symbol X in the above expression is defined as,

$$X = \operatorname{cosec}\beta L \int_{-L}^{L} \sin\beta\left(L - |\ell'|\right)\left[\frac{\cos\beta r(0)}{r(0)} - \frac{\cos\beta r(L)}{r(L)}\right]d\ell'$$

$$= 2(1 + \cos\beta L)C_4(\beta d, \beta L) - 2\cos\beta L C_4(\beta d, 2\beta L) - 2\cot\beta L(1 + \cos\beta L)C_\ell(\beta d, \beta L)$$
$$+ (\cot\beta L\cos\beta L - \sin\beta L)C_\ell(\beta d, 2\beta L), \quad \beta L \le \pi/2$$

$$(3)$$

or

$$X = \int_{-L}^{L} \sin\beta\left(L - |\ell'|\right)\left[\frac{\cos\beta r(L - \lambda/4)}{r(L - \lambda/4)} - \frac{\cos\beta r(L)}{r(L)}\right]d\ell'$$

$$= C_4(\beta d, \pi/2) + (1 + \cos 2\beta L)C_4(\beta d, \beta L - \pi/2) - \cos 2\beta L C_4(\beta d, 2\beta L - \pi/2)$$
$$+ \sin 2\beta L[C_4(\beta d, \beta L) - C_4(\beta d, 2\beta L)] - (1 + \cos 2\beta L)C_\ell(\beta d, \beta L) + \cos 2\beta L$$
$$C_\ell(\beta d, 2\beta L) + \sin 2\beta L[C_\ell(\beta d, \beta L - \pi/2) - C_\ell(\beta d, 2\beta L - \pi/2)], \quad \beta L \ge \pi/2$$

$$(4)$$

or

$$X = 2C_4(\beta d, \pi/2) - C_\ell(\beta d, \pi), \quad \beta L = \pi/2 \tag{5}$$

Here

$$r(\ell) = \sqrt{(\ell - \ell')^2 + d^2} \tag{6}$$

2d   diameter of the antenna.

4   The Far Field Distribution of Wire Antenna in Parallel Polarization

The electric vectors of incident and reflected fields are not entirely vertical and having both the horizontal and vertical components. The electric vector contains horizontal components and vertical components.

The total horizontal component in parallel polarization, E_h' is given by,

$$E_h' = E_i(r_1) - R_v E_i(r_2) \tag{7}$$

The total vertical component in parallel polarization, E_v is given by,

$$E_v = E_i(r_1) + R_v E_i(r_2) \tag{8}$$

Here,

R_v reflection coefficient for vertical polarization

$$= \frac{\cos\theta - \left(\beta/\beta'\right)\left\{1 - \left[\left(\beta/\beta'\right)\sin\theta\right]^2\right\}^{\frac{1}{2}}}{\cos\theta + \left(\beta/\beta'\right)\left\{1 - \left[\left(\beta/\beta'\right)\sin\theta\right]^2\right\}^{\frac{1}{2}}} \tag{9}$$

Applying the field expressions (7, 8 and 9), to the antenna element in Fig. 2, the far field in spherical coordinates is given by,

$$
\begin{aligned}
E_\theta = j30\beta\frac{e^{-j\beta r_1}}{r_1}&\left\{\cos\alpha\cos\left(\varphi - \varphi'\right)\cos\theta\int I(\ell)\exp(j\beta\ell\cos\psi)\right.\\
&[1 - R_v\exp(-j2\beta h_\ell\cos\theta)]d\ell\\
&\left.- \sin\alpha\sin\theta\int I(\ell)\exp(j\beta\ell\cos\psi)[1 + R_v\exp(-j2\beta h_\ell\cos\theta)]d\ell\right\}
\end{aligned} \tag{10}
$$

$$
\begin{aligned}
E_\varphi = -j30\beta\frac{e^{-j\beta r_1}}{r_1}&\left\{\cos\alpha\sin\left(\Phi - \Phi'\right)\times\int I(\ell)\exp(j\beta\ell\cos\psi)\right.\\
&\left.[1 + R_h\exp(-j2\beta h_\ell\cos\theta)]d\ell\right\}
\end{aligned} \tag{11}
$$

Here,

$\exp(j\beta\ell\cos\psi)$ phase advance of current element $d\ell$ at ℓ from the feeding point

h_ℓ height of the current element $d\ell$ above the ground

α angle between the antenna and its projection on the ground

ψ angle between the antenna and the direction of $P(r, \theta, \phi)$

$$\cos\psi = \cos\theta\cos\theta' + \sin\theta\sin\theta'\cos\left(\varphi - \varphi'\right)$$

5   Results

The radiation patterns of vertical dipole are evaluated numerically for different heights above earth. The radiation patterns of a vertical dipole above the earth for the heights of $\lambda/2$, λ, 2λ and 3λ are computed using the field expressions and are presented

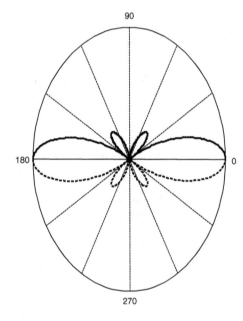

Fig. 3 Radiation pattern of dipole for the height h = $\lambda/2$ from the ground

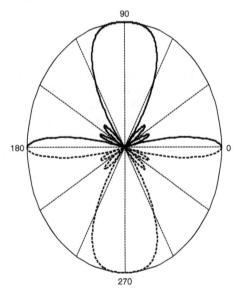

Fig. 4 Radiation pattern of dipole for the height h = λ from the ground

in Figs. 3, 4 and 5. The patterns are presented in 3D form in Figs. 6, 7 and 8 for the realistic view. The earth having high conductivity and behaves as mirror. The image patterns are represented in dotted line.

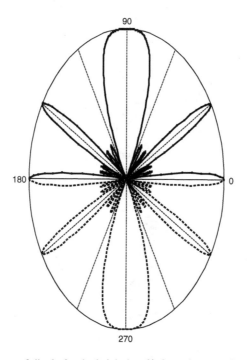

Fig. 5 Radiation pattern of dipole for the height h = 2λ from the ground

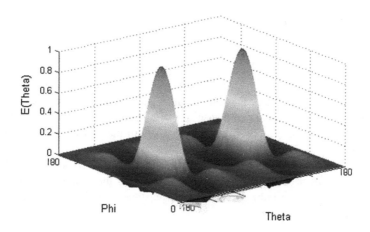

Fig. 6 The 3 dimensional pattern of dipole for the height h = λ/2 from the ground

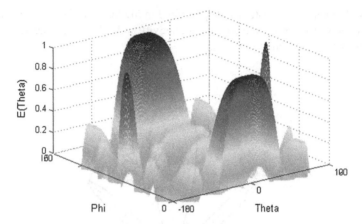

Fig. 7 The 3 dimensional pattern of dipole for the height h = λ from the ground

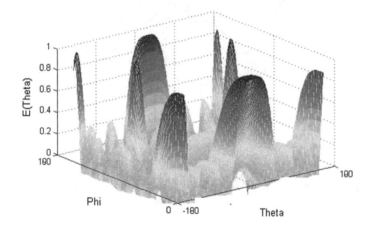

Fig. 8 The 3 dimensional pattern of dipole for the height h = 2λ from the ground

6   Conclusions

From the results, it is found that the radiation patterns are characterized by minor lobes and the main beam has smaller beam width with the influence of earth on the patterns. The patterns are observed that the width of main beam is decreasing with increasing height.

Further, the patterns are found to contain more number of side lobes for larger heights. The effect of earth on the pattern is due to the induced currents in the reflecting surface. The induced currents are dependent on frequency, conductivity and permittivity of the earth.

Acknowledgments The authors of this paper duely acknowledge the encouragement, support and guidance during their thesis work from Prof. G.S.N. Raju, vice-chancellor of Andhra University, Visakhapatnam, Andhra Pradesh, INDIA.

References

1. Marcov, *Antennas* (Progress Publishers, Moscow, USSR, 1965)
2. G.S.N. Raju, *Antenna and Wave Propagation* (Pearson Education PTE Ltd, Singapore, 2005)
3. H. Jasik (ed.), *Antenna Engineering Hand Book* (Mc Graw-Hill, New York, 1961)
4. S. Silver, *Microwave Antenna Theory and Design* (Dover Publications, New York, 1947)
5. R.S. Elliot, Array pattern synthesis. IEEE Antennas Propag. Soc. News Lett.
6. B.D. Steinberg, *Principles of Aperture and Array Systems Design* (Wiley, New York, 1976)
7. D.M. Pozar, *Microwave Engineering*, 4th edn. (Wiley, 2011)
8. C.A. Balanis, *Antenna Theory: Analysis and Design*, 2nd edn. (Wiley, 1997)
9. R.S. Elliott, *Antenna Theory and Design* (Prentice-Hall, USA, 1981)

Non-uniform Circular Array Geometry Synthesis Using Wind Driven Optimization Algorithm

Santosh Kumar Mahto, Arvind Choubey and Sushmita Suman

Abstract This paper describes fast, efficient and global optimization method for pattern synthesis of non-uniform circular array antenna having a minimum side lobe level (SLL) and beam width by controlling the amplitude and position-only using wind driven optimization (WDO) algorithm. The WDO is a new nature-inspired optimization technique based on the movement of air parcel in the earth's atmosphere. It uses a new learning strategy to update the velocity and position of air packets based on their current pressure values. One design example of non-uniform circular array antenna is considered and the results obtained by WDO algorithm is compared with those obtained by other evolutionary algorithms such as GA, PSO, CS, FA, BBO, COA, and MIWO. This algorithm achieves a minimum SLL compared to one of the best results obtained by the cuckoo search algorithm (COA). Also, the learning characteristic shows that WDO algorithm takes less than 50 iterations to determine the optimal excitation amplitude and position of the array element. The simulation results demonstrate the improved performance of the WDO algorithm in terms of directivity, minimum SLL, null control and the rate of convergence compared to other algorithms reported in literature.

Keywords Array antenna · Evolutionary algorithm · Wind driven optimization · Circular array design · Null control · Interference · Sidelobe level

S.K. Mahto (✉) · A. Choubey
Department of Electronics and Communication Engineering,
National Institute of Technology, Jamshedpur, India
e-mail: ec51236@nitjsr.ac.in

A. Choubey
e-mail: achoubey.ece@nitjsr.ac.in

S. Suman
Department of Computer Science Engineering, National Institute
of Technology, Jamshedpur, India
e-mail: sushmita.adm@nitjsr.ac.in

© Springer India 2016
S.C. Satapathy et al. (eds.), *Microelectronics, Electromagnetics*
and Telecommunications, Lecture Notes in Electrical Engineering 372,
DOI 10.1007/978-81-322-2728-1_59

1 Introduction

Several research works have been conducted for the optimal parameters selection of linear and circular arrays for achieving better performance remained as an open research problem. The desired pattern of the antenna array is achieved by controlling the various parameters such as the excitation amplitude-only, the phase-only, the position-only and complex weights (both amplitude and phase) and has its merits and demerits as discussed in [1–14]. Recent studies demonstrate that the circular array antenna performs better than other array geometries such as linear and planar because they provide 360° azimuth coverage and less sensitive to mutual coupling since do not have edge elements [6–14]. It has remarkable applications such as radio direction finding, spatial detection techniques, mobile communication, radar, space navigation, sonar, and other systems. The main design problem in the present context is to determine the optimal selection of parameters for the circular array that provides maximum directivity, minimum SLL while imposing constraints on beam width and other suitable criteria to minimize the interference effect and enhance its performance.

Classical derivative-based techniques for complex multi dimension problems such as linear and non-linear array antenna design are ineffectual, because there is possibility that the solution being trapped in local optima and are unable to determine a global solution. To overcome this problem nature/bio inspired optimization technique is used [2–14]. The population based search algorithms such as Genetic Algorithm (GA) [6] and Particle Swarm Optimization (PSO) is applied in the circular antenna array design problem to ensure maximal SLL suppression subject to constant beam width [6, 7]. A comparative performance analysis of GA, PSO and Differential Evolution (DE) to the above problem is mentioned in [8]. The main objective of his work was to determine the current excitations and phase perturbations of the circular antenna array to provide maximum directivity and SLL suppression. PSO and DE provided comparable results, but were able to outperform GA. Recently Invasive Weed Optimization (IWO) has occupied a special place for solving complex antenna design problems. A modified IWO and an improved version of IWO [9] are used for synthesizing non-uniform circular antenna arrays (NUCAA) with optimized performance. Further, COA performs well compared to others for designing NUCAA in [14].

In this paper wind driven optimization (WDO) is used for designing NUCAA with optimized performance to minimize the interference effect. WDO has recently made a distinct place of its own in solving computational electromagnetic problems as compared to PSO and CLPSO [2]. However, to the best of our knowledge, WDO has not been implemented to optimize circular antenna array geometry. We provide detail simulation results over one design problem of (N = 8) element NUCAA. Comparisons results show that WDO algorithm perform better than other well-known evolutionary optimizers like GA, PSO, CS, FA, BBO, COA, and MIWO in terms of minimum SLL, null control and the rate of convergence.

2 Wind Driven Optimization Algorithm

The WDO is a new global optimization technique which has the capability to employ constraints on the search space similar to particle based algorithm [3–5]. The WDO is inspired by the movement of wind in the earth's atmosphere where it blows to equalize horizontal pressure. Since air density is proportional to temperature. Due to variations in air density and the air pressures at different locations cause the air to move from high to low pressure regions. The pressure gradient (∇P) over a distance is expressed as,

$$\nabla P = \left(\frac{\partial P}{\partial x}, \frac{\partial P}{\partial y}, \frac{\partial P}{\partial z} \right) \tag{1}$$

The net force on the air packet is given as

$$\rho a = \sum F_I \tag{2}$$

where ρ and a is the air density and acceleration of an air parcel respectively. The F_I correspond to all forces acting on the air packets.

The relation between pressure, density and temperature of the air packet are given as

$$P = \rho RT \tag{3}$$

where, P is pressure, R is the universal gas constant and T is temperature.

The pressure gradient force is the vital force that initiates movement of the air parcel, but there are other forces that can also influence its path and speed. The four major forces such as Pressure Gradient Force (*PGF*), Frictional Force (F_f), Gravitational Force (Fg), and Coriolis force (F_C) which describes the motion of the air packet such as velocity and displacement are considered in (2).

2.1 Pressure Gradient Force (PGF)

The *PGF* is the vital force that initiates movement of the air parcel, but there are other forces that can also influence its motion. Considering the fact that air has infinite volume (δV), the *PGF* gradient force can be expressed as,

$$F_{PG} = -\nabla P \delta V \tag{4}$$

2.2 Frictional Force (F_f)

The most obvious force causing the wind motion is the pressure gradient force (PGF). A deterring force that opposes the motion caused by the pressure gradient is the frictional force (F_f), which can be simply written as

$$F_f = -\rho \alpha u \tag{5}$$

where α and u are coefficient of friction and the velocity vector of the wind respectively.

2.3 Gravitational Force (Fg)

The gravitational force can be defined as

$$F_g = \rho \delta V g \tag{6}$$

where g = gravitational constant.

2.4 Coriolis Force

The Coriolis force, arises due to the rotation in the reference time frame caused by the earth's rotation. It is given as

$$F_c = -2\Omega \times u \tag{7}$$

These forces are incorporated in Eq. (2)

$$\rho \frac{\Delta u}{\Delta t} = \rho \delta V g + (-\nabla P \delta V) + (-2\Omega \times u) + (-\rho a u) \tag{8}$$

For simplicity, considering time interval $\Delta t = 1$ and acceleration $a = \frac{\Delta u}{\Delta t}$. Assuming $\delta V = 1$, as the air parcel is infinitesimally small and dimensionless. The velocity update equation [3] is given as

$$u_{new} = (1 - \alpha)u_{cur} - g x_{cur} + \left(RT \left| \frac{1}{i} - 1 \right| (x_{opt} - x_{cur}) \right) + \frac{c * u_{cur}^{otherdim}}{i} \tag{9}$$

where i represent the rank of the particle in the population based on their pressure value at its location, c is a constant that represents the rotation of earth, and u_{cur},

u_{new}, x_{cur} and x_{opt} are initial velocity, new velocity, current position and optimum position respectively.

In Eq. (9) the first term indicates, air packets continues its current trajectory with velocity proportionally trim down by force of friction while neglecting other forces acting on it. The force due to gravity in the second term is proportional to the gravitational constant which continuously drags air packet from its present position in the direction of its center. In Eq. (9) the third term shows the pressure gradient effect. For smaller pressure gradient the air packets with higher ranked will be in a position closer to the optimum position (x_{opt}). The fourth term shows the effect of the Coriolis force.

The velocity of the air packet is updated and then the new updated position is given by Eq. (10)

$$x_{new} = x_{cur} + u_{new} \tag{10}$$

3 Problem Formulation

A circular array antennas having an N-element is shown in Fig. 1. The elements are assumed to be isotropic in nature and un-equally placed on a circle of radius 'r'. The array factor can be written as [11–14]

$$AF(\theta) = \sum_{n=1}^{M} a_n \cdot e^{j(\cos(kr \cdot \cos(\theta - \emptyset_n) + \alpha_n)}, \tag{11}$$

$$kr = \frac{2\pi r}{\lambda_w} = \sum_{i=1}^{N} d_i, \tag{12}$$

where a_n, and α_n are the excitations amplitude, and phase of the nth element, respectively. d_n is the arc distance (arc longitude) between two consecutive elements of array, the wave number is $k = \frac{2\pi}{\lambda_w}$, θ and λ_w the incident angle and the

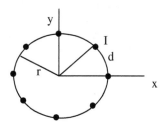

Fig. 1 Circular antenna array geometry

wavelength of the signal respectively. The angular position of nth element is determined by:

$$\emptyset_n = \left(2\pi r/k \cdot r\right) \cdot \sum_{i=1}^{n} d_i \tag{13}$$

For directing the main beam towards the desired angle (θ_0), excitation phase of the nth element is given as $\alpha_n = kr \cos(\theta_0 - \emptyset_n)$.

In this work, the main beam is directed along $\theta_0 = 0°$.

Generally, the main beam of the NUCAA is required to be intended for desired signal, however, signals from other directions to be suppressed.

The power in the main beam increases the corresponding SLL power decreases and vice-versa. The pressure function to control the average SLL suppression is given as [2]

$$Pressure1(\bar{x}) = \sum_{i=1}^{M} \frac{1}{\Delta\theta_i} \int_{\theta_{li}}^{\theta_{ui}} |AF^{\bar{x}}|^2 d\theta \tag{14}$$

where $[\theta_{li}, \theta_{ui}]$ are the regions of SLL suppression, $\Delta\theta_i = \theta_{ui} - \theta_{li}$ and M is the number of regions of SLL suppression.

Due to the increasing population of electromagnetic environment, it is necessary to minimize the unwanted interference by the undesired signal. By controlling the null depth level (NDL) to a desired value, we can minimize the interference effect and pressure function is given as [2]

$$Pressure2 = \sum_{k}^{N} |AF^{\bar{x}}(\theta_{nuk})| \tag{15}$$

The maximum SLL obtained at the desired direction (θ_{msk}) in the specified region in lower bands $[-\pi, \theta_{nuk}]$ [13]

$$Pressure3 = \sum_{k=1}^{N} |AF^{\bar{x}}(\theta_{msk})| \tag{16}$$

where N is number of null and θ_{msk} is the angle at the maximum SLL is obtained in the lower side band $[-\pi, \theta_{nuk}]$.

The size of the circular array antenna is minimized by controlling the perimeter of the antenna and is given as [13]

$$Pressure4 = \sum_{i}^{n} d_i \tag{17}$$

where d_i is the arc distance of ith array element measured from the x-axis.

The main objective of array antenna design is to synthesis array pattern having minimum SLL and beam width by finding an appropriate set of excitation amplitude (a_n) and position of the array element to get the desired array pattern.

The final pressure function is formulated by combining all individual pressure equation given as

$$Pressure = a * Pressure1 + b * Pressure2 + c * Pressure3 + d * Pressure4 \quad (18)$$

where, a, b, c, and d are weight factor, however, by properly selecting the its value the desired radiation pattern can be obtained, and in this paper its values are selected as 1.5, 3, 2 and 0.2 respectively.

4 Numerical Results

To illustrate the simplicity, flexibility and efficacy of the WDO algorithm over the other existing evolutionary optimization methods for circular array pattern synthesis, one design example has been considered. The results are compared with other reported algorithms such as GA, PSO, CS, FA, BBO, COA, and MIWO. The parameters used in WDO algorithm are mentioned in [3]. During initialization, the position of air packet is kept proportional to the excitation amplitude and position of array elements which expedites the convergence of WDO algorithm.

In this example, (N = 8) element circular array antenna is synthesized by optimizing the excitation amplitude and position of array element using WDO algorithm. In order to get the array pattern having minimum SLL in the two spatial region of interest are [180°, 35°] and [−35°, −180°], constant beam width, and desired nulls are placed at −142° and −72° in the lower region of the SLL. The radiation pattern obtained by WDO algorithm is shown in Fig. 2a and the optimized excitation amplitude and position of array elements is given in Table 1. The simulation results of WDO algorithm are compared with other reported algorithms such as GA, PSO, DE, IWO and COA in Table 2 and it demonstrate the improved performance of WDO algorithm compared to others in terms of minimum SLL and directivity.

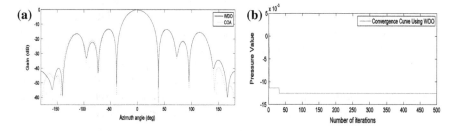

Fig. 2 Normalized pattern and learning characteristics for the 8-elements circular array antenna

Table 1 The amplitude and position of the circular array antenna (N = 8) using WDO algorithm

K	Position (d_n)	Excitation amplitude (a_n)
1	0.7940786	0.32197182
2	0.1407326	0.66894789
3	0.4215873	0.14411802
4	0.8459162	0.78752081
5	0.8963536	0.58480396
6	0.4554067	0.82212964
7	0.8337591	0.80219927
8	0.1640934	0.30264269

Table 2 Optimum results using different algorithms for circular array antenna (N = 8)

Algorithm	MSLL (−dB)	ASLL (−dB)	$\sum d_i(\lambda_\omega)$	Directivity (dB)
GA [6]	09.81	13.70	4.40	10.92
PSO [7]	12.04	20.31	4.43	9.916
MIWO [9]	02.28	10.74	5.95	12.34
COA [10]	13.32	22.60	4.40	14.57
BBO [11]	09.84	15.96	4.43	12.18
FA [13]	12.93	15.48	4.65	11.86
CS [14]	10.31	17.09	4.46	12.78
WDO	**13.61**	**23.23**	4.55	**14.78**

The convergence curve is shown in Fig. 2b. It is evident from the learning characteristics that WDO takes less than 50 iterations to find the optimum sets of controlling parameters (excitation amplitude and position) of array elements.

5   Conclusion

This paper presents a nature based evolutionary computation technique called WDO algorithm for non-uniform circular array antenna design problem having minimum SLL and null control subject to constraints on beam width. The excitation amplitude and position of array elements are optimized using WDO technique. One design example of non-uniform circular array antenna is considered and the results are compared with other algorithms such as GA, PSO, CS, FA, BBO, COA, and MIWO. The simulation results demonstrate improved performance of the WDO algorithm in terms of minimum SLL, directivity, null control and rate of convergence. This algorithm may be considered as a substitute to other evolutionary algorithms for solving electromagnetic and antennas related problems.

References

1. H. Steyskal, R.A. Shore, R.L. Haupt, Methods for null control and their effects on the radiation pattern. IEEE Trans. Antennas Propag. **34**, 404–409 (1986)
2. S.K. Goudos, V. Moysiadou, T. Samaras, K. Siakavara, J.N. Sahalos, Application of a comprehensive learning particle swarm optimizer to unequally spaced linear array synthesis with sidelobe level suppression and null control. IEEE Trans. Antennas Propag. **9**, 125–129 (2010)
3. Z. Bayraktar, M. Komurcu, J.A. Bossard, D.H. Werner, The wind driven optimization technique and its application in electromagnetics. IEEE Trans. Antennas Propag. **61**(3), 771–779 (2013)
4. S.K. Mahto, A. Choubey, A novel hybrid IWO/WDO algorithm for interference minimization of uniformly excited linear sparse array by position-only control, IEEE Antennas Wirel. Propag. Lett. 1–1 (2015)
5. S.K. Mahto, A. Choubey, S. Suman, Linear array synthesis with minimum side lobe level and null control using wind driven optimization, in *IEEE International Conferences SPACES*, India, pp. 191–195
6. M. Panduro, A.L. Mendez, R. Dominguez, G. Romero, Design of non-uniform circular antenna arrays for side lobe reduction using the method of genetic algorithms. Int. J. Electron. Commun. **60**, 713–717 (2006)
7. M. Shihab, Y. Najjar, N. Dib, M. Khodier, Design of non-uniform circular antenna arrays using particle swarm optimization. J. Electr. Eng. **59**(4), 216–220 (2008)
8. M.A. Panduro, C.A. Brizuela, L.I. Balderas, D.A. Acosta, A comparison of genetic algorithms, particle swarm optimization and the differential evolution method for the design of scannable circular antenna arrays. Progr. Electromagnet. Res. B **13**, 171–186 (2009)
9. G.G. Roy, S. Das, P. Chakraborty, P.N. Suganthan, Design of non-uniform circular antenna arrays using a modified invasive weed optimization algorithm. IEEE Trans. Antennas Propag. **59**(1), 110–118 (2011)
10. U. Singh, M. Rattan, Design of linear and circular antenna arrays using cuckoo optimization algorithm. Progr. Electromagnet. Res. C **46**, 1–11 (2014)
11. U. Singh, T.S. Kamal, Design of non-uniform circular antenna arrays using biogeography-based optimization. IET Microwaves Antennas Propag. **5**, 1365–1370 (2011)
12. P. Ghosh, J. Banerjee, S. Das, S.S. Chowdhury, Design of non-uniform circular antenna arrays—an evolutionary algorithm based approach. Progr. Electromagnet. Res. B **43**, 333–354 (2012)
13. S. Ashraf, N.I. Dib, Circular antenna array synthesis using firefly algorithm. Int. J. RF Microwave Comput. Aided Eng. (2013)
14. M. Khodier, Optimization of antenna arrays using the cuckoo search algorithm. IET Microwaves Antennas Propag. **7**(6), 458–464 (2013)

Implementation of ISAR Imaging with Step Frequency and LFM Waveforms Using Gabor Transform

G. Anitha and K.S. Ranga Rao

Abstract Imaging of a target with high resolution is an important task in many Radar applications. In order to obtain high resolution images, Inverse synthetic aperture Radar (ISAR) imaging is implemented. Down range resolution can be obtained by transmitting signals having large bandwidth while cross range resolution is obtained by collecting echo signals obtained from different positions of the target during its rotation. The conventional method in ISAR imaging is Range Doppler (RD) method. In RD method, to obtain the final image either Fourier transform or FFT is used in general. Here in this paper, a new transform named Gabor transform is used which shows clearly the enhancement in the image resolution when compared with the former case. Also a comparison between the images for two different input signals namely Step frequency and LFM waveforms will be shown. The paper is structured as below. In Sect. 1 Introduction is explained. In Sect. 2 Step frequency and LFM waveforms are discussed and Sect. 3 gives the Simulation results. Finally in Sect. 4 Conclusion is given.

Keywords ISAR · RD · Gabor transform · FFT

1 Introduction

ISAR imaging is used to project the reflectivity properties of the target's reflective components onto a two dimensional (2D) plane [1]. The first image dimension is Range or Down-Range (DR). The range axis is the straight line which connects the

G. Anitha (✉) · K.S. Ranga Rao
Department of ECE, JNTU, Kakinada, India
e-mail: anithaguttavelli@gmail.com

K.S. Ranga Rao
e-mail: srrkolluri@gmail.com

K.S. Ranga Rao
NSTL, Visakhapatnam, India

© Springer India 2016
S.C. Satapathy et al. (eds.), *Microelectronics, Electromagnetics
and Telecommunications*, Lecture Notes in Electrical Engineering 372,
DOI 10.1007/978-81-322-2728-1_60

target with the Radar. In order to get high resolution in the DR, pulses having large bandwidth have to be transmitted. This can also be achieved by transmitting waveforms like step frequency signals and LFM waveform. The second image dimension is the Cross-Range (CR). The CR axis is considered to be perpendicular to both range and the target's rotational motion axis. In order to resolve the target points on the CR axis, the target should have relative rotational motion with respect to the Radar. This rotational axis should have at least significant amount of perpendicular component to the range axis.

During target's rotation the reflective points of target will rotate with different relative velocities and reflect the Radar signal with different Doppler shifts. Thus, these Doppler shifts would be evaluated with some kind of signal processing method and virtual CR dimension could be obtained synthetically. The coherent processing of these signals will be done in specific duration of time called coherent integration time. Due to the direct relation in between the CR and the Doppler information it is referred as Range Doppler imaging [2, 3]. Figure 1 shows the image dimensions in 2D.

In general, targets like ship or airplane maneuvers with translational motion (TM) and rotational motion (RM). TM is the motion along the range axis and RM is the motion that causes aspect change of the target from the view of the Radar. In order to generate ISAR images, the system requires some data collection time. Although target's rotation is required for the image generation, in many cases target's rotational and translational motion effects the quality of the image and problems such as blurring or smearing occurs due to coherent intervals. The effect that is caused due to TM is Range offset and Range walk. Range offset causes the range of each scatterer to spread from its actual location to adjacent cells which leads to distorted range profile. Range walk shifts the range profiles by some range bins resulting in an inaccurate position processing of individual scattterers. The

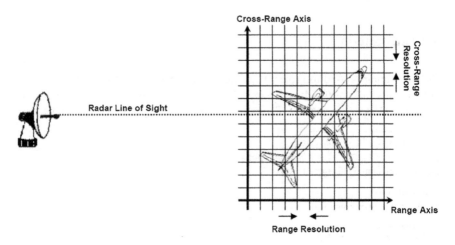

Fig. 1 Image dimensions

effect caused due to RM is cell migration. It occurs by circular motion of the scatterers and leads to distortion of range profile. It can also be said that the radial velocity of the scatterer is not constant throughout the integration time during the rotation [4].

2   ISAR Imaging Using Step Frequency and LFM Waveforms

In HRR Radars, in order to achieve high down range resolution, signals having large bandwidth are required. The selection of Radar signal type is mainly done depending on the application of the Radar. The most commonly used waveforms are Stepped frequency (SF) and linear frequency modulated (LFM) waveform also called Chirp waveform [5].

A Step Frequency waveform

Step frequency waveform is formed by sending a series of single frequency short continuous sub waves and the frequencies between adjacent sub waves is increased by a frequency of Δ_f. For one burst of SF signal, a total number of N continuous wave signals is sent each having a discrete frequency f_n. Each sub wave has time duration of τ and is of distance T_p away from the adjacent sub wave. The total bandwidth, B is given by $B = N. \Delta_f$. The transmitted signal is represented as

$$S(t) = \sum_{m=1}^{M} \sum_{n=1}^{N} \left(\frac{t - t_p}{T_p} \right) \exp(2\pi f_n t) \tag{1}$$

where

$$rect\left(\frac{t - t_p}{T_p} \right) = \left(\begin{matrix} 1, & \frac{t-t_p}{T_p} < 1 \\ 0, & \text{otherwise} \end{matrix} \right) \text{ and } \quad f_n = f_0 + (n - 1) \cdot \Delta f$$

f_n is the carrier frequency of the Nth pulse, f_0 is initial carrier frequency. Here the pulse amplitude is assumed to be normalized to unity during the pulse duration T_p, t_p is the time instant at which the emission of the nth pulse of the Mth burst starts, $n = 1...N$ is the pulse number and $m = 1...M$, the burst number. After compensating both the translational and rotational motion, the final ISAR image is obtained by the expression given below.

$$S_p(m, n) = a_p \, rect\left(\frac{t - t_p}{T_p} \right) \exp\left[j2\pi f_n\left(t - t_p \right) \right] \tag{2}$$

where a_p represents the amplitude of the echo signal.

B LFM waveforms

In Radar systems if short pulses are used it gives wide bandwidth and the resolution is high but limited for short distance detection. Instead if long pulses are used then the detection range is more but resolution is less. In order to achieve wide bandwidth and also long ranges LFM waveforms are used. In LFM waveforms long pulses are frequency modulated so that the resolution is high and is used for long range [5]. This waveform is repeated in every T_{PR} intervals for most common radar applications, especially for localization of targets in the range. The transmitted signal for LFM waveform is given by

$$S\left(\hat{t}, t_m\right) = rect\left(\frac{\hat{t}}{T_P}\right) \exp\left[j2\pi\left(f_c t + \frac{1}{2}Kt^2\right)\right] \tag{3}$$

where

$$rect\left(\frac{\hat{t}}{T_P}\right) = \begin{cases} 1, & \left|\frac{\hat{t}}{T_P}\right| \leq \frac{1}{2} \\ 0, & \left|\frac{\hat{t}}{T_P}\right| > \frac{1}{2} \end{cases}$$

f_c denote the carrier frequency, T_p is pulse width, $\hat{t}$ is fast time and $t_m = n/PRF$ ($n = 0, 1,...N - 1$) is slow time, $t = \hat{t} + t_m$ is full time. N is number of accumulating pulses, PRF is pulse repetition frequency. Suppose that the distance between the Pth scatter and imaging radar at the time of t_m is $R_p(t_m)$, then, the radar echo of this scatter is represented [6] as

$$S_P(\hat{t}, t_m) = a_p rect\left(\frac{\hat{t} - \frac{2R_P(t_m)}{c}}{T_P}\right) \exp\left\{j2\pi\left[f_c\left(t - \frac{2R_P(t_m)}{c}\right) + \frac{1}{2}K(\hat{t} - \frac{2R_P(t_m)}{c})^2\right]\right\}$$

$$\tag{4}$$

In ISAR imaging any of the above two waveforms can be used. Here SF waveforms seems to be technologically more flexible than conventional LFM waveforms, since they consist of single frequency pulses, which are easily generated using a fast, coherent, frequency stepped synthesizer and do not require any wideband processing in the radar receiver whereas, LFM waveforms with large time bandwidth product have high Doppler shift tolerance, though they suffer from the Range Doppler coupling problem [7].

Conventional Fourier Transform is a best method for Range Doppler imaging while obtaining the final ISAR image. But in order to improve the resolution Gabor transform is used. In this paper, Gabor transform is applied in place of Fourier transform. It can be seen that the Gabor transform kernel is the Fourier transform kernel plus a Gaussian function. Since the Gaussian signal is more concentrated than the rectangular function in the frequency domain, the frequency resolution of the Gabor transform is much better than short time Fourier transform [8]. Among all kinds of window functions, the Gabor function is proved to achieve the best analytical resolution in both time and frequency domain. This function is Gaussian modulated by a sinusoidal signal [9]. The generalized function, $G_p(t)$, with normalization of the maximum response in frequency domain is given by

$$G_p(t) = \frac{1}{\sqrt{2\pi\alpha^2}} \exp\left\{ \frac{(t - \tau_p)^2}{2\alpha} \right\} \exp(j2\pi f_p(t - \tau_p)) \tag{5}$$

τ_p is shifting factor, f_p is Gaussian window and α is blurring factor. The above expression represents the Gabor transform which is applied to the received signal for obtaining final ISAR expression given in Eqs. 2 and 4 to get ISAR image.

3 Simulation Results

Here in this section, simulated plane is shown which is constructed with 39 points. The echo signal received from the scatterer is assumed to have equal amplitudes. Figure 2 shows the simulated plane.

Figure 3 shows the ISAR image without any motion compensation i.e., translational and rotational motion. From this image, it can be easily said that the scatterers cannot be identified from this image since it is completely blurred.

A ISAR imaging for Step Frequency waveform

In this section ISAR imaging using SF input is simulated. As already explained in the previous section about the Translational motion, some compensation techniques are applied to remove the effects caused by TM. Once these techniques are applied then the image is obtained as shown in Fig. 4. When Figs. 3 and 4 are compared, then it is clearly observed in Fig. 4 the image quality is improved to some extent

Fig. 2 Simulated plane

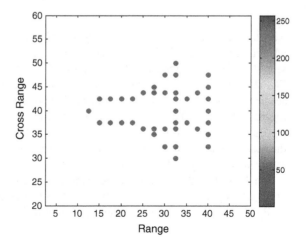

Fig. 3 ISAR image before
motion compensation

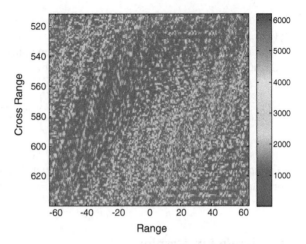

Fig. 4 ISAR image after
TMC for SF

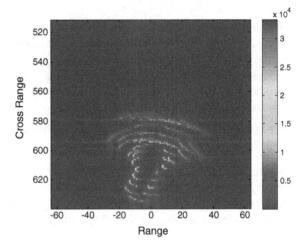

with that from Fig. 3. Because in Fig. 4 TM compensation technique is applied
whereas it is not done in Fig. 3. Figure 5 gives the DR profile. In this figure it can be
seen that there are only 12 scatterers resolved for 39 scattering points as shown in
Fig. 2. This is because some of the scatterers are in the same DR gate. Thus
scatterers that are in the same DR gate can't be resolved by 1D HRR profile. In this
regard only 12 scatterers are resolved. When it comes to cross range Doppler
processing, as shown in Fig. 6, here only 11 scatterers are resolved since some of
the scatterers are in the same CR gate, so there are only 11 scatterers that can be
resolved according to Figs. 6 and 7.

Finally in Fig. 8 the ISAR image using Gabor transform is shown. The image is
obtained only after applying rotational motion compensation. From Fig. 8 one can
easily say the image is having high resolution and good quality.

Fig. 5 Down range profile for SF

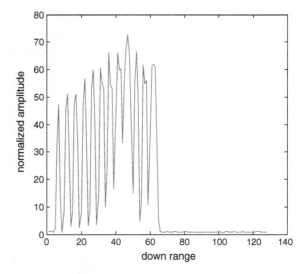

Fig. 6 Cross range profile for SF

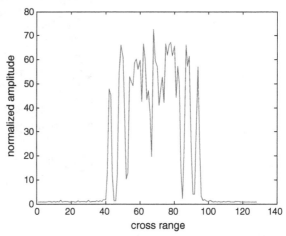

Fig. 7 Down range and cross range profiles for SF

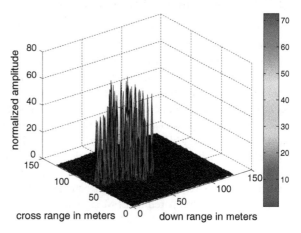

Fig. 8 Final ISAR image for SF

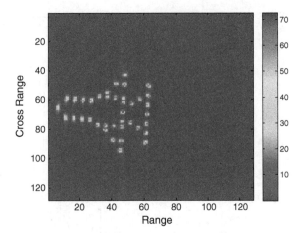

Fig. 9 ISAR image after TMC for LFM

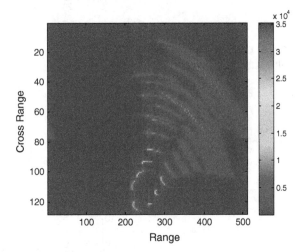

B ISAR imaging using LFM Waveforms

Now the simulation is carried out using LFM waveforms and the figures are shown below in Figs. 9, 10, 11, 12 and 13.

Figure 9 shows the image after TMC and before applying Gabor transform. This image is deteriorated when compared to Fig. 13. Figure 10 shows the DR profile and Fig. 11 the CR profile. In these two images the peaks identified are more when compared to Figs. 5 and 6. Here instead of 12 peaks for DR profile and 11 peaks for CR profile many peaks are identified in these two figures. So depending on the DR and CR profiles, Fig. 12 gives the combined DR and CR profile in 3-D. Number of peaks are spread over throughout the scale for LFM input. The final ISAR image for LFM as shown in Fig. 13 is not with good quality when compared with Fig. 8 which is the ISAR image for step frequency. The image obtained with step frequency is better compared to LFM signal.

Fig. 10 Down range profile for LFM

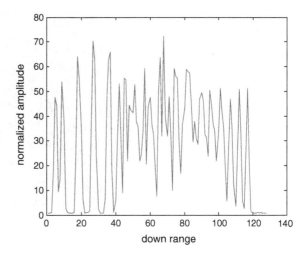

Fig. 11 Cross range profile for LFM

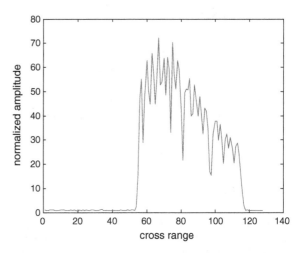

Fig. 12 Down range and cross range profile for LFM

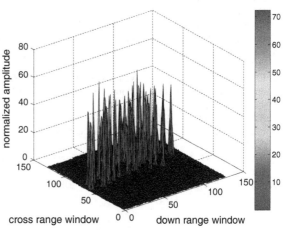

Fig. 13 Final ISAR image for LFM

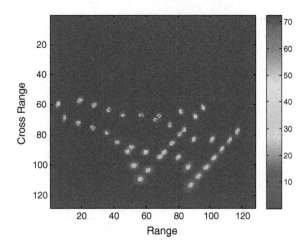

4 Conclusions

In this paper, two Radar waveforms namely Step Frequency and LFM waveforms are used for high DR resolution and Gabor transform is applied for implementing ISAR imaging with high resolution in CR dimension. Here from the simulated waveforms, the ISAR image obtained for the SF waveform is showing better results than for LFM waveform. SF waveforms appear to be more flexible than LFM waveforms.

References

1. D.R. Wehner, *High Resolution Radar* (Artech House Inc, 1995), pp. 367–380
2. Y. Li, R. Wu, M. Xing et al., Inverse synthetic aperture radar imaging of ship target with complex motion. IET Radar Sonar Navig. **3**, 449–460 (2009)
3. G. Zhao Zhao, L. Ya-Chao, X. Mengdao, et al., ISAR imaging of maneuvering targets with the range instantaneous chirp rate technique. IET Radar Sonar Navig. **3**, 449–460 (2009)
4. L. Hongya, *ISAR Imaging With LFM Waveforms* (IEEE Publications, 2007), pp. 729–734
5. Caner Ozdemir.: Inverse Synthetic Aperture Radar Imaging with MATLAB Algorithms (Wiley, 2012), pp. 51–66
6. L. Wu, X. Wei, D. Yang, H. Wang, X. Li, ISAR imaging of targets with complex motion based on discrete chirp fourier transform for cubic chirps. IEEE Trans. Geosci. Remote Sens. **50**, 4201–4212 (2012)
7. A.V. Karakasiliotis, A.D. Lazarov, P.V. Frangos et al., Two-dimensional ISAR model and image reconstruction with stepped frequency-modulated signal. IET Signal Process. **2**, 277–290 (2008)
8. V.C. Chen, H. Ling, *Time-Frequency Transforms for Radar Imaging and Signal Analysis* (Artech House, 2001), pp. 9–11
9. T.S. Lee, Image representation using 2D gabor wavelets. IEEE Trans. Pattern Anal. Mach. Intell. **18**, 959–971 (1996)

Design of Low Power and High Speed Carry Look Ahead Adder (CLAA) Based on Hybrid CMOS Logic Style

Vinay Kumar, Chandan Kumar Jha, Gaurav Thapa
and Anup Dandapat

Abstract Parallel Adders or Ripple Carry Adders (RCA) are the building blocks of the digital processing units. However these RCAs have disadvantage of large propagation delay because of the propagation of carry from the first adder to the last adder. To overcome this drawback, Carry Look Ahead Adders (CLAA) are used. But again these CLAAs consumes more power because of use of large number of gates for generating carry. In this paper, we present a design of CLAA based on hybrid CMOS logic style which consumes less power as compared to the designs of CLAA based on complementary CMOS, Pass Transistor Logic (PTL) and Transmission Gate (TG). The proposed design of CLAA has less propagation delay as compared to its existing designs. Also the number of transistors required for the proposed design of CLAA are less as compared to the existing designs based on other logic style.

Keywords Carry look ahead adder (CLAA) · CMOS · Pass transistor logic (PTL) · Ripple carry adder (RCA) · Transmission gate (TG)

1 Introduction

Parallel Adders or Ripple carry adders (RCA) are most important arithmetic circuits for a digital system [1]. They are widely used in arithmetic logic units and digital signal processors [2]. These parallel adders or RCAs consist of parallel connection of full-adders where carry out of each adder acts as a carry in of the next adder [3].

V. Kumar (✉) · C.K. Jha · G. Thapa · A. Dandapat
Department of ECE, National Institute of Technology, Meghalaya, Shillong
e-mail: vinaynits@gmail.com

C.K. Jha
e-mail: chandanmu11ec06@nitm.ac.in

G. Thapa
e-mail: gauravmu11ec17@nitm.ac.in

A. Dandapat
e-mail: anup.dandapat@gmail.com

© Springer India 2016 645
S.C. Satapathy et al. (eds.), *Microelectronics, Electromagnetics
and Telecommunications*, Lecture Notes in Electrical Engineering 372,
DOI 10.1007/978-81-322-2728-1_61

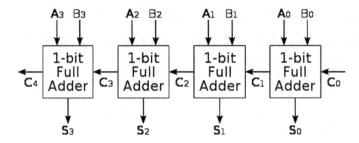

Fig. 1 Diagram of 4-bit parallel adder or RCA

A 4-bit RCA is shown in Fig. 1. RCA suffers from the drawback of large propagation delay because the carry generated from the first adder of RCA propagates through next adders to reach the last adder [4]. This propagation of carry through adders decreases the speed of the RCA. To overcome this drawback, Carry Look Ahead Adders (CLAA) are used where carry of the each adder is generated from the input bits and hence does not depend on the carry out of the previous adders [5]. Thus the carry does not have to propagate from each adder to next adder and thereby increasing the speed of the adder blocks. The CLAA requires large number of AND and OR Gates for generating carry out of each adder as compared to RCA. Hence the design of CLLA requires more number of transistors and thus increasing the layout area as compared to RCA. This also increases the power consumption of CLAA as compared to RCA. This paper presents a design of CLAA based on Hybrid CMOS Logic Style and compares its performance with design of CLAA using different logic styles like complementary CMOS, Pass Transistor Logic (PTL) and Transmission Gate (TG) [6].

1.1 Working of CLAA

The propagation delay of RCA is very large because the carry out of each adder propagates through next adders to reach the last adder of the RCA. But the CLAA eliminates this propagation of carry by generating the carry out of each adder from input bits which acts as a carry in of the next adder and hence the carry does not have to propagate. The CLAA requires two extra functions for each full-adder known as Carry Generated (CG) and Carry Propagated (CP).

The Carry is generated by the full-adder only when both the input bits A and B are high. Thus the carry generated function (CG) is given as AND operation between input bits. The carry in is propagated by the full-adder only when both the input bits A and B are high or either of them is high. So the carry propagated function is given as the OR operation between input bits. The carry-out (C_{out}) of a full-adder is '1' if the CG is '1' or both CP and C_{in} are '1'. Hence the C_{out} function

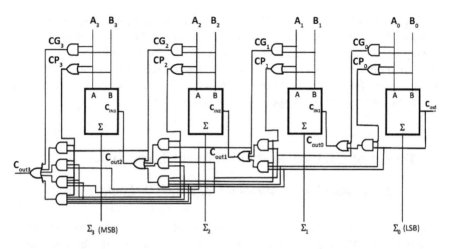

Fig. 2 Diagram of 4-bit CLAA

is given as OR operation of CG with AND operation between CP and C_{in}. The functions CG, CP and C_{out} of nth adder of CLAA are given as:

$$CG_n = A_n \cdot B_n \tag{1}$$

$$CP_n = A_n + B_n \tag{2}$$

$$C_{outn} = CG_n + CP_n \cdot C_{inn} \tag{3}$$

Consider a 4-bit CLAA shown in Fig. 2 in which the carry-out (C_{out}) of each full adder is dependent on its CG, CP and C_{in}. The CG and CP functions of each adder are immediately available as soon as the input bits A and B and C_{in} to the first adder are applied because they are dependent only on these bits. The carry input to each adder is the carry output of the previous adder. The carry-outs C_{out0}, C_{out1}, C_{out2} and C_{out3} of four full-adders FA_0, FA_1, FA_2 and FA_3 of CLAA can be expressed respectively as:

$$C_{out0} = CG_0 + CP_0 \cdot C_{in0} \tag{4}$$

$$C_{out1} = CG_1 + CP_1 \cdot C_{in1} = CG_1 + CP_1 \cdot CG_0 + CP_1 \cdot CP_0 \cdot C_{in0} \tag{5}$$

$$\begin{aligned} C_{out2} &= CG_2 + CP_2 \cdot C_{in2} \\ &= CG_2 + CP_2 \cdot CG_1 + CP_2 \cdot CP_1 \cdot CG_0 + CP_2 \cdot CP_1 \cdot CP_0 \cdot C_{in0} \end{aligned} \tag{6}$$

$$\begin{aligned} C_{out3} &= CG_3 + CP_3 \cdot C_{in3} = CG_3 + CP_3 \cdot CG_2 + CP_3 \cdot CP_2 \cdot CG_1 + CP_3 \cdot CP_2 \\ &\quad \cdot CP_1 \cdot CG_0 + CP_3 \cdot CP_2 \cdot CP_1 \cdot CP_0 \cdot C_{in0} \end{aligned} \tag{7}$$

Expression of C_{out} of each full-adder of CLAA is dependent only on the initial carry-in, C_{in0}, its CG and CP functions and CG and CP functions of the previous adder. Since CG and CP functions can be expressed in terms of inputs A and B of the full adders, carry-out of each adder are immediately available and there is no need to wait for a carry to propagate through all adders before a final result is achieved. Hence the CLAA speeds up the addition Process.

2   Design of CLAA Based on Different Logic Styles

2.1   CLAA Based on Complementary CMOS Logic

The complementary CMOS logic style consists of a block of NMOS and PMOS. The block of NMOS is called Pull-Down Network because it pulls down the output to ground based on corresponding inputs and the block of PMOS is called Pull-UP Network because it pulls up the output to power supply based on corresponding inputs [7]. This type of logic style style has the advantage of full swing as compared to other logic style like pass transistor logic (PTL) and Transmission Gate (TG). But the power consumption of this logic style is large as compared to other logic styles. Also this logic style requires more transistor for implementing the circuits and hence increasing the layout area of the circuit. The sum function of each adder of CLAA based on complementary CMOS logic style is shown in Fig. 3a. It requires 13 NMOS and 13 PMOS and thus a total of 26 transistors. The carry out of each adder of CLAA is obtained from carry generated (CG) and carry propagated function (CP) which are obtained with the help of AND and OR gates. The 2-input AND and OR gates based on complementary CMOS logic style are shown in Fig. 3b, c respectively. The higher input AND and OR gates have been obtained following the logic of 2-input gates.

2.2   CLAA Based on Pass Transistor Logic (PTL)

The Pass transistor logic (PTL) style requires less number of transistor as compared to complementary CMOS logic style and hence consumes less layout area for the circuits [8]. The power consumption for this logic style is comparatively less as compared to other logic style like Complementary CMOS and transmission gates (TG) because there is no direct connection between output and the power supplies. But this logic style suffers from major drawback of drop in the logic swing of the outputs [9]. Hence buffers are used after each stage to restore the logic swing. But the use of buffer after each stage increases the number of transistors and power consumption. The sum function of each adder of CLAA based on PTL is shown in

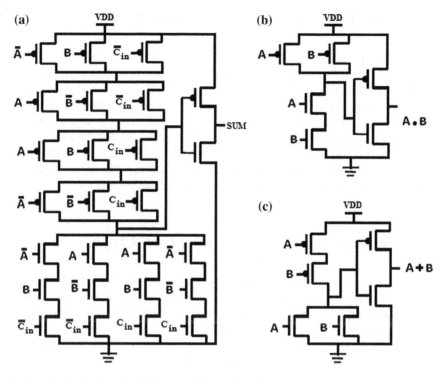

Fig. 3 Circuit of **a** Sum. **b** AND gate. **c** OR gate based on complementary CMOS

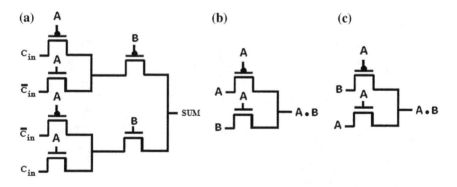

Fig. 4 Circuit of **a** SUM **b** AND gate **c** OR gate based on PTL

Fig. 4a. The 2-input AND and OR gates based on PTL style are shown in Fig. 4b, c respectively. The higher input AND and OR gates have been obtained following the logic of 2-input gates.

2.3   CLAA Based on Transmission Gate (TG)

To overcome the drawback of swing degradation of PTL, a parallel combination of PMOS and NMOS is used which is called Transmission Gate (TG). The NMOS of TG passes strong '0' and PMOS passes strong '1' and thus providing better logic swing as compared to PTL. But TG needs double the number of transistor to implement the same function as compared to PTL and hence more layout area. This also increases the power consumption of TG as compared to PTL. The sum function of each adder of CLAA based on TG is shown in Fig. 5a. The 2-input AND and OR gates based on TG style are shown in Fig. 5b, c respectively. The higher input AND and OR gates have been obtained following the logic of 2-input gates.

2.4   Proposed CLAA Based on Hybrid CMOS Logic

To overcome the drawback of logic styles like complementary CMOS, PTL and TG, we combine the logic styles to design the CLAA. The sum function of each adder of the proposed CLAA is implemented using two XNOR [10]. This XNOR module is based on complementary CMOS logic and PTL as shown in Fig. 6a. This XNOR consists of six transistors and hence less layout area will be required

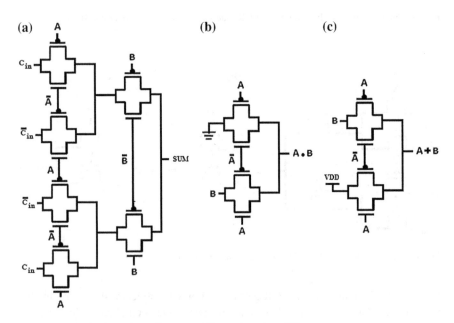

Fig. 5 Circuit of **a** Sum. **b** AND gate. **c** OR gate based on TG

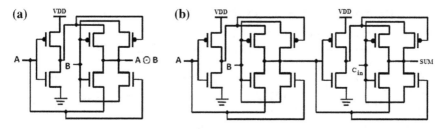

Fig. 6 **a** XNOR module. **b** Adder module

Fig. 7 Circuit of **a** AND gate. **b** OR gate based on hybrid of TG and PTL

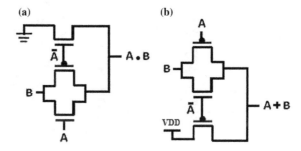

for the adder of CLAA as compared to other logic styles [11]. This also reduces the power consumption and decreases the propagation delay of the proposed CLAA. A combination of PMOS and NMOS is used at the output node of XNOR to restore the output swing. PMOS is used to pull up the output to VDD and NMOS is used to pull down the output to ground whenever there is voltage drop at the output node for high or low output. Now the sum function of each adder of CLAA uses two XNOR to perform XNOR operation between A_n, B_n and C_{in} as shown in Fig. 6b.

Now the carry out of each adder of CLAA consists of AND and OR gates which are implemented by combining Transmission Gate (TG) and Pass Transistor Logic (PTL). AND gate consists of one TG and one Pass Transistor NMOS which is always connected to ground as shown in Fig. 7a. This NMOS is used to pass strong '0' as compared to the PMOS. OR gate consists of one TG and PMOS which is always connected to VDD as shown in Fig. 7b. This PMOS is used to pass strong '1' as compared to the NMOS. Similarly higher input AND and OR gates are implemented using TG and PTL. The hybrid of TG and PTL reduces the number of transistor for gates as compared to TG and also provides better swing than PTL. This also reduces the power consumption for gates and decreases the propagation delay.

Table 1 Comparison between results of various CLAA's

Design	Average power		Delay (ps)		PDP (fj)	Transistor count
	Switching power (μW)	Static power (nW)	Sum	Carry		
CMOS	30.68	50.1	292.99	422.3	10.99	292
PTL	45.8	200.2	400	352.96	17.32	120
TG	23.55	40.01	281.3	237.9	6.12	324
HYBRID	19.36	10.02	163.6	225	3.76	250

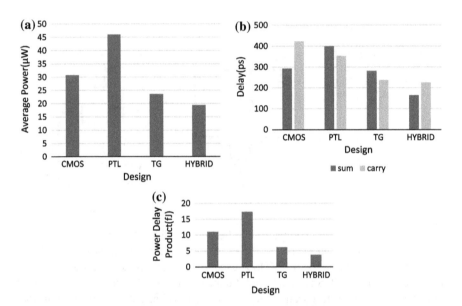

Fig. 8 **a** Average power. **b** Average propagation delay for sum and carry. **c** Power delay product

3 Results and Discussions

The circuit of CLAA based on various logic styles was simulated using Cadence Virtuoso tools in 180-nm technology at 1.8-V power supply. The simulation results are shown in Table 1. The number of transistors required for design of CLAA based on hybrid logic style is less as compared to complementary CMOS and TG. However the number of transistor for design of CLAA based on PTL is less as compared to CLAA based on hybrid CMOS logic style but the design based on PTL suffers from swing degradation and hence buffers are used which increases the number of transistor and power consumption.

The power consumption and propagation delay of proposed CLAA are less as compared to CLAA based on other logic styles and hence the power delay product of the CLAA is reduced significantly. The detailed comparison between CLAAs using different logic styles and proposed CLAA based on average power consumption is shown in Fig. 8a. The comparison for delay and power delay product are shown in Fig. 8b, c respectively.

4 Conclusion

A 4-bit CLAA has been designed and compared with various existing designs of CLAA in this paper. The simulation was carried out using standard Cadence Virtuoso tools in 180-nm technology. The average power consumed by the proposed CLAA is low as compared to the existing designs of CLAA. The proposed design of CLAA based on hybrid CMOS logic style consumes 37, 58 and 18 % less power as compared to designs based on complementary CMOS, PTL and TG respectively. The number of transistors required for design of proposed CLAA are 15 and 23 % less as compared to design of CLAA based on complementary CMOS and TG. The CLAA based on PTL requires less transistors but have a problem of swing degradation and hence buffers are required which increases the number of transistors and power consumption. Generally the power consumption is reduced at the expense of propagation delay but mixing of CMOS logic styles offers low power consumption as well as less propagation delay. Owing to low power consumption and less propagation delay, the PDP of the proposed CLAA is 64, 78 and 43 % less as compared to designs of CLAA based on complementary CMOS, PTL and TG respectively. Similarly this hybrid CMOS logic style can be used to design higher bit CLAA for low power consumption and high speed VLSI applications.

References

1. A.M. Shams, T.K. Darwish, M.A. Bayoumi, Performance analysis of low-power 1-bit CMOS full adder cells. IEEE Trans. Very Large Scale Integr. (VLSI) Syst. **10**, 20–29 (2002)
2. C.H. Chang, J. Gu, M. Zhang, A review of 0.18-μm full adder performances for tree structured arithmetic circuits. IEEE Trans. Very Large Scale Integr. (VLSI) Syst. **13**(6), 686–695 (2005)
3. D. Radhakrishnan, Low-voltage low-power CMOS full adder. IEE Procee. Circuits Devices Syst. **148**(1), 19–24 (2001)
4. M. Zhang, J. Gu, C.H. Chang, A novel hybrid pass logic with static CMOS output drive full-adder cell, in *Proceedings of IEEE International Symposium on Circuits Systems* (2003), pp. 317–320
5. S. Wairya, G. Singh, Vishant, R.K. Nagaria, S. Tiwari, Design analysis of XOR (4T) based low voltage CMOS full adder circuit, in IEEE *Nirma University International Conference Engineering (NUiCONE)* (2011), pp. 1–7
6. J. Samanta, M. Halder, B.P. De, Performance analysis of high speed low power carry look-ahead adder using different logic styles. Int. J. Soft Comput. Eng. (IJSCE) **2**(6), 330–336 (2013). ISSN:2231-2307
7. N.H.E. Weste, D. Harris, A. Banerjee, *CMOS VLSI Design: A Circuits and Systems Perspective*, 3rd edn. (Pearson Education India, India, 2006)

8. R. Zimmermann, W. Fichtner, Low power logic styles: CMOS versus pass-transistor logic. IEEE J. Solid State Circuts **32**(7), 1079–1090 (1997)

9. J.M. Rabey, *Digital Integrated Circuits: A Design Perspective* (Prentice Hall, 1996)

10. J.M. Wang, S.C. Fang, W.S. Feng, New efficient designs for XOR and XNOR functions on the transistor level. IEEE J. Solid-State Circuits **29**, 780–786 (1994)

11. S. Goel, M. Elgamel, M.A. Bayoumi, Novel design methodology for high-performance XOR-XNOR circuit design, in *IEEE Proceedings of 16th Symposium on Integrated Circuits and Systems Design (SBCCI)* (2003), pp. 71–76

TSPC Based Dynamic Linear Feedback Shift Register

**Patel Priyankkumar Ambalal, A. Anita Angeline
and V.S. Kanchana Bhaaskaran**

Abstract A Low power-Linear Feedback Shift register (LP-LFSR) with encoding technique is proposed. The LP-LFSR is a low power pseudo random sequence generator, designed using the True Single Phase Clock (TSPC) and gray encoding technique. It offers very high speed even while working with low power, less area and reduced clock-skew problems. Validation of the proposed design is made through comparison with a True Single Phase Clock (TSPC) based LFSR. The deployment of the LP-LFSR for testing application generates the random pattern with a single bit variation, thus leading to reduction in switching power.

Keywords True single phase clock (TSPC) · Low power pseudo random pattern generator · Linear feedback shift register (LFSR)

1 Introduction

The necessity to design complex integrated circuits is continuously on the rise. Hence it becomes mandatory to adopt to simplified and robust design methods for better performance [1]. One of the key solutions to achieve the same is by adopting a reliable clocking methodology, which is catered to by using a two-phase non overlapping clock [1]. This is further realized in the pseudo-two-phase clock based systems, using the four clock phases of any clock based system [2, 3]. This however leads to more silicon die area. They are also proved to be very sensitive to the clock races, arising due to the skew between the clock phases. Hence, it requires better

P.P. Ambalal (✉) · A. Anita Angeline · V.S. Kanchana Bhaaskaran
School of Electronics Engineering, VIT University, Chennai, India
e-mail: ppriyank287@gmail.com

A. Anita Angeline
e-mail: anitaangeline.a@vit.ac.in

V.S. Kanchana Bhaaskaran
e-mail: kanchana.vs@vit.ac.in

© Springer India 2016   655
S.C. Satapathy et al. (eds.), *Microelectronics, Electromagnetics
and Telecommunications*, Lecture Notes in Electrical Engineering 372,
DOI 10.1007/978-81-322-2728-1_62

controlling of clock timing during the design of the integrated circuits. This is accomplished by the insertion of more "dead time" between the clock phases [2], and this gradually slows down the generation of clock. In the conventional CMOS dynamic technique using N and P blocks, two clock phases are used, without the restriction of any overlaps [2, 4]. In the proposed LP-LFSR, the use of the true single-phase clock avoids any overlapping.

The built-in self-test (BIST) technique, cryptography, and secured communication are some of the areas where the random test pattern generation is necessary. The test patterns are generated using the LFSR. The LFSR is operated in either the normal mode or the test mode. In normal mode of operation, there may be correlation among the inputs. On the other hand, in the test mode, there will be no correlation in the inputs, leading to increased switching activity.

As compared to the normal mode of operation, the circuit under test involves more number of switching activities, which are generated by the BIST circuit [3]. The increase in the switching activities increases the power dissipation and the delay [1]. In order to reduce the switching power, different types of techniques have been proposed [5].

There are two major components of power dissipation in the CMOS integrated circuits. One of them is the static power dissipation, which occur because of the leakage current phenomena into and from the nodes through the devices which are off and such leakage currents are continuously drawn from the power supply. The second component is the dynamic power dissipation, which occurs due to charging and the discharging of the output nodal capacitances and the short circuit or rail to rail current.

In this paper, a single input changing technique is employed for minimizing the switching activities of the generated test pattern. This was aimed at realizing much lower power consumption. The paper is organized as follows. Section 2 deals with the design of TSPC based dynamic D-Flip Flop. Section 3 presents the design of 4-bit LFSR using Dynamic D-Flip Flop and Sect. 4 elaborates the low power LFSR design and Sect. 5 concludes.

2 TSPC Based Dynamic D-Flip FLOP

Figure 1 describes the TSPC based positive edge triggering dynamic D-flip flop [6]. At the node X while the clock signal $clk = 0$, the input inverter samples the inverted D input. This makes the second (dynamic) inverter in the precharge mode, with the device $M4$ charging up the node Y to V_{dd}.

Meanwhile, the third inverter is in the hold mode, as both the devices $M7$ and $M8$ are off. Therefore, while $clk = 0$ its output Q is stable as the input of the final inverter holds its previous value. On the rising edge of the clk, which is the evaluate phase, the dynamic inverter ($M4$-$M6$) evaluates. On the rising edge, if X is high, the node Y discharges. While $clk = 1$ at the node output Q, the value of Y is passed since the third inverter ($M8$-$M9$) is ON. It is to be noted that during the positive phase of

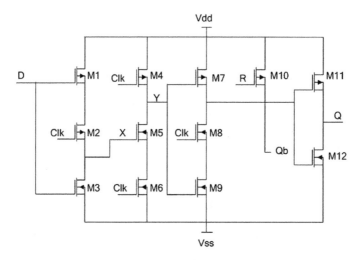

Fig. 1 TSPC based dynamic positive edge triggering D-Flip Flop

clock, when the *D* input transits to logic *HIGH*, the node *X* transits to a LOW. Hence, till the value on the node *X* propagates to *Y*, before the rising edge of the clock, the input must be kept stable. This defines the hold time of the register. In TPSC the sizing of the transistors offers better functionalities.

The glitch problem is corrected by resizing the pull down paths through *M5-M6* and *M8-M9*. This design of dynamic flip flop also enables resetting of the device *M10* which is accomplished by discharging one or two internal nodes. By using dynamic logic, the edge triggered D flip flops are designed. While the device is not in transition mode, the digital output is stored on the parasitic device capacitance [7].

3 LFSR Architecture

The 4-bit linear feedback shift register is implemented as shown in Fig. 2. The current state of the LFSR is a direct computation of its precedent state and this is the combination of different cyclic binary codes. The tap positions are decided by the XOR of previous bit combinations and the shifting property of the particular bits at the tap positions, which permits for serial manipulation until the repetition of the start state [8].

In the LFSR, the strength of different length of precise states depends on the tap positions, which are very useful for generating the feedback bit. The different types of possible states depend upon the maximal tap position. The zero state is an important factor in LFSR because of the fact that it would return the zero state (as XOR operation of the zero value with a zero value can return a zero value only).

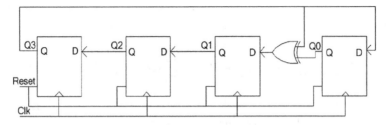

Fig. 2 LFSR using TSPC based dynamic D-flip Flop

Figure 2 presents the internal (type-II) 4-bit LFSR design for the polynomial $x^4 + x + 1$. The advantage of using the Internal LFSR over the Fibonacci configuration is that the XOR functions are placed inside the shift register, instead of placing in the feedback loop in the Fibonacci (type-I) configuration. The significant difference between the two types of LFSRs is that in the Fibonacci configuration, the signal path has two XOR operations, and hence both the functions have to process the signals within the single clock pulse duration. This type of limitation is not there in the Internal LFSR configuration.

Due to the magnificent random property of LFSR, it is very widely used in testing, as a test pattern generator. The LFSR requires small area, which is also an added advantage. As a seed generator circuit, the Internal (Type-II) LFSR is used. The LP-LFSR shown in Fig. 3 consists of an n-bit Gray counter, a seed generator (SG), which is an internal (type-II) LFSR and a sequence of XOR gate. The Gray code counter produces a single bit output changing sequence. References [9] and [10] have employed the designs such as a binary counter and a Gray code converter for the said purpose.

As shown in Fig. 3 of in the LP-LFSR based architecture, the gray counter output C [n − 1:0], XORed with the seed value yields the pseudo random output. The test clock signal (*TCK*) which controls the counter and seed generator plays a vital role in the LP-LFSR. In the n-bit counter, the initial value is set to all zeros. The counter produces different types of 2n continuous binary data periodically. As

Fig. 3 Low power linear
feedback shift generator

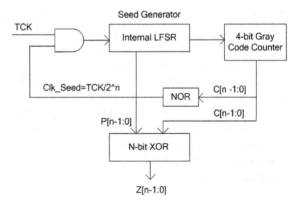

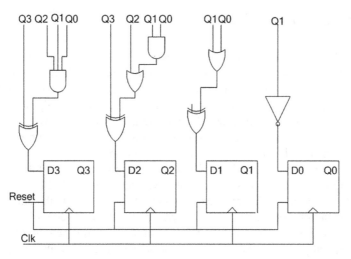

Fig. 4 4-Bit gray code counter

shown in Fig. 4, the single input changing output is generated using a gray code counter. The area necessary for the gray code counter is minimal and the power consumption also is less.

The n-bit output from the counter is given as input to the NOR gate. Depending on the output form NOR gate, the seed generator produces the next seed, since the clock is enabled to the internal LFSR only when the NOR gate output is *HIGH*. All the bits of C[n:0] will become '0' after every 2n outputs, making the output of the NOR gate to go *HIGH*. Hence, the period of the single input changing pattern will be of value 2n. The purpose of the Gray Counter is that two successive values of its output C[n − 1:0] will change only a single bit position. The normal LFSR output cannot be taken as the seed value directly [8], as some seeds values share the same vectors. Hence, the seed generator circuit should affirm that two of the single input changing pattern does not share the same vectors.

The following logic describes the final generated test sequences.

$$Z[0] = P[0] \oplus C[0]$$
$$Z[1] = P[1] \oplus C[1]$$
$$Z[2] = P[2] \oplus C[2]$$
$$\cdots$$
$$Z[n − 1] = P[n − 1] \oplus C[n − 1]$$

Due to the control signal, $TCK/2^n$ is the Seed Generator's clock. The output patterns of the single input changing generator are single input changing sequences, as any vector on performing an XOR operation will be a single input changing pattern only at a random pattern. Hence, it is very useful in large circuits and system-on-chip based circuits for the testing purpose.

4 Simulations and Analysis

The simulation is carried out using Cadence® Virtuoso tool with 180 nm CMOS library. For both the Internal (type-II) LFSR and LP-LFSR, the polynomial $x^4 + x + 1$ is utilized, which counts up to (24-1) i.e. 15 number of stages due to 4-bit LFSR. Figure 5 shows the output waveform of TSPC based D-flip-flop. Figures 6 and 7 show the simulation result of 4-bit Internal (Type-II) LFSR and LP-LFSR respectively. The simulated results obtained proved to have a single bit change in the output, thus reducing the dynamic power consumption.

Table 1 shows the comparison of experimental results between Internal (type-II) LFSR and the LP-LFSR circuit. It is observed that the Galois LFSR circuit consumes 31.12 μW dynamic power and incurs 92.51 ns delay. On the other hand, the LP-LFSR circuit consumes 21.33 μW dynamic power and incurs 67.53 ns delay.

The power consumption mainly depends on the switching activity of the circuit, clock frequency and voltage. The nature of output pattern contributes much to the dynamic power. This measure could be adopted unless there is no constraint imposed on the output. As the LFSR is meant for random pattern generation this measure is more appropriate. To reduce the switching activities, as 2n vectors are inserted between two neighboring seeds, in which each vector has only one bit difference with the last vector.

As there is only a single bit change in the output, it contributes much minimal switching power, and also the random test patterns are generated. As only a single input bit changes for every clock pulse, it leads to less switching power.

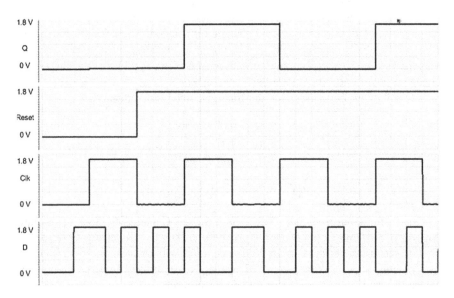

Fig. 5 Output waveform of TSPC based D-flip flop

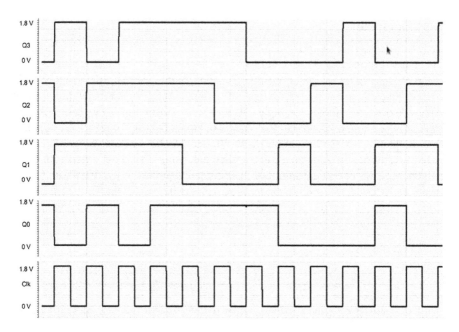

Fig. 6 Output waveform Of 4-bit internal LFSR

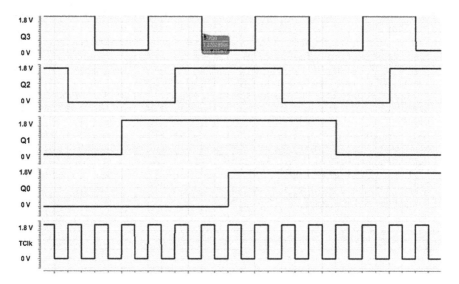

Fig. 7 Output waveform of LP-LFSR

Table 1 Power consumption of LFSR's

Circuits	Dynamic power, µW	Delay, ns	Power-delay-product, fJ
Internal LFSR (Type-II LFSR)	31.12	92.51	2878.91
LP-LFSR	21.33	67.53	1440.43

5 Conclusion

The paper proposed a True Single Phase Clocked LFSR design which ensures reliable output even during transition of seed inputs. The use of the Gray code counter leads to lower power dissipation owing to the reduced switching activity incurred, because of only a single bit change at the output. The modified seed generation circuitry based on the Gray counter's output for every 2n outputs ensures that all possible combinations of the pseudo random generation. The pseudo-random changing sequences generated using LP-LFSR offers 30.87 % reduction in the power consumption than the conventional Type II LFSR. It also incurs 33 % less delay, with 67.53 ns as against the 92.51 ns delay of type II LFSR.

References

1. Y.J. Ran, I. Carlson, C. Svensson, A true single phase clock dynamic circuit technique. IEEE J. Solid State Circuits, **22**(5) (1987)
2. A.K. Mohan, B.P. Shaun, S.S. Mahato, Low power test pattern generator (SICG) for BIST applications. IJVES **4**(1), (2013)
3. C. Mead, L. Conway, *Introduction to VLSI Systems* (Addison-Wesley, Reading, MA, 1980)
4. N. Waste, K. Eshraghian, *CMOS VLSI Design* (Addison-Wesley, Reading, MA, 1985)
5. D.J. Myers, P.A. Ivey, A design style for VLSI CMOS. IEEE J. Solid-State Circuits **SC-20**, 741–745 (1985)
6. N.F. Conclaves, H.J. De Man, NORA: A race free dynamic CMOS technique for pipelined logic structures. IEEE J. Solid-State Circuits **SC-18**, 261–266 (1983)
7. R.S. Kati, X.Y. Roan, H. Chatter, Multiple-output low power linear feedback shift register design. IEEE Trans. Circuits System. I **53**(7), 1487–1495 (2006)
8. Y. Dorian, A distributed BIST control scheme for complex VLSI devices. In *Proceedings of 11th IEEE VLSI Test Symposium*, pp.4–9, Atlantic City, New Jersey, Apr. 1993. 4–9, Apr 1993
9. S.C. Lei, J. Goo, L. Cao, Z. Ye. Liu, X.M. Wang, SACSR: a low power bist method for sequential circuits. Academic journal of Xi'an jiaotong university (English Edition), **20**(3), 155–159 (2008)
10. P. Girard, L. Guilder, C. Landrail, S. Pravossoudovitch, H.J. Wunderlich, A modified clock scheme for a low power BIST test pattern generator. In *19th IEEE Proceedings of VLSI Test Symposium*, CA, pp. 306–311, Apr–May 2001

Design of Wideband Planar Printed Quasi-Yagi Antenna Using Defected Ground Structure

Princy Chacko, Inderkumar Kochar and Gautam Shah

Abstract Conventional Yagi antenna provides a unidirectional radiation pattern but is not preferred for wideband applications. The proposed quasi-Yagi antenna consists of a driver dipole, two directors and the ground plane as the reflector. Four extended stubs are added to the ground plane to improve the bandwidth. A U-shaped defect is also introduced in the ground plane which enhances the bandwidth further. Simulation results show that the proposed antenna provides a 111.31 % bandwidth that ranges from 3.94 to 13.83 GHz. The maximum gain offered by the antenna is 5.92 dBi.

Keywords Quasi-Yagi antenna · Bandwidth enhancement · Defected ground structure

1 Introduction

Wideband antenna design has drawn tremendous attention with the unparalleled development in contemporary wireless communications [1]. In addition to providing a huge impedance bandwidth, antennas are also expected to have a simple structure, low profile and stable radiation patterns. The design of wideband antennas for a portable device is typically tricky since one has to sacrifice the impedance bandwidth for compactness [2]. For instance, alternatives like the corrugated horn and log-periodic dipole antenna provide broadband coverage at the

P. Chacko (✉) · I. Kochar · G. Shah
Electronics and Telecommunication Department, Mumbai University,
St. Francis Institute of Technology, Borivli, India
e-mail: princychacko27@gmail.com

I. Kochar
e-mail: inderkumarkochar@sfitengg.org

G. Shah
e-mail: gautamshah@ieee.org

© Springer India 2016   663
S.C. Satapathy et al. (eds.), *Microelectronics, Electromagnetics
and Telecommunications*, Lecture Notes in Electrical Engineering 372,
DOI 10.1007/978-81-322-2728-1_63

expense of a high profile. By introducing a ground plane it is found that the radiation patterns can be improved, but the antenna structure becomes large and complicated [3]. Yagi antenna has been widely used to obtain unidirectional radiation and high gain due to its simple structure [4]. Printed Yagi antennas not only exhibit directional characteristics of the Yagi arrays but also have a low-profile like microstrip antennas [5]. A quasi-Yagi antenna fed using a microstrip line, first reported in [6] has drawn much attention towards the low profile printed quasi-Yagi antenna owing to its advantages [7].

Various existing techniques for increasing the bandwidth of the low profile quasi-Yagi antennas are described in [8–15]. Using a substrate with a high dielectric constant, a printed uniplanar quasi-Yagi antenna is presented in [8]. In this antenna, metallization at the top of the substrate comprises of the driver and director elements in addition to the microstrip line feed and a broadband microstrip-to-coplanar stripline (CPS) balun. On the other hand, the metallization at the bottom of the substrate is made up of the truncated microstrip ground plane that serves the purpose of a reflector. This antenna achieves bandwidth of only 10–20 % for VSWR ≤ 2 with reasonably high gains of about 6.5 dBi. Using the same antenna, broader bandwidth of 40–50 % is achieved with a sacrifice of approximately 2.5 dBi in gain. This antenna is 93 % efficient and the front-to-back ratio (FBR) exceeds 12 dB. Another way of obtaining a wide impedance bandwidth in quasi-Yagi antenna is to use a balun. However, the balanced to unbalanced transformer increases the electrical size and also has a negative impact on the performance. One of the solutions to this problem is to replace the microstrip line feed with an appropriate coplanar waveguide (CPW) as presented in [9]. The input feed network commonly needed for perfect impedance matching with the quasi-Yagi antenna is dispensed with in this case; while the antenna is electrically small. The antenna covers the X-band from 7.7 to 12 GHz yielding a 10 dB return loss bandwidth of 44 %. The antenna has a peak gain of 7.4 dBi at 10 GHz. The reported radiation efficiency of the structure is 95 % in the entire band. This antenna also provides FBR of 15 dB.

A quasi-Yagi antenna consisting of a couple of radial stubs is presented in [10]. Both the stubs are perpendicular and have different radii. A couple of parallel dipoles are used in lieu of the driver dipole. These dipoles are of different lengths and are tapered, thereby, enhancing the bandwidth. This antenna covers the band from 3.34 to 8.72 GHz with VSWR ≤ 2 and also achieves a gain varying between 6.3 and 7.5 dBi across the bandwidth, low cross-polarizations (≤17 dB), and FBR > 15 dB. A UWB quasi-Yagi antenna employing a dual-resonant driver is presented in [11]. This antenna consists of two parts i.e. the UWB balun and the dual-resonant driver. This antenna works well between 4.7 and 10.4 GHz corresponding to a fractional bandwidth of 75 % with VSWR ≤ 2. The measured gain lies between 3.6 and 4.5 dBi. A fully-differential millimeter wave Yagi-Uda antenna is presented in [12]. This antenna uses a CPS feed. A folded dipole feed is used in this design to increase the input impedance from 18 to 150 Ω. The antenna results in medium gain of around 8–10 dBi. Another feature of this antenna is that the cross-polar components have low magnitude (−22 dB) in the 22–26 GHz range. A two-element quasi-Yagi array which consists of two folded dipole drivers is presented in [13]. The drivers are fed by a coplanar stripline. A quarter radial stub matches the coplanar stripline with a 50 Ω microstrip line. A 50 Ω T-junction power divider consisting of a 50–25 Ω wideband microstrip line transformer is used as the feeding network. The folded dipole reduces the mismatch between the CPS feedline and the antenna driver. A microstrip-stub which is added in the middle of the two folded-dipole-driven quasi-Yagi antennas on the ground plane helps to improve mutual coupling. Experimental results have shown a bandwidth of 1.67 GHz for this two element quasi-Yagi array. The antenna also yields a stable radiation pattern with a FBR > 17 dB. The measured gain of the two-element array shows ripples between 6.14 and 7.12 dBi.

A planar quasi-Yagi antenna, consisting of a driver dipole, a microstrip line to slotline transition feed, and a director is presented in [14]. A microstrip circular stub at the end of the microstrip line and a slot circular stub at the end of the slotline are used to improve the impedance transformation. A pair of rectangular apertures is created in the bottom ground plane to enhance the radiation characteristics in the low frequency band. A stepped connection structure is utilized in order to achieve a wide bandwidth. A 92.2 % fractional bandwidth ranging from 3.8 to 10.3 GHz is achieved. The minimum and maximum gain for this UWB antenna is found to be 4.1 dBi at 5.2 GHz and 7 dBi at 8.8 GHz respectively. A quasi-Yagi antenna using a driver dipole and two parasitic strips and a feeding network that employs a microstrip line which slowly transits to a slotline structure is presented in [15]. The coplanar stripline (CPS) connects the driver dipole to the slotline. Two extended stubs positioned symmetrically in the ground plane are used to minimize the lateral dimensions. The width of the antenna is reduced by approximately 16.7 % compared with the original antenna. This antenna operates over an impedance bandwidth from 3.6 to 11.6 GHz. The antenna gain fluctuates between 4.1 and 6.8 dBi over the entire bandwidth.

In this paper, a planar printed quasi-Yagi antenna with the use of defected ground structure (DGS) is proposed. This is the first quasi-Yagi antenna using a DGS for bandwidth enhancement. The driver dipole, a reflector using the ground-plane and a couple parasitic strip elements make the antenna. The antenna has a U-shaped defect symmetrically etched on both sides of the ground plane for improving the bandwidth. It also consists of four stubs placed symmetrically on both the sides of the ground plane which helps to enhance the bandwidth further. The antenna operates over an impedance bandwidth from 3.94 to 13.83 GHz.

In the next section, the structure and the performance of the proposed quasi-Yagi antenna is discussed. Simulation results are presented in the penultimate section. Conclusions are drawn towards the end.

2 Antenna Structure and Performance

As an improvement to existing techniques mentioned above, a planar printed quasi-Yagi antenna using DGS is proposed to further enhance the bandwidth. The geometric structure of the printed quasi-Yagi antenna is shown in Fig. 1. The

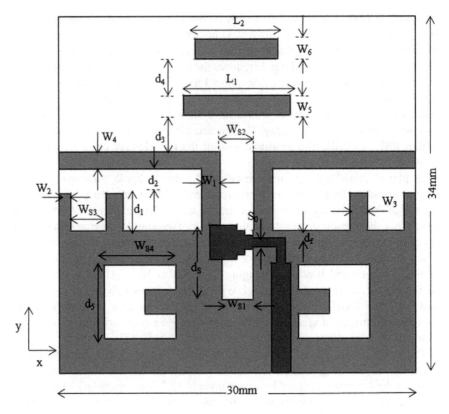

Fig. 1 Geometric structure of the proposed planar printed quasi-Yagi antenna

proposed antenna is printed on a FR-4 substrate with $\varepsilon_r = 4.4$ and substrate height of 0.8 mm. The size of the substrate is 30 mm × 34 mm (W × L).

The feed network begins with a microstrip line and tapers into a slotline structure. A driver dipole and two optimally placed directors are used. While the feed network is attached on the top of the substrate, the driver dipole and the optimally placed directors are printed on the bottom of the substrate. The driver dipole is connected to the ground plane through coplanar stripline (CPS). A couple of symmetrically etched U-shaped slots on the ground plane enhance the bandwidth. Four stubs are extended from the ground plane which further improves the bandwidth and also reduces the lateral dimensions. A four stage Chebyshev transformer was used initially as the feedline, however it has a large size due to which it was not preferred. Thus, a stepped microstrip feedline was adopted in order to obtain a wideband impedance match [15].

Impedance match and gain at higher frequencies within the operating band is improved because of two directors. Figure 2 shows the variation in surface current

(a)   **(b)**

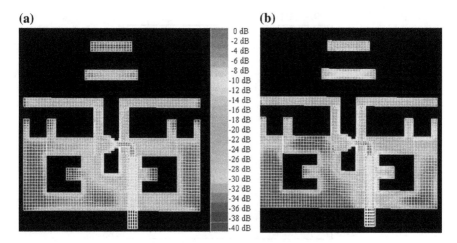

Fig. 2 Density of surface currents distribution at **a** 4.5 GHz and **b** 10.28 GHz

density along the proposed structure at 4.5 and 10.28 GHz. As shown in Fig. 2a, the surface current intensity on the directors is poor at 4.5 GHz whereas on the metallic strips, a greater amount of surface current flows at 10.28 GHz as shown in the Fig. 2b. Therefore the directors play an important role in the high frequency band. It is also found that increasing the number of directors further reduces the input impedance as every additional director results in a loading effect, thereby worsening the gain and the bandwidth.

Achieving a wide impedance bandwidth is the main objective in the optimization procedure. The dimensions of the U-shaped slots in the ground plane and the stubs are optimized to achieve a wide bandwidth. The final optimized dimensions are specified as follows:

L1 = 9 mm, L2 = 7 mm, WS1 = 2.6 mm, WS2 = 2.8 mm, WS3 = 3 mm, W1 = 1.6 mm, W2 = 1 mm, W3 = 1.5 mm, W4 = 1.4 mm, W5 = 1.6 mm, W6 = 1.6 mm, d1 = 1 mm, d2 = 2 mm, d3 = 3 mm, d4 = 3 mm, dS = 5.7 mm, df = 1 mm, S0 = 0.7 mm.

3 Simulation Results

The proposed antenna with the optimized parameters was designed and simulated using Hyperlynx 3D EM software. Figure 3 shows the magnitude of reflection coefficient of the proposed quasi-Yagi antenna. The antenna operates satisfactorily with VSWR < 2 between 3.94 and 13.83 GHz. This antenna achieves the widest

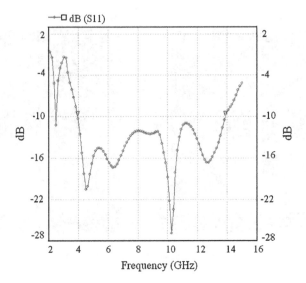

Fig. 3 Plot of reflection coefficient of the proposed quasi-Yagi antenna

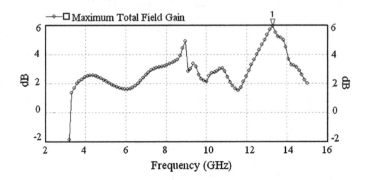

Fig. 4 Gain of the proposed quasi-Yagi antenna

impedance bandwidth compared to all the existing techniques. It can thus be clearly seen that this antenna works in majority of the UWB band as well as the X-band. The resonance at 4.5 GHz is due to the dipole while the resonance at 6.36, 10.28, and 12.78 GHz is due to the first director. The second director is added to improve the gain further.

The variation in realized gain with frequency is shown in Fig. 4. The gain varies from 1.59 to 4.9 dBi in the lower frequency range, while it increases suddenly in the high frequency range. The maximum gain achieved is 5.92 dBi at 13.28 GHz. The

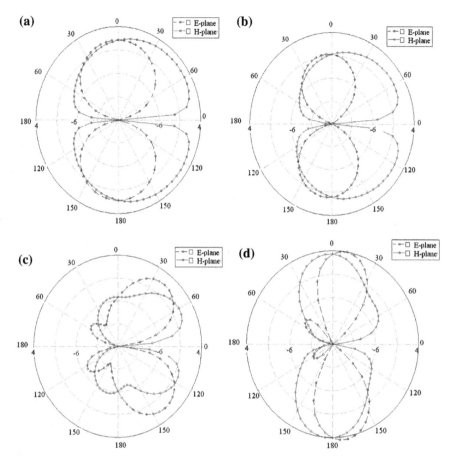

Fig. 5 Radiation patterns of the proposed quasi-Yagi antenna in E-plane and H-plane at
a 4.5 GHz, **b** 6.36 GHz, **c** 10.28 GHz, and **d** 12.78 GHz

radiation patterns are plotted to demonstrate the radiation characteristics of the
quasi-Yagi antenna. The E-plane and H-plane radiation patterns at 4.5, 6.36, 10.28
and 12.78 GHz are depicted in Fig. 5. The E-plane pattern is devoid of sidelobes.
Sidelobes in the H-plane pattern at higher frequencies are due to the surface wave
effect.

The shape of the radiation pattern obtained by the proposed antenna can be
compared with a standard Yagi-Uda antenna for which the far-zone electric field
generated by M modes of the nth element oriented parallel to the z-axis is given
as [16]

$$E_{\theta_n} \cong j\omega \frac{\mu e^{-jkr}}{4\pi r} \sin\theta \int\limits_{-l_{n/2}}^{+l_{n/2}} I_n e^{jk(x_n \sin\theta \cos\Phi + y_n \sin\theta \sin\Phi + z'_n \cos\theta)} dz'_n \qquad (1)$$

4 Conclusion

A low profile, planar quasi-Yagi antenna with enhanced bandwidth is designed. The antenna consists of two U-shaped slots in the ground plane which acts as the defected ground structure and helps to improve the bandwidth. These slots also improve the gain in the higher frequency region. Four stubs are added in the ground plane which helps to enhance the bandwidth further. The proposed antenna achieves a percentage bandwidth of 111.31 % ranging from 3.94 to 13.83 GHz. It also provides a gain ranging from 1.59 to 5.92 dBi in the desired band.

Acknowledgments The authors thank the Electronics and Telecommunication Department at St. Francis Institute of Technology for providing an environment conducive for research.

References

1. G. Lei, K.M. Luk, A wideband magneto-electric dipole antenna. IEEE Trans. Antennas Propag. **60**, 4987–4991 (2012)
2. W.X. Liu, Y.Z. Yin, W.L. Xu, S.L. Zuo, Compact open-slot antenna with bandwidth enhancement. IEEE Lett. Antennas Wirel. Propag. **10**, 850–853 (2011)
3. S. Wang, Q. Wu, D. Su, A novel reversed T-match antenna with compact size and low profile for ultra wideband applications. IEEE Trans. Antennas Propag. **60**, 4933–4937 (2012)
4. S. Lim, H. Ling, Design of a closely spaced, folded Yagi antenna. IEEE Lett. Antennas Wirel. Propag. **5**, 302–305 (2006)
5. T.T. Thai, G.R. DeJean, M.M. Tentzeris, Design and development of a novel compact soft-surface structure for the front-to-back ratio improvement and size reduction of a microstrip Yagi array antenna. IEEE Lett. Antennas Wirel. Propag. **7**, 369–373 (2008)
6. J. Huang, A.C. Demission, Microstrip Yagi array antenna for mobile satellite vehicle application. IEEE Trans. Antennas Propag. **39**, 1024–1030 (1991)
7. S.H. Sedighy, A.R. Mallahzadeh, M. Soleimani, J. Rashed-Mohassel, Optimization of printed Yagi antenna using invasive weed optimization (IWO). IEEE Lett. Antennas Wirel. Propag. **9**, 1275–1278 (2010)
8. N. Kaneda, W.R. Deal, Y. Qian, R. Waterhouse, T. Itoh, A broadband planar quasi-Yagi antenna. IEEE Trans. Antennas Propag. **50**, 1158–1160 (2002)
9. H.K. Kan, R.B. Waterhouse, A.M. Abbosh, M.E. Bialkowski, Simple broadband planar CPW-fed quasi-Yagi antenna. IEEE Lett. Antennas Wirel. Propag. **6**, 18–20 (2007)
10. S.X. Ta, I. Park, Wideband double dipole quasi-Yagi antenna using a microstrip-to-slotline transition feed. J. Electromagnet. Eng. Sci. **13**, 22–27 (2013)
11. A. Abbosh, Ultra-wideband quasi-Yagi antenna using dual-resonant driver and integrated balun of stepped impedance coupled structure. IEEE Lett. Antennas Wirel. Propag. **61**, 3885–3888 (2013)

12. R.A. Alhalabi, G.M. Rebeiz, Differentially-fed millimeter-wave Yagi-uda antennas with folded dipole feed. IEEE Trans. Antennas Propag. **58**, 966–969 (2010)
13. S.X. Ta, S. Kang, I. Park, Closely spaced two-element folded-dipole-driven quasi-Yagi array. J. Electromagnet. Eng. Sci. **12**, 254–259 (2012)
14. J. Wu, Z. Zhao, Z. Nie, Q.H. Liu, Design of a wideband planar printed quasi-Yagi antenna using stepped connection structure. IEEE Trans. Antennas Propag. **62**, 3431–3435 (2014)
15. J. Wu, Z. Zhao, Z. Nie, Q.H. Liu, Bandwidth enhancement of a planar printed quasi-Yagi antenna with size reduction. IEEE Trans. Antennas Propag. **62**, 463–467 (2014)
16. C.A. Balanis, *Antenna Theory: Analysis and Design*, 3rd edn. (Wiley, 2005)

ECG Signal Preprocessing Based on Empirical Mode Decomposition

L.V. Rajani Kumari, Y. Padma Sai and N. Balaji

Abstract Most of the Heart diseases can be diagonised with the help of ECG signal. If the ECG signal is degraded by noise, then accurate analysis is not possible. So it is most important to extract the features of ECG signal clear without noise. In this paper, an effective method named Empirical Mode Decomposition (EMD) is implemented for removing the noise from the ECG signal corrupted by non stationary noises. EMD is widely used because of its Adaptive nature and its high efficient decomposition for which any kind of complex signal could be decomposed into a limited number of Intrinsic Mode Functions (IMFs). In the proposed method, we implemented Empirical Mode Decomposition by using low pass filters for removing the noise efficiently. Also, removal of Baseline Wander and Power Line Interference which are the dominant artifacts present in ECG recordings can be done. The results show that above method is capable to remove the noises effectively. We have taken the input signals from MIT-BIH arrhythmia database. The performance is calculated in terms of Signal to Noise ratio improvement in dB. Simulations and Synthesis were carried in Modelsim and Xilinx ISE Environment.

Keywords Electrocardiogram (ECG) · EMD · SNR · Verilog · MIT-BIH database

L.V. Rajani Kumari (✉) · Y. Padma Sai
Department of ECE, VNR-VJIET, Hyderabad, India
e-mail: rajanikumari_lv@vnrvjiet.in

Y. Padma Sai
e-mail: padmasai_y@vnrvjiet.in

N. Balaji
JNTUK, ECE, University College of Engineering JNTU, Vizianagaram, India
e-mail: narayanamb@rediffmail.com

© Springer India 2016
S.C. Satapathy et al. (eds.), *Microelectronics, Electromagnetics and Telecommunications*, Lecture Notes in Electrical Engineering 372,
DOI 10.1007/978-81-322-2728-1_64

1   Introduction

ECG is the vital tool which is used to record the bioelectric potentials generated by
heart. Pathological conditions of the heart [1] are examined by the physicians using
ECG. During the recordings of ECG signals, various types of noises will interfere
with the ECG signal. Literature survey indicates obtaining the original characteristics [2] of an ECG signal by eliminating the AC interference of 50/60 Hz is a
challenging problem. Recently an efficient method for discarding the power line
interference and baseline wander was reported [3].

The basic principle of EMD is to divide a signal into a finite group of oscillatory
modes (AM-FM components) called Intrinsic mode Functions (IMFs). Partial
reconstruction of the signal by removing noisy IMFs [4, 5] is followed by most of
the EMD based denoising methods. EMD method is used for eliminating 60 Hz
noise [6] and a notch filter is used for filtering. The basic principle used in this
process is filtering the 60 Hz noise, present in the first IMF by using notch filter.

The baseline wander can be removed by dividing the ECG into Intrinsic mode
functions (IMFs) and reconstructing the ECG signal by eliminating the final IMF
which contains the base line wander. In paper [7], EMD is used for eliminating the
Baseline wander from the ECG signal. The baseline removal is done by eliminating
the low frequency component of the ECG signal.

A software tool called Xilinx ISE is used for synthesis and analyzes the HDL
designs, enables to synthesize the designs, examine RTL diagrams, and configure
the targeted device.

2   Theoretical Background

2.1   Empirical Mode Decomposition Method (EMD)

The empirical mode decomposition (EMD) method is most popular for the study of
nonlinear and non stationary signals [8]. In EMD method, the noisy ECG signal s
(t) can be decomposed as

$$s(t) = \sum_{i=1}^{n} imf_i(t) + r(t) \tag{1}$$

where n = the number of IMFs, $imf_i(t)$ = ith IMF [9] and $r(t)$ = the final residue of s
(t) respectively.

(1) Recognize the extrema [10] of the original signal $s(t)$. Find the lower envelope
     and upper envelope by passing cubic spline through the extrema.
(2) Find the mean envelope $m(t)$ by taking the average of the envelopes of the
     extrema.

(3) The candidate IMF $h_j(t)$ is calculated i.e.

$$h_j(t) = s(t) - m(t) \tag{2}$$

where j is the iteration index.

As the number of iterations increases, there will be no useful information in the obtained signals. To present this problem, some boundary conditions are adapted. The sifting process is stopped by confining the normalized standard deviation (NSTD) [11]. Where

$$SD = \sum_{t=0}^{L-1} \frac{\left| h_{j-1}(t) - h_j(t) \right|^2}{h_{j-1}^2(t)} \tag{3}$$

For having proper results, The Standard Deviation is set between 0.2 and 0.3 [12]. L is the signal length.

(4) Verify whether $h_j(t)$ satisfies the conditions which defines an IMF.

(5) The residue function $r(t)$ is obtained,

$$r(t) = s(t) - imf_1(t) \tag{4}$$

(6) To find other IMFs [13] $imf_1(t)$, $imf_2(t)$,... $imf_n(t)$, repeat the Steps (1–5) by considering $r(t)$ as new set of data $s(t)$.

(7) If the residue function $(t) = s(t) - \sum_{i=1}^{n} imf_i(t)$ is a monotonic function, we can not extract IMF's, then stop the sifting process. Thus, the original signal s (t) is divided into the n IMFs $imf_1(t)$, $imf_2(t)$,...$imf_n(t)$ and a final residue function.

3   Proposed Methodology

Using EMD, the proposed denoising method is illustrated in Fig. 1 and various steps are discussed below

1. The source of ECG signals are from MIT BIH arrhythmia data base [14]. The noisy ECG signal s(t) is obtained as sum of original ECG signal x(t) and the 60 Hz noise signal n(t), s(t) = x(t) + n(t).

2. The noisy ECG signal s(t) obtained is decomposed into IMFs using EMD algorithm. The Butter worth low pass filter is used to filter first decomposed IMFs with cut off frequency of 60 Hz to remove noise.

3. In the high order IMFs, the Baseline Wander is present. To remove the BW, We deduct the sum of higher order IMFs from the ECG

4. Thus the original ECG signal (Denoised signal) is obtained by adding the remaining IMFs.

Fig. 1 Block diagram of
proposed ECG denoising
method

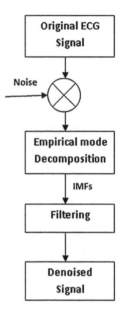

4   Simulation and Synthesis Results

For the experimental study we have received the records from MIT-BIH database.
Some ECG data records from 100 to 230 are verified to evaluate the performance of
the algorithm in terms of SNR.

The Simulation results of records 100, 116 and 111 are shown below which are
obtained using Modelsim software (Fig. 2).

An ECG Record 116 has been chosen from the database. Figure 3 shows the
overall simulation result of removal of Power Line Interference.

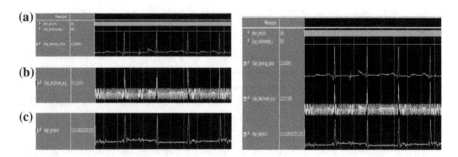

Fig. 2 Overall simulation result of denoising using EMD. Record 100: **a** input ECG sinal, **b** noisy
ECG signal, **c** denoised output using EMD

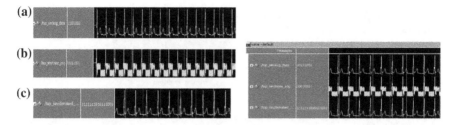

Fig. 3 Overall simulation result of removal of power line interference. Record 116: **a** Original signal. **b** Original signal with noise. **c** Denoised signal

Fig. 4 Overall simulation result of BW and muscle (EMG) noise removal. Record 111: **a** input ECG signal with baseline shift and EMG noise, **b** denoised signal using EMD

An ECG Record 111 has been chosen from the ECG signal database affected with baseline noise. This signal is applied as the input signal to the EMD method. Figure 4 shows the overall denoising result of ECG signal removing Baseline Wander and muscle noise. Figures 5 and 6 shows synthesized RTL schematic of top level of our proposed design and final synthesis report.

Fig. 5 RTL schematic generated by the synthesis tool

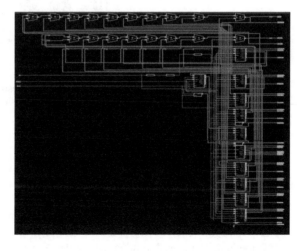

ecg2 Project Status (07/25/2014 - 15:42:15)			
Project File:	ecg2.ise	Current State:	Synthesized
Module Name:	topnn	• Errors:	No Errors
Target Device:	xc3s500e-5fg320	• Warnings:	135 Warnings
Product Version:	ISE 10.1 - Foundation	• Routing Results:	
Design Goal:	Balanced	• Timing Constraints:	
Design Strategy:	Xilinx Default (unlocked)	• Final Timing Score:	

ecg2 Partition Summary	⊟
No partition information was found.	

Device Utilization Summary (estimated values)				⊟
Logic Utilization	Used	Available	Utilization	
Number of Slices	1393	4656		29%
Number of Slice Flip Flops	280	9312		3%
Number of 4 input LUTs	2628	9312		28%
Number of bonded IOBs	280	232	120%	
Number of MULT18X18SIOs	11	20		55%
Number of GCLKs	1	24		4%

Fig. 6 Synthesis report

Fig. 7 SNR values for the EMD denoising algorithm

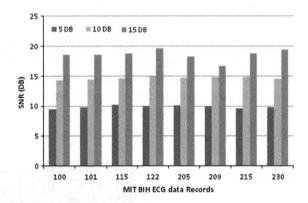

The Fig. 7 shows the enhanced ECG signal SNRs for input SNR values varying from 5, 10, and 15 dB of noisy ECG signals. These results evidently show us that as the input signal SNR goes on increasing from 5 to 15 dB enhanced ECG signal SNR increases.

5 Conclusion

In this paper, the ECG signal is preprocessed and enhanced by eliminating major noise power line interference and base line drift along with EMG noise by using EMD method. Compared to other techniques, the results show that EMD method is capable to eliminate the noises from the ECG signal in one step. The experimentation is performed on the standard data records of MIT-BIH arrhythmia data base. Simulation is done using Verilog language and synthesis in Xilinx ISE Environment. EMD method is best suited for ECG signal analysis as the computation time is less.

References

1. O. Sayadi, M.B. Shamsollahi, ECG denoising and compression using a modified extended Kalman filter structure. IEEE Trans. Biomed. Eng. **55**(9), 2240–2248 (2008)
2. M. Kaur, B. Singh, Power line interference reduction in ECG using combination of MA method and IIR notch. Int. J. Recent Trends Eng. **2**(6) (2009)
3. Z.-D. Zhao, Y.-Q. Chen, A new method for removal of baseline wander and power line interference in ECG signals. In *International Conference on Machine Learning and Cybernetics*. doi:10.1109/ICMLC.2006.259082, 04 March 2009
4. A.O. Boudraa, J.-C. Cexus, EMD-based signal filtering. IEEE Trans. Instrum. Meas. **56**(6), 2196–2202 (2007)
5. P. Flandrin, P. Con calves, G. Rilling, Detrending and denoising with empirical mode decomposition. In *Proceedings EUSIPCO*, Vienna, Austria, pp. 1581–1584 (2004)
6. A.J. Nimunkar, W.J. Tompkins, EMD-based 60-Hz noise filtering of the ECG. In *Proceedings 29th IEEE EMBS Annual International Conference*, pp. 1904–1907 (2007)
7. M. Blanco-Velasco, B. Weng, K.E. Barner, ECG signal denoising and baseline-wander correction based on the empirical mode decomposition. Comput. Biol. Med. **38**, 1–13 (2007)
8. N.E. Huang, Z. Shen, S.R. Long, M.C. Wu, H.H. Shih, Q. Zheng, N.-C. Yen, C.-C. Tung, H. H. Liu, The empirical mode decomposition and the Hilbert spectrum for nonlinear and nonstationary time series analysis. Proc. Roy. Soc. Lond. A, Math. Phys. Sci. **454**, 903–995 (1998)
9. M.-H. Lee, K.-K. Shyu, P.-L. Lee, C.-M. Huang, Y.-J. Chiu, Hardware implementation of EMD using DSP and FPGA for online signal process. IEEE Trans. Indus. Electron. (2011)
10. G. Tang, A. Qin, ECG denoising based on empirical mode decomposition. In *9th International Conference for Young Computer Scientists*, pp. 903–906
11. H. Ling, Q.-H. Lin, J.D.X. Chen, Application of the empirical mode decomposition to the analysis of esophageal reflux disease. IEEE Trans. Biomed. Eng. **52**(10) (2005)
12. V. Almenar, A. Albion, A new adaptive scheme for ECG enhancement. Signal Process. **75**(3), 253–263 (1999)
13. A. Karagiannis, P. Constantinou, Noise-assisted data processing with empirical mode decomposition in biomedical signals. IEEE Trans. Inform. Technol. Biomed. **15**(1), 11–18 (2011)
14. M. Kaur, B. Singh, Seema, Comparisons of different approaches for removal of baseline wander from ECG signal. In *Proceedings IJCA 2nd International Conference and Workshop on Emerging Trends in Technology* (2011)

Design and Analysis of Magnetic Lenses for High Energy Proton Accelerators

Vikas Teotia, Sanjay Malhotra and P.P. Marathe

Abstract High Energy Proton beams have application in scientific, industrial and Medical fields. High energy proton accelerators mainly consist of ion source and array of RF accelerating cavities and focusing magnets. Low energy section of accelerator deploys solenoid focusing magnet as they focusses the beam simultaneously in both the axis but are less efficient then a quadrupole focusing magnets in focusing strength. This paper discusses design of an Electromagnetic Quadrupole for transverse focusing of 200 MeV section of High Energy Proton Accelerator. Method used for optimization of magnetic pole shape for obtaining better than 1000 ppm uniformity in Good field region is described. Paper details the studies carried out on influence of Magneto motive forces on figure of merits of the magnet in terms of uniformity and magnetic field gradient. Paper describes the uniformity, linearity and higher order modes achieved in the design.

Keywords Accelerators · Proton · EMQ · Good field region · HEPA · Emittance

1 Introduction

The charged particle beams in particle accelerators tends to defocuses due to Columbic repulsions and transverse kicks due to fringe E-fields, strength of which depends on the synchronous phase and EM design of RF accelerating cavities [1, 5]. This transverse blowing up of beams tends to increase the emittance of the beam. Increase in emittance degrades the spatial current density which is undesirable. Electromagnetic forces are required to annul this transverse defocusing. Among the available options of using Electric field or magnetic field for charged particle focusing, magnetic fields are preferred since generation of equivalent

V. Teotia (✉) · S. Malhotra · P.P. Marathe
Accelerator Control Division, Bhabha Atomic Research Centre,
Trombay Mumbai 400085, India
e-mail: vteotia@barc.gov.in

© Springer India 2016
S.C. Satapathy et al. (eds.), *Microelectronics, Electromagnetics
and Telecommunications*, Lecture Notes in Electrical Engineering 372,
DOI 10.1007/978-81-322-2728-1_65

B-field is convenient than generation of equivalent E-field [2]. However at low particle energy, E-fields are preferred as magnetic forces are low owing to low particle velocity. For high energy beams, B-field focusing is natural choice for transverse focusing. Magnetic Quadrupole are used for focusing of charged particle beams. Depending on design, these quadrupoles could be permanent magnet based [3, 4] or electromagnet based, later provides advantage of ease in tuning while former is more efficient in terms of power consumption during operations [5]. The magnetic field strength of quadrupole magnet is given in terms of integral magnetic field gradient denoted as *integral Gdl*. The required integral Gdl depends on beam emittance at entry of the magnetic lens, magnetic quadrupoles are therefore operated normally from 50 to 100 % of their rated strength. Electromagnetic Quadrupole becomes the obvious choice for such applications. This paper describes design and analysis of an EMQ for 200 MeV section of a proton accelerator. Since quadrupoles provides alternate gradient focusing [5], the focusing and de-focusing quadrupoles are always used in pair which is called a doublet assembly.

2 System Specifications

2.1 Layout

The high energy section of proton accelerator consists of array of accelerating cavities and focusing elements. Depending on particle β, the accelerating cavities could be normal conducting DTL, half wave resonators, Spoke resonators or elliptical cavities. At 200 MeV elliptical cavities are more efficient than other families of resonators. HEPA (High Energy Proton Accelerator) have three five cell elliptical cavities followed by a doublet assembly. This arrangement is periodic, number of which depends on desired output energy.

2.2 Specifications

The beam envelope along one of the sections of the HEPA is shown in Fig. 1. The input and output phase space of the beam is shown in Fig. 2. The left and middle figure shows how quadrupole magnets limits the beam emittance and when quadrupoles are off the beam emittance blow ups as shown in right most part of Fig. 2. The integral Gdl in good field region of 24 mm (diameter) with uniformity better than 1000 ppm is a critical requirement. The required value of Integral Gdl is 3 T. The sum of higher order multipoles normalized to quadrupole component shall be less than 0.1 %.

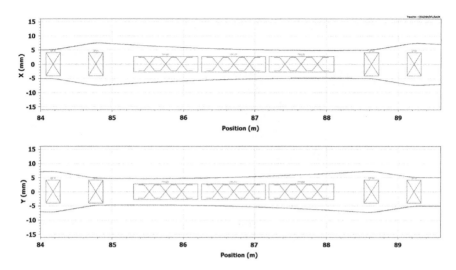

Fig. 1 Beam envelope in one section of the HEPA with doublet assembly and RF cavities

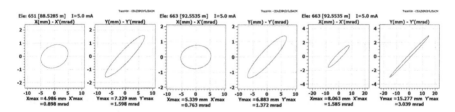

Fig. 2 Phase space of beam at entry (*left*) at exit (*middle*) of the doublet magnet assembly. The *right hand side figure* shows phase space in absence of the quadrupole doublet

3 Design and Analysis

The magnetic design of EMQ meeting above specifications is carried out using TOSCA/OPERA-3D from Vector Fields [6]. A perfect Quadrupole have hyperbolic pole, however due to engineering constraints, a perfect hyperbola is truncated which results in systematic multipoles which are odd multiples of quadrupole (n = 2). The errors in mechanical fabrication results in non-systematic multipoles which are even multiples of the quadrupole [7]. The hyperbola pole shape is modified to a customized pole shape which gives high integral magnetic field uniformity. The design consists of optimization of the magnetic pole contour to achieve the required uniformity and minimal higher order multipoles. Second order splines are used for magnetic pole design and the coordinates of the constituent points are optimized for high integral Gdl uniformity. The magnetic field distribution in the yoke and histogram of the magnetic field in the beam aperture is

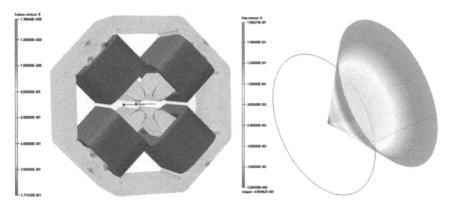

Fig. 3 3-D model of the magnet with magnetic field profile in the yoke (*left*) and histogram of the magnetic field magnitude in the good field region (*right*)

shown in Fig. 3. The primary figure of merit of the EMQ are strength, uniformity and linearity of integral Gdl and also the higher order multipoles in the GFR.

3.1 Integral Magnetic Field Gradient

The strength of integral Gdl determines the focal length of the EMQ and thereby the phase space of the beam along the axis. Ideally the integral Gdl shall remain constant in entire good field region. Non uniform integral Gdl results in aberration in the beam. This quantity is therefore studied as function of radial and azimuthal axis to determine the points of maxima and minima for evaluation of uniformity in the designed EMQ. Figure 4 shows the variation of integral Gdl as function of azimuthal axis for different values of the radius. The linearity of integral Gdl along the radius is shown in Fig. 5.

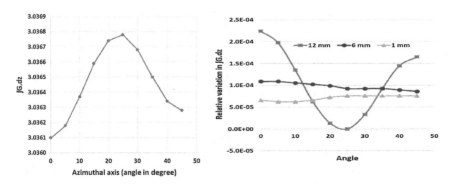

Fig. 4 Integral Gdl as function of azimuthal axis for radius of 12 mm (*left*) and relative variation of integral Gdl as function of azimuthal axis of different values of radius (*right*)

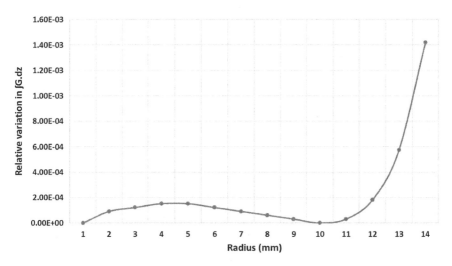

Fig. 5 Linearity of magnetic field gradient as function of radius

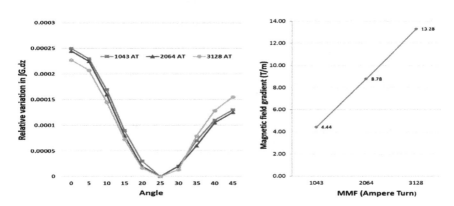

Fig. 6 Relative variation in integral Gdl as function of azimuthal axis for different MMF (*left*) and variation of magnetic field gradient as function of MMF (*right*)

The uniformity in integral Gdl shall remain within specified value for the range of operations which is normally 50–100 %. The magnetic analysis was conducted for different values of MMF and is shown graphically in Fig. 6.

3.2 Good Field Region

The uniformity of integral Gdl is inverse function of radius in the beam aperture. At low radius, uniformity is high. The required Good field region is decided on basis of

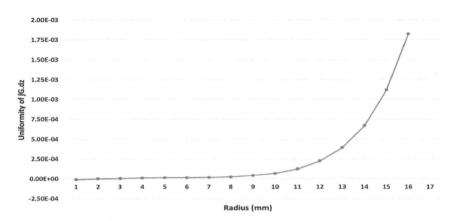

Fig. 7 Uniformity in integral Gdl as function of radius

the beam size and uniformity affects the output beam emittances. The uniformity of integral Gdl as function of radius is shown in Fig. 7.

3.3 Higher Order Multipoles

The sum of amplitudes of the higher order multipoles (from $n = 3$ to $n = 8$) shall be less than 0.1 % of the quadrupole components ($n = 2$). Figure 8 gives spectrum of multipoles in the designed EMQ. The achieved sum of HoMs normalized to quadrupole component is 3.9e-4.

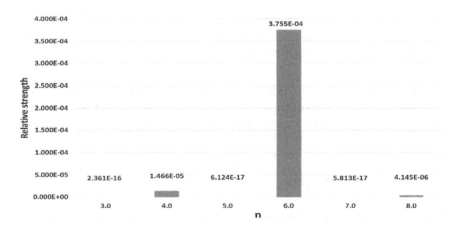

Fig. 8 Spectrum of HoMs

4 Conclusion

The magnetic design of EMQ is completed and desired performance is achieved. Uniformity of 225 ppm in GFR is obtained which is better by factor of four from desired uniformity of 1000 ppm. There is sufficient margin for deviations due to fabrication errors. It is planned to fabricate full scale prototype of the EMQ in next 1 year.

References

1. H. Samy, *RF Linear Accelerators for Medical and Industrial Application*, 1st edn. ISBN-13:978-1608070909
2. W. Helmut, *Particle Accelerator Physics*, 3rd edn. ISBN-13:978-3540490432
3. M. Sanjay et al., Electromagnetic design and development of quadrupole focusing lenses for drift tube Linac. In *Proceedings of International Topical Meeting on Nuclear Research Applications and Utilization of Accelerators*, AT/P5-18, 4–8 May 2009, Vienna
4. T. Vikas et al., Focusing magnets for drift tube Linac. In *Proceedings of Indian Particle Accelerator Conference-2011*, February 12–15, 2011, New Delhi, India
5. T.P. Wangler, *RF Linear Accelerators*, 2nd edn. ISBN:978-3-527-40680-7
6. User manual of OPERA-3D (TOSCA), Vector Fields
7. S.Y. Lee, *Accelerator Physics*, 2nd edn. ISBN-13:9789812562005

Skeletonization of Players in Dynamic Backgrounds Using Discrete Curve Evolution

Narra Dhanalakshmi, Y. Madhavee Latha and A. Damodaram

Abstract Skeletal part extraction of the human or player becomes important in applications like developing the gaming consoles, event prediction in sports, gait based human recognition and classification of human activity etc. The efficiency of the said applications depend on how efficiently the skeletal part is extracted and the extraction of skeletal part is influenced by dynamic background of the video, occlusion and resolution of the video. So in this work, we proposed a method to extract the skeletal part of the sports man with a varying background thereby facilitating the subsequent analysis. Histogram of Oriented Gradients (HOG) is used to detect the human region by making use of Support Vector Machines (SVMs) and then Graph Cut technique is applied to remove the background to extract only the foreground in the form of silhouette. Finally a skeletal pruning method is applied which is based on contour partitioning method such as Discrete Curve Evaluation (DCE) technique. The proposed method is tested on sports video like cricket and shows effectiveness of the method in extracting the skeletal shape from the video with dynamic backgrounds.

Keywords Histogram of Oriented Gradients (HOG) · Graph-cut segmentation · Silhouette · Skeletonization · Discrete Curve Evaluation (DCE)

N. Dhanalakshmi (✉)
ECE Department, V.N.R. Vignana Jyothi Institute
of Engineering & Technology, Hyderabad, India
e-mail: dhanalakshmi_n@vnrvjiet.in

Y. Madhavee Latha
ECE Department, Malla Reddy Engineering College for Women, Hyderabad, India
e-mail: madhuvsk2003@yahoo.co.in

A. Damodaram
Sri Venkateswara University, Tirupati, Andhra Pradesh, India
e-mail: damodarama@rediffmail.com

© Springer India 2016
S.C. Satapathy et al. (eds.), *Microelectronics, Electromagnetics and Telecommunications*, Lecture Notes in Electrical Engineering 372,
DOI 10.1007/978-81-322-2728-1_66

689

1 Introduction

Improvements in video processing technology and computation technology motivated researchers to use computer vision and pattern recognition methods for analysis of videos. Now a days, this kind of research tremendously increased in sports videos in indexing, retrieval and event detection and even prediction on real time. Prediction of an event will be happened by analyzing the player in successive frames from a video. This analysis will be divided into different sections (1) Detection of a player, (2) Subtracting the background from the player (3) Skeletonization, (4) Prediction of an Event. So, accuracy of the prediction depends primarily on player detection, detection of player is a challenging task due to great variability in appearances, poses and also due to large variations in the background.

The important step in skeletonization is detection of human region i.e. human silhouette. So far, many methods have been developed in this area. Different authors used different approaches to detect the players. A player can be detected using background subtraction methods [1]. Pascual [2] used a player separation algorithm without background subtraction. Some of the techniques use color models with special templates to calculate likelihood maps for color templates matching to extract an object of interest [3]. Cloud System Model (CSM) framework [4] is designed to handle 2D articulated bodies in order to segment humans in the video. This methods works when the player facing towards the camera. Matthias Grundmann [5] uses hierarchical graph-based algorithm to segment the human. The next step in Skeletonization process is removing the background in order to precisely extract the player. Removing the background can be done using different color space approaches like Hybrid Color Space (HCS), RGB color space, and $L^*U^*V^*$ color space [6]. Most of the existing methods used low level features. But these methods are not able to segment the player precisely.

After achieving human silhouette, skeleton part should be extracted. To achieve this, the methods [7, 8] use iterative algorithm to extract skeleton part, while keeping the topological structure of pixels. Accuracy and smoothness of the skeleton may be achieved by thinning.

Towards this, we presented a novel and effective way for automatic retrieval of skeletal part of the player from the sports video sequences like cricket for further processing like prediction of an event or classifying the event. We proposed a system for skeletonization of the sports man, this uses the combination of HOG and SVM for the detection of a player, Graph-Cut segmentation to subtract the background and Discrete Curve Evolution method in order to extract the skeletal part. The updated model provides more accurate skeleton for further video analysis like even prediction or classification.

This paper has been organised into following sections. Section 2 gives background information of the methods. Section 3 discusses the proposed method and Sect. 4 explains experimental results. In Sect. 5, we draw conclusions and discuss about future work.

2 Background

The proposed system employs gradient based algorithm such as Histograms of Oriented Gradients (HOG) descriptors are trained using SVM for detecting human region and Graph cut method for segmenting player and non player regions in the sports video sequence and Discrete Curve Evaluation for skeletonization. We improved our work [9] and extended the work to extract skeletal parts.

The HOG descriptor [10] was focused on the detection of pedestrian (human) by calculating gradients (G_x, G_y) in both the horizontal and vertical directions for all the pixels in the frame on overlapping basis in order to improve the performance. These HOG descriptors of both player and non player are trained using SVM [11]. However, the detected region of the player alone may not be directly suitable for action analysis of a player, since it also captures gradients of non human neighboring pixels. As a result of this, the accuracy of evaluating human action may be reduced.

Graph cut [12] method represents each frame as a graph with nodes and edges, where edges are assigned with some non negative weight or cost based on edge weighting functions. This function clusters the pixels that possess similar characteristics. Greig et al. [13] discovered the powerful optimized graph cut algorithm for solving many problems in the field of computer vision. But, the Graph Cut method alone is not enough to segment only foreground i.e., player, when the video contains signification motion in successive frames and variation in pixel intensities. Due to this background pixels tend to be segmented as foreground. So, we eliminated the problems occurred when HOG and Graph Cut individually applied by combining both HOG and Graph-cut methods in our paper.

Most of the skeletonization methods work well when proper silhouette is given. So, we have used the Discrete Curve Evaluation (DCE) method used in [8] for skeletonization. This DCE simplifies the contour of the segmented human silhouette and then pruned the skeleton by contour partitioning.

3 Proposed Method

We used the combination of above three methods to achieve our goal of skeletonization. An overview of the major steps involved in our method is shown in Fig. 1.

Fig. 1 Block diagram of the proposed work

HOG is used to extract only players as it is popular as human detection algorithm. HOG descriptor is derived by splitting each frame in terms of blocks and then each block into cells. Gradients (G_x, G_y) are computed in both the horizontal and vertical directions for all the pixels of all the cells in the frame. Then gradient magnitudes (G) and directions (θ) are computed by using the following expressions Eqs. 1 and 2:

$$G = \sqrt{G_x^2 + G_y^2} \tag{1}$$

$$\theta = \tan^{-1}\left(\frac{G_x}{G_y}\right) \tag{2}$$

Once the HOGs are computed, we do provide the training the SVM classifier with positive class HOGs of the players and negative class HOGs of non-player or human. A player detection of the frame is done by scanning a detection window across each frame at multiple positions and scales, in each position runs SVM classifier. It results in multiple overlapping detections in 3D position and scale space, around each player object in a frame and these are combined to get the final player position using Non-Maximum suppression with mean shift seeking algorithm [14]. But this HOG alone results in a human part with some background. Example results are shown in Fig. 2.

So, next is to segment the precise human part and eliminate background. Typically, segmentation is termed as a binary labeling problem where pixels are assigned to either foreground or background by the set of labels and can be optimally solved by an execution of min-cut/max-flow such as Graph-cuts which acts like a powerful energy minimization tool. The graph cut based segmentation approach of Boykov et al. [15] is adopted in this work with an energy function of the form:

$$E(A) = \lambda \cdot R(A) + B(A) \tag{3}$$

Fig. 2 Human detection using HOG

The coefficient $\lambda \geq 0$ specifies a relative importance of the region properties term $R(A)$ (penalties for assigning a pixel p to Foreground (F)) versus the boundary properties term background $B(A)$.

$$R(A) = \sum_{p \in P} R_p(A_p) \tag{4}$$

and

$$R_p(F) = -\ln Pr(I_p|F) \tag{5}$$

$$R_p(B) = -\ln Pr(I_p|B) \tag{6}$$

where negative log-likelihoods is motivated by the MAP-MRF formulations in [13]. In this work, the boundary penalties are set based on the following function Eq. 7:

$$B(A) = \sum_{p \in P} \sum_{\{p,q\} \in N} B_{\{p,q\}} \cdot \delta(A_P, A_q) \tag{7}$$

where

$$\delta(A_p, A_q) = \begin{cases} 1 & A_p \neq A_q \\ 0 & A_p = A_q \end{cases}$$

$$B_{\{p,q\}} \propto \exp\left(-\frac{(I_p - I_q)^2}{2\sigma^2}\right) \cdot \frac{1}{dist(p,q)}$$

where I_p and I_q are the intensities of pixel p and q with a penalty discontinuities between functions σ, when $|I_p - I_q| < \sigma$ the penalty is large, when $|I_p - I_q| > \sigma$ the penalty is small. Finally, the best cut that would give an "optimal" segmentation with minimum cost among all possible cuts as the sum of costs of all edges that go from S (source) to T (sink). This assigns each pixel either foreground i.e., player or background i.e., black.

$$C(S, T) = \sum_{(p,q) \in P, p \in S, q \in T} w(p,q) \tag{8}$$

When we used Graph-cut alone without HOG, this results in a foreground that comprises of player and non-player objects as well. This is due to variations in pixel intensities and camera movement (motion). We got the impressive results when HOG output fed to graph cut method. Figure 3 shows the results with Graph-Cut alone and Fig. 4 shows Graph-CUT with HOG.

On the extracted human body in the form of silhouette, we applied DCE method for skeletonization as a final step. The extracted skeleton has more accurate for further applications.

Fig. 3 Graph-cut alone

Fig. 4 Graph-cut with HOG

4   Implementation and Result

The proposed system uses a combination of HOG and Graph-cut for getting region based silhouette for players in the sports video sequence and then DCE is used for skeletonization for further processing in event detection and classification. We selected a cell size (8 × 8) and block size (16 × 16) for HOG computation. In this paper, the SVM is trained with sports players as positive class and non human as negative class on the HOG and S-T graph cut algorithm is used for segmenting the player from the background. Results of HOG and Graph-Cut for players detection is shown in Fig. 5a. DCE algorithm has been applied on the results of HOG and

(a)   **(b)**

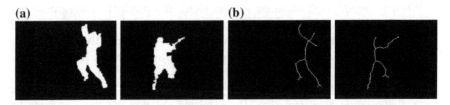

Fig. 5 From *left* to *right*: extracted silhouette using HOG-graph-CUT and extracted skeleton by DCE

Graph-cut methods and achieved the accurate skeleton as shown in Fig. 5b. This extracted skeleton further can be used for applications like event recognition, gait based human recognition, event classification and so on.

5　Conclusions and Future Work

We proposed a method to extract the skeleton of the player, which is based on a combination of Histogram of Oriented Gradients (HOG) features, Graph-cut method and DCE. This method precisely extracts human silhouette even in the varying backgrounds and then skeleton. These results also show that combination of HOG and Graph cut produces improved performance than applied individually. Silhouette features can adequately represent the movements performed by players in a video and DCE is applied on silhouette to extract skeleton for future work. Which includes, improving the accuracy of HOG-Graph-Cut further and then developing a system which extracts features from skeleton like joint features and train them to predict human action based on skeleton feature points or to classify the action or to detect the event. The actions could be bowling, batting, fielding etc.

References

1. J. Sullivan, S. Carlsson, Tracking and labelling of interacting multiple targets, in *Proceedings of the 9th European Conference on Computer Vision (ECCV 2006)* (2006)
2. P.J. Figueroa, N.J. Leite, Ricardo M.L. Barros, Tracking soccer players aiming their kinematical motion analysis. Comput. Vis. Image Underst. **101**(2), 122–135 (2006)
3. S. Gedikli, J. Bandouch, N. v. Hoyningen-Huene, B. Kirchlechner, M. Beetz, *An Adaptive Vision System for Tracking Soccer Players from Variable Camera Settings*
4. T.V. Spina, M. Tepper, A. Esler, V.M.N. Papanikolopoulos, A.X. Falcao, G. Sapiro, *Video Human Segmentation Using Fuzzy Object Models and Its Application to Body Pose Estimation of Toddlers for Behavior Studies*, 29 May 2013. arXiv:1305.6918v1 [cs.CV]
5. M. Grundmann, V. Kwatra, M. Han, I. Essa, Efficient hierarchical graph-based video segmentation, in *IEEE Conference on Computer Vision and Pattern Recognition (CVPR)*, San Francisco, USA, June 2010
6. N. Vandenbroucke, L. Macaire, J.G. Postaire, Color pixel classification in an color hybrid color space, in *Proceedings of International Conference on Image Processing* (1998), pp. 176–180
7. W. Xie, R.P. Thompson, R. Perucchio, A topology-preserving parallel 3D thinning algorithm for extracting the curve skeleton. Pattern Recogn. **36**(7), 1529–1544 (2003)
8. X. Bai, L. Longin, L. Wenyu, Skeleton pruning by contour partitioning with discrete curve evolution. IEEE Trans. Pattern Anal. Mach. Intell. **29**(3), 449–462 (2007)
9. N.D. Lakshmi, Y.M. Latha, A. Damodaram, Silhouette extraction of a human body based on fusion of HOG and graph-cut segmentation in dynamic backgrounds, in *Computational Intelligence and Information Technology, 2013*. CIIT 2013. Third International Conference on, 18–19 Oct 2013, pp. 527–531

10. N. Dalal, B. Triggs, Histograms of oriented gradients for human detection, in *Proceedings of the 2005 IEEE Computer Society Conference on Computer Vision and Pattern Recognition (CVPR'05)*, vol. 1 (2005), pp. 886–893
11. T. Joachims, Making large-scale SVM learning practical, in *Advances in Kernel Methods— Support Vector Learning*, ed. by B. Schölkopf, C. Burges, A. Smola (MIT-Press, 1999)
12. X. Lin, B. Cowan, A. Young, Model-based graph cut method for segmentation of the left ventricle, in *Proceedings of the 2005 IEEE Engineering in Medicine and Biology 27th Annual Conference Shanghai*, China, September 1–4 2005
13. D. Greig, B. Porteous, A. Seheult, Exact maximum a posteriori estimation for binary images. J. Roy. Stat. Soc. B **51**(2), 271–279 (1989)
14. D. Comaniciu, Nonparametric information fusion for motion estimation, in *Proceedings of the Conference on CVPR 2003*, vol. I, Madison, Wisconsin, USA, 2003, pp. 59–66
15. Y. Boykov, V. Kolmogorov, An experimental comparison of min-cut/max-flow algorithms for energy minimization in vision. IEEE Trans. (PAMI) **26**(9), 1124–1137 (2004)

FPGA Implementation of Test Vector Monitoring Bist Architecture System

J.L.V. Ramana Kumari, M. Asha Rani, N. Balaji and V. Sirisha

Abstract Test pattern generation Built-In Self Test (BIST) system is used to carryout testing operation of a circuit. We apply input vectors to the circuit based on logic implementation. A sequence of vectors are applied to the dadda multiplier circuit as inputs, and outputs can be observed in the examined window. The testing analysis of the Device under test (DUT) or Circuit Under Test (CUT) is monitored. The fault conditions and fault free conditions can be observed in the normal mode and test mode. The design and clocking analysis of BIST system is analyzed. In this system, Dadda multiplier circuit can be used as Device under test (DUT). The vector monitoring system is verified by Modelsim simulator, synthesized using Xilinx ISE tool and implemented in Spartan3 FPGA.

Keywords Built-in self-test (BIST) · Response verifier (RV) · Circuit under test (CUT) · Test vectors

1 Introduction

With the advancements in VLSI technology, the speed of digital circuits is growing rapidly. The performance improvement and testing issues have become the most analytical challenges for memory manufacturing. Built-in self-test (BIST) is an

J.L.V. Ramana Kumari (✉)
Department of ECE, VNRVJIET, Hyderabad, India
e-mail: ramanakumari_jlv@vnrvjiet.in

M. Asha Rani
Department of ECE, JNTU College of Engineering, Hyderabad, India
e-mail: ashajntu1@yahoo.com

N. Balaji
Department of ECE, JNTUK College of Engineering, Vizianagaram, India
e-mail: narayanamb@rediffmail.com

V. Sirisha
VLSI System Design, VNRVJIET, Hyderabad, India
e-mail: vemireddysirisha@gmail.com

© Springer India 2016   697
S.C. Satapathy et al. (eds.), *Microelectronics, Electromagnetics*
and Telecommunications, Lecture Notes in Electrical Engineering 372,
DOI 10.1007/978-81-322-2728-1_67

efficient method of design of a circuit used to test the circuit itself. BIST represents a combination of the concepts of built-in test (BIT) [1, 2] and self-test. The related term built-in-test equipment (BITE) refers to the hardware and software integrated into a unit to provide BIST or DFT capability.

Built-in self-test BIST [1, 3] technique can be classified into two types, namely off-line BIST, which includes functional and structural approaches and on-line BIST, which includes both concurrent and non-concurrent techniques. In on-line BIST, testing occurs during normal functional operating conditions, i.e., the Device Under Test (DUT) is not placed into a test condition where normal functional operation is stopped. In Concurrent on-line BIST, testing occurs simultaneously with normal functional operation. Non-concurrent on-line BIST is a method in which testing is carried out when the system is in idle state. Off-line [4, 5] BIST technique carried out, testing a system when it is not performing its normal operating condition. Test data and hardware of a DUT are completely controlled by offline-BIST. And the time it takes for testing the DUT is also relatively short. It can be applied at the manufacturing, field and operational levels. Detecting errors in real time cannot be done by Off-line testing. Linear Feedback Shift Registers (LFSRs) component can be used as most popular pseudo-random test pattern generation in BIST systems. There have the advantage of low hardware. However, for circuits with random pattern vectors, high fault coverage cannot be obtained within an acceptable test length. By analyzed the previous techniques [4, 6], a concurrent BIST architecture is developed.

2   CBIST

2.1   Device Under Test (DUT)

In this paper, the dadda multiplier is considered as combinational DUT for testing. A combinational DUT with x input lines can generate 2^x possible input vectors. The enabling of the RV is controlled by comparing the output of the DUT with that of logic block.

2.2   Concurrent BIST Unit (CBU)

The device under test circuit and self testing hardware are employed in the same architecture, hence the name Built In Self Test (BIST) system. The name concurrent can support simultaneous operations of testing as well as normal functional operation of dadda multiplier. No boundary scan or internal scan paths need to be used to a centralized BIST architecture. The logic is developed by using decoder and test generator (TG) circuits as shown in the Fig. 2. This system consists of two phases.

Fig. 1 CBIST

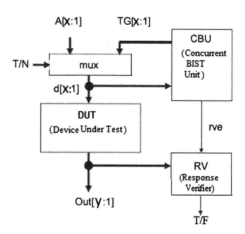

In the first phase, the system is used in the normal mode when the T/N control signal line has logic HIGH value. In the second phase the control signal T/N has logic LOW value to operate the system in test mode. During normal operating condition of the system, the input vector A[x:1] shown in Fig. 1 from the multiplexer are transferred to the dadda multiplier circuit (DUT). The response of the dadda multiplier circuit can be observed. When the system is in Test operating condition, the inputs to the DUT (denoted by d[x:1] in Fig. 1 are fed from TG[x:1].

2.3 Response Verifier (RV)

The response verifier RV compares the value of response verifier generator (rve) from the logic block with the output generated by dadda multiplier (DUT). The response of RV tells whether the fault has been occurred or not by generating its True/False T/F signal. If the response verifier generates True/False T/F signal logic LOW value that indicates that a fault in the dadda multiplier circuit otherwise no fault in the circuit. Since C-BIST have less number of components that indicates low hardware, but the time it takes to generate the input vector is more. In the scan methods for testing, test vectors are shifted out serially using scan registers to check DUT response, special hardware is needed for Automatic Test Equipment (ATE) systems.

3 Bist Architecture

The input vector [7] sequence of the dadda multiplier that can be used as DUT are divided into two different sequences of sets that are indicated with z set of bits and k set of bits respectively, such that z and k together gives the value of x. The k-most

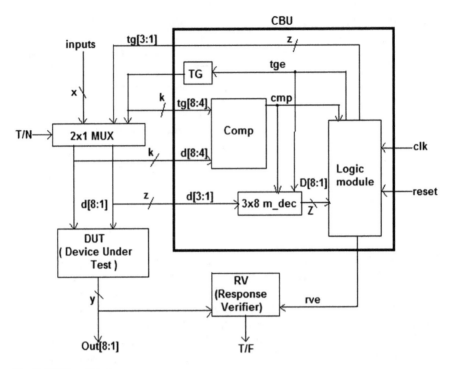

Fig. 2 BIST architecture

significant sequence of bits in the input vector is given to the comparator to compare the k-most significant sequence of bits in the TG. The BIST architecture system uses a 3 × 8 decoder shown in Fig. 3 used as modified decoder denoted as m_dec in Figs. 2 and . The result from the comparator is given to both modified decoder and logic module. The z sequence of bits is given to the modified decoder to generate the response verifier generator rve. There are two different cases which prevents the modified decoder for performing its normal function. The first case can be when tge is set to logic 1 which enables all the decoder outputs. In the second case, the decoder outputs are disabled for a logic LOW value on the TGE and comparator. When tge is logic LOW and comparator output is set to logic HIGH, modified decoder module performs its normal decoding function. The True/False T/F signal from the Response Verifier explains fault and fault free conditions of the DUT.

3.1 Dadda Multiplier

In this system 4-bit Dadda multiplier [8, 9] used as the DUT. Normally, the multiplication is carried out by forming partial products from the multiplication operation of the multiplicand with each bit in the multiplier. Finally, the product is

Fig. 3 Decoder

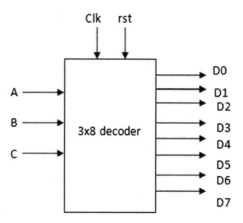

derived by adding all the partial products. In this scheme dadda performs the different process and gets the exact product terms, using the AND gates, Half Adders (HA) and Full Adders (FA).

3.1.1 Dadda Multiplier Partial Product Partitioning

A two 8-bit binary numbers a0a1a2…..a7 and b0b1b2…..b7 for the n by n Dadda multiplier [5], partial products axby of two n-bit numbers where x, y are ordering from 0, 1,….7. The partial products form n rows and 2n − 1 columns of a matrix. For each product term the numbers are represented as shown in Fig. 4a. E.g. a0b0 is given as an index 0, a0b1 is the index 1 and so on. For the second representation, rearrange the partial product terms as shown in Fig. 4b. The lengthy column in the

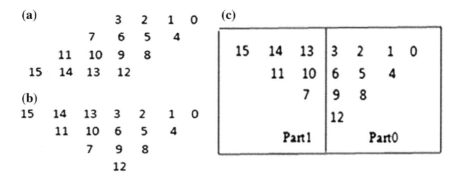

Fig. 4 Partial products separation: **a** partial product array diagram for 4 * 4 dadda multiplier, **b** second representation, **c** partitioned structure of the multiplier showing part 1 and part 0

middle of the partial products PP is the maximum delay in the Partial Product Summation Tree PPST. We break-up the PPST into two parts as shown in the Fig. 4c, in which the part 1 and part 0 consists of n columns. We proceed the addition of each column to the two parts in parallel. The addition procedure is described in next section. In this, partial products are partitioned differently part 0 and part 1, this multiplier operation is different from the other multiplier operations.

3.1.2 Dadda Multiplier Methodology

Two 4-bit binary numbers, multiplier and multiplicand, those two sequences perform multiplication process. That numbers divide equally and 2nd part terms are reversed shown in Fig. 4. Next (2, 5) are apply to the HA it produce the S0 and C0, (3, 6, 9) are apply to the FA it generate S1, C1 then (13, 10) are apply to the HA it generate S2 and C2 every time the carry will be generated that carry is added to next stage. Then second stage to add (C0, S1) require HA and to add (C1, S2, 7) require FA, next (C2, 14, 11) also use FA all addition operations are performed accordingly. Final stage every two terms to perform carry save adder each time the carry will generate and that carry is propagated to next and then added to FA. Finally only first two are HA and remaining all are FA's to get the final product terms p0, p1, p2, p3, p4, p5, p6 and p7 and partial product terms shown in Fig. 5. Reduction of the partial products for the Dadda multiplier shown in Fig. 5 based on

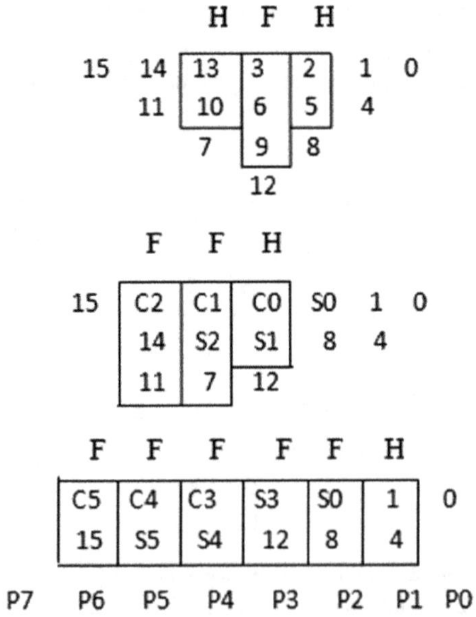

Fig. 5 Reduction of the partial products of dadda multiplier

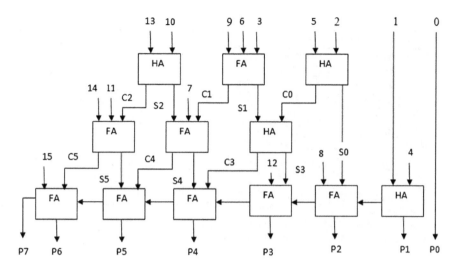

Fig. 6 The dadda based implementation

the Dadda multiplier implementation shown in Fig. 6, where totally four half adders and eight full adders are used. And the product terms are p0 to p7 with carry. Every time the multiplication operation is performed through the HA and FA. The position of partial product is denoted by the numerals residing on the HA and FA.

4 Simulation Results and Analysis

4.1 Normal Mode

If T/N = 1, circuit operates in normal mode, without fault and upon giving the binary value for multiplicand and multiplier is 10101101(10&13), results in direct multiplication and output from DUT is 1000010(130) and it is correct hence RV goes high as shown in Fig. 7a.

If T/N = 1, circuit operates in normal mode, with fault, fault is included in the DUT, then we apply the same inputs but the output is 00010010(18) it is wrong then RV shows low as shown in Fig. 7b.

4.2 Test Mode

Operation of the circuit: If T/N = 0, it operate in test mode without fault case, we give the binary value 10101101(10&13). In this case decoder, comp, TG and logic

(a)

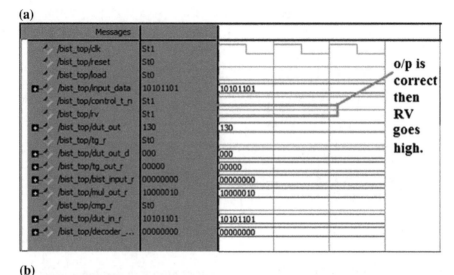

(b)

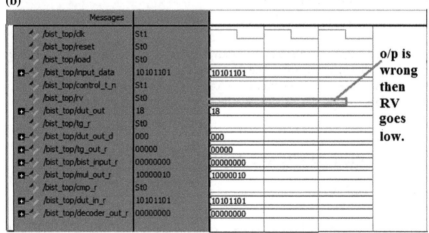

Fig. 7 **a** RV is high in fault free condition. **b** RV is low in faulty case

blocks are activated. Initially 5-bits of TG is all zero's and the 3-bits of decoder is 0 to 7 then logic block output can go from 000 to 111 and every time logic value is appended to TG value, if logic value is 000 to 111 completed one bit of TG is incremented, then 8-bits are given to DUT it operate the multiplication operation. EX: initially 00000000 up to 00000111 one cycle is completed next one bit of TG is incremented 00001000 and it continue up to 11111111 are performed the multiplication operation, if output is correct then RV goes high as shown in Fig. 8a.

Operation of the circuit: If T/N = 0, it operates in test mode with fault case, include the fault in DUT and now by applying the same inputs as earlier, to the

(a)

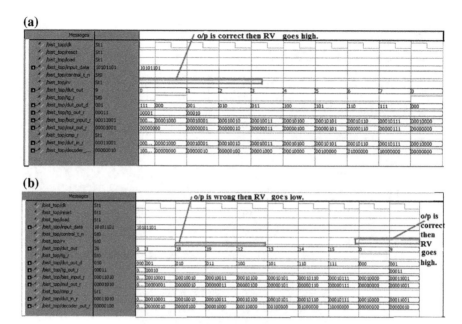

(b)

Fig. 8 **a** RV is high in fault free condition of test mode. **b** RV is high and low in faulty condition of a test mode

DUT. It causes RV to go high if output matches with the actual output (i.e. fault free output). Otherwise as shown in Fig. 8b gives faulty output, it shows RV low.

4.3 Hardware Implementation Results Using Spartan3 FPGA

Normal mode operation: The inputs to the FPGA's are 0100(4) and 0010(2) i.e. multiplier and multiplicand, cntrl = 1, rst = 0, load = 0 and output obtained is 00001000(8) which is correct and can be verified by observing RV which is high as shown in Fig. 9a.

Test mode operation: The inputs to the FPGA's are 0100(4) and 0010(2) i.e. multiplier and multiplicand, cntrl = 0, rst = 1, load = 1 and outputs obtained is 11111111(255). All the outputs (0–255) are shown in test mode with in fraction of seconds but we can't observe all the outputs only final output is shown. If the output is correct then RV shows high as shown in Fig. 9b.

(a)   **(b)**

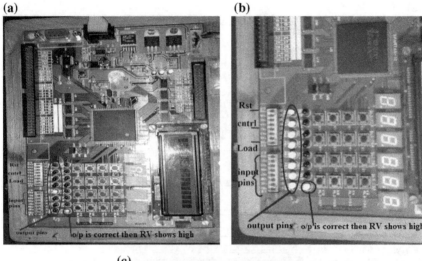

(c)

Fig. 9 a Normal mode on FPGA. **b**. Test mode on Spartan3 FPGA. **c** Fault condition in normal mode

Fault condition operation: In normal mode rst, load pins are 0 and cntrl pin is 1, inputs are 00111010(3&10) but the output shows 00100110(38) which is wrong output and can be verified by observing RV which is not glowing as shown in Fig. 9c. Synthesys report and clocking analysis is shown in Tables 1 and 2 respectively.

Table 1 Synthesis report using FPGA

Module Name:	bist_top	• Errors:	No Errors
Target Device:	xc3s400-5pq208	• Warnings:	4 Warnings
Product Version:	ISE 9.2i	• Updated:	Wed Jul 8 17:32:16 2015

KRISH Partition Summary					
No partition information was found.					

Device Utilization Summary					
Logic Utilization	Used	Available	Utilization	Note(s)	
Number of Slice Flip Flops	9	7,168	1%		
Number of 4 input LUTs	57	7,168	1%		
Logic Distribution					
Number of occupied Slices	31	3,584	1%		
Number of Slices containing only related logic	31	31	100%		
Number of Slices containing unrelated logic	0	31	0%		
Total Number of 4 input LUTs	57	7,168	1%		
Number of bonded IOBs	21	141	14%		
Number of MULT18X18s	1	16	6%		
Number of GCLKs	1	8	12%		
Total equivalent gate count for design	4,449				
Additional JTAG gate count for IOBs	1,008				

Table 2 Clocking analysis

Minimum period: 4.191 ns (Maximum frequency: 238.592 MHz)
Minimum input arrival time before clock: 4.480 ns
Maximum output required time after clock: 17.710 ns
Maximum combinational path delay: 18.504 ns

5 Conclusion and Future Scope

In this paper, test vector monitoring concurrent BIST scheme is designed. Testing operation for the combinational circuit dadda multiplier has been implemented. This testing operation is monitored under normal and test mode, the fault and fault free conditions have been observed. The simulation results of the system are verified using Modelsim simulator, synthesized using Xilinx ISE and implemented on Spartan3 FPGA. This work can be carried on to various sequential circuits.

References

1. B. Konemann, J. Mucha, G. Zwiehoff, Buit-in test for complex digital integrated circuits. IEEE J. Solid State Circuits **SC-15**, 315–318 (1980)
2. V. Sreedeep, Harish, B. Ram kumar, M. Kittur, Member, A design technique for faster dadda multiplier. IEEE, (2005)
3. E.J. McCluskey, Built-in self-test techniques. IEEE Design Test Compote. **2**(2), 21–28 (1985)
4. M.G. Buehler, M.W. Sievers, Off-line, built-in test techniques for vlsi circuits. Computer **18**, 69–82 (1982)
5. J. Savir, W.H. McAnney, P.H. Bardell, *BIT (Built-in Test) for VLSI: Pseudorandom Techniques* (John Wiley and Sons, New York, 1987)
6. I. Voyiatzis, C. Efstathiou, Input vector monitoring concurrent BIST architecture using SRAM cells. IEEE **22**(7) (2014)
7. I. Voyiatzis, A. Paschalis, D. Gizopoulos, N. Kranitis, C. Halatsis, A self testing RAM is based on a concurrent BIST architecture. IEEE Trans. Rel. **54**(1), 69–78 (2005)
8. L.T. Wang, E.J. McCluskey, Condensed linear feedback shift register (lfsr) testing—a pseudo exhaustiv test technique. IEEE Trans. Comput. **35**(4), 367–370 (1986)
9. J.P. Hayes, Transition Count Testing of Combinational Logic Circuits. IEEE Trans. Comput. **25** (6), 613–620 (1976)

Weighting Multiple Features and Double Fusion Method for HMM Based Video Classification

Narra Dhanalakshmi, Y. Madhavee Latha and A. Damodaram

Abstract In this paper we present an effective and innovative way of classifying videos into different genres based on Hidden Markov Model (HMM) thereby facilitating subsequent analysis like video indexing, retrieval and so on. In particular, this work focuses on weighting Multiple Features and also on the challenging task of fusion technique at two different levels. The multiple features are used based on the observation that no single feature can provide the necessary discriminative information to better characterize the given video content in different aspects for distinguishing large video collections. Hence, the features such as 3D-color Histogram, Wavelet-HOG, and Motion are extracted from each video and a separate HMM is trained for each feature of video class. All the classifiers are grouped into sections such that each section contains classifiers with different features of the same genre. These features are evaluated in terms of weights based on Fuzzy Comprehensive Evaluation (FCE) technique for finding the degree of use of each feature in identifying the class. For classification, Double Fusion strategy is applied in terms of Intra section fusion and Inter section fusion methods. Intra section Fusion i.e. weighted-sum method is applied at the outputs of classifiers within the section of each genre. These weights represent the relative importance which is assigned to each feature vector in finding that particular class. Then an Inter section fusion i.e. Arg-Max method is applied to fuse the scores of all sections to make final decision. We tested our scheme on video database having videos such as Sports, Cartoons, Documentaries and News and the results are compared with other methods. The results show that multiple features, double fusion and also the use of

N. Dhanalakshmi (✉)
Department of ECE, VNRVJIET, Hyderabad, India
e-mail: dhanalakshmi_n@vnrvjiet.in

Y. Madhavee Latha
Department of ECE, MRECW, Hyderabad, India
e-mail: madhuvsk2003@yahoo.co.in

A. Damodaram
Department of CSE, Sri Venkateswara University, Tirupati, Andhra Pradesh, India
e-mail: damodarama@rediffmail.com

© Springer India 2016
S.C. Satapathy et al. (eds.), *Microelectronics, Electromagnetics and Telecommunications*, Lecture Notes in Electrical Engineering 372,
DOI 10.1007/978-81-322-2728-1_68

fuzzy logic enhance video classification performance in terms of Accuracy Rate (AR) and Error Rate (ER).

Keywords Hidden markov model (HMM) · Fuzzy logic · Video classification · HOG · 3D-color histogram · Classifier fusion

1 Introduction

The rapid growth of multimedia content requires more effective video organization to increase the efficiency of video retrieval. This paper addresses an algorithm for the video classification to organize the video database. One solution to the problem of multimedia content management is through video classification and it's labeling. This would facilitate retrieval, efficient browsing, management, manipulation and analysis of visual data. Different modalities can be used for performing the automatic video classification from multimedia database. They include text-based approaches, audio-based approaches, visual-based approaches and combination of these approaches. The general process in all approaches for automatic video classification is as follows: First, features are extracted from the video to analyze the video. Then based on these features, a classifier classifies video into different categories. Text based approach is used in [1] for the classification of News video. Liu et al. [2] uses audio based approach for the classification of TV programs. Girgensohn and Foote [3] used color features for the classification of video into News, Commercials, Basketball games, and Football games. Lu [4] proposed new approach for motion detection. Roach et al. [5] performs classification only using object and camera motion. In this work visual-based approach is used. Both spatial and temporal features are used for the classification. Commonly used classifiers include Hidden Markov Model (HMM), Artificial Neural Network (ANN), Support Vector Machine (SVM), Bayesian Network, C4.5 decision tree [6], and others. Roach and Mason [7] extracted the audio features from video and used Gaussian Mixture Model (GMM) as a classifier. Motion and Color features are used in [8] and C4.5 decision tree is used for the classification. In [9] video genre classification is done using a set of visual features and SVM is used as a classifier. [10, 11] uses HMM as a classifier. A survey of different techniques for automatic indexing and retrieval of video data can be found in [12, 13]. They deal the problem of video genre classification for three classes: Sports, Cartoons and News. Hidden Markov Models (HMMs) [14] are statistical models which are used as classifiers in this work. The use of HMMs is their better learning capability for both static and dynamic features. This paper presents an innovative fusion technique based on Fuzzy Comprehensive Evaluation (FCE). A FCE method is applied in the assessment of algorithms for reconstructed remote sensing images [15]. Some research has been done on finding ways to fuse various classifiers [16]. In [17], presented different combination schemes to fuse classifiers decisions.

The paper is organized as follows. Section 2 describes the method of extracting multiple visual features from the videos. In Sect. 3, methodology is given. Experimental results are presented in Sect. 4, and in Sect. 5 are drawn conclusions.

2 Feature Extraction

The method of extracting features in this work is shown in Fig. 1. During this process, first the video sequence is converted into number of frames as the further processing depends on frames only. This process is also called Frame Grabbing. The method of extracting keyframes and different feature representations from the videos are described below.

2.1 Keyframe Extraction

The inclusion of automatic keyframes extraction method leads to efficient implementations in terms of processing time and energy consumption. Hence, Keyframes are extracted by comparing consecutive frames based on Pearson Correlation Coefficient (PCC) [18]. It would be noticed that the keyframes are identified based on the content of video but not on the size of the video.

2.2 Feature Representation

2.2.1 3D-Color Histogram

RGB color space is used to form 3D-color histogram [19] from the sequence of Keyframes. This 3D-color histogram is used to describe the color distribution of the

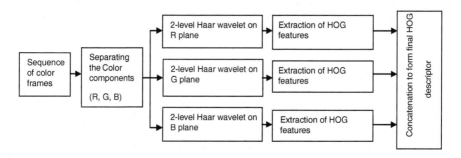

Fig. 1 Block diagram for deriving wavelet-HOG feature vector

video during the classification. The RGB color space is viewed as $8 \times 8 \times 8$ bins in each keyframe. The 3D-color histogram is particularly well suited for the problem of recognizing an object of unknown position and rotation within a scene.

2.2.2 Wavelet-HOG Features

Wavelet-HOG features are extracted in the RGB color space. HOG descriptor was first introduced by Dalal and Triggs [20], but this work includes some enhancements over the original approach. It includes, enhancing the horizontal and vertical local information in the frame by applying wavelets. It results in upgrading the shape information coded in the descriptor. All the coefficients of both low-frequency and high-frequency bands are considered to derive the feature vector. The procedure to extract the Wavelet-HOG feature vector is depicted in Fig. 1. Haar decomposition is performed on each keyframe before applying HOG. For the task of video classification, the frame is divided into 4×4 blocks. Then weighted histogram is generated for each block of cells with a total of 10 orientation bins. Then, the histograms from each block are normalized and concatenated to form the HOG descriptors. Finally, Wavelet-HOG features are formed by concatenating HOG descriptors which are derived in each color channel from 2-level DWT. The length of the Wavelet-HOG feature vector is $3 \times (7 \times (4 \times 4 \times 10))$ i.e. 3360. This process leads to enhancing the object shape information coded in the Wavelet-HOG features.

2.2.3 Motion Features

A novel motion detection algorithm is applied that integrates background subtraction and temporal differencing methods to achieve better performance [21]. For this, foreground objects are first extracted by removing background information from the video sequence [22]. The result image, which contains background area in black color and foreground objects in their original colors, is converted into gray image. It is robust even without knowing the background and also in the presence of dynamic backgrounds. Then the motion is estimated [21] using temporal differencing method which is a two-frame differential method by considering consecutive keyframes.

3   Proposed Method

An innovative approach for HMM based Video classification by weighting multiple features and double Fusion is shown in Figs. 2 and 3. The first objective of this approach is to design a multi-feature and multi-class system for the classification of videos into four different genres using three different features. Color features are extracted from the RGB color space which is a 3-channel web environment supported more than most of other color spaces. Wavelet-HOG feature is extracted for representing shape

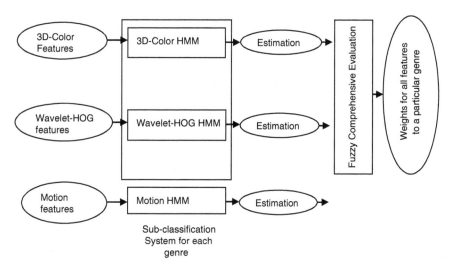

Fig. 2 Block diagram for weighting multiple features using FCE technique

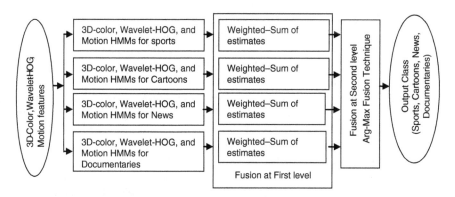

Fig. 3 Block diagram for HMM based video classification based on double fusion

information in the video. The motion vectors from the sequence of frames are extracted via background subtraction and a difference technique. The HMM classifiers are grouped into sections depending on the class. Each section contains different feature-HMMs corresponds to a particular class and forms a HMM based sub-classification system which is shown in shown in Fig. 2. The idea to combine different multi-feature HMMs in the HMM based sub-classification system is based on the assumption that the combination of multiple features gives better results than single ones. This work also focuses the analysis on the fusion of the estimations obtained from the multi-feature HMMs belongs to the same class according to their degree of use in terms of weights for the task of identifying that class (Fig. 4).

There are two major phases to meet the first objective in our framework. During the first phase, the framework includes the process of extracting the features to better

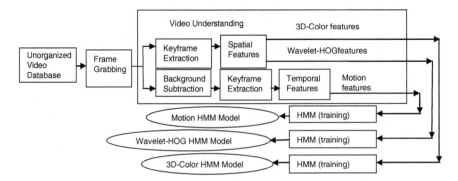

Fig. 4 Block diagram for feature extraction and modelling HMMs

represent the video. During the second phase, it includes identifying the class of unknown video and weighting the estimates from each feature-HMM corresponding to that particular class. Figure 4 shows sequence of steps required for deriving the features and modeling HMMs. As a first step, keyframes are extracted for 3D-color histogram and Wavelet-HOG but keyframes are extracted after background subtraction for motion feature to preserve the video temporal order. Then, the same process is repeated for all classes. As HMM is a supervised classifier, the classification requires testing as well as training. During the training phase of HMM, three different features which are derived from the same video are given as input to the respective HMMs in the corresponding section. Then, a model is built for each feature-HMM for all classes using Baum-Welch algorithm during training by finding HMM parameters given by $\lambda = (A, B, \pi)$. Then, Forward algorithm is used during testing to compute the probability of observation sequence $P(O|\lambda)$ for finding the class of unknown video. The database containing videos from that class and also videos of other classes are applied to each feature-HMM in all sections for testing. Then, the test video is classified to a particular predefined class depending on the value of the likelihood as an estimation from the feature-HMM. After classification, all feature-HMMs decisions are evaluated independently to know the effectiveness of a feature-HMM on the classification. For this, FCE has been used with the metrics such as Accuracy Rate (AR) and Error Rate (ER) and thus generating the weights for each feature-HMM. The level of use of each feature on the performance of the classification system is assessed by considering the fraction of relevant videos which are correctly classified, fraction of relevant videos which are incorrectly classified and the fraction of irrelevant videos which are incorrectly classified. The first two factors are explained by a metric called AR and the last is explained by the metric called ER. The metrics AR and ER are defined using Eqs. 1 and 2 as follows

$$AR = \text{total no. of videos correctly classified/total no. of relevant videos } (V_x) \quad (1)$$

$$ER = \text{total no. of videos incorrectly classified/total no. of relevant videos } (V_x) \quad (2)$$

Fig. 5 Membership functions for the ER and AR

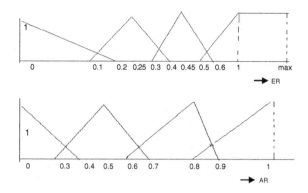

A system is good classifier when it is achieved with AR = 1 and ER = 0. AR ranges from 0 to 1 and an ER ranges from 0 to max (i.e. $V/V_x = 4$). Where, V is the total number of videos in the test database. In a fuzzy set, each element of universe of disclosure is awarded a degree of membership value in a fuzzy set using a membership function. There are four linguistic terms (Very Good (VG), Good (G), Average (A) and Poor (P)) to represent the effectiveness of the video classification. The Accuracy and Error Rates of testing database are fuzzified based on the triangular like member functions which are shown in Fig. 5. The procedure to compute the weight for each feature-HMM based on FCE is as follows. Factor set (U) is formed with the factors such as AR and ER. As these both factors are equally important equal weights are assigned to determine weight set W = [0.5, 0.5]. The Evaluation set (V) includes the scale factor for AR and ER, such as Very Good (VG), Good (G), Average (A) and Poor (P). Fuzzification technique is applied to determine the values of these scale factors for both AR and ER. A fuzzy relation matrix R on U × V is determined by taking the values of membership degrees of all four factors in V from the set U. Then, the elements in the matrix R are to be normalized so that the sum of the elements in each row is 1. The results of evaluation (b) is calculated by using b = W · R. Finally the weight (W_i) of each Feature-HMM is found by multiplying b with fuzzy transformation function T(f). A Fuzzy transformation T(f) is used to transform Factor set U into the evaluation set V. T(f) is formed as a column matrix with elements 1, 0.7, 0.3, and 0.1. The coefficients in this vector are corresponds to the credibility of each factor in the evaluating process. The process is repeated to determined the weight of each feature-HMM technique for all classes. The weights W_1, W_2, and W_3 are calculated for 3D color Histogram, Wavelet-HOG, and Motion features respectively. These weights are separately calculated for each genre of the video. Then the second objective of this work is to implement double fusion. Double fusion is applied for making a final decision on the class using weighted sum rule fusion as intra-section fusion method at first level and Arg-max fusion as inter-section fusion method at second level. The weights which are required at first fusion level are calculated by using FCE technique during the off-line classification phase which will be done only once. During the on-line testing phase, these weights are used for the fusion of decisions which are given by three individual classifiers in the HMM based sub-classification system. The first level of fusion can solve the problem of combining estimates given by three feature-HMMs in the same section. These results are compared with that of single feature alone and also multiple features without fuzzy logic. One fusion technique is to use sum method with fixed weights which is equivalent to sum method using without weights. The following are Eqs. 3 and 4 to combine estimates of three feature-HMMs in each section for sum without weights and weighted-sum techniques respectively.

$$SUM(Ws) = \sum_{i=1}^{n} S_i \qquad (3)$$

$$Weighted - SUM(Ws) = \sum_{i=1}^{n} w_i S_i \qquad (4)$$

where W_s is the weighted score derived from each section and n represents number of estimations to be fused. In our case, n value is three which intern corresponds to number features used to characterize the video. The symbols S_i and w_i represent the value of estimation from the feature-HMM and the weight for each estimation within the section respectively. By using the above procedure, the weighted scores such as W_{ss}, W_{sc}, W_{sn}, and W_{sd} are derived from the sections such as Sports, Cartoon, News and Documentaries respectively. At the second level of fusion, a fusion technique that is Arg-max as intersection fusion is applied in order to make a better judgment on the class from the weighted scores which are obtained from each section. We are finding the best possible class Class* using Arg-max and is defined by using the Eq. 5.

$$Class^* = \frac{arg - max}{C} E(C/UnknownVideo) \qquad (5)$$

where $C = \{Sports, News, Cartoon, Documentaries\}$ and E(C) is the decision derived from each section. We are finding that Class* which maximizes E(C) and returns that particular argument as a final class Class*.

4 Results

In this section, the results of the proposed algorithm for video classification and its labeling are discussed. The videos collected differ in color, length, motion and category (e.g. Cartoons, News, Sports, and Documentaries videos). The results are organized as follows. The results from Video classification with single feature are shown first. Then the classification performance using non fuzzy based method that is SUM fusion with equal weights and also the impact of the fuzzy logic on classification performance for combining multiple features are discussed.

In Table 1, the results derived using individual features such as 3D-color histogram, Motion, and Wavelet-HOG are presented. In case of 3D-color histogram, Cartoon videos as well as Sports are well classified because of the use of bright colors and unique playground colors respectively. The motion feature has achieved good AR that is 0.9 for Documentaries but at a cost of ER equal to 0.6. The Wavelet-HOG got AR equal to 0.9 for News and 0.86 for Documentaries. The Documentaries which contain animations in most of their part are misclassified into Cartoon. Table 2 gives the degree of memberships for Sports genre after fuzzification of all features. These are used to get variable weights for each feature based on FCE. The weight represents the level of ability using a feature in classifying an unknown video into its correct class. Similarly, weights for remaining genres for all features are also calculated which are shown in Table 3. Fusion is applied for combining estimation from different features in order to make a better judgment. From the Table 4, some of the News videos are miss-classified into the Sports and Documentaries in case of weighted-sum with fixed weights. This will be the case when a large part of the News video covers the Sports and Documentaries. But,

Table 1 Mean accuracy and error rates for various features

	3D-color histogram		Motion		Wavelet-HOG	
	AR	ER	AR	ER	AR	ER
News	0.5	0.62	0.8	0.3	0.9	0.2
Sports	0.86	0.5	0.89	0.2	0.6	0.5
Cartoon	0.9	0.2	0.85	0.4	0.6	0.62
Documentaries	0.7	0.7	0.9	0.6	0.86	0.4
Mean AR and ER	0.74	0.5	0.86	0.37	0.74	0.43

Table 2 A numerical example for fuzzification of accuracy and error rates for sports genre

	Degree of membership for AR				Degree of membership for ER			
	VG	G	A	P	VG	G	A	P
3D-color histogram AR = 0.86, ER = 0.5	0.3	0.4	0	0	0	0	0.66	0
Motion AR = 0.89, ER = 0.2	0.45	0.1	0	0	0	0.66	0	0
Wavelet-HOG AR = 0.6, ER = 0.5	0	0	0.5	0	0	0	0.66	0

Table 3 Weights for different genres of videos for various techniques

	3D-color histogram	Motion	Wavelet-HOG
Sports	0.6	0.81	0.4
Cartoon	0.85	0.5	0.25
News	0.25	0.7	0.85
Documentaries	0.4	0.55	0.6

Table 4 Mean accuracy and error rates for various techniques

	Multiple features without fuzzy (weighted-sum with equal weights)		Multiple features with FCE (weighted-sum using variable weights)	
	AR	ER	AR	ER
News	0.7	0.4	0.9	0.1
Sports	0.8	0.2	0.9	0.1
Cartoon	0.8	0.2	1.0	0
Documentaries	0.7	0.4	0.8	0.2
Mean AR and ER	0.75	0.3	0.9	0.1

using weighted-Sum with variable weights method this percentage is reduced to 0.1 which is shown in Table 4. The Cartoon videos are classified with 100 % classi-fication AR using FCE and Weighted-Sum methods because of its unique features in terms of color, object shapes and motion. Sports and News genres each got 0.9 as AR using FCE and weighted-Sum combination. This may be caused by properly weighting the estimates of the three techniques (motion, identification of players and playing field.). Documentaries are misclassified due to the use of diverse colors, different unpredictable objects with different shapes and unlikeness of object's motion in the scenes. The AR of FCE weighting method along with weighted-sum technique is as high as 0.9, which is higher than the accuracy rate of weighted-sum method using equal weights or sum method using without weights.

In this work, the performance of the Video Classification system is measured by Mean AR and Mean ER. From the experiments we can say that the double fusion improves the classification performance with high AR and almost low tolerable ER. Experimental results from Table 4 show that weighted-sum fusion scheme along with FCE method as a statistical technique performs better than non statistical techniques such as sum rule without weights fusion method.

5   Conclusions

The proposed method uses HMM as a classifier to classify videos using multiple features into four different categories such as News, Cartoon, Documentaries, and Sports. The proposed method yields an average classification performance with AR = 0.9 and ER = 0.1 by using double fusion. At one level of fusion, the combination FCE and Weighted-Sum is used. The use of FCE is to assess each feature-HMM and evaluate them by assigning weights. At second level of fusion, Arg-max is used for making a final decision on the classification. It was shown that combining estimates of various feature-HMMs in a diplomatic manner results in an improvement of the performance of classification.

References

1. W. Zhu, C. Toklu, R. Lion, Automatic news video segmentation and categorization based on closed-captioned text, in *Proceedings IEEE International Conference Multimedia Expo (ICME2001)*, pp. 829–832
2. Z. Liu, J. Huang, Y. Wang, Classification of TV programs based on audio information using hidden markov model, in *Proceedings IEEE Signal Processing Society Workshop Multimedia Signal Processing* (1998), pp. 27–32
3. C. Lu, M.S. Drew, J. Au, Classification of summarized videos using hidden Markov models on compressed chromaticity signatures, in *Proceedings 9th ACM International Conference Multimedia* (2001), pp. 479–482
4. N. Lu, J. Wang, Q.H. Wu, L. Yang, An improved motion detection method for real-time surveillance, IAENG Int. J. Comp. Sci. (IJCS) **35**(1), 16 (2008)
5. M. Roach, J. Mason, M. Pawlewski, Video genre classification using dynamics, IEEE Int. Conf. Acoust. Speech Signal Process. (ICASSP) **3**, 1557–1560 (2001)
6. B. Truong, S. Venkatesh, C. Dorai, Automatic genre identification for content-based video categorization, in *Proceedings 15th International Conference on Pattern Recognition* (2000), pp. 230–233
7. X. Gibert, H. Li, D. Doermann, Sports video classification using HMMs, in *Proceedings International Conference Multimedia Expo (ICME 2003)*, vol. 2. pp. 345–348
8. B.T. Truong, S. Venkatesh, C. Dorai, Automatic genre identification for content based video categorization. Int. Conf. Pattern Recogn. **4**, 230–233 (2000)
9. V. Suresh, C. Krishna Mohan, R. Kumaraswamy, B. Yegnanarayana, Content-based video classification using SVMs, in *International Conference on Neural Information Processing*, Kolkata, Nov 2004, pp. 726–731
10. C. Lu, M.S. Drew, J. Au, An automatic video classification system based on a combination of HMM and video summarization. Int. J. Smart Eng. Syst. Design **5**(1), 33–45 (2003)
11. X. Gibert, H. Li, D. Doermann, Sports video classification using HMMs, in *Proceedings of ICME 2003* (2003), pp. 345–348
12. R. Brunelli, O. Mich, C. Modena, A survey on the automatic indexing of video data. J. Vis. Commun. Image Represent. **10**(2), 78–112 (1999)
13. Y. Wang, Z. Liu, J.-C. Huang, Multimedia content analysis using both audio and visual clues. IEEE Signal Process. Mag. **17**, 12–36 (2000)
14. L.R. Rabiner, A tutorial on hidden markov models and selected applications in speech recognition, in *Proceedings of the IEEE*, vol. 77. (1989), pp. 257–286
15. L. Zhai, X. Tang, Fuzzy comprehensive evaluation method and its application in subjective quality assessment for compressed remote sensing images, in *Proceedings of the 4th International Conference on Fuzzy Systems and Knowledge Discovery (FSKD 2007)*, vol. 1. Haikou, China, 2007, pp. 145–148
16. L. Kuncheva, J.C. Bezdek, R. Duin, Decision templates for multiple classifier fusion: an experimental comparison. Pattern Recogn. **34**, 299–314 (2001)
17. R. Benmokhtar, B. Huet, Classifier fusion: combination methods for semantic indexing in video content, in *Proceedings of ICANN*, vol. 2. (2006), pp. 65–74
18. A. Miranda Neto, L. Rittner, N. Leite, D.E. Zampieri, R. Lotufo, A. Mendeleck, Pearson's correlation coefficient for discarding redundant information in real time autonomous navigation system, in *IEEE Multi-conference on Systems and Control (MSC)*, Singapura (2007)
19. N. Dhanalakshmi, Y. Madhavee Latha, A. Damodaram, Implementation of HMM based automatic video classification algorithm on the embedded platform, in *IEEE International Advance Computing Conference (IACC)* (2015), pp. 1263–1266

20. N. Dalal, B. Triggs, Histograms of oriented gradients for human detection, in *IEEE Conference on Computer Vision and Pattern Recognition (CVPR)* (2005)
21. N. Dhanalakshmi, Y. Madhavee Latha, A. Damodaram, Motion features for content-based video classification in dynamic backgrounds, in *ICSPCOMSD* (2015)
22. N. Dhanalakshmi, Y. Madhavee Latha, A. Damodaram, Silhouette extraction of a human body based on fusion of HOG and graph-cut segmentation in dynamic backgrounds. Comput. Intell. Inf. Tech. 527–531 (2013)

A Suitable Approach in Extracting Brain Source Signals from Disabled Patients

Solomon Gotham and G. Sasibushana Rao

Abstract Brain is the central processing unit of human body. Analyzing brain signals plays vital role in diagnosis and treatment of brain disorders. Brain signals are obtained from electrodes of Electroencephalogram (EEG). These are linear mixture of evoked potentials (EVP) of some number of neurons. Earlier work considered processing these mixed signals for analyzing brain functioning of brain disabled patients. But processing the original signals gives better result. Hence original signals have to be separated from linear mixture of source signals. This work will suggest a suitable approach in extracting evoked potentials of neurons.

Keywords Electroencephalogram (EEG) · Non gaussianity · Evoked potentials (EVP)

1 Introduction

Brain is collection of huge number of Neurons. Each neuron is an interconnecting segment in the network of the nervous system. Since brain controls overall functioning of mind and body, brain disorders have deep impact on pleasantness of human life. Communication between humans is affected severely with these brain disorders. Brain disorders cause different diseases. Some of them are briefed here. Cerebral Palsy (CP) is disease [1, 2] caused by events that happen before or during birth. Cerebral palsy (CP) occurs in 1.4–3.0 per 1000 live births children. Lou Gehrig's disease is a disease that is known to lead to the locked-in syndrome and is

S. Gotham (✉) · G. Sasibushana Rao
Department of Electronics and Communication Engineering,
Andhra University College of Engineering, Visakhapatnam,
Andhra Pradesh, India
e-mail: Solomongotham1@gmail.com

G. Sasibushana Rao
e-mail: Sasi_gps@yahoo.co.in

© Springer India 2016
S.C. Satapathy et al. (eds.), *Microelectronics, Electromagnetics and Telecommunications*, Lecture Notes in Electrical Engineering 372,
DOI 10.1007/978-81-322-2728-1_69

otherwise called as amyotrophic lateral sclerosis (ALS) disease [3]. These patients are fully conscious and aware of what is happening in their environment but are not able to communicate or move. Other kinds of brain disorder diseases are lack of learning, seizures [4], Attention Deficit/Hyperactivity Disorder ADHD [5] etc., Analyzing the functioning of brain can help working on solutions to the problems mentioned above.

Evoked potential (EVP) variations of neurons in brain can be recorded through various means like EEG, Magneto encephalogram (MEG) etc., Obtained signals from EEG recordings are linear combinations of original neuron EVP. The studies of the electrical signals produced by the brain are addressed both to the brain functions and to the status of the full body. EEG signals of different subjects, recorded from scalp sensors are not the original electrical potentials evoked from the neurons of brain. Instead an EEG recorded from a sensor is a sum of the large number of brain cell (neuron) potentials. Earlier work [1–5] considered processing these mixed signals for analyzing brain functioning of brain disabled patients. The basic requirement is not the potentials on the scalp, but the potentials of the sources inside the brain. So it is needed to extract the original source signals of brain from the EEG signals obtained from the scalp.

By applying digital signal processing methods to the original brain signals extracted from recordings like EEG [6] or MEG rather than electrode signals directly, it is possible, for example, to obtain patterns for diagnosis and treatment of brain disorders.

2 Brain Signals

Electroencephalography (EEG) uses the electrical activity of the neurons inside the brain. When the neurons are active, they produce an electrical potential. The combination of this electrical potential of groups of neurons can be measured outside the skull, which is done by EEG. A signal at the electrode in electroencephalogram (EEG) is a measure of brain electrical activity. This gives changes in the potential difference between two points on the scalp. Exact location of the activity can't be estimated since there is some tissue and even the skull itself between the neurons and the electrodes. For EEG measurements an array of electrodes is placed on the scalp. Many available caps use 19 electrodes, although the number of caps using more electrodes is rising. The electrodes are placed according to the international 10–20 system [7], as is depicted in Fig. 1. The 10–20 system is an internationally adopted procedure to describe the placement of sensors on brain scalp as a standard for better comparisons between different measurements. 10 and 20 refer to distances between adjacent electrodes in the order of percentage of total distance between nasion (N) and inion (I) as shown in Fig. 1. Each sensor has a letter to identify lobe and a number to identify the hemisphere location.

The letters F, T, C, P and O stand for frontal, temporal, central, parietal, and occipital lobes, respectively as shown. The letter "C" is used only for identification

Fig. 1 Distance between electrodes of 10–20 system [7]

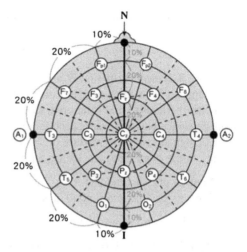

central part. A "z" (zero) refers to an electrode placed on the midline. Even subscript numbers (2, 4, 6, 8) refer to electrode positions on the right hemisphere, whereas odd numbers (1, 3, 5, 7) refer to those on the left hemisphere. In addition, the letter codes A, Pg and Fp identify the ear lobes, nasopharyngeal and frontal polar sites respectively. The magneto encephalogram (MEG) is EEGs magnetic counterpart.

3   Methodology

The obtained signals are weighed sums of the cell potentials, where the weights depend on the signal path from the neurons to the electrodes. Because the same potential is recorded from more than one electrode, the signals from the electrodes are highly correlated. The potential variations in the sources could be computed from a sufficient number of electrode signals through a mathematical tool Independent Component Analysis [8]. Considering the sensor signals as weighted sums of original source signals, let signals of sensors be taken as

$$
\begin{aligned}
x_1(t) &= a_{11}s_1(t) + a_{12}s_2 + a_{13}s_3 + \cdots\cdots\cdots \\
x_2(t) &= a_{21}s_1(t) + a_{22}s_2 + a_{23}s_3 + \cdots\cdots\cdots \\
x_3(t) &= a_{31}s_1(t) + a_{32}s_2 + a_{33}s_3 + \cdots\cdots\cdots
\end{aligned}
\tag{1}
$$

{x} is set of sensor signals A is mixing matrix and {s} is set of original signals. Equation (1) can be written in matrix form as

$$
X = AS
\tag{2}
$$

S, the original source signals vector is found by

$$S = A^{-1}X \tag{3}$$

Now it is needed to find S and A from the known X. The solution to such problem can be treated as optimization problem. Blind source separation can be one of better methods to find solution to such problems. This method assumes statistical independence between sources, no more than one source has Gaussian distribution, sensor signals are linear mixture of sources and number of sensors and sources are same. In the algorithm sources are separated in two steps as

1. Sources are uncorrelated
2. Sources are maximum non Gaussian

The above two steps are formulated as

1. Apply linear transformation to mixed signals X to reduce correlation

   a. Centering

   $$\hat{x} = X - E(X) \tag{4}$$

   b. Sphere the centered data using Eigen decomposition and diagonal transformation

   $$E\{XX^T\}\Phi = \Phi\Lambda$$
   $$\Lambda^{-\frac{1}{2}}\Phi^T E\{XX^T\}\Phi\Lambda^{-\frac{1}{2}} = I \tag{5}$$

Eigen value matrix

$$\Lambda = diag(\lambda_1, \lambda_2, \lambda_3, \ldots \ldots \lambda_m)$$

Orthogonal eigenvector matrix

$$\Phi = [\phi_1, \phi_2, \phi_3, \ldots \ldots \ldots \phi_m]$$

where $\Phi^{-1} = \Phi^T$

$$P = VD^{-\frac{1}{2}}V^T$$
$$\tilde{x} = P\hat{x} \tag{6}$$

2. Transform the Whitened X to get $W = A^{-1}$ such that non-Gaussianity is maximized. An estimated source signal $y_i = w_i^T x$ where w_i is corresponding row of

W for y_i. Non-Gaussianity can be measured through parameters like relative entropy [9], negentropy, etc., Negentropy is given by

$$N(y) = (E(G(y)) - E(G(v)))^2 \tag{7}$$

$G(y)$ is a non-quadratic function. v is a Gaussian variable with zero mean and unit variance, so the term $E\{G(v)\}$ is a constant. The best choice of $G(y)$ depends on the problem, Commonly used functions are

$$G_1(y) = -e^{\left(-\frac{1}{2}y^2\right)}, \; G_2(y) = \tanh(y), \; G_3(y) = y^3, \; G_4(y) = y^2 \tag{8}$$

It is needed to find $\mathbf{w}_i$ that maximizes $N(w_i^T \tilde{x})$.

From optimization theory we have that extrema of $E\{G(y)\}$ are found where the gradient of the Lagrange function is zero. As the constraint $||w_i|| = 1$ is equivalent to $w^T w - 1 = 0$, the Lagrange function is given by:

$$N(w, \lambda) = E(G(w^T x)) - \lambda(w^T w - 1) \tag{9}$$

The gradient with respect to $\mathbf{w}$ is

$$N_w \prime(w, \lambda) = E(xg(w^T x)) - 2\lambda w \tag{10}$$

where $g(y)$ is the gradient of $G(y)$, so the solution is obtained by taking

$$E(xg(w^T x)) - \beta w = 0 \text{ where } \beta = 2\lambda$$

The Lagrange multiplier β can be calculated with

$$\beta = E(w^{*T} xg(w^{*T} x))$$

here $\mathbf{w}^*$ is the optimum w. The gradients of the G-functions are derivatives of G-functions, for example Gradient of $G_1(y)$ is

$$g_1(y) = ye^{\left(-\frac{1}{2}y^2\right)}$$

Finding W is an optimization problem which is found through the following steps, till W converges.

The characteristic equation for the optimization step is

$$w_i = E(\tilde{x}g(w^T \tilde{x})) - E(g\prime(w^T \tilde{x}))w_i \tag{11}$$

w_i is normalized through

$$w_i = \frac{w_i}{||w_i||}$$

Transform to original axis is done through

$$w_i = w_i - \sum_{i=1}^{j-1} w_i^T w_j w_j \qquad (12)$$

w_i is again normalized as in the second step.

In this algorithm the best choice of Gaussian estimate $G(y)$ in (7), depends on the problem. This paper investigates the suitable Gaussian estimate for mentally disabled subjects.

4   Data Recording

The EEG was recorded from 32 electrodes (scalp) at a sampling rate of 2048 Hz. The system was tested with five disabled patients. The following variables are contained in the data files. Data matrix contains the raw EEG with $34 \times 115,507$ number of samples dimension. Each of the 34 rows corresponds to one electrode. The ordering of electrodes is Fp1, AF3, F7, F3, FC1, FC5, T7, C3, CP1, CP5, P7, P3, Pz, PO3, O1, Oz, O2, PO4, P4, P8, CP6, CP2, C4, T8, FC6, FC2, F4, F8, AF4, Fp2, Fz, Cz, MA1, MA2. Each column corresponds to one temporal sample.

5   Simulation Results and Discussions

For the disabled subjects movement of body parts will be relatively lesser than healthy subjects. Extracted components will have more number of low frequency components than high frequency components. Raw EEG of Subject 5 which has the least movement and higher disability, is shown in Fig. 2. High frequency components that may result from power lines, EMG or EOG will be mixed with electrode signals. Hence source signals must have less no of high frequency components.

The EVPs were extracted through the algorithm mentioned in Sect. 3. Four different Gaussian estimates shown in (8), were applied in the algorithm. Applying Gaussian function $G_1(y)$ in (8), 'Gauss' type, last 4 components are not original signals as they are high frequency signals (Fig. 3). Remaining 3 signals have lower frequency components. For a Gaussian estimate of $G_2(y)$ (Fig. 4) low frequency

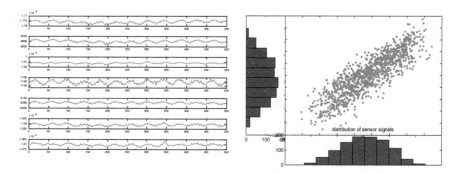

Fig. 2 Original sensor signals and spread of two signals

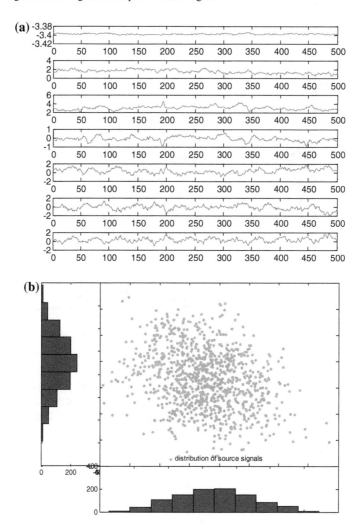

Fig. 3 **a** Extracted signals with $G(y) = G_1(y)$. **b** Spread of two signals of **a**

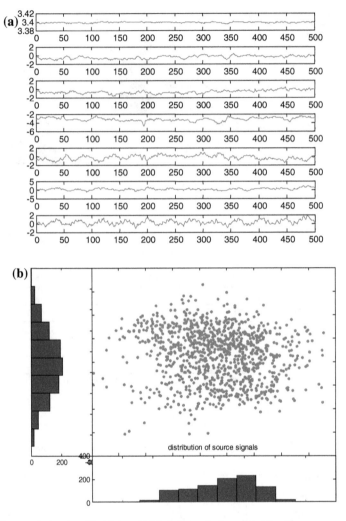

Fig. 4 **a** Extracted signals with $G(y) = G_2(y)$. **b** Spread of two signals of **a**

components are more than 3 in number as shown in Fig. 6. Guass estimate types 'Pow3' (Fig. 5) and 'Skew' (Fig. 6) have less number of components than number of sensors and mostly dominated by high frequency components. From Figs. 3, 4, 5 and 6 the data distributions are shown for two extracted signals as part 'b' of the respective figure.

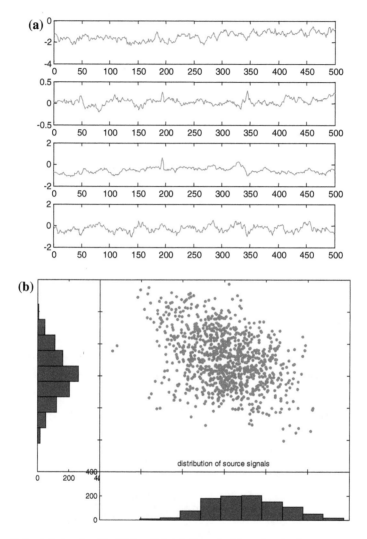

Fig. 5 **a** Extracted signals with $G(y) = G_3(y)$. **b** Spread of two signals of **a**

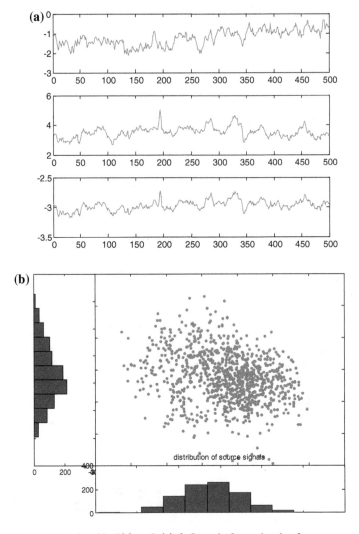

Fig. 6 **a** Extracted signals with $G(y) = G_4(y)$. **b** Spread of two signals of **a**

6 Conclusions

The signals recorded from EEG are linear mixture of originated brain evoked potentials. Applying digital processing techniques directly on recorded EEG signals cannot be expected to give right results for any category of data. Hence original neuron EVPs need to be extracted from EEG. In this process this paper has suggested a suitable approach in retrieving EVP of neurons from EEG of disabled patients. The maximum number of iterations and maximum value of 'w_i difference'

(Eq. 12) between two iterations for stopping criteria is considered to be 1000 and 0.0001.

Gaussian function	No. of sensors	Extracted signals	HF signals	Select
$G_1(y)$ 'Gauss'	7	7	3	YES
$G_2(y)$ 'Tanh'	7	7	1	YES
$G_3(y)$ 'Pow3'	7	4	3	NO
$G_4(y)$ 'Skew'	7	3	3	NO

From the results and analysis it can be concluded the better Gaussian estimate can be chosen between 'Gauss' or 'Tanh' types. Since the visual inspection of extracted components depict that both have lesser high frequency components and are giving same number of source signals as the number of EEG signals chosen. It can also be observed that the spread of data in the extracted signals through 'Gauss' or 'tanh' is relatively less Gaussian.

References

1. B.M. Faria et al., *Cerebral Palsy EEG signals Classification* (Springer, 2012)
2. L.C. Fonseca1 et al., Quantitative EEG in children with learning disabilities, analysis of band power. Arq Neuropsiquiatr **64**(2-B), 376–381 (2006)
3. U. Hoffmann, in *Bayesian Machine Learning Applied in a Brain-Computer Interface for Disabled Users*, Bernhard Karl's University, Thesis (2007)
4. J. Echauz et al., Monitoring, signal analysis, and control of epileptic seizures, in *Meditarrean IEEE Conference on Control and Automation* (2007)
5. B. Hillard, in *Anlysys of EEG Rhythems Using Custom Made Matlab for Processing ADHD Subjects University of Louisville*, Thesis (2012)
6. S. Sanei, J. Chambers, *EEG Signal Processing* (Wiley, 2007)
7. *10–20 System Positioning Manual* (Trans Cranial Technologies pvt. Ltd., 2012)
8. A. Hyvärinen, E. Oja, *Independent Component Analysis: Algorithms and Applications* (Wiley, 2001)

Design and Analysis of Multi Substrate Microstrip Patch Antenna

R. Prasad Rao, Budumuru Srinu and C. Dharma Raj

Abstract The idea of patch antenna raised from utilization of printed circuit technology not only to the circuit components and also transmission lines but the radiating elements of an electronic system. The major consideration of this work is to increase the bandwidth of a microstrip patch antenna fabricated with multilayer substrate. The narrow bandwidth of a microstrip antenna increases with increasing in thickness of the substrate. Alternatively, the bandwidth of antenna can be enhanced with use multilayer substrate. In this paper, analysis is carried out for determining bandwidth of the proposed microstrip with substrate having two layers of different materials and thicknesses in various frequency bands.

Keywords Bandwidth · Multilayer substrate · Microstrip patch

1 Introduction

Microstrip patch antenna consists support by a metallization over a ground plane by a thin dielectric substrate and fed to the ground at an accurate location. The patch structure can be arbitrary; in practice-rectangle, circle, equilateral triangle, annular-ring etc [1]. There are various methods of feeding the antenna patch for example, coaxial, microstrip, aperture coupled and proximity types. Electromagnetic energy is coupled into the substrate region between the patch and the ground plane. This substrate under the patch has similar characteristics to that of a resonant cavity. The microstrip antennas are very compact, planar and can be designed for any shape and array patterns which make them very suitable for applications like aircrafts, space crafts and missiles.

R. Prasad Rao (✉)
Department of Electronics and Communications Engineering,
Avanthi Institute of Technology, Narsipatnam, A.P., India
e-mail: prasadrao.rayavarapu@yahoo.com

B. Srinu · C. Dharma Raj
Department of Electronics and Communications Engineering, GITAM University,
Visakhapatnam 530045, A.P., India

© Springer India 2016
S.C. Satapathy et al. (eds.), *Microelectronics, Electromagnetics and Telecommunications*, Lecture Notes in Electrical Engineering 372,
DOI 10.1007/978-81-322-2728-1_70

733

Microstrip antenna typically having bandwidth less than 5 %. However, various band width widening techniques have been developed. Bandwidths up to a 50 % have been reported. Wider the bandwidth, larger is the size of the antenna. The major limitation of the microstrip antenna is that it has very low power handling capability due to thin substrate layer thickness. The average power that can be handled by the antenna is of order of tens of watts. These antennas have large dissipation losses like Ohmic loss compared to that of the antennas having equivalent aperture.

2 Microstrip Antenna

Substrate Material The metallic patch in a microstrip antenna is normally constructed by fabricating thin copper foil on the substrate. The substrate material provides mechanical support for the radiating patch elements. The required spacing between the patch and its ground plane can be achieved by choosing proper dimensions of the substrate [2]. The substrate thickness for the general microstrip antenna is of the range between 0.01 and 0.05 free-space wavelength. The dielectric constant of the substrate ranges from 1 to 10 and can be separated into 3 categories. Materials like air, polystyrene foam or dielectric honeycomb have dielectric constant in the range of 1–2. Fiber-glass reinforced Teflon has the dielectric constant between 2 and 4. The materials with dielectric constant somewhere between 4 and 10 are ceramic, quartz or alumina.

Microstrip Feed Line A microstrip patch connected directly to a microstrip ransmission line, at the edge of a patch. The impedance is normally much higher than 50 Ω (of order of 200 Ω). The mismatch between to microstrip patch and the feed line can be avoided by introducing quarter wave transformer. One more method of matching the antenna impedance to the feed line is by tapering the microstrip line smoothly into the patch. In microstrip-line feed method of feed, an array of microstrip patch elements are fabricated on the same substrate along with the feed lines which reduces the size, volume and also the cost. [3]

Microstrip Line The feed line is generally microstrip line which is built on a monolithic substrate. This has the greatest advantage that the matching this line with the microstrip is very easy since altering the shape of the line can lead to design of the line impedance of required value.

Design The major intention of the work is to enhance the BW of a antenna by using multilayer substrate, viz., the two substrates selected are of FR4 (lossless) each with equal width (Fig. 1) [4, 5].

- Design the substrate 1 with required dimensions
- On the surface of the substrate 1 casacde the substrate 2 with required dimensions
- Extrude the ground plane to the substrate 1
- Fabricate the metallic patch element to the surface of the substrate 2

Fig. 1 Multilayer substrate patch antenna

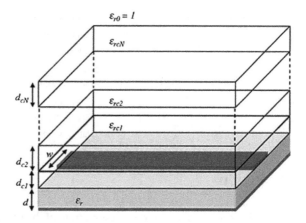

- Connect the strip line feed to the patch
- Design a port to provide the required potential
- Simulate the design and test the results.

The model (MSA) represented by two slots of width is (W) and height is (h) separated by transmission line of finite length (L). The width of the patch normally determined from the given equation [6, 7].

$$W = \frac{c}{2f_0\sqrt{\frac{(\varepsilon_r + 1)}{2}}} \tag{1}$$

Due to the fact that the fringing field around the periphery of the patch is not confined to the dielectric, the effective dielectric constant (ε_{eff}) is less than (ε_r).

$$\varepsilon_{eff} = \frac{\varepsilon_r + 1}{2} + \frac{\varepsilon_r - 1}{2}\sqrt{\frac{1}{\left[1 + 12\frac{h}{w}\right]}} \tag{2}$$

For TM_{10} Mode of operation in a microstrip, the length of the patch must be less than half the wavelength ($\lambda/2$). This difference in the length (ΔL) can be evaluated to be

$$\Delta L = 0.412h\frac{\left(\varepsilon_{eff} + 0.3\right)\left(\frac{W}{h} + 0.264\right)}{\left(\varepsilon_{eff} - 0.258\right)\left(\frac{W}{h} + 0.813\right)} \tag{3}$$

$$L_{eff} = \frac{c}{2f_r\sqrt{\varepsilon_{eff}}} \tag{4}$$

where c = light speed, L_{eff} = effective length. F_r = resonant frequency, ε_{eff} = effective dielectric constant (Table 1).

Table 1 Data sheet of micro strip antenna	S. no	Type of element	Physical length (mm)
	1	Patch length	38
	2	Substrate 1	5 mm
	3	Substrate 2	5 mm
	4	Ground plate thickness	0.5
	5	Feed line	12.5
	6	Gap of feed line	1
	7	Patch width	51

3 Results

Analysis is carried out for the estimation of bandwidth of a micro strip patch antenna for single substrate with different thickness at different frequencies, that can be compared with same parameters of multi layer substrate. The thickness of the substrate considered is 5 and 8 mm for both single and multi layer substrates with 1 and 2 GHz frequencies (Figs. 2, 3, 4, 5, 6, 7, 8 and 9).

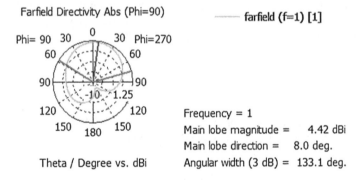

Frequency = 1
Main lobe magnitude =　　4.42 dBi
Main lobe direction =　　8.0 deg.
Angular width (3 dB) =　133.1 deg.

Fig. 2 Substrate with width h = 8 mm and frequency = 1 GHz

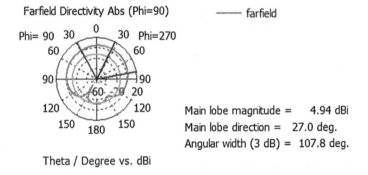

Main lobe magnitude =　　4.94 dBi
Main lobe direction =　27.0 deg.
Angular width (3 dB) =　107.8 deg.

Fig. 3 Substrate with width h = 8 mm and frequency = 2 GHz

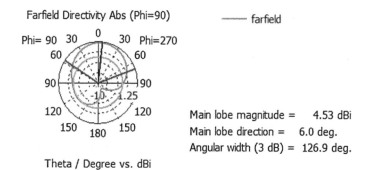

Main lobe magnitude =   4.53 dBi
Main lobe direction =   6.0 deg.
Angular width (3 dB) = 126.9 deg.

Fig. 4 Substrate with width h = 5 mm and frequency = 1 GHz

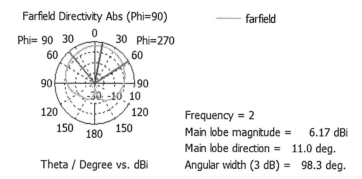

Frequency = 2
Main lobe magnitude =   6.17 dBi
Main lobe direction =   11.0 deg.
Angular width (3 dB) =   98.3 deg.

Fig. 5 Substrate with width h = 5 mm and frequency = 2 GHz

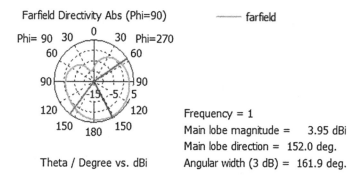

Frequency = 1
Main lobe magnitude =   3.95 dBi
Main lobe direction = 152.0 deg.
Angular width (3 dB) = 161.9 deg.

Fig. 6 Double substrate with width h = 5 mm and frequency = 1 GHz

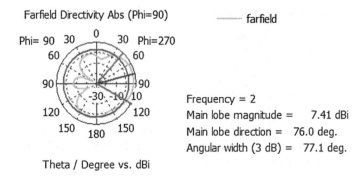

Fig. 7 Double substrate with width h = 5 mm and frequency = 2 GHz

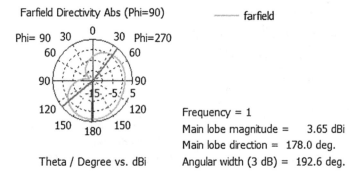

Fig. 8 Double substrate with width h = 8 mm and frequency = 1 GHz

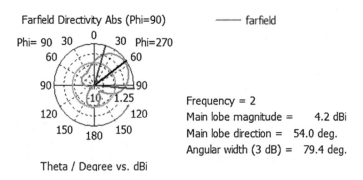

Fig. 9 Double substrate with width h = 8 mm and frequency = 2 GHz

Table 2 Comparison of band widths single layer with multi layer substrate MSA

Thickness	Single layer substrate		Multi layer substrate	
h = 5 mm	Frequency (GHz)	Bandhwidth (deg)	Frequency (GHz)	Bandhwidth (deg)
	1	126.9	1	161.9
	2	98.3	2	77.1
h = 8 mm	Frequency (GHz)	Bandhwidth (deg)	Frequency (GHz)	Bandhwidth (deg)
	1	133.2	1	192.6
	2	107.8	2	79.2

4 Conclusion

According to the above results it can be concluded that the bandwidth of a micro strip patch antenna is enhanced by increasing the thickness of the substrate up to 7 dB for a single layer substrate and it can be enhanced up to 31 dB in case of two layered substrate. There is a tradeoff between the directivity and the bandwidth of a micro strip patch antenna. At 2 GHz frequency by increasing the thickness for single layer substrate from 5 to 8 mm it is observed that the bandwidth in enhanced approximately by 9 dB, while that for multi layered substrate up to 3 dB. The band width further can be increased by using greater than three or more substrates (Table 2).

References

1. W.L. Stutzman, G.A. Thiele, *Antenna Theory and Design*, 2nd edn. (New York, Wiley, 1998)
2. G. Kumar, K.P. Ray *Broadband Micro Strip Antennas* (Artech House, Inc., 2003)
3. C.A. Balanis, *Advanced Engineering Electromagnetics* (Wiley, New York, 1989)
4. J.A. Ansari, *Analysis of Multilayer Rectangular Patch Antenna for Broadband Operation* © (Springer Science+Business Media, LLC., 2010)
5. S.D. Gupta, Multilayer microstrip antenna quality factor optimization for bandwidth enhancement. J. Eng. Sci. Tech. (JESTEC) **7**(6), 756–773 (2012)
6. R. Garg, P. Bhartia, I. Bahl, A. Ittipiboon, *Micro Strip Antenna Design Handbook* (Artech House, Boston, 2001)
7. A. Thomas, Milligan, 2nd edn. *Modern Antenna Design*

Design and Analysis of Reversible Binary and BCD Adders

A.N. Nagamani, Nikhil J. Reddy and Vinod Kumar Agrawal

Abstract Reversible logic in recent times has attracted a lot of research attention in the field of Quantum computation and nanotechnology due to its low power dissipation capability. Adders are one of the basic components in most of digital systems. Optimization of these adders can improve the performance of the entire system. In this work we have proposed designs of reversible Binary and BCD adders. Ripple carry adder, conditional adders for binary addition and regular and flagged adders for BCD addition. The proposed adder designs are optimized for quantum cost, Gate count and delay. The effectiveness of the negative control Toffoli and Peres gates in reducing quantum cost, delay and gate count is explored. Due to this the adder performance increases along with area optimization which will make these designs useful in future low power Reversible computing.

Keywords Binary adders · BCD adders · Reversible computing · Negative controlled Toffoli · Quantum cost

1 Introduction

In conventional logic circuits after the computations the input data cannot be retrieved from the outputs resulting in loss of input data. For every bit of information lost, there is heat dissipation of the order of $kT\ln 2$ Joules [1]. Bennett illustrated that if a computation were carried out using reversible logic so that there

A.N. Nagamani (✉) · N.J. Reddy
Department of Electronics and Communication Engineering,
PES Institute of Technology, Bangalore 560085, India
e-mail: nagamani@pes.edu

N.J. Reddy
e-mail: nikhiljreddi@gmail.com

V.K. Agrawal
Department of Information Science and Engineering, PES Institute
of Technology, Bangalore 560085, India
e-mail: vk.agrawal@pes.edu

© Springer India 2016   741
S.C. Satapathy et al. (eds.), *Microelectronics, Electromagnetics
and Telecommunications*, Lecture Notes in Electrical Engineering 372,
DOI 10.1007/978-81-322-2728-1_71

is no loss of information, kTln2 Joules energy dissipation would not occur [2]. If an operation does not convert energy to heat and produces no entropy then it is said to be physically reversible [3]. A logic circuit is said to be reversible if it computes a bijective (one-to-one and onto) logic function [4]. Many design constraints are imposed by the reversible logic for designing a combinational block or optimization of a Boolean function for any particular application. Firstly, fan out and loops are not allowed in reversible logic [5]. However, both can be achieved with some additional gates. Secondly, with the use of reversible logic additional functions are generated at the output along with the required outputs which are called as Garbage.

In this paper we have discussed implementation of various BCD and binary adders in reversible logic implementation. The proposed designs are improved performance parameters compared to existing designs available in literature. A 1-bit full adder is designed using 2 Peres Gate (PG), resulting in reduced gate count, quantum cost and delay compared to existing designs. Such 1-bit adders are used in designing ripple carry adders which are used for designing the BCD and Binary adders proposed in this paper. The BCD adder is designed using Negative Controlled Toffoli (NCT) and other standard reversible logic gates, which has resulted in reduced quantum cost and delay than the existing designs.

Rest of the paper is organized as follows; literature survey in Sect. 2, proposed designs in Sect. 3, results and discussions in Sect. 4, Conclusion in Sect. 5 and reference follows. The comparison of parameters such as quantum cost, delay, ancilla input and garbage output are considered and compared with the existing designs. The design parameters are generalized for n-bit operands.

2 Literature Survey

2.1 Reversible Logic

A reversible gate is having a unique input–output pattern, i.e., they have same number of output as the inputs [6, 7]. The fan out or feedback facilities are not applicable to the reversible circuits and these are made up of only reversible gates. When it comes to the use of this reversible hardware, they find their importance in some of the crucial fields such as low power design, nanotechnology, quantum computing, bio-informatics and optical information processing. An $n \times n$ reversible gate means it is having n-input pins and n-output pins.

Some of the basic reversible gates used in this work are CNOT/Feynman gate (FG), Fredkin gate, Toffoli gate (TG) and Peres gate (PG). These gates are also synthesizable.

One of the variants of TG is the Negative Controlled Toffoli (NCT) which can have one or more negative control lines. It works complimenting to that of TG i.e. the target bit toggles if the negative control bit is zero. This gate can be used with a quantum cost of 6 and delay of 6Δ.

The output function, quantum cost and delay of the basic reversible gates used in this work are as listed below:

1. Feynman Gate (a, b) = (a, a xor b); Quantum cost = 1; Delay = 1Δ.
2. Toffoli Gate (a, b, c) = (a, b, ab xor c); Quantum cost = 5; Delay = 5Δ.
3. Fredkin Gate (a, b, c) = (a, ab + a'c, ac + a'b); Quantum cost = 5; Delay = 5Δ.
4. Peres Gate (a, b, c) = (a, a xor b, ab xor c); Quantum cost = 4; Delay = 4Δ.
5. Negative Toffoli Gate (a, b, c) = (a, b, a'b' xor c); Quantum cost = 6; Delay = 6Δ.

Various design techniques and synthesis methods have been proposed for Toffoli gate with negative control lines. A post—synthesis DD based optimization technique for NCT has been proposed in [5]. Usage of NCT optimizes the circuit design because of its capability to replace series of gates which reduces the area, gate count and power consumption. Synthesis algorithm proposed for negative control lines are used to optimize the circuit for a given function as mentioned in [7] with compromise in computational time. Further optimization with the use of NCT in place of Toffoli according to a set of rules defined based on template matching technique in [8] is another method which reduces gate count, area and power consumption.

2.2 Existing Adder Designs

Design of combinational blocks using the concept of reversible gates is one of the trending topics in VLSI domain. A reversible BCD adder circuit is proposed in [9]. This circuit has a quantum cost of 113 and delay of 97Δ. Various attempts and designs have been proposed for BCD among which [10] concentrates on a modular synthesis method to design a reversible BCD adder circuit. A BCD adder design with the help of NEW gates and FG gates has been proposed in [11], in which they have mainly concentrated on reducing the garbage outputs, neglecting the delay, ancilla inputs and quantum cost. In [12] the BCD adder is designed with the goal of optimizing the number of ancilla inputs and garbage outputs compared to [11] using HNG gates [10] and SLC gates which are new kind of gates. This design also results in reduced gate count, but the quantum cost and delay of new gates are not discussed. The flagged BCD adder which works by generating flag bits has shown that it is better when compared to that of all the existing work in terms of power dissipation and area [11, 13, 14].

In this work we have proposed reversible implementations of basic adders such as the conditional sum adder and conditional carry adder [15], flagged BCD adder [16]. Conditional Carry adder [15] which calculates sum and carry faster than other adder variants which finds its application in high speed designs. These adders have a better power-delay product compared to other adders for high-speed applications [15]. However, design complexity increases with the number of bits.

The BCD adders along with Carry Look-Ahead (CLA) adder proposed in [13] does faster calculation compared to conventional one, but uses additional area, power and has higher latency. Hence, to further reduce power and latency in BCD addition, flagged binary addition [16] is proposed for the correction logic.

In this work, a 2 PG model of a 1-bit adder proposed in [17] is used, with one ancilla input and 2 garbage outputs. This design is used to build the ripple carry adder and BCD adder.

3 Proposed Designs

3.1 Binary Adders

A common and yet very useful type of combinational circuit that can be realized by simple logic operations and used in the addition of binary numbers are called as binary adders. These are designed using basic AND and XOR operations. The simplest type of adders is a single bit full adder. Few of the binary adder architectures are designed in this section.

3.1.1 Reversible Ripple Carry Adder

Only the PG is used to design a 1-bit full adder with 1 ancilla input and 2 garbage values. The design consists of only 2 Peres gates, as shown in Fig. 1a. The 1-bit full adder discussed is used to design a 4-bit ripple carry adder as shown in the Fig. 1b. Four 1-bit full adders are connected in cascade to get the 4-bit ripple carry adder. This design has one Ancilla input and two garbage for each input bit pair. The 4-bit adder has a quantum cost of 32 and a delay of 20Δ. It is evident that there is

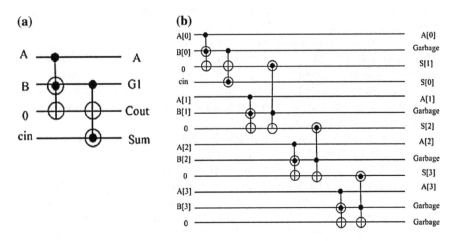

Fig. 1 **a** 1-bit full adder [17], **b** 4-bit ripple carry adder

(a)   **(b)**

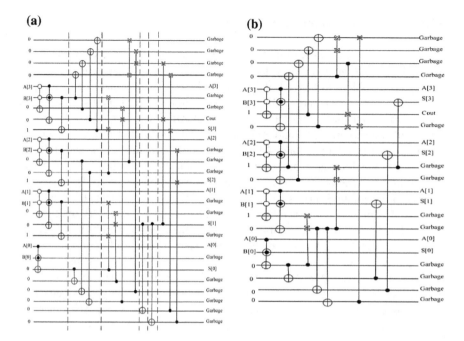

Fig. 2 **a** Quantum diagram conditional sum adder, **b** quantum diagram of conditional carry adder

improvement in performance as there is reduction of quantum cost and delay when compared to the ripple carry adder proposed in [9].

3.1.2 Reversible Conditional Sum Adder

The problem of carry propagation is solved by conditional sum addition rules. All possible carries are generated from the adder, which are simultaneously used to select the true sum outputs from the two provisional sums. The schematic diagram of the conditional sum adder (CSA) for 4-bit is shown in the Fig. 2a. The true sum outputs are selected from various sequences at all possible sum outputs. The selection of the sum bits are dependent on the carry bits. When $C_{i-1} = 0$, the sum bit can be can be obtained by the expression $S_i = A_i \text{ xor } B_i \text{ xor } 0 = A_i \text{ xor } B_i$, else if $C_{i-1} = 1$, then the sum bit is given as $S_i = A_i \text{ xor } B_i \text{ xor } 1 = A_i \text{ xnor } B_i$. The total number of multiplexers that are used in a 2^N bit CSA is obtained by the expression

$$\sum_{n=1}^{N} \left(2^{N-n} + 1\right)\left(2^n - 1\right) = \sum_{n=1}^{N} 2^N - 2^{N-n} + 2^n - 1 \tag{1}$$

3.1.3 Reversible Conditional Carry Adder

The basic process by which Conditional Carry Adder (CCA) works is same as that of the CSA, but involves only the selection of carries to overcome the carry propagation problem in Carry Look Ahead Adders [18]. Distinct carries are generated using which true carries are generated simultaneously from two provisional carries. The provisional carries are obtained under different carry input conditions. The simultaneous carry generations are all done independently for each bit pair. The conditional carry addition for a 4-bit will be completed in 3 steps with an extra XOR of the previous carry and sum bit to generate final sum bits. CCA [15] is a method involving selection of carry output of every bit of the operands. It uses less number of 2:1 multiplexers as compared to conditional sum adders [15]. The CCA for a 4-bit addition only generates C_0, C_1 and C_3. Hence, some 2:1 multiplexers have to be supplied by the carry unit in order to select the other carry bit C_2. As the conventional CSA, CCA still uses the multiplexer for carry selection. To explain the process, if $C_{i-1} = 0$, the carry signal is expressed as $C_i = A_iB_i + (A_i + B_i) \cdot 0 = A_iB_i$. If carry signal $C_{i-1} = 1$, the next carry is given by the expression $C_i = (A_iB_i) \cdot 0 + (A_i + B_i) \cdot 1 = A_i + B_i$. Thus every carry generation module is designed with a two input OR gate and a two input AND gate. To generalize, the total number of multiplexers in a 2^N bit CCA is given by

$$\sum_{n=1}^{N} 2^{N-n}(2^n - 1) = \sum_{n=1}^{N} 2^N - 2^{N-n} \tag{2}$$

where, $N = \log_2 n$. It involves 22 reversible gates for a 4-bit operation with quantum cost and delay of 73Δ and 25Δ respectively. The quantum diagram of the Conditional carry Adder is given in Fig. 2b.

3.1.4 BCD Adders

BCD adders are also a type of binary adders in which the out sum is constrained between 0 and 9 and any bit above this will be detected by excess 9 detectors and corrected in correction logic. Two designs of reversible BCD adders are proposed in this section.

Reversible 4-Bit BCD Adder

Full adder circuits along with an overflow detection circuit are used in the construction of the conventional BCD adders [19]. A 4-bit reversible BCD adder has been proposed using the 4-bit ripple carry adder. 3 * 3 NCT and a 3 * 3 Toffoli are used to realize the overflow detection circuit. The use of NCT in the design reduces

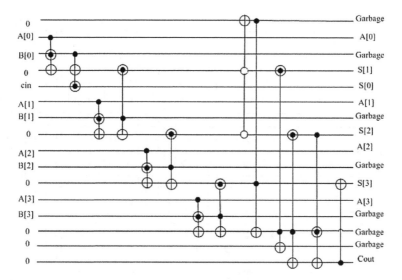

Fig. 3 Quantum diagram of 4-bit BCD adder

the gate count [20] and hence aids the requirement. In the correction circuit a 1-bit adder has been used along with a Feynman gate and a PG which has reduced gate count and delay when compared to that of the existing designs. The BCD adders are used for addition of two numbers say A $(A_3 A_2 A_1 A_0)$, B $(B_3 B_2 B_1 B_0)$ and produce an output in the range 0–9. If the output is greater than 9 it will be detected by excess 9 detector and generates a correction bit which is nothing but the carry bit. This activates the correction block where 0110 (6 in decimal) will be added to the sum bits. The excess 9 detector uses a correction logic which generates Cout given by

$$C_{out} = S_3S_2 + S_3S_1 + C_3 \tag{3}$$

This Boolean expression is realized by using an NCT and a Toffoli. In this design the quantum cost is found to be 56 with gate count of 14 and a delay of 40Δ. Figure 3 shows the quantum diagram of the reversible BCD adder.

Reversible 4-Bit Flagged BCD Adder

A flagged BCD adder [16] is designed with a 4-bit ripple carry adder discussed in section along with excess-9 detector, flag bit computation block and four 2:1 multiplexers. The multiplexers design is realized by using a 3 * 3 Fredkin gate. The excess-9 detector is same as that of reversible BCD adders made up of a 3 * 3 Toffoli and Negative Toffoli gate. A flagged BCD adder is an improved design of the conventional BCD adder which uses the flag inversion logic [16]. Flag inversion

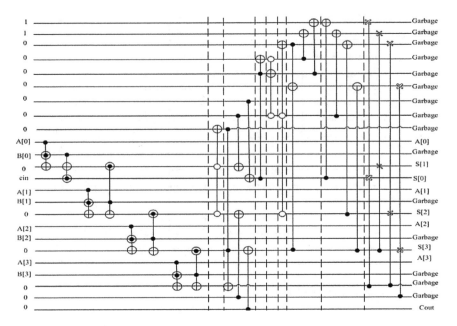

Fig. 4 Quantum diagram of flagged BCD adder

logic realizes the addition of constant (0110) with lesser number of gates and also the delay of the proposed model is comparatively lower than the existing designs. Figure 4 gives the quantum diagram of the proposed flagged BCD adder. The input operands A (A_3 A_2 A_1 A_0), B (B_3 B_2 B_1 B_0) are given to the 4-bit full adder. Sum bits S (S_3 S_2 S_1 S_0) and carry out (Cout) are used in excess-9 detector circuit to get correction bit. If the correction bit is 0, then the sum bits are passed through the multiplexers. If it is 1, then the following steps are to be followed to calculate sum bits.

Step 1: $d_1 = d_0$ and S_0; $d_2 = d_1 + S_1$; $d_3 = d_2 + S_2$; $d_4 = d_3$ and S_3

The flag bits are generated using these bits.

Step 2: $f_1 =$ not d_1; $f_2 =$ not d_2; $f_3 = d_3$; $f_4 = d_4$

In the next step the flag bits f (f_3 f_2 f_1 f_0) and sum bits S (S_3 S_2 S_1 S_0) are passed through flag inversion logic to get the BCD output m (m_3 m_2 m_1 m_0) which is one of the inputs to the multiplexer, obtained as output when Cout value is 1.

Step 3: $m_0 = f_0$ xor S_0; $m_1 = f_1$ xor S_1; $m_2 = f_2$ xor S_2; $m_3 = f_3$ xor S_3

4 Results and Discussions

In this section, comparisons of various Reversible circuit performance parameters with respect to the existing designs in literature are discussed and improvements of the proposed designs are analysed.

4.1 Binary Adders

The proposed 4-bit reversible Ripple carry adder using only PG has a delay of 20Δ, quantum cost of 32, gate count of 8, ancilla input of 4 and garbage output of 8. This when compared with [9], the quantum cost will reduce by 16 for a 4-bit adder. The delay also has reduced compared to [12, 21]. The comparison is given in Table 1.

Among all types of adders ripple carry adder is the simplest form of adder and has longest delay of all. To avoid this linear increase in the delay of carry generation as bits increase. In case of CLA carry is calculated parallel using carry generators [23]. The speed of the adders can be increased at the cost of area using the carry select adders. In this type of adders two versions of carry are calculated and the right one will be selected. These pairs are repeated in a 4-bit adder module. Thus the propagation of such carry select adders is reduced to that of n/4 adder modules. In carry select design the previous carry selects the sum using a multiplexer gate. Thus each module is having a delay time equal to that of a multiplexer delay and hence will be faster than the CLA unit. The speed of the carry select adders can be further improved by using more number of adders and multiplexers. Where the speed increases at the cost of increase in area [24].

Table 1 Comparison of N bit reversible RCA adder

Bits	[12]	[9]	Proposed	% Improvement w.r.t [12]	% Improvement w.r.t [9]
4	56	48	32	42.85	33.33
8	114	96	64	43.85	33.33
16	234	192	128	45.29	33.33
32	474	384	256	45.99	33.33
64	954	768	512	46.33	33.33
128	1914	1536	1024	46.49	33.33
256	3834	3072	2048	46.58	33.33
512	7674	6144	4096	46.62	33.33

Designs	Ancilla input	Garbage output	Quantum cost	Delay
[22]	0	0	$17n - 6$	$10n + 6$
[21]	0	0	$17n - 22$	$15n - 6$
[12]	0	0	$15n - 6$	$9n + 1$
[9]	4n	0	12n	10n
Proposed	N	2n	8n	$4n + 4$

Table 2 Comparison of 4-bit binary adders

Adder type	Ancilla input	Garbage output	Quantum cost	Delay
RCA	4	8	32	40
CSA	19	19	91	25
CCA	13	13	73	24

The conditional carry design based on carry select adder [25, 26] is improved version of CSA which is still restricted to the carry select propagation. The conditional carry adder has achieved highest performance than any other adders used in high speed applications [27, 28]. The number of multiplexers is more in case of CSA but it has been reduced in the CCA design and hence the area also reduces. The comparison table of CSA and CCA are tabulated Table 2.

4.2 BCD Adders

When compared with the previous works of the reversible BCD adders the proposed design has shown improvements in various parameters tabulated in Table 3. It has mainly reduced quantum cost by 20 % and delay by 29.82 % when compared to the most optimized design proposed in [12].

In flagged BCD adders multiplexers are used for addition of correction logic, thus the circuit has lower delay compared to the adder with correction bits. The flagged BCD adders are better than the normal BCD adders in terms of delay if the sum is less than or equal to 9. They are just passed through the multiplexers to get the sum bits. The delay for a 4-bit flagged BCD adder in this case will be 48Δ and if the sum is greater than 9 and the flagged correction logic is used to correct them, then the delay will go up to 80Δ. But this delay is less when compared to other

Table 3 Comparison of 4-bit reversible BCD adder and flagged BCD adders

References design	Ancilla	Garbage output	Quantum cost	Delay
[13]	28	24	220	Not discussed
[29]	16	16	676	Not discussed
[10]	56	64	336	Not discussed
[30]	8	24	412	Not discussed
[12]	4	3	70	57
[31]	6	10	Not discussed	Not discussed
[11]	17	22	Not discussed	Not discussed
[9]	13	10	113	97
Proposed BCD	7	7	56	40
Flagged BCD with C_{out}	10	17	90	80
Flagged BCD without C_{out}	10	17	90	48

Table 4 Comparison of n-bit reversible BCD Adders

	Ancilla Input	Garbage Output	Quantum Cost	Delay
[13]	7n	6n	55n	Not discussed
[29]	4n	4n	169n	Not discussed
[10]	14n	16n	84n	Not discussed
[30]	2n	6n	103n	Not discussed
[12]	N	N − 1	70N	57N
[9]	3n + n/4	2n + n/2	28n + n/4	24n + n/4
[Proposed]	2n − 1	2n − 1	14n	10n

BCD adders. Last two rows of the Table 3 give the parameters of the flagged BCD and the normal BCD adder with and without overflow in addition.

From the Table 3 we can say that when compared to the existing work, the proposed work has good amount of improvement in all the parameters. In [12] author has designed a 1-digit BCD addition which is 4-bit and in [9] the data for this design is taken as 1-bit design. In this work, we have addressed the parameters for 4-bit addition. From Table 3 we can see that the quantum cost of the proposed design has reduced to noticeable levels. In the Table 4 comparisons of BCD adders for n-bit design are tabulated.

5 Conclusion

From this work we conclude that, a single adders cannot satisfy all the constraints of the design parameters. Each adder will have their own better feature than others. Conditional adders when compared to the other adders have less delay and hence perform faster at the cost of increase in quantum cost. A maximum of 40 % delay reduction is achieved in CCA compared to RCA and CSA offers 37 % reduction in delay with approximately 2× and 3× increase in quantum cost respectively. The BCD adder proposed is better when compared to previous work with reduction of 20 % in quantum cost and 29.82 % in delay.

References

1. R. Landauer, Irreversibility and heat generation in the computational process. IBM J. Res. Develop. **5**, 183–191 (1961)
2. C.H. Bennett, Logical reversibility of computation. IBM J. Res. Develop. **17**(6), 525–532 (1973)
3. D.P. Vasudevan, P.K. Lala, J.P. Parkerson, A novel approach for on-line testable reversible logic circuit design. In *Proceedings of the 13th Asian Test Symposium (ATS 2004)*
4. I. Polian, J.P. Hayes, Advanced modeling of faults in reversible circuits. IEEE (2010). 978-1-4244-9556-6/10

5. R. Barnes Earl, G. Oklobdzija Vojin, New multilevel scheme for fast carry-skip addition. IBM Techn. Disclosure Bull. **27**, 133–158 (2009)
6. B. Parhami, Fault tolerant reversible circuits. In *Proceedings of 40th Asimolar Conference Signals, Systems, and Computers*, Pacific Grove, CA, pp. 1726–1729, Oct. 2006
7. M.M. Rahman, Md. Saiful Islam, Z. Begum, M. Hafiz, Synthesis of fault tolerant reversible logic circuits. IEEE (2009). 978-1-4244-2587- 7/09
8. K. Datta, G. Rathi, R. Wille, *Exploiting Negative Control Lines in the Optimization of Reversible Circuits* (Springer-Verlag, Berlin Heidelberg, 2013)
9. A.N. Nagamani, S. Ashwini, V.K. Agarwal, Design of optimized reversible binary and BCD adders. In *2015 International VLSI Systems, Architecture, Technology and Applications (VLSI-SATA)*, pp. 1–5, 8–10 Jan. 2015
10. M. Mohammadi, M. Eshghi, M. Haghparast, A. Bahrololoom, Design and optimization of reversible bcd adder/subtractor circuit for quantum and nanotechnology based systems. World Appl. Sci. J. **4**(6), 787–792 (2008)
11. H. Hasan Babu, A. Raja Chowdhury, Design of a reversible binary coded decimal adder by using reversible 4-bit parallel adder. In *Proceedings of the 18th International Conference on VLSI Design and 4th International Conference on Embedded Systems Design, 1063-9667/05*. IEEE (2005)
12. H. Thapliyal, N. Ranganathan, Design of efficient reversible logic based binary and BCD adder circuits. ACM J. Emerg. Technol. Comput. Syst. V(N), Month, 20YY
13. A.K. Biswas, M. Hasan, A.R. Chowdhury, Md Hafiz, H. Babu, Efficient approaches for designing reversible binary coded decimal adders. Microelectron. J. **39**(12), 1693–1703 (2008)
14. B. Ramkumar, H.M. Kittur, Low-power and area-efficient carry select adder. IEEE Trans. Very Large Scale Integr. VLSI Syst. **20**(2), 371–375 (2012)
15. K.-H. Cheng, S.-W. Cheng, Improved 32-bit conditional sum adder for low-power high-speed applications. J. Inform. Sci. Eng. **22**, 975–989 (2006)
16. K.N. Vijeyakumar, V. Sumathy, FPGA implementation of low power hardware efficient flagged binary coded decimal adder. Int. J. Comput. Appl. **46**(14), 0975–8887 (2012)
17. V. Devi, H. Lakshmisagar, FPGA implentation of reversible floating point multiplier using CSA. (IJIRSE) Int. J. Innov. Res. Sci. Eng. ISSN (Online) 2347–3207
18. H. Thapliyal, H.V. Jayashree, A.N. Nagamani, H.R. Arabnia, A novel methodology for reversible carry look-ahead adder. In *Transactions on Computational Science XVII, pp 73–97 Part I* (Springer-Verlag Berlin Heidelberg, 2013)
19. H.M.H. Babu, A.R. Chowdhury, Design of a compact reversible binary coded decimal adder circuit. Elsevier J. Syst. Archit. **52**, 272–282 (2006)
20. R. Wille, M. Soeken, N. Przigoda, R. Drechsler, Exact synthesis of Toffoli gate circuits with negative control lines. In *IEEE 42nd International Symposium on Multiple Valued Logic* (2012)
21. K.-W. Cheng, C.-C. Tseng, Quantum full adder and subtractor. Electron. Lett. **38**(22), 1343–1344 (2002)
22. K.V.R.M. Murali, N. Sinha, T.S. Mahesh, M.H. Levitt, K.V. Ramanathan, A. Kumar, Quantum information processing by nuclear magnetic resonance: experimental implementation of half-adder and subtractor operations using an oriented spin-7/2 system. Phys. Rev. A **66**(2), 022313 (2002)
23. K. Hwang, *Computer Arithmatic: Principles, Architecture, and Design* (Wiley, 1976)
24. O.J. Bedrij, Carry-select adder. IRE Trans. Electron. Comput. EC **11**, 340–346 1962
25. S. Perri, P. Corsonello, G. Cocorullo, A high-speed energy-efficient 64-bit reconfigurable binary adder. IEEE Trans. Very Large Scale Integ. (VLSI) Syst. **11**, 939–943 (2003)
26. Y.M. Huang, J.B. Kuo, A high-speed conditional carry select(CCS) adder circuit with a successively incremented carry number block (SICNB) structure for low-voltage VLSI implementation. IEEE Trans. Circuit Syst. II: Analog Digital Signal Process. **47**, 1074–1079 (2000)

27. K.H. Cheng, S.M. Chiang, S.W. Cheng, The improvement of conditional sum adder for low power applications. In *Proceedings of the IEEE International Application Specific Integrated Circuits Conference*, pp. 131–134 (1998)

28. J. Sklansky, Conditional-sum addition logic. IRE Trans. Electron. Comput. EC **9**, 226–231 (1960)

29. M.K. Thomsen, R. Glck, Optimized reversible binary-coded decimal adders. J. Syst. Archit. **54**(7), 697–706 (2008)

30. M. Mohammadi, M. Haghparast, M. Eshghi, K. Navi, Minimization optimization of reversible bcd-full adder/subtractor using genetic algorithm and don't care concept. Int. J. Quant. Inform. **7**(5), 969–989 (2009)

31. H.R. Bhagyalakshmi, M.K. Venkatesha, Optimized reversible BCD adder using new reversible logic gates. J. Comput. **2**(2), 2151–9617 (2010)

Magnetic Field Analysis of a Novel Permanent Magnetic Suspension Pole Shoe Based Bearing Less Switched Reluctance Motor Using Finite Element Method

P. Nageswara Rao, G.V.K.R. Sivakrishna Rao
and G.V. Nagesh Kumar

Abstract For high speed control of Bearing Less Switched Reluctance Motor (BSRM), the torque and radial force has to be decoupled. A novel Permanent magnet pole shoe type based suspension poles is proposed for a 12/14 BSRM, to decouple suspension force and radial torque and also radial forces generated in x and y-axis. A pole shoe type permanent magnet made of Neodymium Magnet, which is an alloy of Neodymium, iron and boron material (NDEFB) is placed on the all four suspension poles. The selection of pole shoe is taken according to force required to levitate the rotor mass in air gap. The width and arc are taken without disturbing the self starting phenomena of rotor. When the motor is excited, the flux distribution in the air gap under phase A is short flux path and there is no flux reversal in the stator core. This will decrease saturation effect and the MMF required producing motoring torque which decreases core losses. The decoupling of radial force and motoring torque are achieved and the decoupling of radial force in positive and negative x and y directions is achieved. The performance of the proposed motor is compared with 12/14 conventional BSRM model and the results are presented and analysed.

Keywords Magnetic field · Bearingless motor · Stress

1 Introduction

BSRM has good performance under special environments such as fault tolerance, robustness, tolerance of high temperature or in intense temperature variations [1]. BSRM has two types of stator windings consists of motor main windings or Torque

P. Nageswara Rao · G.V. Nagesh Kumar
Department of EEE, GITAM University, Visakhapatnam, A.P., India

G.V.K.R. Sivakrishna Rao (✉)
Department of EEE, Andhra University, Visakhapatnam, A.P., India
e-mail: drgvnk@rediffmail.com

© Springer India 2016
S.C. Satapathy et al. (eds.), *Microelectronics, Electromagnetics
and Telecommunications*, Lecture Notes in Electrical Engineering 372,
DOI 10.1007/978-81-322-2728-1_72

755

windings and radial force or suspension force windings in the stator to produce suspension force that can realize rotor shaft suspension without mechanical contacts [2–5]. Recently, several structures of bearing less switched reluctance motor (BSRM) have been proposed. According to size of suspension pole and torque pole to get the self starting of motor and improvements in decoupling between suspension force and motor torque. All the above performances are achieved by implementing novel 12/14 BSRM [6–8]. With the novel winding arrangement and pole structures, the short flux paths with no reversal of flux paths are achieved compared with conventional 12/8 BSRM pole structure [9].

In 12/14 BSRM suspension winding the main tasks is levitation of rotor in air gap at the time of motor starts. It requires the DC excitation on all four suspension poles. Second thing is control the rotor position to its axis point when the rotor center axis point is changed due to change in loads and disturbances. With the single winding on suspension pole it has to do the above mentioned tasks at the same instant, so it is very difficult to get independent control of position and levitation [10–13].

A pole shoe type permanent magnet an alloy of Neodymium, iron and boron material (NDEFB) material is placed on the all four suspension poles. The performance of the proposed model is compared with novel 12/14 BSRM at same excitation levels without changing the winding pattern and switching supply. When the motor is excited, the flux distribution in the air gap under phase A is short flux path and there is no flux reversal in the stator core. This will decrease saturation effect and the MMF required producing motoring torque these leads to lower core losses. The radial force and motoring torques are taken for all positive and negative directions of both X-axis and Y-axis. As compare with 12/14 BSRM model, the proposed model shows better uniform flux patterns and simple controlling of rotor position and levitation of rotor with less value of currents. The decoupling of radial force and motoring torque are achieved and the decoupling of radial force in positive x and negative x directions similarly radial force in positive y and negative y directions are achieved.

2   Proposed Bearingless Motor

The hybrid stator pole structure of 12/14 BSRM is shown in Fig. 1. In general, any BSRM model consists of both torque and suspending force poles on the stator only. The rotor does not consist of any windings; it is a simple solid iron core with the salient poles. Figure 1 shows four suspension poles S_{x1}, S_{x2}, S_{x3} and S_{x4} are individually excited. Four torque poles i.e. T_{A1}, T_{A2}, T_{A3} and T_{A4} are connected in series to construct phase A, and remaining four torque poles T_{B1}, T_{B2}, T_{B3} and T_{B4} are connected in series to construct phase B. X-directional suspending force is controlled by the winding currents in S_{x2} and S_{x4} stator radial force poles. Similarly

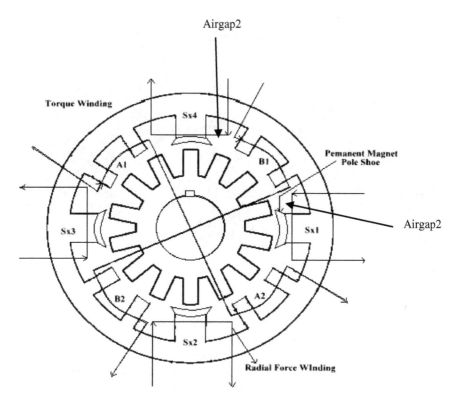

Fig. 1 Winding diagram of 12/14 BSRM

Y-directional suspending force is controlled by controlling the winding currents in S_{x1} and S_{x3}. In order to get a continuous suspending force, the suspending force pole arc is selected to be not less than one rotor pole pitch.

The magnetic flux is generated when x-axis suspension winding current is excited. It can be seen that the flux density in the air gap 1 is increased; because of the direction of the Permanent Magnet flux from pole shoe is in the same direction as the suspension magnetic flux. At the same instant the flux density in the airgap2 is decreased, as the direction of Permanent Magnet flux is opposite to that of the suspension magnetic flux. Therefore, this overlaid magnetic field results in the radial suspension force Fx acting on the rotor toward the positive direction in the x-axis. A radial suspension force toward the negative direction in the x-axis can be produced with a negative current. Similarly, the principle can be applied to the Y-axis also.

A pole shoe type permanent magnet is placed on all suspension poles to levitate the rotor. The selection of pole shoe is taken according to force required to levitate the rotor mass in air gap. The width and arc are taken without disturbing the self starting phenomena of rotor.

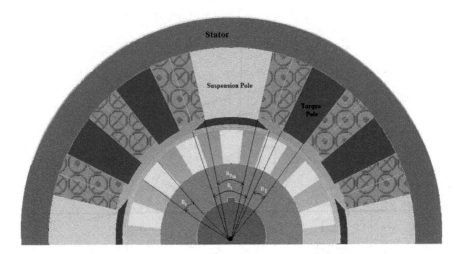

Fig. 2 3D fem model

This hybrid stator model exhibits linear characteristics with respect to rotor position, because of the independent controlling between motor torques and suspending force poles. This model can effectively decouple the torque from the suspension force. The output torque is considerably enhanced because of the small flux path without any flux reversal, and also it is simple to control the air-gap. The inductance of suspending winding is almost constant due to that pole arc of suspending stator is equal to one rotor pole pitch. Therefore, for the toque design, it can refer to the conventional single-phase SRM. The 3-D FEM Model is shown in Fig. 2. When ignoring fringing effect and saturation effects, the inductance may be explained as:

$$L = \frac{\mu_0 N^2 L_{STK} R}{g}(\theta_0 + K_{fr}) \tag{1}$$

Consequently, the torque T in this pole can be explained as follows:

$$T = \frac{1}{2}i^2 L \tag{2}$$

Based on the FEM analysis, the net suspending force produced by proposed BSRM and is given in Eq. (3).

$$\begin{bmatrix} F_x \\ F_y \end{bmatrix} = \begin{bmatrix} K_{Fx} & 0 \\ 0 & K_{Fy} \end{bmatrix} \begin{bmatrix} i_x \\ i_y \end{bmatrix} \text{ where } i_x = \begin{bmatrix} i_1^2 \\ i_3^2 \end{bmatrix} i_y = \begin{bmatrix} i_2^2 \\ i_4^2 \end{bmatrix} \tag{3}$$

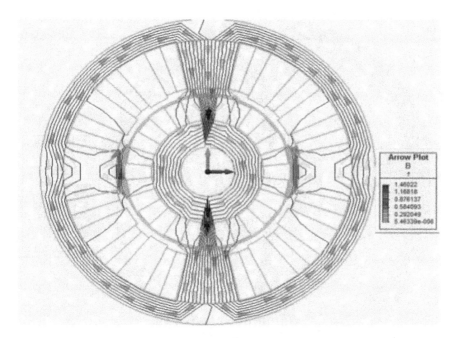

Fig. 3 Magnetic flux distribution pattern from PM suspension poles

3   Results and Discussions

Figure 3 shows the flux distribution of the permanent magnet pole shoes on suspension poles. The magnetic force produced from the permanent magnet to levitate the rotor is =56N. The peak value of flux density in air gap from permanent magnet is 0.879T.

Figure 4a shows the Field distributions of four pole suspension poles produced by the four pole suspension windings. Figure 4b shows the instantaneous energy and co energy productions by suspension windings. By Finite–Element method, the suspension force computed is Fx = 5N and Fy = 106N, and the maximum value of flux density in air gaps is about 1.2T. The maximum magneto motive force from each suspension poles is about 160AT.

Figure 5 shows the flux distributions of phase A with short flux paths, without reversal of flux paths at constant load. In order to get easy analysis only phase A is considered.

The Fig. 6, shows that at constant load, the suspension force produced from suspension stator windings S_{x1} and S_{x2} windings is independent to each other in X

(a)

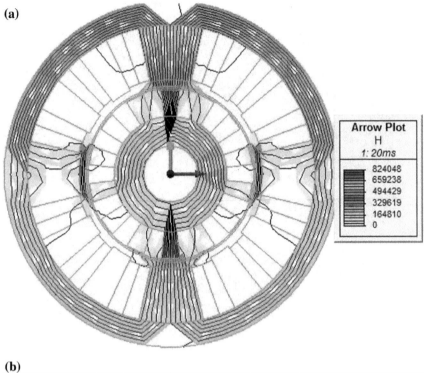

(b)

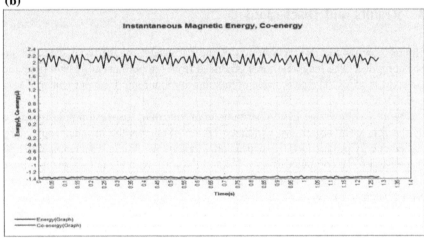

Fig. 4 **a** Magnetic flux distributions of excited suspension poles. **b** Instantaneous energy and co energy due to suspension poles with PM pole shoes

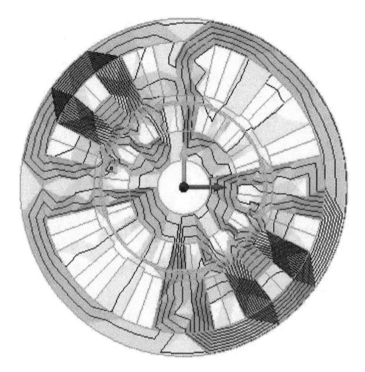

Fig. 5 Short flux paths of phase A

and Y directions i.e. the decoupling between x-axis suspension force and y-axis suspension force is possible. Figure 7 shows the flux linkages of motoring coils and Fig. 8 shoes the flux linkages of suspension coils. Fig. 9 shows the variation of instantaneous magnetic energy and co energy. From Fig. 10, the force produced from suspension coils and torque produced from torque windings are independent at any time of operation of motor. Table 1 shows the comparison of conventional and proposed BSRM.

From the Fig. 10, the force produced from suspension coils and torque produced from torque windings are independent at any time of operation of motor. Because of the saliency of both stator and rotor of the BSRM, the distribution of the magnetic flux and electric field in air gap is varied with the position of the rotor. For the proposed BSRM, Several important positions are chosen to compare the FEM analysis results with conventional 12/14 BSRM. The distribution of magnetic flux in both the models has the short flux paths and there is no flux reversal in air gap. Due to short flux paths produced from Phase and phase B, the saturation and core losses are reduced. As compare with conventional model the proposed model shows better independent control of radial force in x-axis and y-axis directions. The force required for the levitation of rotor is achieved at just 2A of suspension winding currents due to PM Pole shoe on suspension poles. But in case of conventional 12 × 14 BSRM the current required is 4A to levitate the rotor in air gap.

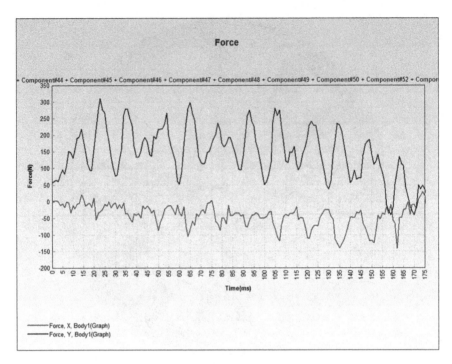

Fig. 6 Suspension force produced in x and y directions

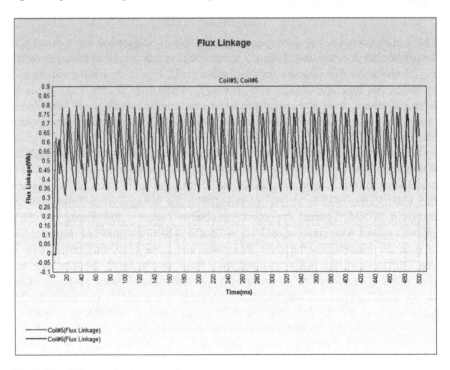

Fig. 7 Flux linkages of motoring coils

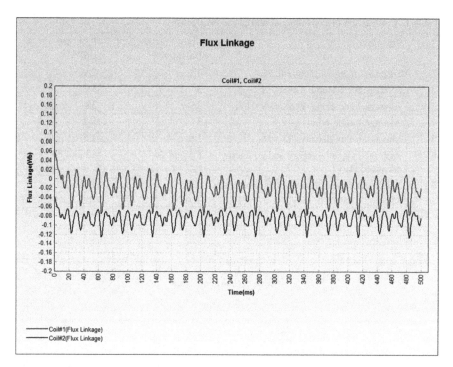

Fig. 8 Flux linkages of suspension coils

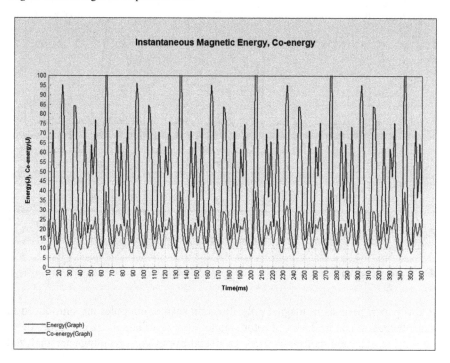

Fig. 9 Instantaneous magnetic energy and co energy

Table 1 Comparison of conventional and proposed BSRM

S. no	Parameter	Conventional 12/14 BSRM	Proposed 12/14 BSRM
1	Suspension winding current, A	4A	2A
2	Average force along X-Axis, N	35	38
3	Average force along Y-Axis, N	35	38
4	Average torque N-mt	8	8.2
5	Maximum flux density (T)	1.9	1.6
6	Average magnetic energy and co-energy produced (J)	8.5 and 24	15 and 35
7	Maximum MMF in suspension coils (AT)	320	160
8	Maximum MMF in torque coils (AT)	360	360

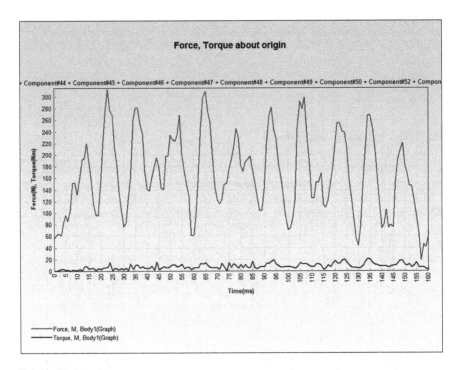

Fig. 10 Force and torque

4 Conclusion

In this paper, permanent magnet pole shoes on suspension poles are introduced to make levitation and positions of rotor simple. Flux distributions, radial forces of all suspension coils and motoring torque produced from torque windings are analyzed based on finite element method (FEM). The saturation effect is reduced due to short flux path and there is no reversal of flux under both the phases. Decoupling of positive x and negative x directions of radial force are achieved. Only 2A is sufficient to levitate the rotor for suspension windings with the aid of PM Pole shoe flux.

References

1. R. Krishnan, R. Arumugan, J.F. Lindsay, Design procedure for switched-reluctance motors. IEEE Trans. Ind. Applicat. **24**, 456–461 (1988)
2. M. Takemoto, K. Shimada, A. Chiba, T. Fukao, A design and characteristics of switched reluctance type bearing less motors, in *Proceedings of 4th International Symposium Magnetic Suspension Technology*, vol. NASA/CP-1998-207654, May 1998, pp. 49–63
3. Z. Xu, D.-H. Lee, J.-W. Ahn, *Modeling and Control of a Bearingless Switched Reluctance Motor with Separated Torque and Suspending Force Poles*
4. Z. Xu, F. Zhang, J.-W. Ahn, *Design and Analysis of a Novel 12/14 Hybrid Pole Type Bearingless Switched Reluctance Motor*
5. H. Wang, Y. Wang, X. Liu, J.-W. Ahn, Design of novel bearingless switched reluctance motor. IET Electr. Power Appl. **2**, 73–81 (2012)
6. Z. Xu, D.-H. Lee, J.-W. Ahn, Control characteristics of 8/10 and 12/14 bearingless switched reluctance motor, in *The 2014 International Power Electronics Conference*
7. Z. Xu, D.-H. Lee, J.-W. Ahn, Comparative analysis of bearingless switched reluctance motors with decoupled suspending force control. IEEE Trans. Ind. Appl. **51**(1), (2015)
8. J. Bao, H. Wang, J. Liu, B. Xue, Self-starting analysis of a novel 12/14 type bearingless switched reluctance motor, in *2014 IEEE International Conference on Industrial Technology (ICIT)*, Busan, Korea, 26 Feb–1 Mar 2014)
9. J. Bao, H. Wang, B. Xue, The coupling characteristics analysis of a novel 12/14 type bearingless switched reluctance motor, in *17th International Conference on Electrical Machines and Systems (ICEMS)*, Hangzhou, China, 22–25 Oct 2014)
10. H. Wang, J. Bao, B. Xue, J. Liu, *Control of Suspending Force in Novel Permanent Magnet Biased Bearingless Switched Reluctance Motor* (IEEE, 2013)
11. B. Xue, H. Wang, J. Bao, Design of novel 12/14 bearingless permanent biased switched reluctance motor, in *2014 17th International Conference on Electrical Machines and Systems (ICEMS)*, Hangzhou, China, 22–25 Oct 2014
12. H. Jia, C. Fang, T. Zhang, Finite element analysis of a novel bearingless flux switching permanent magnet motor with the single winding, in *2014 17th International Conference on Electrical Machines and Systems (ICEMS)*, Hangzhou, China, 22–25 Oct 2014
13. H. Yang, Z. Deng, X. Cao, L. Zhang, Compensation strategy of short-circuit fault for a single-winding bearingless switched reluctance motor, in *2014 17th International Conference on Electrical Machines and Systems (ICEMS)*, Hangzhou, China, 22–25 Oct 2014

Performance Analysis of a Field-Effect-Transistor Based Aptasensor

Md. Saiful Islam, Eugene D. Coyle and Abbas Z. Kouzani

Abstract Prostate cancer is one of the most diagnosed cancers which leads to a considerable number of deaths due to the lack of early and sensitive detection. This paper presents an aptamer functionalized field effect (FET) based biosensor for the detection of prostate cancer. Prostate specific antigen (PSA) is considered as the biomarker for prostate cancer whose detection is confirmed by attaching aptamers onto the sensor surface. Through the modelling and numerical simulation, the paper aims to evaluate and predict the performance parameters such as sensitivity, settling time, and limit of detection (LOD) of a label-free FET based electronic biosensor. Various sensor parameters such as structure (i.e., geometry), type of the FET (e.g., nanowire FET, spherical FET, ion-selective FET, and magnetic particle) radius of the FET channel and incubation time are optimized and analyzed. In addition, concentration of analyte biomolecules, diffusion coefficients and affinity to the receptor molecules are also investigated to determine the optimize performance parameters.

Keywords Field-effect-transistor · Biosensor · Aptamer · Aptasensor

1 Introduction

Prostate specific antigen (PSA) is the most widely used biomarker for early diagnosis of prostate cancer whose detection can be conducted by using deoxyribonucleic acid (ssDNA) aptamers and a sensitive transducing method. The selection

Md. Saiful Islam (✉) · E.D. Coyle
Military Technological College, PO Box 262, 111 Muscat, Oman
e-mail: sirajonece@yahoo.com

E.D. Coyle
e-mail: Eugene.Coyle@mtc.edu.om

A.Z. Kouzani
School of Engineering, Deakin University, Geelong, Victoria 3216, Australia
e-mail: kouzani@deakin.edu.au

© Springer India 2016
S.C. Satapathy et al. (eds.), *Microelectronics, Electromagnetics and Telecommunications*, Lecture Notes in Electrical Engineering 372,
DOI 10.1007/978-81-322-2728-1_73

of the transducing approach for aptasensors, aptamer-based biosensors, remains a bottleneck for sensitive and selective detection of target biomolecules [1–3]. A silicone nanowire (Si-NW) FET based biosensor offers high surface-to-volume ratio of Si-NW, and enables carrier distribution over the entire cross-sectional conduction pathway to be modulated by the presence of a few charged biological macro-molecules on its surface which results in increased sensitivity compared to one dimensional (1D) planar ion-sensitive FET (ISFET) [4]. Recently, Knopfmacher et al. [5] published an article presenting an organic FET sensor to overcome some existing barriers in rapid traction for sensing applications.

Sensitivity is a parameter in biosensors which can be tuned in the absence of an external gate by precisely controlling the dopant type and concentration in Si-NW FET based biosensors. Several articles have already demonstrated the improvement of sensitivity in Si-NW FET based biosensors for DNA and proteins detection [4, 6]. However, very few of them have focused on prostate cancer detection using this similar approach. On the other hand, although sensitivity of Si-NW FET based biosensors has been investigated and studied extensively, limit of detection (LOD) which provides an important performance parameter is yet to be fully explored in terms of the device geometry and other design parameters. In this paper, the per-formance of a FET based DNA aptasensor is investigated. The investigation of the performance analysis of FET based biosensors has been accompanied through four different approaches, including ISFET, cylindrical Si-NW FET, nanosphere FET and magnetic particle (MP). The impact of different parameters such as radius of Si-NW, incubation time, analyte concentration, buffer ion concentration and DNA base pair on settling time and LOD have been investigated. The outcome of this research has been explored using analytical model and is supported by detailed numerical simulation using nanohub and MATLAB [4].

2 Modeling of FET Based Biosensors

In the FET based biosensing approach, the gate is replaced with a functionalized layer of receptor biomolecules on the sensor surface. When selective target bio-molecules are bonded to the probe biomolecules, the distribution of charge in the liquid-transducer interface of the boundary layer changes which results in the modulation of the conductance of the transducer that enables the detection of the target biomolecules. In this paper, a modeling and simulation exercise has been carried out for the FET based biosensor for DNA and prostate cancer detection using the PSA biomarker. A schematic of the FET based aptasensor model is shown in Fig. 1. Four different approaches including ISFET, cylindrical Si-NW FET, nanosphere FET and magnetic particle have been considered in the model description. The device is comprised with a microfluidic channel to facilitate liquid analyte and three electrodes (source, drain, and gate). The goal of the source and drain electrodes is to link the semiconductor channel and the gate electrode which

Fig. 1 Schematic
reperestattion of a FET based
biosensor

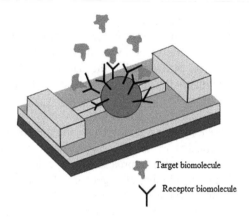

Target biomolecule

Receptor biomolecule

are responsible for modulating the channel conductance. Electrodes immersed in the analyte solution are protected by oxide layer.

To avoid parasitic response, the channel of the FET is immobilized with specific biological receptors (e.g., aptamers for prostate cancer detection) that recognizes and binds only to the target biomolecules. In the modelling of the FET based biosensor, the device is modelled using self-consistent solutions of the Diffusion-Capture model, Poisson-Boltzmann and Drift-Diffusion equations. In the definition of the proposed device, it considers liquid analyte to flow through a microfluidic channel and a top oxide layer protects the front-end complementary metal oxide semiconductor process from biological solution. Throughout the simulation, the dimension of the device is kept unchanged; it is then varied to optimize the performance of the biosensor. To determine the electrical response of the binding of target biomolecules, the surface of the sensor is functionalized with specific receptors that recognizes and binds only to the target biomolecules. In this process, target biomolecules introduced in the liquid solution are diffused and captured by the receptor molecules. Target biomolecules such as DNA in liquid solution carry a negative charge. Thus the electrostatic interaction between the charge of the target DNA (cDNA) and the sensor surface results in the modulation of the sensor characteristics. By measuring the change of the electrical characteristics such as drain current and conductance, the existence of the target biomolecules in the sample is detected. Throughout the modelling and simulation, it aims to optimize the design parameters so as to enhance the performance of the biosensors in terms of sensitivity, LOD and settling time.

3   Theory

The structure of the FET based biosensor is divided into three regions: (i) Si-NW channel, (ii) insulating native oxide around the NW with certain thickness, and (iii) electrolyte containing the target biomolecules and various ions generating the necessary buffer ions [4, 7, 8]. The drift-diffusion equation is used to describe the model solution of the carrier transport through the Si-NW channel. The potential in the dielectric oxide in the presence of an electrolyte is described using Poisson's equation provided that the native oxide is defect free. Finally, the electrostatic potential in the analyte solution is modeled using the nonlinear Poisson-Boltzmann equation. The presence of target charged biomolecules (e.g., DNA) provides an electrostatic interaction between receptor and target biomolecules which results a diffusion of the target DNA throughout the sensor volume and modulates the electrical characteristics such as drain current and conductance of the FET channel. A semi-classical approach is used to describe the conductance change of the Si-NW due to the binding of target biomolecules to its surface. The change of conductance or drain provides the necessary signal of the presence of desired target biomole-cules. The binding of the target molecules by the receptor molecules on the sensor surface is described by two steps [8]: (i) transport of target biomolecules, and (ii) conjugation with the receptor molecules. The target-receptor conjugation is con-sidered as a first-order chemical reaction and expressed by [7]:

$$\frac{d\rho}{dt} = D\nabla^2\rho \tag{1}$$

$$\frac{dN}{dt} = k_F(N_0 - N)\rho_s - k_R N \tag{2}$$

where D is the diffusion coefficient and ρ is the concentration of target biomolecules in the analyte solution. N represents the density of the conjugated receptors, and N_0 denotes the total receptor density on the sensor surface. k_F and k_R are the capture and dissociation constants, and ρ_s is the concentration of analyte particles at the sensor surface. The capture and diffusion of the target biomolecules are described by Eqs. (1) and (2), respectively. Solution of Eq. (1) determines the density of the captured biomolecules as follows [7]:

$$\frac{k_F N_0 + E - k_F N_{equi}}{k_{Fp0} + k_R} \log\left(1 - \frac{N}{N_{equi}}\right) + \frac{k_F}{k_{Fp0} + k_R} N = Et \tag{3}$$

where N_{equi} is the equilibrium concentration of the conjugated biomolecules, and is defined as:

$$N_{equi} = \frac{k_F N_0 \rho_s}{k_{Fp0} + k_R} \tag{4}$$

$E = (N_{avo}C_D(t))/A_D$, N_{avo} is Avogadro's number, $C_D(t)$ is the time dependent diffusion equivalent capacitance, A_D is the dimension dependent area of the sensor.

4   Numerical Simulations

The design of four different FET based biosensors is considered in a simulation investigation. The structure is first defined in a modelling platform with inclusion of various boundary conditions based on the requirement of the solution to be conducted. A modelling and simulation exercise of the biosensors is performed as a means of detection of target DNA and prostate cancer. To investigate the performance of the FET based biosensor through the electrical response for the existence of DNA, the sensor surface is immobilized with ssDNA for the detection of corresponding cDNA molecules. A liquid solution that contains the target cDNAs is introduced into the sensor surface which in turns diffused and captured by the receptor DNA molecules. The electrostatic interaction between the charge of target cDNAs and the sensor surface constitutes the change of the sensor characteristics such as the drift of drain current and shift of channel conductance. Measuring the shift of the electrical characteristics including drain current and conductance, the presence of the target cDNAs is identified. Similarly, to detect the prostate cancer, the sensor surface is functionalized with the specific DNA biomarker. The sensor surface is immersed into the analyte solution that contains PSA. Unlike DNA, the charge of PSA relys on the pH of the analyte solution which can be approximated using the Henderson Hasselbalch equation and the dissociation constants of the amino acids that constitute the protein [4, 9].

In the case of the magnetic particle based biosensor, the receptor is functionalized with iron oxide MP. Throughout the entire detection process based on the MP biosensor, the following three steps occur. The target captured by the receptor molecules goes for a subsequent incubation with gold nanopartilces functionalized with MP to produce MP-target-nanoparticle conjugate [10]. The localization of MP-target-nanoparticle is accompanied by the magnetic field and finally releases the nanoparticle and detects the MP based target biomolecules. During the entire simulation, the device dimensions are kept constant unless otherwise stated. The device structure is defined as: length = 40 nm, width = 20 nm, radius of Si-NW channel = 30 nm and oxide thickness = 2 nm. The biological conjugation parameters are selected as: forward reaction coefficient = 3×10^6 and reverse reaction coefficient = 1 and receptor density = 1×10^4 μm^{-2}. The diffusion coefficient of DNA is determined using the DNA diffusion model where the DNA diffusion coefficient = $A \times (bp)^{-n}$ [11] where A is the pre-factor = 4.9×10^{-6}, n is the exponent = 0.72 and bp is the base pair = 20 for initial simulation. The parameter of diffusion coefficient of DNA is considered at room temperature and thus the simulation environment is considered at 300 K ambient temperature.

5   Results and Discussions

Limit of detection (LOD) and sensitivity are two important parameters in the design of a biosensor determining the performance of the biosensors. For the best performance of the FET based biosensor, LOD should be as small as possible whereas the sensitivity is expected to be as high as possible. Both of these parameters depend on a number of factors such as device parameters (e.g., ISFET, Si-NW FET etc.), biological and surface parameters, ambient conditions and pH of the buffer ion. The below results give some of the optimized performance parameters.

5.1   Limit of Detection of FET Based Biosensor

Limit of detection refers to the lowest amount of sample analyte that can be reliably detected by a sensor. It is the determination of the smallest concentration of the target biomolecules that the sensor can clearly distinguish from its response to a bulk solution (solution containing no target biomolecules). For example, if a sensor has a limit of detection of 10 fM, it cannot generate any signal in response to the analyte that has concentration below 10 fM. To determine the LOD, the first step is to investigate the settling time of the biosensor. Settling time is defined as the time required by the FET based biosensor to produce a stable signal change upon the introduction of the sample analyte. The settling time is entirely dependent on the concentration of the sample analyte (e.g., less time required to generate stable signal with higher concentration of the analyte), and the diffusion coefficient and association constants. Settling time also depends on the radius of the Si-NW channel.

To determine the LOD, the settling time is first plotted against the variation of the target analyte concentration. Thereafter, a horizontal line corresponding to the incubation time is superimposed into the plot. The intersection of the incubation time line and settling time indicates the LOD. For this simulation, four different FET structures have been considered as shown in Fig. 2a. LODs for different incubation time and for different FET structures have been found from Fig. 2a. The black squares in this figure indicates the LOD for 1 s incubation time. It is obvious from Fig. 2a that the nanosphere FET provides the lowest LOD which is further reduced by using an MP biosensor. The results indicate that a 1D planar ISFET can produce LOD in the few nM range, while a 2D cylindrical Si-NW FET can produce LOD around 110 fM for a standard incubation time of 100 s. While Si-NW FET can detect four to five orders magnitude lower analyte concentration compared to ISFET, the MP sensor however can provide a lower LOD compared to both ISFET and Si-NW FET. Figure 2a also suggests that LOD is affected by the incubation time and the analyte concentration. To acquire the better LOD, it is required to have a higher incubation time. However, to facilitate ultrafast detection (less incubation time), the magnitude of LOD increases. Therefore, it is necessary to establish a trade-off between LOD and incubation time. Figure 2b illustrates that LOD

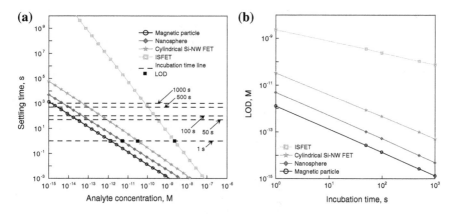

Fig. 2 **a** Determination and comparison of LOD for four different biosensors. The intersection point between the settling time line and the incubation time line (*black dotted horizontal line*) indicates LOD. **b** Dependency of LOD on incubation time for 4 biosensors structures

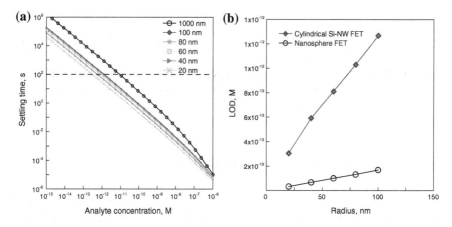

Fig. 3 **a** Determination of LOD for different radii of cylindrical Si-NW FET at 100 s incubation time. **b** Illustration of LOD as a function of radius of nanosphere and cylindrical Si-NW FET

decreases with the increase of the incubation time. The best LOD has been achieved by MP sensor.

The goal is to achieve fast detection at femtomolar analyte concentration. However, this is quite challenging as the settling time is inversely proportional to the analyte concentration. Consequently, an alternative approach has been investigated to achieve a better LOD where the radius of the Si-NW channel is varied to determine the minimum LOD. As indicated in Fig. 3a, the settling time declines with decreasing radius. Figure 3b suggests that a minimum LOD can be obtained by scaling down the radius of the FET channel (cylindrical or spherical). This is expected theoretically as the minimum detectable concentration is inversely

Fig. 4 Density of the captured target biomolecules at various times, and a comparison between the cylindrical NW-FET and the nanosphere biosensors

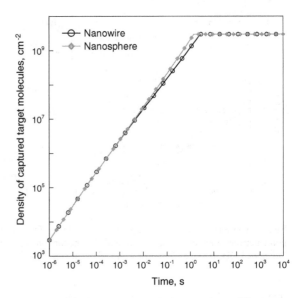

proportional to the square of the radius of the FET channel. Investigation of the density of the captured target molecules against time is shown in Fig. 4. It is clear from the figure that the density of the captured biomolecules increases linearly at the beginning of the reaction, however it becomes saturated and remains constant after a ceratin period of the reaction started.

5.2   Sensitivity of FET Based Biosensor

Sensitivity of a biosensor is defined as the ratio of the sensor output (e.g., conductance change) to the change in the measurand (e.g., concentration of target biomolecule). The sensitivity of a FET based biosensor depends on a number of factors including geometry, coated materials, oxide thickness, doping density of the sensor, target biomolecule as well as characteristics of the fluidic environment. The enhancement of the sensitivity can be obtained if the sensor surface is functionalized with aptamers that contain higher DNA base pairs.

6   Conclusion

This paper described the design and analysis of a FET based aptasensor to estimate the performance parameters of the sensor. Aptamers were functionalized onto the sensor surface for the detection of prostate cancer. It was demonstrated that the sensor response is affected by a number of factors including analyte concentration, buffer ion concentration, incubation time and geometry of the sensor. The simulation results show that it is possible to optimize LOD of FET biosensor by varying the diameter of cylindrical/spherical channel of the FET. They also suggest that better LOD is obtained with smaller diameters and MP sensor. In addition, the best sensitivity can be achieved with lower buffer ion concentration which ensures the minimum interference of undesired molecules in the analyet solution.

References

1. M.S. Islam, A.Z. Kouzani, Simulation and analysis of a sub-wavelength grating based multilayer surface plasmon resonance biosensor. J. Lightwave Technol. **31**, 1388–1398, 2013/05/01 (2013)
2. Y.C. Lim, A.Z. Kouzani, W. Duan, Aptasensors: a review. J. Biomed. Nanotechnol. **6**(2), 93–105 (2010)
3. Y.C. Lim, A.Z. Kouzani, W. Duan, X.J. Dai, A. Kaynak, D. Mair, A surface-stress-based microcantilever aptasensor. IEEE Trans. Biomed. Circuits Syst. **8**(1), 15–24 (2014)
4. P.R. Nair, M.A. Alam, Design considerations of silicon nanowire biosensors. IEEE Trans. Electron Devices **54**, 3400–3408 (2007)
5. O. Knopfmacher, M.L. Hammock, A.L. Appleton, G. Schwartz, J. Mei, T. Lei, et al., Highly stable organic polymer field-effect transistor sensor for selective detection in the marine environment. Nat Commun. **5**, 01/06/online (2014)
6. S.M. Kwon, G.B. Kang, Y.T. Kim, Y.H. Kim, B.K. Ju, In-situ detection of C-reactive protein using silicon nanowire field effect transistor. J. Nanosci. Nanotechnol. **11**, 1511–1514 (2011)
7. P.R. Nair, M.A. Alam, Performance limits of nanobiosensors. Appl. Phys. Lett. **88**, (2006) June 5
8. P.R. Nair, J. Go, G.J. Landells, T.R. Pandit, M.A. Alam, BioSensorLab, ed (2008)
9. P.D. Constable, A simplified strong ion model for acid-base equilibria: application to horse plasma. J. Appl. Physiol. **1985**(83), 297–311 (1997)
10. P.R. Nair, M.A. Alam, Theoretical detection limits of magnetic biobarcode sensors and the phase space of nanobiosensing. Analyst **135**, 2798–2801 (2010)
11. G.L. Lukacs, P. Haggie, O. Seksek, D. Lechardeur, N. Freedman, A.S. Verkman, Size-dependent DNA mobility in cytoplasm and nucleus. J. Biol. Chem. **275**, 1625–1629 (2000)

Index

© Springer India 2016
S.C. Satapathy et al. (eds.), *Microelectronics, Electromagnetics
and Telecommunications*, Lecture Notes in Electrical Engineering 372,
DOI 10.1007/978-81-322-2728-1

CPSIA information can be obtained
at www.ICGtesting.com
Printed in the USA
LVHW01*1427110318
569449LV00010B/1231/P